T0214235

Lecture Notes in Computer Science 9527

Commenced Publication in 1973
Founding and Former Series Editors:
Gerhard Goos, Juris Hartmanis, and Jan van Leeuwen

More information about this series at http://www.springer.com/series/7407

Moreno Falaschi (Ed.)

Logic-Based Program Synthesis and Transformation

25th International Symposium, LOPSTR 2015
Siena, Italy, July 13–15, 2015
Revised Selected Papers

 Springer

Editor
Moreno Falaschi
Università di Siena
Siena
Italy

ISSN 0302-9743   ISSN 1611-3349   (electronic)
Lecture Notes in Computer Science
ISBN 978-3-319-27435-5   ISBN 978-3-319-27436-2   (eBook)
DOI 10.1007/978-3-319-27436-2

Library of Congress Control Number: 2015956368

LNCS Sublibrary: SL1 – Theoretical Computer Science and General Issues

Printed on acid-free paper

This Springer imprint is published by SpringerNature
The registered company is Springer International Publishing AG Switzerland

Preface

This volume contains a selection of the papers presented at LOPSTR 2015, the 25th International Symposium on Logic-Based Program Synthesis and Transformation held during July 13–15, 2015, at the University of Siena, Italy. It was co-located with PPDP 2015, the 17th International ACM SIGPLAN Symposium on Principles and Practice of Declarative Programming.

Previous LOPSTR symposia were held in Canterbury (2014), Madrid (2013 and 2002), Leuven (2012 and 1997), Odense (2011), Hagenberg (2010), Coimbra (2009), Valencia (2008), Lyngby (2007), Venice (2006 and 1999), London (2005 and 2000), Verona (2004), Uppsala (2003), Paphos (2001), Manchester (1998, 1992, and 1991), Stockholm (1996), Arnhem (1995), Pisa (1994), and Louvain-la-Neuve (1993). More information about the symposium can be found at: http://alpha.diism.unisi.it/lopstr15/.

The aim of the LOPSTR series is to stimulate and promote international research and collaboration on logic-based program development. LOPSTR is open to contributions in all aspects of logic-based program development, all stages of the software life cycle, and issues of both programming-in-the-small and programming-in-the-large. LOPSTR traditionally solicits contributions, in any language paradigm, in the areas of synthesis, specification, transformation, analysis and verification, specialization, testing and certification, composition, program/model manipulation, optimization, transformational techniques in SE, inversion, applications, and tools. LOPSTR has a reputation for being a lively, friendly forum for presenting and discussing work in progress. Formal proceedings are produced only after the symposium so that authors can incorporate this feedback in the published papers.

In response to the call for papers, 30 contributions were submitted from 17 different countries. The Program Committee accepted six full papers for immediate inclusion in the formal proceedings, and 14 full papers plus one short paper presented at the symposium were accepted after a revision and another round of reviewing. Each submission was reviewed by at least three Program Committee members or external reviewers. In addition to the 21 contributed papers, this volume includes the abstract of the talks by two outstanding invited speakers: Patrick Cousot (Courant Institute of Mathematical Sciences, New York University, USA), whose talk was shared with PPDP, and Gilles Barthe (IMDEA Software Institute, Madrid, Spain). This volume also includes the full paper by a third outstanding invited speaker: Dale Miller (Inria and LIX/Ecole Polytechnique, Palaiseau, France), whose talk was shared with PPDP.

We want to thank the Program Committee members, who worked diligently to produce high-quality reviews for the submitted papers, as well as all the external reviewers involved in the paper selection. We are very grateful to the LOPSTR 2015 local organizers for the great job they did in managing the symposium. Many thanks also to Elvira Albert, the Program Committee chair of PPDP, with whom we often interacted to coordinate the two events. We would also like to thank Andrei Voronkov for his excellent EasyChair system that automates many of the tasks involved in chairing a conference. Special thanks go to all the authors who submitted and presented their papers at LOPSTR 2015, without whom the symposium would have not been possible. We also thank the Dipartimento di Ingegneria dell'Informazione e Scienze Matematiche, the Italian Chapter of the European Association for Computer Science, and the Association for Logic Programming for their cooperation and support in the organization of the symposium.

October 2015   Moreno Falaschi

Organization

Program Committee

Slim Abdennadher	German University in Cairo, Egypt
María Alpuente	UPV, Spain
Demis Ballis	University of Udine, Italy
Olaf Chitil	University of Kent, UK
Michael Codish	Ben-Gurion University of the Negev, Israel
Moreno Falaschi	DIISM, University of Siena, Italy
Jérôme Feret	Inria/Ecole normale supérieure, France
Maurizio Gabbrielli	University of Bologna, Italy
Jürgen Giesl	RWTH Aachen, Germany
Miguel Gómez-Zamalloa	Complutense University of Madrid, Spain
Arnaud Gotlieb	SIMULA Research Laboratory, Norway
Gopal Gupta	University of Texas at Dallas, USA
Manuel V. Hermenegildo	T.U. Madrid (UPM) and IMDEA Software, Spain
Viktor Kuncak	EPFL Lausanne, Switzerland
Luigi Liquori	Inria, France
Alexei Lisitsa	University of Liverpool, UK
Narciso Martí-Oliet	Universidad Complutense de Madrid, Spain
Jorge A. Navas	NASA Ames Research Center, USA
Kazuhiro Ogata	JAIST, Japan
Carlos Olarte	Universidade Federal do Rio Grande do Norte, Brazil
Catuscia Palamidessi	Inria and LIX, Ecole Polytechnique, France
Maurizio Proietti	IASI-CNR, Rome, Italy
Albert Rubio	Universitat Politècnica de Catalunya, Spain
Wim Vanhoof	University of Namur, Belgium

Steering Committee

Elvira Albert	Complutense University of Madrid, Spain
María Alpuente	Universidad Politecnica de Valencia, Spain
Gopal Gupta	University of Texas at Dallas, USA
Alberto Pettorossi	University of Rome Tor Vergata, Italy
Maurizio Proietti	IASI-CNR, Italy
Hirohisa Seki	Nagoya Institute of Technology, Japan
Germán Vidal	Universidad Politecnica de Valencia, Spain

General Chair

Moreno Falaschi   DIISM, University of Siena, Italy

Organizing Committee

Andrea Machetti   DIISM, Siena, Italy
Simonetta Palmas   DIISM, Siena, Italy
Guillermo Román Díez   Technical University of Madrid, Spain
Alessandro Rossi   DIISM, Siena, Italy
Elisa Tiezzi   DIISM, Siena, Italy
Sara Ugolini   University of Pisa, Italy

Additional Reviewers

Aguirre, Luis   Echahed, Rachid   Madhavan,
De Gouw, Stijn   Kuraj, Ivan   Ravichandhran
Kafle, Bishoksan   Mauro, Jacopo   Meo, Maria Chiara
Marple, Kyle   Palomino, Miguel   Pettorossi, Alberto
Morales, Jose Francisco   Pita, Isabel   Riesco, Adrian
Peña, Ricardo   Rueda, Camilo   Salazar, Elmer
Rojas, José Miguel   Sharaf, Nada   Ströder, Thomas
Sapiña, Julia   Zaki, Amira   Zunino, Roberto
Valencia, Frank   De Angelis, Emanuele
Dalla Preda, Mila   Haemmerle, Remy

Invited Talks

Verification by Abstract Interpretation, Soundness and Abstract Induction

Patrick Cousot

Courant Institute of Mathematical Sciences, New York University, USA
pcousot@cims.nyu.edu

Abstract. Automatic program verification tools have to cope with programming language and machine semantics, undecidability, and mathematical induction, and so are all complex and imperfect. The ins and outs of automatic program verification will be discussed in light of the theory and practice of abstract interpretation [1–3].

References

1. Cousot, P., Cousot, R.: Abstract interpretation: a unified lattice model for static analysis of programs by construction of approximations of fixed points. In: Graham, R.M., Harrison, M. A., Sethi, R. (eds.) POPL 1977, pp. 238–252. ACM (1977)
2. Cousot, P., Cousot, R.: Systematic design of program analysis frameworks. In: Aho, A.V., Zilles, S.N., Rosen, B.K. (eds) POPL 1997, pp. 269–282. ACM (1979)
3. Cousot, P., Cousot, R.: Abstract interpretation: past, present and future. In: Henzinger, T.A., Miller, D. (eds) Joint Meeting of the Twenty-Third EACSL Annual Conference on Computer Science Logic (CSL) and the Twenty-Ninth Annual ACM/IEEE Symposium on Logic in Computer Science (LICS), CSL-LICS 14, pp. 2:1–2:10, Vienna, Austria, 14–18 July 2014. ACM

An extended version appears in the proceedings of the 17th International Symposium on Principles and Practice of Declarative Programming (PPDP 2015), July 14–16 2015, Siena, Italy. ACM Press.

Towards Verified Cryptographic Implementations

Gilles Barthe

IMDEA Software Institute, Madrid, Spain
gjbarthe@gmail.com

Abstract. Building secure cryptographic software is difficult. I will discuss some language-based methods for analyzing and improving the security of implementations against cache attacks, differential power attacks, and fault attacks.

Keywords: Description logic • Programming paradigm • Smart environments

It is notoriously difficult to build secure cryptographic software. On the one hand, security goals for cryptographic constructions are often mathematically elaborate, as they quantify over the probability that an adversary with bounded computational resources can win a security experiment. On the other hand, mathematical proofs of security are carried in an ideal model that elides many attack vectors, and thus cannot guarantee that deployed software is secure. Indeed, there are two main classes of attacks that are generally not considered by mainstream provable security: side-channel attacks and fault attacks.

Side-channel attacks exploit signals emitted by a program's execution to recover information about the confidential data it manipulates, and are surprisingly effective to recover key material and other secrets almost instantaneously from cryptographic implementations. Over the last twenty years, there has been a continuous stream of devastating side-channel attacks, using execution time, power consumption, or cache usage.

Fault attacks are attacks in which an adversary with physical access to a cryptographic device, say a smartcard, tampers with the execution of an algorithm to retrieve secret material. Since the seminal Bellcore attack on modular exponentiation, there has been extensive work to discover new fault attacks against cryptographic schemes and develop countermeasures against such attacks.

In this talk, I will present some language-based methods (type systems and program logics) used for analyzing the security of implementations against cache attacks, differential power attacks, and fault attacks. I will also report on implementations and applications of our methods both to produce secure implementations and to discover attacks in existing implementations.

Proof Checking and Logic Programming

Dale Miller

Inria and LIX/École Polytechnique
dale.miller@inria.fr

Abstract. In a world where trusting software systems is increasingly important, formal methods and formal proof can help provide trustable foundations. Proof checking can help to reduce the size of the *trusted base* since we do not need to trust an entire theorem prover if we can check the proofs they produce by a trusted (and smaller) checker. Many approaches to building proof checkers require embedding within them a full programming language. In most many modern proof checkers and theorem provers, that programming language is a functional programming language, often a variant of ML. In fact, parts of ML (e.g., strong typing, abstract datatypes, and higher-order programming) were designed to make ML into a trustworthy "meta-language" for checking proofs. While there is considerable overlap in the foundations of logic programming and proof checking (both benefit from unification, backtracking search, efficient term structures, etc), the discipline of logic programming has, in fact, played a minor role in the history of proof checking. I will argue that logic programming can have a major role in the future of this important topic.

Contents

Program Transformation

Constraint Solving and Programming

Invited Talk

Proof Checking and Logic Programming

Dale Miller[✉]

Inria and LIX/École Polytechnique, Palaiseau, France
dale.miller@inria.fr

Abstract. In a world where trusting software systems is increasingly important, formal methods and formal proof can help provide trustable foundations. Proof checking can help to reduce the size of the *trusted base* since we do not need to trust an entire theorem prover if we can check the proofs they produce by a trusted (and smaller) checker. Many approaches to building proof checkers require embedding within them a full programming language. In most many modern proof checkers and theorem provers, that programming language is a functional programming language, often a variant of ML. In fact, parts of ML (e.g., strong typing, abstract datatypes, and higher-order programming) were designed to make ML into a trustworthy "meta-language" for checking proofs. While there is considerable overlap in the foundations of logic programming and proof checking (both benefit from unification, backtracking search, efficient term structures, etc.), the discipline of logic programming has, in fact, played a minor role in the history of proof checking. I will argue that logic programming can have a major role in the future of this important topic.

1 Introduction

There are a number of theorem provers used by academics and industry and the kinds of formalisms that they take on are becoming increasingly complex and important. For example, computer systems such as Coq, HOL/Lite, and Isabelle have been used to help formally prove the four color theorem [18], the Kepler conjecture [20], and the correctness of a compiler [26] and a micro kernel [24]. For theorem provers to be part of our approach to trusting formalized mathematics and software systems, such provers must be trusted as well. However, trusting modern theorem provers is proving difficult since over time they become increasingly more complex: they evolve to allow for stronger inference rules and for the integration of specialized proving technology. Furthermore, one might not wish to require that theorem provers are formally trusted since anything that is formally validated has stopped evolving and is not generally subject to innovation and improvements.

1.1 Validate Proofs, Not Provers

One method for addressing the correctness of theorem provers is to move from formally validating an entire theorem prover to simply validating individual

M. Falaschi (Ed.): LOPSTR 2015, LNCS 9527, pp. 3–17, 2015.
DOI: 10.1007/978-3-319-27436-2_1

proofs that are emitted by provers. With such a move, we need to trust the proof checker instead of the entire theorem prover: presumably a checker is much simpler and does not need to evolve frequently. Some provers, like Coq, separate the activity of theorem proving from proof checking by providing its own *kernel* which checks all proposed proofs. There are, however, several reasons why it is desirable to move the proof checking operation to be *outside* a theorem prover.

- One of the philosophically motivated aspects of proofs must surely be their ability to communicate across time and space the reason to trust that a formula is indeed true [30]. Moving proof checkers outside a prover emphasizes that proofs are meant to be *communicated*, at least from prover to checker.
- When the kernel is part of a prover, there is a great tendency for the kernel to provide exactly what the prover needs: such a communication would evolve to just involve two entities (the kernel and prover) instead of the many possible other actors which might also want to check, trust, and use a proof.
- Finally, when the checker is formally separated from the prover, the structure of emitted proofs and the semantics of the kernel would become independent of the technology of the prover. Anyone could, therefore, reimplement the kernel and check the emitted proofs. Having several kernels by several different teams of implementers provides a well recognized path to having increased trust in software.

An example of an architecture for moving proof checking outside of theorem provers is currently being explored and implemented within the Dedukti system [11]. Dedukti is based the $\lambda\Pi$-modulo formal framework of Cousineau and Dowek [9] which mixes two well-known and powerful frameworks, one for hypothetical reasoning (via the dependently typed λ-calculus known as LF [21] and $\lambda\Pi$) and one for functional programming style computations (via confluent rewriting). The Dedukti project has recently developed software that allows several existing theorem provers—e.g., Coq, HOL, Matita—to output proofs into a format that Dedukti can check independently from those provers [2].

1.2   Proof Checking Vs Proof Reconstruction

What is typically called proof checking generally contains elements of *proof reconstruction*: that is, the process of checking whether or not a given document is a proof might require computing some details that are not present explicitly in the document. For example, even for rather low-level and detailed notions of proof, it is seldom the case that one would expect to have every detail present within a formal proof object. For example, in order to check that the assumptions $(p \supset p \supset q) \supset (p \supset q)$ and $(p \supset p \supset q)$ and the rule of *modus ponens*, written schematically as "from A and $A \supset B$, conclude B", infers $p \supset q$, it is not likely that one needs to provide *in the proof itself* the explicit ordering of assumption and the binding of schematic variables $[A \mapsto (p \supset p \supset q), B \mapsto (p \supset q)]$. Such an ordering and binding can easily be computed, leaving less to store in the proof document.

Existing theorem provers often contain much more significant notions of proof reconstruction. For example, the Boyer-Moore theorem prover attempts to fill in significant gaps between lemmas using various proof procedures parametrized by the collection of previously proved lemmas [5]. If the proof procedure is not able to fill in a gap in a sequence of lemmas, the user must design some additional lemmas to split up that gap into more manageable parts. Other theorem provers, particularly those based on the LCF framework [19], allow functional programs (usually in a variant of ML) to be executed in order to compute ways to complete the gaps between lemmas (or between hypotheses and goal). In other systems, an interactive theorem prover might call a completely automated prover in order to complete a step of inference: for example, Isabelle can call the Vampire theorem prover to close a gap in a user's proof attempt [28].

1.3   The Community of Logic Programming

Oddly enough, the community that has invested a great deal of energy into providing effective implementations of logic—namely, the logic programming community—has not traditionally been involved with proof checking. There are at least three likely reasons for this mismatch between that community and those interested in theorem proving and proof checking.

Efficiency Versus Soundness. In proof checking, logical soundness is *everything*: there is no reason to be doing proof checking if one is not confident that the underlying logic engine is logically sound. The logic programming community has often emphasized efficiency instead of logical soundness: for example, many Prolog systems have not supported the occurs-check in unification since that was seen as a feature only needed for toy examples [34, Sect. 3.3]. Experience with automated theorem proving shows, however, that soundness of inference critically depends on the presence of occurs-check in unification. Of course, a programming language like Prolog can still be used to implement proof checkers even when unification is unsound since *any programming language* can be used to build, in principle, any programming task. One would suspect, however, that logic programming languages should have a much more immediate and transparent ways to support the effective implementation of logic.

Lack of Logical Expressiveness. Another aspect of most logic programming languages is that they do not have direct support for quantified formulas and the concomitant operations of substitution into and unification of expressions containing bindings. While individual Prolog clauses are interpreted as universally quantified, such quantification is implicit. Furthermore, no direct and logical support for bindings is available within Prolog even though the interplay between formula-level bindings (quantifiers) and term-level bindings (λ-abstractions) has been well understood since Church's Simple Theory of Types [8]. Thus, while Prolog provides many logical principles that could be used to implement proof checkers, that support does not extend to much of logic itself.

Lack of Abstractions. There is still at least one other reason that logic programming can be a poor match with proof checking: most Prolog systems do not support rich forms of *abstractions*, such as abstract datatypes and procedural (higher-order) abstraction. Each of these features were explicitly introduced into the first design and implementation of the functional programming language ML since they were seen as important for building the LCF proof system [19]. Abstract datatypes help provide guarantees that, for example, there are only certain ways to build objects of type thm (the type for theorems): once thm is established, it is made into an abstract datatype and the design of ML's type system enforces that theorems can only arise from the originally provided constructors and functions. Similarly, higher-order programming in ML was introduced to allow for certain "kernel" operations (the tacticals) to have their trusted code separated from the "clients" code (the tactics).

The lack of expressiveness and abstraction can be addressed within logic programming if one is willing to move beyond first-order Horn clauses for fragments of higher-order, intuitionistic logic [29]. In fact, λProlog [12,31,33] and Twelf [35] are two logic programming languages that treat bindings in expressions and proofs directly as part of their logical foundations. Furthermore, λProlog also exploits features of its underlying logic to provide logically sound notions of modules and abstract datatypes as well as higher-order programming (such as is found in most functional programming languages).

Where should we begin in looking for connections between the logic programming paradigm and proof checking? Remarkably, one has to look no further than the recent literature of proof theory (see references in Sect. 3) to find a framework where relations and not functions dominate, where bounded backtracking search has obvious and immediate applications, and where formula-level and term-level abstractions occur naturally together. In addition, the topic of proof theory presents a mathematically and not a technologically notion of proof.

2   Proof Theory as a Framework

If a modern theorem prover outputs a proof as a (persistent) document, that proof document is usually based on specific technology built into the prover. On some other occasions, provers output documents meant for tracing and debugging. It is the exceptional theorem prover that outputs a document that is intended to outlast the (version of the) prover itself.[1] Given the existence of the mathematical literature on *proof theory* initiated by Frege and Gentzen, proofs-as-documents can be as *eternal* as Peano numerals: 0, $s(0)$, $s(s(0))$, $s(s(s(0)))$, etc.

Once proofs are liberated from the technology that produces them, then they can be checked by independently constructed checking programs which can be written by anyone keen to develop their own trusted base of code instead of adopting someone else's code. Once checked, such proofs can be placed in

---

[1] Some specialized theorem provers related to SAT solving have adopted standards of outputting their proof evidence (see, for example, [36]).

libraries that survive changes in theorem proving and proof checking technologies. Thus, proofs can be used to communicate trust between different provers and across changes in technology. In order to reach such a status in the sharing and trusting of proofs, we probably need to design a framework based on a mathematically well defined and sophisticated notion of proof. Examining the literature on proof theory, however, reveals a number of formally defined proof structures. In the earliest days, Frege and Hilbert proposed rather simple, linear proof structures; Gentzen introduced both the sequent calculus as well as natural deduction; later, resolution refutations and tableaux proof systems were also introduce, in part, to support automation of theorem proving. Still more structures can be found that can be accepted as proofs, such as proof nets, matings, deep inference, and winning strategies.

In this paper, I will outline a multi-year effort that proposes to use the sequent calculus as the *assembly language* of proof and to describe how to compile many other higher-level notions of proof into that assembly language. The formal devices for making such definition of proof languages will be based on the notion of *focused proofs* for first-order classical and intuitionistic logics. We illustrate such a proof system in the next section.

3   Focused Versions of Sequent Calculi

The sequent calculus of Gentzen provides an appealing form of formal proof structure since they can be used to describe proofs in both classical, intuitionistic, and linear logics. They also support propositional, first-order, and higher-order logics and do so in a modular and clear fashion. The cut-elimination theorem [15] also reveals that this notion of proof supports sophisticated manipulations (such as substitution and composition of proofs). On the other hand, the proofs in the sequent calculus can be chaotic: if a proof of a given sequent exists, many trivial and not-so-trivial variations of that proof also exists. All these variants work to hide structure. Relying on the small inference steps that are part of Gentzen's presentation of the sequent calculus not only makes finding sequent calculus proofs difficult, it also makes communicating them challenging. Consider the following example (taken from [4]). Attempting to prove the sequent $\Gamma \vdash \exists x \exists y [(p\ x\ y) \vee ((q\ x\ y) \vee (r\ x\ y))]$, where Γ contains, say, a hundred formulas. The search for a (cut-free) proof of this sequent can confront the need to choose from among a hundred-and-one introduction rules. If we choose the right-side introduction rule, we will then be left with, again, a hundred-and-one introduction rules to apply to the premise. Thus, reducing this sequent to, say, $\Gamma \vdash (q\ t\ s)$ requires picking one path of choices in a space of 101^4 choices.

One of the first attempts to use sequent calculus in computer science needed to develop a normal form of sequent proof that discarded a great deal of those variants. The *uniform proofs* of [29,32]—with its notion of alternating phases of goal-reduction and backchaining—was used to provide a general framework for defining proof search in logic programming. With the advent of linear logic [16], that two phase structure was extended to all of linear logic using Andreoli's *focused proof system* [1] and the notion of *polarity* [1,17]. Soon afterwards, various focused proof systems for intuitionistic logic [6,13,22,23] and classical logic [10,17,25] appeared. The *LJF* and *LKF* frameworks of [27] can be seen as offering a general framework and generalization to these various classical and intuitionistic focused proof systems.

We limit our attention here to first-order classical logic. A similar development holds for intuitionistic logic as well.

Polarizing Connectives. The emphasis on focused proofs is an emphasis on proof structure and not provability. For example, consider the following different ways to write the introduction rules for disjunction and conjunction in a one-sided sequent system.

$$\frac{\vdash \Gamma, B_1 \quad \vdash \Gamma, B_2}{\vdash \Gamma, B_1 \wedge B_2} \quad \frac{\vdash \Gamma_1, B_1 \quad \vdash \Gamma_2, B_2}{\vdash \Gamma_1, \Gamma_2, B_1 \wedge B_2} \quad \frac{\vdash \Gamma, B_1, B_2}{\vdash \Gamma, B_1 \vee B_2} \quad \frac{\vdash \Gamma, B_i}{\vdash \Gamma, B_1 \vee B_2} \ i \in \{1,2\}$$

Given that the structural rules of weakening and contractions are available in classical logic, the first pair of rules and the second pair of rules are inter-admissible inference rules: any sequent provable with one element of the pair is provable also with the second member of the pair. Notice also that the first member of each pair is *invertible* while the second member is *not invertible*. People presenting proof systems for classical logic or who are implementing such systems generally pick one member from each pair and that choice is usually the invertible rule. Given our interest here in proof structures (an not just provability), our eventual focused proof system will contain all four of these introduction rules. They will be distinguished from each other by having them introduce different *polarized* versions of disjunction and conjunction.

$$\frac{\vdash \Gamma, B_1 \quad \vdash \Gamma, B_2}{\vdash \Gamma, B_1 \wedge^- B_2} \quad \frac{\vdash \Gamma_1, B_1 \quad \vdash \Gamma_2, B_2}{\vdash \Gamma_1, \Gamma_2, B_1 \wedge^+ B_2} \quad \frac{\vdash \Gamma, B_1, B_2}{\vdash \Gamma, B_1 \vee^- B_2} \quad \frac{\vdash \Gamma, B_i}{\vdash \Gamma, B_1 \vee^+ B_2} \ i \in \{1,2\}$$

The introduction rules for the negative polarized connectives ($\wedge^-$ and $\vee^-$) are invertible while the introduction rules for the positive polarized connectives ($\wedge^+$ and $\vee^+$) are not invertible. The units for these connectives are also polarized similarly: t^-, t^+, f^-, and f^+, and these have the following introduction rules.

$$\frac{}{\vdash \Gamma, t^-} \quad \frac{}{\vdash t^+} \quad \frac{\vdash \Gamma}{\vdash \Gamma, f^-}$$

(There is no introduction rule for the positive false f^+).

Some connectives have fixed polarity: universal quantification is negative and its de Morgan dual, existential quantification, is positive. Atoms can be either positive or negative: this choice can be made in an arbitrary but fixed fashion. The negated atom $\neg A$ has the opposite polarity to A. A formula has positive or negative polarity depending only on its top-level logical connective (if it has one) or on the polarity as a literal.

Grouping Don't-Care and Don't-Know Non-determinism. If several invertible rules can be applied to yield a given sequent then those rules can be applied in any order and, in fact, in all possible orderings to yield a proof. In order to factor away such don't-care non-determinism, we introduce the notion of the $\Uparrow$ phase in focused proof construction display the invertible rules as follows:

$$\frac{}{\vdash \Theta \Uparrow t^-, \Gamma} \qquad \frac{\vdash \Theta \Uparrow A, \Gamma \quad \vdash \Theta \Uparrow B, \Gamma}{\vdash \Theta \Uparrow A \wedge^- B, \Gamma} \qquad \frac{\vdash \Theta \Uparrow \Gamma}{\vdash \Theta \Uparrow f^-, \Gamma} \qquad \frac{\vdash \Theta \Uparrow A, B, \Gamma}{\vdash \Theta \Uparrow A \vee^- B, \Gamma}$$

$$\frac{\vdash \Theta \Uparrow [y/x]B, \Gamma}{\vdash \Theta \Uparrow \forall x.B, \Gamma} \; \dagger \qquad \frac{\vdash \Theta, C \Uparrow \Gamma}{\vdash \Theta \Uparrow C, \Gamma} \; store$$

Here, sequents are of the form $\vdash \Theta \Uparrow \Gamma$, where Θ is a schematic variable ranging over *multisets* of formulas and Γ is a schematic variable ranging over *lists* of formulas. A list is used here instead of a multiset as a way to reduce the don't-care non-determinism: we only need to consider introduction rules on the first formulas of that list. In the $\forall$-introduction rule, the $\dagger$ proviso is the usual one: the variable y is not free in the lower sequent. In the store rule, C is a positive formula or negative literal: this rule is responsible for recognizing that the first formula in the right-hand context cannot be introduced by an invertible inference. In general, the context Θ contains only positive formulas and negative literals.

Another phase contains sequents of the form $\vdash \Theta \Downarrow B$ where Θ is as before and B is a formula. The introduction rules associated to this phase are written as follows.

$$\frac{}{\vdash \Theta \Downarrow t^+} \qquad \frac{\vdash \Theta \Downarrow B_1 \quad \vdash \Theta \Downarrow B_2}{\vdash \Theta \Downarrow B_1 \wedge^+ B_2} \qquad \frac{\vdash \Theta \Downarrow B_i \quad i \in \{1, 2\}}{\vdash \Theta \Downarrow B_1 \vee^+ B_2} \qquad \frac{\vdash \Theta \Downarrow [t/x]B}{\vdash \Theta \Downarrow \exists x.B}$$

Structural and Identity Rules. The following two "structural" rules are needed to move between these two phases.

$$\frac{}{\vdash \neg P_a, \Theta \Downarrow P_a} \; init \qquad \frac{\vdash \Theta \Uparrow B \quad \vdash \Theta \Uparrow \neg B}{\vdash \Theta \Uparrow \cdot} \; cut$$

$$\frac{\vdash \Theta \Uparrow N}{\vdash \Theta \Downarrow N} \; release \qquad \frac{\vdash P, \Theta \Downarrow P}{\vdash P, \Theta \Uparrow \cdot} \; decide$$

Here, P is a positive formula; N a negative formula; P_a a positive literal; C a positive formula or negative literal; and $\neg B$ is the negation normal form of the negation of B.

Synthetic Inference Rules. One of the purposes of introducing a focused proof system is to make the following identification: the phases introduce new, synthetic inference rules. Gentzen's introduction and structural rules form the assembly instructions of proof and the synthetic inference rules form the higher-level notions of proof. For example, assume that Θ contains the formula $a \wedge^+ b \wedge^+ \neg c$ and that we have a derivation that Decides on this formula.

$$
\cfrac{
\cfrac{\vdash \Theta \Downarrow a}{} \, Init
\quad
\cfrac{\vdash \Theta \Downarrow b}{} \, Init
\quad
\cfrac{
\cfrac{
\cfrac{\vdash \Theta, \neg c \Uparrow \cdot}{\vdash \Theta \Uparrow \neg c} \, Store
}{\vdash \Theta \Downarrow \neg c} \, Release
}{} \, \wedge^+
}{
\cfrac{\vdash \Theta \Downarrow a \wedge^+ b \wedge^+ \neg c}{\vdash \Theta \Uparrow \cdot} \, Decide
}
$$

This derivation is possible if and only if Θ is of the form $\neg a, \neg b, \Theta'$. Thus, the "macro-rule" is

$$
\cfrac{\vdash \neg a, \neg b, \neg c, \Theta' \Uparrow \cdot}{\vdash \neg a, \neg b, \Theta' \Uparrow \cdot}
$$

Soundness and Completeness of Focusing. The formulas used in *LK* are unpolarized while those in *LKF* are polarized. In order to state soundness and completeness of focusing, we must introduce the notion of polarizing a formula. Given the unpolarized formula B, let $\hat{B}$ be one of the exponentially many formulas that result by placing $+$ or $-$ on the occurrences of $\vee$ and $\wedge$ as well as attributing polarization to the atoms in B. (The quantifiers have fixed polarities: $\forall$ is negative and $\exists$ is positive.) The soundness theorem for *LKF* is immediate: Assume that $\vdash \cdot \Uparrow \hat{B}$ has an *LKF* proof. We can recover an *LK* proof by simply replacing the $\Uparrow$ and $\Downarrow$ with commas, deleting some repetitions of sequents, and dropping the $+$ and $-$ annotations on the propositional connectives. Conversely, completeness (which is proved in [27]) states that if B is a first-order theorem and $\hat{B}$ is any polarization of B then $\vdash \cdot \Uparrow \hat{B}$ is provable in *LKF*. A consequence of soundness and completeness implies that if any polarization of B is provable in *LKF* then every polarization is provable in *LKF*. Clearly, polarization is not relevant to *provability* but is relevant to the structure of proofs.

4 Foundational Proof Certificates

The two phases in *LKF* are strikingly different. The invertible $\Uparrow$ phase can be built from the bottom up as a purely deterministic computation: one just applies the inference rules in a straightforward fashion. On the other hand, the $\Downarrow$ phase is not straightforward since some of the inference rules need information that is lacking from the conclusion: in particular, the $\exists$ introduction rule requires a substitution term and the $\vee^+$ introduction rule requires an indicator of whether to select the left or right disjunct. If this information is lacking, then the construction of the proof can be seen as a non-deterministic computation, where choices and substitution terms are guessed.

It is now easy to see that the focused proof system in Sect. 3 provides a *communication protocol* between an entity that possesses some evidence that a formula is a theorem and a low-level tool attempting to build a sequent calculus proof from the bottom-up of that proposed theorem.

$$\frac{}{\Xi \vdash \Theta \Uparrow t^-, \Gamma} \qquad \frac{\Xi_1 \vdash \Theta \Uparrow A, \Gamma \quad \Xi_2 \vdash \Theta \Uparrow B, \Gamma \quad \wedge_c(\Xi, \Xi_1, \Xi_2)}{\Xi \vdash \Theta \Uparrow A \wedge^- B, \Gamma} \qquad \frac{\Xi' \vdash \Theta \Uparrow \Gamma \quad f_c(\Xi, \Xi')}{\Xi \vdash \Theta \Uparrow f^-, \Gamma}$$

$$\frac{\Xi' \vdash \Theta \Uparrow A, B, \Gamma \quad \vee_c(\Xi, \Xi')}{\Xi \vdash \Theta \Uparrow A \vee^- B, \Gamma} \qquad \frac{\Xi' \vdash \Theta \Uparrow [y/x]B, \Gamma \quad \forall_c(\Xi, \Xi')}{\Xi \vdash \Theta \Uparrow \forall x.B, \Gamma} \; \dagger$$

$$\frac{true_e(\Xi)}{\Xi \vdash \Theta \Downarrow t^+} \qquad \frac{\Xi_1 \vdash \Theta \Downarrow B_1 \quad \Xi_2 \vdash \Theta \Downarrow B_2 \quad \wedge_e(\Xi, \Xi_1, \Xi_2)}{\Xi \vdash \Theta \Downarrow B_1 \wedge^+ B_2}$$

$$\frac{\Xi' \vdash \Theta \Downarrow B_i \quad i \in \{1,2\} \quad \vee_e(\Xi, \Xi', i)}{\Xi \vdash \Theta \Downarrow B_1 \vee^+ B_2} \qquad \frac{\Xi' \vdash \Theta \Downarrow [t/x]B \quad \exists_e(\Xi, \Xi', t)}{\Xi \vdash \Theta \Downarrow \exists x.B}$$

$$\frac{\Xi_1 \vdash \Theta \Uparrow B \quad \Xi_2 \vdash \Theta \Uparrow \neg B \quad cut_e(\Xi, \Xi_1, \Xi_2, B)}{\Xi \vdash \Theta \Uparrow \cdot} \; cut$$

$$\frac{\Xi' \vdash \Theta \Uparrow N \quad release_e(\Xi, \Xi')}{\Xi \vdash \Theta \Downarrow N} \; release \qquad \frac{init_e(\Xi, l) \quad \langle l, \neg P_a \rangle \in \Theta}{\Xi \vdash \Theta \Downarrow P_a} \; init$$

$$\frac{\Xi' \vdash \Theta \Downarrow P \quad decide_e(\Xi, \Xi', l) \quad \langle l, P \rangle \in \Theta \quad positive(P)}{\Xi \vdash \Theta \Uparrow \cdot} \; decide$$

$$\frac{\Xi' \vdash \Theta, \langle l, C \rangle \Uparrow \Gamma \quad store_c(\Xi, \Xi', l)}{\Xi \vdash \Theta \Uparrow C, \Gamma} \; store$$

Fig. 1. The augmented LKF proof system LKF^a (Color figure online).

One way to implement such a protocol would be to instrument the focused proof system LKF with certain augmentations as shown in Fig. 1. The augmentation is accomplished in three simple steps: (i) a proof certificate term, denoted by the syntactic variable Ξ is added to every sequent; (ii) every inference rule of LKF is given an additional premise using either an *expert predicate* or a *clerk predicate*; and (iii) the multiset of formulas to the left of the arrows $\Uparrow$ and $\Downarrow$ is replaced with a multiset of pairs of an *index* and a formula. (Viewing this figure in color shows the augmentations in blue.) Clearly, the LKF proof system can be recovered from LKF^a by removing all occurrences of the syntactic variable Ξ and by removing all premises with a subscripted e or c as well as replacing all occurrences of tuples such as $\langle l, B \rangle$ with just B.

Notice that the extra premise added to invertible rules are called *clerks*: in such inference rules, only simple computations are done and, in general, no information in a certificate term needs to be consumed. On the other hand, the extra premise added to non-invertible rules are called *experts*: these predicates are responsible for examining certificates and, possibly, extracting information from them. We also allow for experts to invoke non-determinism: that is, they can guess additional information (such as substitution terms).

Depending on exactly how one defines and uses certificate terms, indexes, and the clerk and expert predicates (e.g., $\wedge_c(\cdot, \cdot, \cdot)$, $\exists_e(\cdot, \cdot, \cdot)$, etc.), the LKF^a inference rules can be directed to build widely varying proof structures. Collecting together such definitions yields what we call an *FPC*, that is, a *foundational proof certificate* definition. We shall present an example shortly.

5   Proof Checking as Logic Programming

It is possible to see the inference rules in Fig. 1 as describing a logic program: in particular, Fig. 2 contains part of a λProlog specification for several of those rules. The first five lines of Fig. 2 declare the types, constants, and predicates used to encode first-order polarized (*LKF*) formulas; the next seven lines declare the types and predicates of clerks and expert predicates; the next three lines provide the type declaration of the predicates that form the core of the proof checking kernel; and the remaining lines contain six clauses that are a direct specification of six inference rules from Fig. 1: specifically of the inference rules for introducing $\wedge^-$, $\forall$, and $\exists B$, and the rules for store, decide, and initial. All the remaining augmented inference rules can be specified in a similar fashion.

We have used λProlog here instead of Prolog for the following two major reasons.

Bindings in Formulas and Proofs. λProlog encodes bindings in formulas (quantifiers) and in proofs (eigenvariables) directly and implements them into its unification and substitution mechanisms. Achieving a Prolog implementation is possible but would require implementing such binding structures and associated logical operations, all rather difficult things to get right.

Context Management and its Dynamics During Proof Search. Focused and augmented sequents contain a context denoted by the Θ: this is intended to be a multiset of pairs of indexes and formulas. This multiset only needs to support the following operations: add a pair to the multiset (the store rule) and select a formula (nondeterministically) from the multiset by providing an index (the decide and initial rules). Notice that we do not need to know how many members there are in Θ nor anything about possible orderings between pairs. Note also that no functional dependency is assumed to hold between indexes and formula: that is, many formulas may be associated to the same index within a given Θ context. For all these reasons, the hypothetical context on λProlog serves as an interesting and direct implementation of this aspect of these inference rules. (See, for example, the fourth clause in Fig. 2.)

By forging such a direct link between a proof checking kernel and a logic program, that kernel has access to backtracking search and unification, which means that it can be used to support *proof reconstruction*: if a proof certificate does not contain all the details necessary to complete a (sequent calculus) proof, then it should be possible to allow backtracking and unification to discover some of them.

The full process of defining and checking a certain format of proof evidence can now be done as follows.

1. Pick some discipline to polarize a given classical logic formula into a polarized (*LKF*) formula.
2. Provide the signature (term constructors) for certificate (terms of type `cert`) and indexes (terms of type `index`). Any term structure possible in λProlog are allowed for these structures.

```
kind form, i          type.
type nand, por        form -> form -> form.
type all, some        (i -> form) -> form.
type complem          form -> form -> o.
type pos_or_lit       form -> o.

kind cert, index      type.
type andC             cert -> cert -> cert -> o.
type allC             cert -> (i -> cert)  -> o.
type storeC           cert -> cert -> index -> o.
type initE            cert ->          index -> o.
type someE            cert -> cert -> i      -> o.
type decideE          cert -> cert -> index -> o.

type uparrow          cert -> list form -> o.
type downarrow        cert ->      form -> o.
type store            index ->     form -> o.
```

```
uparrow Cert ((nand A B)::Gamma) :- andC Cert Cert1 Cert2,
     uparrow Cert1 (A::Gamma), uparrow Cert2 (B::Gamma).
uparrow Cert ((all B)::Gamma) :- allC Cert Cert',
     pi x\ uparrow (Cert' x) ((B x)::Gamma).
downarrow Cert (some B) :- someE Cert Cert' T,
     downarrow Cert' (B T).
uparrow Cert (C::Gamma) :- pos_or_lit C,
     storeC Cert Cert' Idx, store Idx C => uparrow Cert Gamma.
uparrow Cert nil :- decideE Cert Cert' Index, store Index B,
     downarrow Cert' B.
downarrow Cert B :- initE Cert Idx, store Idx C, complem B C.
```

Fig. 2. A λProlog implementation of part of Fig. 1 (Color figure online)

3. Provide logic program specifications of the clerk and expert relations.
4. Prove the goal **uparrow Cert (B::nil)** where B is the polarized form of the proposed theorem and **Cert** is the supplied certificate term (proof evidence).

6   Non-determinism in Proof Checking

In general, clerk predicates are intended to be functional: in the case of the conjunctive-clerk, this means that for every Ξ there exists at most one Ξ_1 and at most one Ξ_2 such that $\wedge_c(\Xi\,\Xi_1\,\Xi_2)$ is provable.

One could insist that this is also the case with experts. For example, if the $\exists$ expert predicate is functional then for every Ξ for which $\exists_e(\Xi,\Xi',t)$ is provable, the continuation certificate Ξ' and the substitution term t are uniquely determined. This can be achieved if, for example, the certificate simply stores this substitution term inside itself. While such proof certificates are clearly possible to design and check, one might want some flexibility in the design of certificates:

for example, storing all substitution instances might require certificates to be huge. If such substitution terms could be inferred from context using, say, the unification mechanism of logic programming, then the size of a proof certificate might be much smaller. Thus, allowing the proof checker to also reconstruct details of a proof allows proof certificates to be possibly much smaller in size.

For a specific example of the trade-off between proof size and proof checking, consider the following example (taken from [7]). It is possible to convert some decision procedures into proof certificates. For example, consider the procedure for determining whether or not a given propositional formula is a tautology: first compute the conjunctive normal form of the formula and then check that all the resulting clauses contain complementary literals. It is an easy matter to define a proof certificate encapsulating this procedure. Following the four steps mentioned above, we choose to polarize the connectives in a propositional classical formula using the negative (invertible) connectives. We then arrive at the following code (which specifies the result of the second and third steps).

```
type  lit        index.
type  cnf        cert.

andC      cnf cnf cnf  &  initE         cnf lit.
orC           cnf cnf  &  decideE    cnf cnf lit.
falseC        cnf cnf  &  storeC     cnf cnf lit.
releaseE      cnf cnf.
```

Here, there is exactly one proof certificate—just the token cnf. Similarly, there is just one index—the token lit—which is used to index all stored formulas. Note that the only formulas stored in this way are (both positive and negative) literals. Thus, the association between indexes and formulas is not functional and, as a result, the decide rule will be asked to chose some formula with index lit for which a complement is found. Such a step works perfectly in a logic programming setting where the decide rule (on index lit) is immediately followed by the initial rule (on index lit): thus the decide rule will generate positive literals and the initial rule will test those against available negative literals. Notice that if we required the indexing mechanism as well as the decide and initial experts to be functional, we would need to insert into the certificate a great deal of indexing information: since there can be exponentially many clauses in the conjunctive normal form of a propositional formula, such certificates would be huge, in contrast to the description of the cnf certificate that is described here.

7 Conclusion

Proof checking has been part of the history and development of high-level languages, starting, in particular, with the LCF system and the ML meta-language for it [19]. I have argued here that proof checking can be closely linked to logic programming. Such a close linkage should benefit both communities. There are, however, some challenges ahead for such a close relationship to actually occur.

One such challenge is that implementers and designers of logic programming languages have often favored efficiency and expressiveness over logical soundness: witness the presence of unification without the occurs-check, negation-as-failure, assert/retract, etc. Proof checking is a setting where logical soundness is paramount. While soundness can be delivered using unsound, quasi-logical processing, the logic programming community should certainly be able to deliver much more interesting and ambitious approaches to the implementation of logical deduction.

A second such challenge would be to have powerful techniques for reasoning about logic programming specifications. It is possible to view provability from Horn clauses as a simple inductive definition (a feature that a number of theorem provers support), but more direct support seems desirable since provability has more properties than just any inductive definition. In the case of logic programming with hypothetical reasoning and with bindings, as with λProlog, the simple inductive style approach is problematic. Fortunately, initial steps to address this challenge have been made in the design and implementation of the Abella theorem prover [3,14] which is capable of reasoning directly with specifications like those found in λProlog. Thus, it should be possible to develop formal proofs of correctness for λProlog based proof checkers using Abella and related tools.

Acknowledgments. Zak Chihani provided comments on an earlier version of this paper. The work presented here has been funded by the ERC Advanced Grant ProofCert.

References

1. Andreoli, J.-M.: Logic programming with focusing proofs in linear logic. J. Log. Comput. **2**(3), 297–347 (1992)
2. Assaf, A.: A framework for defining computational higher-order logics. Ph.D. thesis, École Polytechnique, September 2015
3. Baelde, D., Chaudhuri, K., Gacek, A., Miller, D., Nadathur, G., Tiu, A., Wang, Y.: Abella: a system for reasoning about relational specifications. J. Formalized Reason. **7**(2), 1–89 (2014)
4. Blanco, R., Miller, D.: Proof outlines as proof certificates: a system description. Draft available online, September 2015
5. Boyer, R.S., Moore, J.S., San Diego: A Computational Logic. Academic Press, San Diego (1979)
6. Chaudhuri, K., Pfenning, F., Price, G.: A logical characterization of forward and backward chaining in the inverse method. J. Autom. Reason. **40**(2–3), 133–177 (2008)
7. Chihani, Z., Miller, D., Renaud, F.: Foundational proof certificates in first-order logic. In: Bonacina, M.P. (ed.) CADE 2013. LNCS, vol. 7898, pp. 162–177. Springer, Heidelberg (2013)
8. Church, A.: A formulation of the simple theory of types. J. Symb. Log. **5**, 56–68 (1940)
9. Cousineau, D., Dowek, G.: Embedding pure type systems in the lambda-Pi-calculus modulo. In: Della Rocca, S.R. (ed.) TLCA 2007. LNCS, vol. 4583, pp. 102–117. Springer, Heidelberg (2007)

10. Danos, V., Joinet, J.B., Schellinx, H.: LKT and LKQ: sequent calculi for second order logic based upon dual linear decompositions of classical implication. In: Girard, J.-Y., Lafont, Y., Regnier, L., (eds.) Advances in Linear Logic. London Mathematical Society Lecture Note Series, vol. 22, pp. 211–224. Cambridge University Press (1995)
11. The Dedukti system (2013). https://www.rocq.inria.fr/deducteam/Dedukti/index.html
12. Dunchev, C., Guidi, F., Coen, C.S., Tassi, E.; A fast interpreter for λProlog. In: International Conference on LPAR-20: Logic Programming and Automated Reasoning (2015) (to appear)
13. Dyckhoff, R., Lengrand, S.: Call-by-value λ-calculus and LJQ. J. Logic Comput. **17**(6), 1109–1134 (2007)
14. Gacek, A., Miller, D., Nadathur, G.: A two-level logic approach to reasoning about computations. J. Autom. Reason. **49**(2), 241–273 (2012)
15. Gentzen, G.: Investigations into logical deduction. In: Szabo, M.E. (ed.) The Collected Papers of Gerhard Gentzen, pp. 68–131, North-Holland, Amsterdam (1935)
16. Girard, J.-Y.: Linear logic. Theo. Comput. Sci. **50**, 1–102 (1987)
17. Girard, J.-Y.: A new constructive logic: classical logic. Math. Struct. Comp. Sci. **1**, 255–296 (1991)
18. Gonthier, G.: The four colour theorem: engineering of a formal proof. In: Kapur, D. (ed.) ASCM 2007. LNCS (LNAI), vol. 5081, pp. 333–333. Springer, Heidelberg (2008)
19. Gordon, M.J., Milner, A.J., Wadsworth, C.P.: Edinburgh LCF: A Mechanised Logic of Computation. LNCS, vol. 78. Springer, Heidelberg (1979)
20. Hales, T.C.: A proof of the Kepler conjecture. Ann. Math. **162**(3), 1065–1185 (2005)
21. Harper, R., Honsell, F., Plotkin, G.: A framework for defining logics. J. ACM **40**(1), 143–184 (1993)
22. Herbelin, H.: Séquents qu'on calcule: de l'interprétation du calcul des séquents comme calcul de lambda-termes et comme calcul de stratégies gagnantes. Ph.D. thesis, Université Paris 7 (1995)
23. Howe, J.M., Proof search issues in some non-classical logics. Ph.D. thesis, University of St Andrews. Available as University of St Andrews Research Report CS/99/1, December 1998
24. Klein, G., Elphinstone, K., Heiser, G., Andronick, D., Cock, P., Derrin, D., Elkaduwe, K., Engelhardt, R., Kolanski, M., Norrish, T., Tuch, S.T., Winwood, S.: seL4: Formal verification of an OS kernel. In Proceedings of the 22nd Symposium on Operating Systems Principles (22nd SOSP 2009), Operating Systems Review (OSR), pp. 207–220, Big Sky, MT. ACM SIGOPS, October 2009
25. Laurent, O.: Etude de la polarisation en logique. Ph.D. thesis, Université Aix-Marseille II, March 2002
26. Leroy, X.: Formal verification of a realistic compiler. Commun. ACM **52**(7), 107–115 (2009)
27. Liang, C., Miller, D.: Focusing and polarization in linear, intuitionistic, and classical logics. Theo. Comput. Sci. **410**(46), 4747–4768 (2009)
28. Meng, J.: The integration of higher order interactive proof with first order automatic theorem proving. Ph.D. thesis, University of Cambridge, Computer Laboratory (2015)
29. Miller, D.: Abstractions in logic programming. In: Odifreddi, P. (ed.) Logic and Computer Science, pp. 329–359. Academic Press, San Diego (1990)

30. Miller, D.: Communicating and trusting proofs: the case for broad spectrum proof certificates. In: Proceedings of the Fourteenth International Congress on Logic, Methodology, and Philosophy of Science, pp. 323–342. College Publications (2014)
31. Miller, D., Nadathur, G.: Programming with Higher-Order Logic. Cambridge University Press, Cambridge (2012)
32. Miller, D., Nadathur, G., Pfenning, F., Scedrov, A.: Uniform proofs as a foundation for logic programming. Ann. Pure Appl. Logic **51**, 125–157 (1991)
33. Nadathur, G., Mitchell, D.J.: System description: Teyjus - a compiler and abstract machine based implementation of λProlog. In: Ganzinger, H. (ed.) CADE 1999. LNCS (LNAI), vol. 1632, pp. 287–291. Springer, Heidelberg (1999)
34. Pereira, F.: C-Prolog User's Manual, Version 1.5, June 1988
35. Pfenning, F., Schürmann, C.: System description: Twelf - a meta-logical framework for deductive systems. In: Ganzinger, H. (ed.) CADE 1999. LNCS (LNAI), vol. 1632, pp. 202–206. Springer, Heidelberg (1999)
36. Wetzler, N., Heule, M.J.H., Hunt Jr, W.A.: DRAT-trim: efficient checking and trimming using expressive clausal proofs. In: Sinz, C., Egly, U. (eds.) SAT 2014. LNCS, vol. 8561, pp. 422–429. Springer, Heidelberg (2014)

Semantics of Logic Languages

On Dual Programs in Co-Logic Programming

Hirohisa Seki[✉]

Department of Computer Science, Nagoya Institute of Technology,
Showa-ku, Nagoya 466-8555, Japan
seki@nitech.ac.jp

Abstract. Co-logic programming is an extension of the conventional
logic programming language, by allowing each predicate to be annotated
as either inductive or coinductive. To define its procedural semantics as
well as an alternating fixpoint semantics, the *stratification restriction*, a
condition on predicate dependency in programs, has been imposed on co-
logic programs (co-LPs). In this paper, we first consider *dual* programs
in co-logic programming: Given a program P, its dual program P^* is a
program such that it defines the "complement" of P, i.e., for any ground
atom $p(\bar{t})$, it computes its negation $\neg p(\bar{t})$. When we consider co-LPs
with negation, we show that the stratification restriction becomes too
restrictive in general, and that the Horn μ-calculus by Charatonik et al.
can be used as an extension of co-logic programming for handling "non-
stratified" co-LPs. We then consider some applications of non-stratified
co-LPs to Answer Set Programming (ASP) and the well-founded semantics (WFS). In particular, we give new iterated fixpoint characterizations
of answer sets as well as the WFS via dual programs. We also discuss
some applications of non-stratified co-LPs to program transformation
such as partial deduction, and a proof procedure for the WFS.

1 Introduction

Co-logic programming, proposed by Gupta et al. [11] and Simon et al. [26,27], is
an extension of logic programming, where each predicate in definite programs is
annotated as either *inductive* or *coinductive*. To define its semantics, the *stratification restriction*, a condition on predicate dependency in co-logic programs
(co-LPs), is assumed, and the declarative semantics by an alternating fixpoint
semantics has been given: the least fixpoints for inductive predicates and the
greatest fixpoints for coinductive predicates. A top-down procedural semantics,
co-SLD resolution, has also been proposed, and recent SWI Prolog [28] has added
support for coinduction.

As a result, co-logic programming provides a powerful computational framework, where many interesting applications such as modelling ω-automata [8],
model checking [6], non-monotonic reasoning and SAT solvers can be easily

H. Seki — This work was partially supported by JSPS Grant-in-Aid for Scientific
Research (C) 15K00305.

M. Falaschi (Ed.): LOPSTR 2015, LNCS 9527, pp. 21–35, 2015.
DOI: 10.1007/978-3-319-27436-2_2

expressed and computed (see, e.g., [12]). Recently, there has been reported some work [16,17] on applying co-LP techniques to Answer Set Programming using *dual programs.*

In this paper, we first consider *dual programs* in co-logic programming: Given a program P, its dual program P^* defines the "complement" of P, i.e., for any ground atom $p(\bar{t})$, $SEM(P) \models \neg p(\bar{t})$ iff $SEM(P^*) \models not\_p(\bar{t})$, where not_p is a new predicate symbol and $SEM(P)$ is a semantics of P. The notion of dual programs has been studied in the literature; among others, Sato and Tamaki [22] has introduced a technique for program transformation called the *negation technique*. It has also been utilized in partial evaluation (or *partial deduction*), program transformation, implementation of proof procedures (see, e.g., [1,4,20, 21,24]).

Considering dual programs of co-LPs with negation requires us to handle "non-stratified" co-LPs, and we show that the Horn μ-calculus by Charatonik et al. [5] can be used as an extension of co-logic programming. We then consider some applications of non-stratified co-LPs to Answer Set Programming (ASP) [10] and the well-founded semantics (WFS). In particular, we give new iterated fixpoint characterizations of answer sets as well as the WFS through dual programs. To the best of our knowledge, this is the first result of giving fixpoint characterizations of answer sets/WFS using dual programs. Some applications of non-stratified co-LPs to program transformation such as partial deduction, and a proof procedure for the WFS are also discussed.

The organization of this paper is as follows. In Sect. 2, we summarise some preliminary definitions on co-LPs and dual programs. In Sect. 3, we explain non-stratified co-logic programs and the Horn μ-calculus. In Sect. 4, we consider non-stratified co-LPs in the well-founded semantics. Finally, we discuss about the related work and give a summary of this work in Sect. 5.[1]

Throughout this paper, we assume that the reader is familiar with the basic concepts of logic programming, which are found in [3,15].

2   Preliminaries

In this section, we first recall some basic definitions and notations concerning co-logic programs (co-LPs). The details and more examples are found in [11,26,27]. Then, we also explain some preliminaries on deriving dual programs by negation elimination.

A *co-logic program* (co-LP) is a definite program, where predicate symbols are annotated as either inductive or coinductive. There is one restriction on co-LP, referred to as the *stratification restriction*: Inductive and coinductive predicates are not allowed to be mutually recursive. An example which violates the stratification restriction is $\{p \leftarrow q;\ q \leftarrow p\}$, where p is inductive, while q is coinductive.

When a co-LP P satisfies the stratification restriction, it is possible to decompose the set $\mathcal{P}$ of all predicates in P into a collection (called a *stratification*) of

---

[1] Due to space constraints, we omit most proofs and some details, which will appear in the full paper.

mutually disjoint sets $\mathcal{P}_0, \ldots, \mathcal{P}_r$ $(0 \leq r)$, called *strata*, so that, for every clause $p(\tilde{x}_0) \leftarrow p_1(\tilde{x}_1), \ldots, p_n(\tilde{x}_n)$ in P, we have that $\sigma(p) \geq \sigma(p_i)$ if p and p_i have the same inductive/coinductive annotations, and $\sigma(p) > \sigma(p_i)$ otherwise, where $\sigma(q) = i$, if the predicate symbol q belongs to $\mathcal{P}_i$. σ is called a *stratification function*.

The following is an example of co-LPs due to Simon et al. [26], which shows that co-logic programming can handle infinite terms such as infinite lists or trees like $f(f(\ldots))$ as well as finite ones.

Example 1 [26]. Suppose that predicates *member* and *drop* are annotated as inductive, while predicate *comember* is annotated as coinductive.

$member(H, [H|\_]) \leftarrow$ $drop(H, [H|T], T) \leftarrow$

$member(H, [\_|T]) \leftarrow member(H, T)$ $drop(H, [\_|T], T_1) \leftarrow drop(H, T, T_1)$

$comember(X, L) \leftarrow drop(X, L, L_1), comember(X, L_1)$

The definition of *member* is a conventional one; its meaning is defined in terms of the least fixpoint, since it is an inductive predicate. So, the prefix ending in the desired element H must be finite. The same applies to predicate *drop*.

On the other hand, predicate *comember* is coinductive, whose meaning is defined in terms of the greatest fixpoint. Therefore, it is true if and only if the desired element X occurs an infinite number of times in the list L. Hence it is false when the element does not occur in the list or when the element only occurs a finite number of times in the list.

For example, the query $X = 1, L = [0, 1|L], comember(X, L)$ is true, while the query $X = 1, L = [0, 1, 0, 1], comember(X, L)$ is false. Note that $L = [0, 1|L]$ represents an infinite list L consisting of 0 s and 1 s.   □

A meta-interpreter for co-logic programming has been developed and available [14], and recent SWI-Prolog (version 6.5.1) has also offered a module for supporting coinduction.[2]

The declarative semantics of a co-logic program is a stratified interleaving of the least fixpoint semantics and the greatest fixpoint semantics. To handle infinite terms, we consider the *complete* (or *infinitary*) Herbrand base [13,15], denoted by HB_P^*, where P is a program.[3]

Let P be a co-logic program with a stratification $\mathcal{P}_0, \ldots, \mathcal{P}_r$ $(0 \leq r)$. Let Π_i $(0 \leq i \leq r)$ be the set of clauses whose head predicates are in $\mathcal{P}_i$. Then, $P = \Pi_0 \cup \ldots \cup \Pi_r$. Let Π (resp., S) be a set of clauses (resp., ground atoms). Similarly to the "immediate consequence operator" T_P in the literature, our operator $T_{\Pi,S}$ assigns to every set I of ground atoms a new set $T_{\Pi,S}(I)$ of ground atoms as

---

[2] http://www.swi-prolog.org/pldoc/doc/swi/library/coinduction.pl.

[3] In the following sections, we will restrict ourselves to propositional programs for the ease of exposition, thus the "standard" Herbrand base HB_P will suffice. In this section, however, we explain some of the general basics of co-LPs for readers unfamiliar with them.

$$T_{\Pi,S}(I) = \{A \in HB_{\Pi}^* \mid \text{there is a ground instance of some clause in } \Pi$$
$$A \leftarrow B_1, \cdots, B_n, \ n \geq 0, \text{ such that, for every } 1 \leq i \leq n,$$
$$\text{either } B_i \in I \text{ or } B_i \in S\}.$$

In the above, the atoms in S are treated as facts. S is intended to be a set of atoms whose predicate symbols are in lower strata than those in the current stratum Π. We consider $T_{\Pi,S}$ to be the operator defined on the set of all subsets of HB_{Π}^*, ordered by standard inclusion. Then, $T_{\Pi,S}$ admits a least and a greatest fixpoint denoted by $lfp(T_{\Pi,S})$ and $gfp(T_{\Pi,S})$, respectively.

Finally, the model $M(P)$ of a co-logic program $P = \Pi_0 \uplus \ldots \uplus \Pi_r$ is defined inductively as follows: Let $M(\Pi_{-1}) = \emptyset$. For $k \geq 0$, $M(\Pi_k) = lfp(T_{\Pi_k, M_{k-1}})$ if $\mathcal{P}_i$ is inductive; $gfp(T_{\Pi_k, M_{k-1}})$ if $\mathcal{P}_i$ is coinductive, where M_{k-1} is the model of lower strata than Π_k, i.e., $M_{k-1} = \cup_{i=-1}^{k-1} M(\Pi_i)$.

Then, the *model* of P is $M(P) = \cup_{i=0}^{r} M(\Pi_i)$, the union of all models $M(\Pi_i)$.

Example 2. Let $P_0 = \{p \leftarrow q; \ q \leftarrow q; \ r \leftarrow r\}$ be a set of clauses. In the traditional logic programming, the meaning of P_0 is given in terms of the least fixpoint semantics, $lfp(T_{P_0}) = \emptyset$.

In co-logic programming, on the other hand, assume that p and r are inductive predicates, while q is a coinductive predicate. Then, since P_0 satisfies the stratification restriction, its meaning is defined in co-logic programming, i.e., $M(P_0) = \{p, q\}$.   □

Dual Programs in Co-Logic Programming. Our approach to handling negation in (co-)LPs is based on *negation elimination* (*NE* for short), a familiar program transformation technique [22], tailored to co-logic programs [24]. Given a (co-)LP P, the NE transformation derives from P a set P^* of definite clauses, called the *dual program* of P, by replacing negative literals $\neg p(\tilde{t})$ by $not\_p(\tilde{t})$, where not_p is a newly introduced predicate symbol.

In the following, we explain NE for a program P such that a clause in P might contain negative literals in its body for later use, and we assume that P is a propositional program for the ease of exposition, although it is applicable to programs without existential variables.[4] NE consists of the following two steps:

(step 1) for each clause in P, we replace each occurrence $\neg p$ of negative literals (if any) by not_p, where not_p is a new predicate not appearing elsewhere;
(step 2) for each predicate p, let $comp(p)$ be its completed definition in P. We then derive the definition of not_p from $comp(p)$ as follows:

(i) [Definition Derivation] Suppose that $comp(p)$ is of the form $p \leftrightarrow B_1 \vee \cdots \vee B_n$. Then, negating both sides of $comp(p)$, and replacing every negative occurrence $\neg p$ by not_p, we obtain $not\_p \leftrightarrow \neg(B_1 \vee \cdots \vee B_n)$.
Next, transforming the right-hand side in the above to a disjunctive form,

---

[4] A variable in a clause is *existential* if it appears in the body of the clause, but not in the head.

using De Morgan's laws, replacing each occurrence of $\neg\neg q$ by q, and each occurrence of $\neg q$ by not_q, we obtain the completed definition of not_p, i.e., $not\_p \leftrightarrow NB_1 \vee \cdots \vee NB_{n'}$, where each NB_i is a conjunction of positive literals. Finally, we transform $comp(not\_p)$ to a set of clauses: $\{not\_p \leftarrow NB_1; \ldots ; not\_p \leftarrow NB_{n'}\}$.

(ii) [Annotation Inversion] Annotate the derived predicate not_p as "coinductive" (resp. "inductive") if the annotation of the original predicate p is inductive (resp. coinductive).

Let P be a (definite) co-LP with the stratification restriction σ, and P^* be the set of all clauses obtained by applying the above NE transformation. We define the stratification function σ^* for P^* as follows: $\sigma^*(p) = \sigma(p)$ for all predicates defined in P, and $\sigma^*(not\_p) = \sigma(p) + 1$ for all predicates not_p newly introduced in NE. Then, we can show that P^* satisfies the stratification restriction w.r.t. σ^*.

Proposition 1. Correctness of Negation Elimination [24]
Let P be a *definite* co-logic program. If every clause in P has no existential variable, then the procedure of negation elimination gives a complementary co-logic program P^*, i.e., for any ground term $\tilde{t}$,

$$M(P) \models \neg p(\tilde{t}) \text{ iff } M(P^*) \models not\_p(\tilde{t}).$$

Example 3 (Continued from Example 2). Consider again $P_0 = \{p \leftarrow q; q \leftarrow q; r \leftarrow r\}$ in Example 2. Then, its dual program $P^* = P_0 \cup \{not\_p \leftarrow not\_q; not\_q \leftarrow not\_q; not\_r \leftarrow not\_r\}$.

Recall that p and r are inductive predicates, while q is a coinductive predicate. Therefore, not_p and not_r are coinductive, while not_q is inductive. Thus, $M(P^*) = \{p, q, not\_r\}$. We note that $M(P) \models \neg r$ and $M(P^*) \models not\_r$. □

In the following, we restrict ourselves to propositional programs, where the condition of NE (Proposition 1) is always satisfied. When a program has existential variables, NE will be still applicable if a certain condition [22] is satisfied. Another way is to use grounding by lparse in smodels [19], which allows us to deal with more general classes of programs such as range-restricted programs. Moreover, our applications of dual programs here include AS and abduction, where it is often the case that datalog programs are considered.

3   Non-stratified Co-Logic Programs and Horn μ-calculus

Proposition 1 deals with *definite* co-logic programs, thus satisfying the stratification restriction. However, the stratification restriction becomes too restrictive, when we consider dual programs of (co-)LPs with negation. In this section, we show that the Horn μ-calculus by Charatonik et al. [5] can be used as a framework for handling "non-stratified" co-LPs.

3.1   Dual Programs of Non-Stratified Programs

The notion of dual programs has been studied in logic programming; we have already explained the program transformation technique, called the *negation technique*, by Sato and Tamaki [22] in Sect. 2. Aravindan and Dung [4] have then proposed partial evaluation (or *partial deduction*) of logic programs in the well-founded semantics (WFS) using dual programs. Dual programs have also been used for implementing proof procedures; ABDUAL by Alferes et al. [1] and its successors such as TABDUAL [20,21] are abductive reasoning systems for the WFS. Marple and Gupta [16] have proposed a proof procedure for answer set programs using their dual programs.

In particular, Aravindan and Dung [4] have shown the following result on partial deduction in the WFS, which is given in the case of propositional programs for the sake of simplicity.

Proposition 2 (Aravindan and Dung). Partial Deduction in the WFS [4]
Let P be a program whose well-founded model is complete[5] and $P^{*-\mathrm{ai}}$ a negative partial deduction of P obtained by replacing selected negative literals $\neg p$ by not_p and adding new definitions for not_p. Then, for every goal $G: \leftarrow A$ which contains no not_p predicate, $WFS(P) \models A$ iff $WFS(P^{*-\mathrm{ai}}) \models A$.   □

In the above, when applied to propositional programs, a *negative partial deduction* of P, denoted by $P^{*-\mathrm{ai}}$, is the same as the dual program P^* explained in Sect. 2, except that annotation inversion is not employed. However, the above proposition is not always correct as the following example shows. Given a program P, we denote by $WFS(P) = \langle T; F \rangle$, where T (resp., F) is the set of atoms true (resp., false) in the WFS of P. The truth value of the remaining atoms in $U = H_P \setminus (T \cup F)$ is undefined, where H_P is the Herbrand base of P.

Example 4. Consider the following program $P = \{p \leftarrow \neg q;\ q \leftarrow q\}$. Since P is a stratified program, P has a unique answer set, i.e., the perfect model, $PERF(P) = \{p\}$, which coincides with $WFS(P) = \langle \{p\}; \{q\} \rangle$ and thus it is complete.

Consider the negative partial deduction $P^{*-\mathrm{ai}} = \{p \leftarrow not\_q;\ not\_p \leftarrow q;\ not\_q \leftarrow not\_q;\ q \leftarrow q\}$, where all the atoms are annotated as inductive.

Then, $WFS(P^{*-\mathrm{ai}}) = \langle \emptyset; \{p, q, not\_p, not\_q\} \rangle$. In particular, p is thus false.

On the other hand, the dual program P^* is the same set of clauses as $P^{*-\mathrm{ai}}$ with different predicate annotations: not_p and not_q are coinductive, while p and q are inductive. P^* satisfies the stratification restriction. Then, $M(P^*) = \{p, not\_q\}$; this means that P^* with its co-LP semantics $M(P^*)$, exactly captures the semantics of P in the WFS.   □

When we consider conventional general (i.e., non-stratified) programs, all predicates are supposed to be annotated as *inductive*. Then, the resulting dual programs do not satisfy the stratification restriction in general. For example,

---

[5] Well-founded model of a program P is complete when it classifies all the elements of the Herbrand base as 'true' or 'false'.

let $P = \{p \leftarrow \neg q; \ q \leftarrow \neg p\}$. Then, its dual program $P^* = \{p \leftarrow not\_q; \ q \leftarrow not\_p; \ not\_p \leftarrow q; \ not\_q \leftarrow p\}$, which does not satisfy the stratification restriction.

3.2   Horn μ-Calculus and Its Fixpoint Semantics

Charatonik et al. [5] have proposed the *Horn μ-calculus*; it is an extension of logic programs by allowing nesting of least and greatest fixpoints, in terms of a priority of each predicate for specifying whether its semantics has to be computed as a least or a greatest fixpoint. They have given to the Horn μ-programs the semantics based on ground proof trees as well as the nested fixpoints semantics.

A *Horn μ-program* (P, Ω) is a set of definite clauses in which every predicate symbol p in P is associated with a non-negative number $\Omega(p)$, called the *priority* of p.

Charatonik et al. [5] give an iterated fixpoint characterization of the semantics $[\![(P, \Omega)]\!]$ of a program P, which we will use in the following.

First we recall the familiar T_P operator of logic programming (see [15]); for any set A of ground atoms, we define $T_P(A)$ in the standard manner. Next, for all sets A and B of ground atoms, and non-negative priority k, we define $A[k := B]$ to be the set of ground atoms such that $p \in A[k := B]$ if either the priority of p is different from k and $p \in A$ or the priority of p equals k and $p \in B$. We now define the operator T_P^k such that $T_P^k(A) = A[k := T_P(A)]$; The operator T_P^k is thus analogous to T_P except that T_P^k only updates predicates of priority k. Then, for each value of k, we take fixpoints of the operators T_P^k. To do that, for any integer k we define $F_P^k(A)$ as follows. First, for negative integer k, we define $F_P^k(A) = A$.

$$F_P^k(A) = \begin{cases} \nu B.T_P^k(F_P^{k-1}(A[k := B])) & \text{if } k \text{ is even,} \\ \mu B.T_P^k(F_P^{k-1}(A[k := B])) & \text{if } k \text{ is odd,} \end{cases}$$

Then, $[\![(P, \Omega)]\!] = F_P^n(\emptyset)$, where $n = \max(\Omega(P))$, i.e., n is the maximal priority of any predicate in P.

It is easy to show that the Horn μ-calculus is an extension of co-logic programming. In fact, let P be a co-LP with a stratification σ. Then, we call a priority function Ω *consistent* with σ, if it satisfies the following: (i) $\sigma(p) \leq \sigma(q)$ iff $\Omega(p) \geq \Omega(q)$ for any predicates p and q, i.e., a predicate in lower stratum has a higher priority, and (ii) $\Omega(p)$ is even (odd) if p is a coinductive (inductive) predicate, respectively. Then, we have the following:

Proposition 3. Let P be a co-logic program with a stratification function σ. Then, $M(P) = F_P^n(\emptyset)$ with priority Ω, where Ω is consistent with σ, and $n = \max \Omega(p)$ for any predicate p in P. $\square$

Example 5. Consider again $P_0 = \{p \leftarrow q; \ q \leftarrow q; \ r \leftarrow r\}$ in Example 2, where p and r are inductive predicates, while q is a coinductive predicate. Then, its stratification function σ is defined as: $\sigma(q) = 0, \sigma(p) = \sigma(r) = 1$.

Consider a priority Ω consistent with σ such that $\Omega(p) = \Omega(r) = 1, \Omega(q) = 2$. Then, we have that $[\![(P_0, \Omega)]\!] = F_{P_0}^2(\emptyset) = \{p, q\} = M(P_0)$. $\square$

On the other hand, the Horn μ-calculus is more general than stratified co-LP.

Example 6 (Adapted and Simplified from [5]). Let $P_0 = \{p \leftarrow p; \ p \leftarrow q; \ q \leftarrow p\}$ be a set of clauses, where p is an inductive predicate, while q is a coinductive predicate. Since P_0 does not satisfy the stratification restriction, its meaning is not given in co-logic programming.

In the Horn μ-calculus, however, the semantics of P_0 can be determined in terms of priorities assigned to the predicates. Suppose, for example, that the coinductive predicate q has a higher priority than the inductive predicate p. We thus define: $\Omega(p) = 1$ and $\Omega(q) = 2$. Then, $[\![(P_0, \Omega)]\!] = \{p, q\}$. $\square$

The framework for unfold/fold transformation of co-logic programs is proposed in [23], where a program is assumed to satisfy the stratification restriction. We note that unfolding does not preserve the meaning of a Horn μ-program in general, as the following example shows.

Example 7. Let $P_0 = \{p \leftarrow q; \ q \leftarrow p\}$ be a set of clauses, where $\Omega(p) = 1$ (i.e., p is an inductive predicate), while $\Omega(q) = 2$ (i.e., q is a coinductive predicate). Then, $[\![(P_0, \Omega)]\!] = \{p, q\}$. Note that P_0 does not satisfy the stratification restriction.

$$P_0 : p \leftarrow q \xrightarrow{\text{unfolding}} P_1 : p \leftarrow p$$
$$q \leftarrow p \qquad\qquad\qquad q \leftarrow p$$

$$P_0 : p \leftarrow q \xrightarrow{\text{unfolding}} P_2 : p \leftarrow q$$
$$q \leftarrow p \qquad\qquad\qquad q \leftarrow q$$

In the above, the atoms in bold letters are the ones on which unfolding is applied. Then, $M(P_1) = \emptyset$, while $M(P_2) = \{p, q\}$. Therefore, when P_0 does not satisfy the stratification restriction, a simple application of unfolding will yield programs with different meanings. $\square$

We will consider unfolding in the Horn μ-calculus in the following section.

Answer Set Programs and Horn μ-programs. We are now in a position to give the relationship between answer sets and Horn μ-programs. Let M be a set of atoms and M^* a set of atoms including ones of the form not_p for some atom p. In the following, we denote by $M \equiv M^*$ if, for any ground atom p, $p \in M$ iff $p \in M^*$, and $p \notin M$ iff $not\_p \in M^*$.

Example 8. Consider again the following program $P = \{p \leftarrow \neg q; \ q \leftarrow \neg p\}$ and its dual program $P^* = \{p \leftarrow not\_q; \ q \leftarrow not\_p; \ not\_p \leftarrow q; \ not\_q \leftarrow p\}$. P is a non-stratified program, and it has two answer sets $M_1 = \{p\}$ and $M_2 = \{q\}$.

On the other hand, P^* does not satisfy the stratification restriction. First, we consider a Horn μ-program (P^*, Ω_1), where we define the priority Ω_1 as

$\Omega_1(not\_q) = 2$ and $\Omega_1(not\_p) = 0$, while $\Omega_1(p) = \Omega_1(q) = 1$. Then, $[\![(P, \Omega_1)]\!] = \{p, not\_q\}$, thus $M_1 \equiv [\![(P, \Omega_1)]\!]$.

On the other hand, we consider a Horn μ-program (P^*, Ω_2) in a symmetric fashion; we define the priority Ω_2 as $\Omega_2(not\_p) = 2$ and $\Omega_2(not\_q) = 0$, while $\Omega_2(p) = \Omega_2(q) = 1$ as before. Then, $[\![(P, \Omega_2)]\!] = \{q, not\_p\}$, thus $M_2 \equiv [\![(P, \Omega_2)]\!]$.   □

In general, we can show the following relationship between the answer sets of P and the fixpoints of $F_{P^*}^2(\emptyset)$.

Proposition 4. AS is a fixpoint of a dual program
Let P be a logic program and $AS(P)$ the set of its answer sets. If $M \in AS(P)$, then there exists a priority Ω such that $M \equiv F_{P^*}^2(\emptyset)$.

In particular, when P is a stratified program, $PERF(P) \equiv F_{P^*}^2(\emptyset) = M(P^*)$, where priority Ω is defined to be consistent with the stratification function of P.   □

Example 4 is a special case of the above proposition.

Unfolding Dual Programs. Next, we consider unfolding of dual programs.

Example 9. Consider again the dual program P^* in Example 8 and recall that the priority Ω_1 is defined as $\Omega_1(not\_q) = 2$ and $\Omega_1(not\_p) = 0$, while $\Omega_1(p) = \Omega_1(q) = 1$. Then, $[\![(P, \Omega_1)]\!] = \{p, not\_q\}$. We consider the following two cases of applying unfolding to P^*, where the atoms in bold letters are the ones on which unfolding are applied.

(i)
$$P^* : p \leftarrow not\_q \xrightarrow{\text{unfolding}} P_1^* : p \leftarrow not\_q$$
$$q \leftarrow \mathbf{not\_p} \qquad\qquad q \leftarrow q$$
$$not\_p \leftarrow q \qquad\qquad not\_p \leftarrow q$$
$$not\_q \leftarrow p \qquad\qquad not\_q \leftarrow p$$

Then, $[\![(P_1^*, \Omega_1)]\!] = \{p, not\_q\}$, thus unfolding preserves the semantics. We note that, in the unfolded clause, $\Omega_1(q) \geq \Omega_1(not\_p)$.

(ii)
$$P^* : p \leftarrow \mathbf{not\_q} \xrightarrow{\text{unfolding}} P_2^* : p \leftarrow p$$
$$q \leftarrow not\_p \qquad\qquad q \leftarrow not\_p$$
$$not\_p \leftarrow q \qquad\qquad not\_p \leftarrow q$$
$$not\_q \leftarrow p \qquad\qquad not\_q \leftarrow p$$

Then, $[\![(P_2^*, \Omega_1)]\!] = \emptyset$, thus unfolding does not preserve the semantics. We note that, in the unfolded clause, $\Omega_1(p) < \Omega_1(not\_q)$.   □

The following proposition explains the above applications of unfolding.

Proposition 5. Unfolding of Horn μ-programs
Let (P, Ω) be a Horn μ-program and P' a program derived by applying unfolding to P. Let $p \leftarrow q_1, \ldots, q_n$ $(n > 1)$ be the unfolded clause in P and q_i $(1 \leq i \leq n)$ the atom upon which unfolding is applied. If $\Omega(p) \geq \Omega(q_i)$, then the semantics of (P, Ω) is preserved, i.e., $[\![(P, \Omega)]\!] = [\![(P', \Omega)]\!]$. □

Aravindan and Dung [4] also studied unfolding of dual programs in the well-founded semantics, where unfolding is defined as usual, i.e., no condition is imposed on applying unfolding. In contrast, unfolding in our framework requires the above-mentioned condition for its application, since each atom in a Horn μ-program is assigned its priority and the semantics (P, Ω) is defined based on the priorities of atoms.

4   Dual Programs in the Well-Founded Semantics

Finally, we consider a fixpoint characterization of the well-founded semantics (WFS) via dual programs, which is based on the Horn μ-calculus.

Proposition 6. Fixpoint Characterization of the WFS
Let P be a logic program and v the truth valuation in the well-founded semantics $WFS(P)$. Let P^* be its dual program and Ω be a priority defined as $\Omega(p) = 1$ (resp., $\Omega(not\_p) = 2$) for any predicate p in P. Then, we have

- $v(p) = \mathbf{t}$ iff $p \in F_{P^*}^2(\emptyset)$ and $not\_p \notin F_{P^*}^2(\emptyset)$,
- $v(p) = \mathbf{f}$ iff $not\_p \in F_{P^*}^2(\emptyset)$ and $p \notin F_{P^*}^2(\emptyset)$, and
- $v(p) = \mathbf{u}$ iff $p \in F_{P^*}^2(\emptyset)$ and $not\_p \in F_{P^*}^2(\emptyset)$. □

Example 10. Consider the following non-stratified program $P = \{p \leftarrow q;\ q \leftarrow \neg p\}$, and its dual program $P^* = \{p \leftarrow q;\ q \leftarrow not\_p;\ not\_p \leftarrow not\_q;\ not\_q \leftarrow p\}$. Then, $WFS(P) = \langle \emptyset; \emptyset \rangle$, i.e., the truth value of every atom in P is undefined: $v(a) = \mathbf{u}$ for $a \in \{p, q\}$.

On the other hand, we consider a Horn μ-program (P^*, Ω), where we define the priority Ω as $\Omega(not\_a) = 2$ and $\Omega(a) = 1$ for $a \in \{p, q\}$. Then, $[\![(P, \Omega)]\!] = \{p, q, not\_p, not\_q\}$. □

A Proof Procedure of Dual Programs in the WFS. From Proposition 6, we will propose a simple proof procedure, which is based on the notion of *P-derivation* of the Horn μ-calculus [5].

The semantics of a Horn μ-program is given in terms of ground derivations [5]. Given a logic program P, r is called a *P-derivation* if for each node n in r, labeled by some ground atom h, there exists a ground instance $h \leftarrow b_1, \ldots, b_m$ $(m \geq 0)$ of a clause in P, and n has m children nodes, each of which is labeled by b_i $(0 \leq i \leq m)$. When $m = 0$, the node n has no children nodes and is a *success node*. If there are no such clauses in P, then n has no children nodes and is a *failure node*. When the root node of r is labeled by p, r is a P-derivation *of p* (or p has a P-derivation r).

Given a P-derivation r, let w be an infinite sequence $w_0 w_1 w_2 \ldots$ of nodes in r such that w_{i+1} is a child of w_i. Such a sequence w is called an *infinite path*. For an infinite path π in a P-derivation, we denote by $Inf(\pi)$ the set of all priorities occurring infinitely often on the path π. We say that a path w in r is *accepting* if the largest element of $Inf(\pi)$ is even. A P-derivation r is *accepting* if every infinite path in r is accepting and every finite path ends with a success node. A P-derivation r is *not accepting* if there exists either an infinite path in r which is not accepting or a finite path ending with a failure node.

From the equivalence of the procedural semantics and the iterated fixpoint semantics of the Horn μ-calculus [5], we have the following characterization of procedural semantics of a dual program P^* in the WFS.

- $v(p) = \mathbf{t}$ iff p (resp. not_p) has an (resp. no) accepting P^*-derivation,
- $v(p) = \mathbf{f}$ iff not_p (resp. p) has an (resp. no) accepting P^*-derivation, and
- $v(p) = \mathbf{u}$ iff both p and not_p have accepting P^*-derivations.

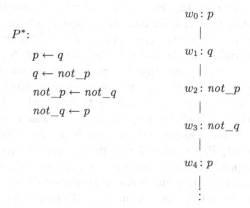

Fig. 1. The P^*-Derivation of p (Example 10)

Since we consider here propositional dual programs in the WFS, we can have a simple procedure for detecting whether $v(p) = \mathbf{u}$ for a ground atom p. To do that, we will make the notion of *(non)-acceptance* more detailed.

Given a P^*-derivation r of an atom p_0, suppose that there is an infinite path $\pi = w_0 w_1 w_2 \ldots$ in r. We denote the label of node w_i $(i \geq 0)$ by $w_i.l$. Since dual programs are propositional, there exist some node w_n $(n \geq 0)$ and $k \geq 0$ such that w_n and w_{n+k+1} have the same label, i.e., $w_n.l = w_{n+k+1}.l$. Then, π is of the form: $w_0 w_1 \ldots (w_n \ldots w_{n+k})^\omega$.

Definition 1. Let r be a P^*-derivation. Then,

- r is called *successful* if every finite path in r ends with a success node and for every infinite path π in r, $\Omega(w_i.l) = 2$ for all i $(n \leq i \leq n+k)$.

– r is called *failed* if there exists either a finite path in r ending with a failure node or an infinite path π in r such that $\Omega(w_i.l) = 1$ for all i ($n \leq i \leq n+k$).

– r is called *undefined* if it is neither successful nor failed. In this case, r has an infinite path π such that $\Omega(w_n.l) \neq \Omega(w_m.l)$ for some m ($n \leq m \leq n+k$). □

Example 11. Figure 1 shows a P^*-derivation of p in Example 10. Recall that $v(p) = \mathbf{u}$. p has a single P^*-derivation r; it consists of a single infinite path $\pi = (w_0 \ldots w_3)^\omega$, where $\Omega(w_0.l) = \Omega(p) = 1$ and $\Omega(w_2.l) = \Omega(not\_p) = 2$. The priorities of labels in the infinite loop are thus alternating. □

Proposition 7. Procedural Semantics of Dual Programs in the WFS
Let P be a logic program and v the truth valuation in the well-founded semantics $WFS(P)$. Let P^* be its dual program and Ω be a priority defined as $\Omega(p) = 1$ (resp., $\Omega(not\_p) = 2$) for any predicate p in P. Then, we have

– $v(p) = \mathbf{t}$ iff p has a successful P^*-derivation,
– $v(p) = \mathbf{f}$ iff all of P^*-derivations of p are failed, and
– $v(p) = \mathbf{u}$ iff all of P^*-derivations of p are neither successful nor failed.   □

We note some implementation issues on the above procedural semantics. ABDUAL [1] has some problems in handling loops involving negative atoms, referred to as "negative loops over negation" (*NLoN*) [20]. In Example 10, for example, goals ← p and ← not_p succeed unexpectedly in ABDUAL. In TABD-UAL [20], such problems have been remedied by introducing some mechanisms for dealing with NLoN. However, its implementation for handling NLoN is dependent on the XSB built-in predicates such as tnot/1 and call_tv/2 (see Fig. 2 (below)) together with an auxiliary predicate over/1, which would make it difficult to perform program analysis and its possible optimization.

On the other hand, Proposition 7 will give a simple proof procedure for the WFS. For example, predicate solve/1 in DRA interpreter [14] (see Fig. 2 (above)) has an argument which maintains information about the current path of ancestors (stack). Using this mechanism, it would be simple to check whether a path currently stored on the stack is successful, failed or undefined, since it is enough to examine the labels of atoms in the stack.

5   Related Work and Concluding Remarks

The notion of dual programs has been proposed and utilized in various fields in logic programming. Techniques using dual programs have been proposed for performing partial evaluation (or partial deduction), program transformation and implementation of reasoning systems and proof procedures (see, for example, [1,4,16–18,20–22,24]). In this paper we have extended to non-stratified (co-)LPs negation elimination with the operation of annotation inversion, which was proposed for definite co-LPs [24].

The main contributions of this paper are the following. (i) We have shown that the Horn μ-calculus can be utilized as a framework for handling non-stratified co-LPs. In fact, the Horn μ-calculus is an extension of co-logic programming (Proposition 3). Gupta et al. [12] have also proposed an extension

```
% solve/1 in DRA interpreter
solve( + sequence of goals, + stack, + coinductive hypotheses, + level):
%% Solve the sequence of goals, maintaining information about the current chain
%% of tabled ancestors (stack) and the chain of coinductive ancestors
%% (coinductive hypotheses). The level is the level of recursion, and is used
%% only for tracing.

% Tbdual Implementation of Ex. 10
% the predicate over(G)/1 is defined as over(G) :- tnot(G).
1. :- table q_ab/1, over/1, not_p/1, p_st/3.
2. q_ab(E) :- tnot p_ab([]), not_p_ab([],E).
3. not_p_ab(I,O) :- call_tv(tnot over(not_p(I)), V),
        (V=undefined, O=I, undefined;
          inspect(p_st(I,O,[]))).
4. not_p(I) :- p_st(I,O,[]).
5. ... (omitted)
```

Fig. 2. Predicate solve/1 in DRA interpreter [14] (above) and implementation in TABDUAL (below)

of co-LPs to handle non-stratified co-LPs. To do that, they have introduced *strong/weak_inductive* annotations, which play a similar role of priorities in the Horn μ-calculus. However, they have not discussed the relationship of their extension with the Horn μ-calculus, and its declarative semantics is not known either. (ii) We have given new iterated fixpoint characterizations of answer sets as well as the WFS for propositional programs via dual programs. A lot of work has been done on the fixpoint semantics for logic programming (see, e.g., an excellent survey in [9]). Denecker et al. [7], for example, have proposed a fixpoint theory as a uniform framework of major semantics of general logic programs. In contrast, our approach has focused on the use of dual programs, and to the best of our knowledge, this is the first result of giving fixpoint characterizations of answer sets/WFS through dual programs. In [25], the relationship between co-LP and the Horn μ-calculus has been studied from the procedural point of view. (iii) Finally, we have proposed an unfolding rule for Horn μ-programs (Proposition 5) and a procedural semantics of dual programs in the WFS (Proposition 7). The unfolding rule in this paper is more general than that for co-LPs [23] in that the latter is applicable only when the stratification restriction of a given program is satisfied. Furthermore, the proof procedure for the WFS in this paper is much simpler than that in [25] in that the latter requires checking a well-founded ordering among ground atoms, while such checking is replaced here by simply examining whether priorities of labels are alternating or not.

In this paper, we have restricted ourselves to propositional programs. This would be reasonable, since co-logic programming has some computational difficulty [2]. This restriction is due to the condition of NE (Proposition 1), i.e., no existential variables in every clause in a given co-LP. One direction for future work is thus to extend the current framework to handle a more general class of co-LPs. Some approaches have already been mentioned in the end of

Sect. 2. Another approach will be to allow arbitrary first order logic formulas in the body of a clause, as in the work by Denecker et al. [7].

We have proposed the proof procedure for the WFS, and compared it with ABDUAL and TABDUAL (Fig. 2). It will be interesting to extend proof procedure to allow *abducibles* for performing abduction. We also have a plan to implement our proof procedure for the WFS in Proposition 7.

Acknowledgement. The author would like to thank anonymous reviewers for their constructive and useful comments on the previous version of the paper. The idea of using co-LP techniques for a proof procedure for the WFS in Sect. 4 came from the discussions with Gopal Gupta at LOPSTR'13 in Madrid.

References

1. Alferes, J.J., Pereira, L.M., Swift, T.: Abduction in well-founded semantics and generalized stable models via tabled dual programs. Theor. Pract. Log. Program. **4**, 383–428 (2004)
2. Ancona, D., Dovier, A.: co-LP: Back to the Roots. Theory and Practice of Logic Programming 13(4–5) (2013)
3. Apt, K.R.: Introduction to logic programming. In: Handbook of Theoretical Computer Science, pp. 493–576. Elsevier (1990)
4. Aravindan, C., Dung, P.M.: Partial deduction of logic programs wrt well-founded semantics. New Gener. Comput. **13**(1), 45–74 (1994)
5. Charatonik, W., McAllester, D., Niwinski, D., Podelski, A., Walukiewicz, I.: The Horn Mu–calculus. In: LICS 1998, pp. 58–69. IEEE Computer Society (1998)
6. Clarke, E.M., Grumberg, O., Peled, D.A.: Model Checking. MIT Press, Cambridge (1999)
7. Denecker, M., Bruynooghe, M., Vennekens, J.: Approximation fixpoint theory and the semantics of logic and answers set programs. In: Erdem, E., Lee, J., Lierler, Y., Pearce, D. (eds.) Correct Reasoning. LNCS, vol. 7265, pp. 178–194. Springer, Heidelberg (2012)
8. Farwer, B.: ω-Automata. In: Grädel, E., Thomas, W., Wilke, T. (eds.) Automata, Logics, and Infinite Games. LNCS, vol. 2500, pp. 3–21. Springer, Heidelberg (2002)
9. Fitting, M.: Fixpoint semantics for logic programming a survey. Theoret. Comput. Sci. **278**(1–2), 25–51 (2002)
10. Gelfond, M., Lifschitz, V.: The stable model semantics for logic programming. In: Proceedings of the Fifth International Conference and Symposium on Logic Programming, pp. 1070–1080. MIT Press (1988)
11. Gupta, G., Bansal, A., Min, R., Simon, L., Mallya, A.: Coinductive logic programming and its applications. In: Dahl, V., Niemelä, I. (eds.) ICLP 2007. LNCS, vol. 4670, pp. 27–44. Springer, Heidelberg (2007)
12. Gupta, G., Saeedloei, N., DeVries, B., Min, R., Marple, K., Kluźniak, F.: Infinite computation, co-induction and computational logic. In: Corradini, A., Klin, B., Cîrstea, C. (eds.) CALCO 2011. LNCS, vol. 6859, pp. 40–54. Springer, Heidelberg (2011)
13. Jaffar, J., Stuckey, P.: Semantics of infinite tree logic programming. Theoret. Comput. Sci. **46**, 141–158 (1986)

14. Kluźniak, F.: Meta-interpreter supporting tabling and coinduction (2009). http://www.utdallas.edu/gupta/meta.html
15. Lloyd, J.: Foundations of Logic Programming, 2nd edn. Springer, Heidelberg (1987). Extended edition
16. Marple, K., Bansal, A., Min, R., Gupta, G.: Goal-directed execution of answer set programs. In: PPDP 2012. pp. 35–44 (2012)
17. Marple, K., Bansal, A., Min, R., Gupta, G.: Dynamic consistency checking in goal-directed answer set programming. Theory Pract. Log. Program. **14**(4–5), 415–427 (2014)
18. Min, R.: Predicate Answer Set Programming with Coinduction. Ph.D. thesis, Universityof Texas at Dallas (2010)
19. Niemelä, I., Simons, A.: Smodels — an implementation of the stable model and well-founded semantics for normal logic programs. In: Fuhrbach, U., Dix, J., Nerode, A. (eds.) LPNMR 1997. LNCS, vol. 1265, pp. 420–429. Springer, Heidelberg (1997)
20. Pereira, L.M., Saptawijaya, A.: Abductive logic programming with tabled abduction. In: Proceedings of the Seventh International Conference on Software Engineering Advances ICSEA 2012, pp. 548–556 (2012)
21. Saptawijaya, A., Pereira, L.M.: Tabled abduction in logic programs. Theory Pract. Log. Program. **13**, 4–5 (2013)
22. Sato, T., Tamaki, H.: Transformational logic program synthesis. In: FGCS 1984 Tokyo, pp. 195–201 (1984)
23. Seki, H.: Proving properties of co-logic programs by unfold/fold transformations. In: Vidal, G. (ed.) LOPSTR 2011. LNCS, vol. 7225, pp. 205–220. Springer, Heidelberg (2012)
24. Seki, H.: Proving properties of co-logic programs with negation by program transformations. In: Albert, E. (ed.) LOPSTR 2012. LNCS, vol. 7844, pp. 213–227. Springer, Heidelberg (2013)
25. Seki, H.: Extending co-logic programs for branching-time model checking. In: Gupta, G., Peña, R. (eds.) LOPSTR 2013, LNCS 8901. LNCS, vol. 8901, pp. 127–144. Springer, Heidelberg (2014)
26. Simon, L., Mallya, A., Bansal, A., Gupta, G.: Coinductive logic programming. In: Etalle, S., Truszczyński, M. (eds.) ICLP 2006. LNCS, vol. 4079, pp. 330–345. Springer, Heidelberg (2006)
27. Simon, L.E.: Extending Logic Programming with Coinduction. Ph.D. thesis. University of Texas at Dallas (2006)
28. Wielemaker, J., Schrijvers, T., Triska, M., Lager, T.: SWI-Prolog. Theory Pract. Log. Program. **12**(1–2), 67–96 (2012)

Equational Formulas and Pattern Operations in Initial Order-Sorted Algebras

José Meseguer and Stephen Skeirik[(⊠)]

Department of Computer Science,
University of Illinois at Urbana-Champaign, Champaign, USA
skeirik2@illinois.edu

Abstract. A *pattern* t, i.e., a term possibly with variables, denotes the set (language) $[\![t]\!]$ of all its *ground instances*. In an untyped setting, symbolic operations on finite sets of patterns can represent Boolean operations on languages. But for the more expressive patterns needed in declarative languages supporting rich type disciplines such as subtype polymorphism untyped pattern operations and algorithms break down. We show how they can be properly defined by means of a signature transformation $\Sigma \mapsto \Sigma^{\#}$ that enriches the types of Σ. We also show that this transformation allows a systematic reduction of the first-order logic properties of an initial order-sorted algebra supporting subtype-polymorphic functions to equivalent properties of an initial many-sorted (i.e., simply typed) algebra. This yields a new, simple proof of the known decidability of the first-order theory of an initial order-sorted algebra.

Keywords: Pattern operations · Initial decidability · Order-Sorted logic

1 Introduction

Term patterns are used everywhere in functional and logic programming: to define predicates and functions, to perform automated deduction tasks like rewriting, matching, unification, resolution, and Knuth-Bendix completion, and also as a *symbolic notation* to describe *languages* as sets of term instances, and *language operations* by corresponding symbolic operations on the term patterns defining them. Such pattern operations, first systematically studied by Lassez and Marriott in [12] and further studied in, e.g., [6,11,18,19] have many applications to, e.g., machine learning, negation in logic programming, sufficient completeness of function definitions, inductive theorem proving, and automated model building.

For greater expressiveness many declarative languages support rich type disciplines. This holds true for both higher-order functional languages and rule-based languages. For example, OBJ [10], CafeOBJ [7], and Maude [1] all support types, subtypes, subtype polymorphism, and–through their parameterized types—polymorphic and dependent types. Obviously, all the above-mentioned applications of pattern operations are also needed for these languages. What is not at all

© Springer International Publishing Switzerland 2015
M. Falaschi (Ed.): LOPSTR 2015, LNCS 9527, pp. 36–53, 2015.
DOI: 10.1007/978-3-319-27436-2_3

obvious—and to the best of our knowledge does not seem to have been investigated so far—is whether the algorithms defining the Boolean algebra of pattern operations for the *untyped* case in, e.g., [6,11,12,18,19] extend in a straightforward way to the more expressive patterns now available in these richer type disciplines. The example below clearly shows that they do not.

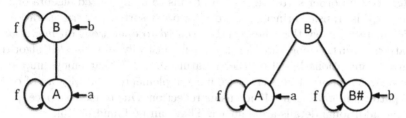

The graph on the left describes an *order-sorted signature* [9] with two types, A and B, and a subtype inclusion $A < B$ depicted by the vertical bar. f is *subtype polymorphic*, with two typings: $f : A \rightarrow A$, and $f : B \rightarrow B$. We have constants a, b of respective types A, B. A *pattern* t, i.e., a term possibly with variables, denotes the set (language) $\llbracket t \rrbracket = \{t\sigma \mid \sigma \ ground\}$ of all its *ground instances*. The symbolic pattern difference $t - t'$ denotes the language $\llbracket t - t' \rrbracket = \llbracket t \rrbracket - \llbracket t' \rrbracket$. In the untyped case, it is well-known [12] that when t and t' are *linear patterns* (have no repeated variables), the symbolic difference $t - t'$ always denotes a language expressible as $\llbracket u_1 \rrbracket \cup \ldots \cup \llbracket u_k \rrbracket$, for $\{u_1, \ldots, u_k\}$ a finite set of patterns. If this were to hold in the order-sorted case, it should hold, in particular, for $\llbracket x{:}B - y{:}A \rrbracket$, with $x{:}B, y{:}A$ variables of sorts A, B. Adopting the convention $f^0(x) = x$, we have, $\llbracket y{:}A \rrbracket = \{f^n(a) \mid n \geq 0\}$, and $\llbracket x{:}B \rrbracket = \llbracket y{:}A \rrbracket \cup \{f^n(b) \mid n \geq 0\}$. Therefore, $\llbracket x{:}B - y{:}A \rrbracket = \{f^n(b) \mid n \geq 0\}$. But *there is no finite set of patterns* $\{u_1, \ldots, u_k\}$ such that $\llbracket u_1 \rrbracket \cup \ldots \cup \llbracket u_k \rrbracket = \{f^n(b) \mid n \geq 0\}$. Indeed, the only possible choice for a u_i is $u_i = b$. All other choices: $u_i = a$, $u_i = x'{:}B$, $u_i = y'{:}A$, $u_i = f^{n+1}(x'{:}B)$, or $u_i = f^{n+1}(y'{:}A)$, $n \geq 0$, are impossible.

Is all lost? Not if we make our signature more expressive: the graph on the right adds a new subtype $B^\# < B$, lowers the typing of b to $B^\#$, and adds the typing $f : B^\# \rightarrow B^\#$. Now $\llbracket z{:}B^\# \rrbracket = \{f^n(b) \mid n \geq 0\}$, and we can *symbolically compute* the difference $x{:}B - y{:}A = z{:}B^\#$. This example shows that the problem is insoluble *as formulated*, but it can be solved by a *signature transformation* extending the original signature Σ. In Sect. 3 we formally define such a transformation $\Sigma \mapsto \Sigma^\#$ that enriches a finite order-sorted signature Σ with additional sorts like the sort $B^\#$ above. This is a key step for obtaining a Boolean algebra of *order-sorted* patterns in Sect. 5.

But the $\Sigma \mapsto \Sigma^\#$ transformation has other far-reaching consequences. Since it is well-known that pattern operations are intimately connected with first-order logic formulas and with negation elimination in such formulas [6,11,18,19], we should first of all ask what light can the $\Sigma \mapsto \Sigma^\#$ transformation shed on the *validity of formulas in initial order-sorted algebras*. As we show in Sect. 4, it sheds a lot of light: it makes the validity of a first-order formula in an initial *order-sorted* algebra equivalent to the validity of an associated formula in an associated *many-sorted* initial algebra. Since the first-order theory of a many-sorted initial algebra is well-known to be decidable [2,13,14], this proves the decidability of the first-order theory of an initial order-sorted algebra. This result goes back to [3,4], but the proof obtained through the $\Sigma \mapsto \Sigma^{\#}$ transformation is considerably simpler. Furthermore, it provides a new, general *transfer principle* to *reduce* certain order-sorted algebra problems to many-sorted algebra ones.

We put this transfer principle to work for order-sorted pattern operations in Sect. 5, where we show that they can be *reduced* to operations on *many-sorted* $\Sigma^{\#}$-patterns. Furthermore, we develop an intrinsically *order-sorted* algorithm for pattern operations based on the signature $\Sigma \cup \Sigma^{\#}$ that enjoys important advantages. As reported in Sect. 6, we have implemented this algebra of order-sorted pattern operations in Maude using reflection. Due to lack of space, proofs and some additional details are omitted. They can be found in [15].

2   Preliminaries on Order-Sorted Algebra

The following material is adapted from [16], which generalizes [9]. It summarizes the basic notions of order-sorted algebra needed in the rest of the paper. It assumes the notions of many-sorted signature and many-sorted algebra, e.g., [5].

Definition 1. *An* order-sorted signature *is a triple* $\Sigma = (S, \leq, \Sigma)$ *with* $(S, \leq)$ *a poset and* (S, Σ) *a many-sorted signature.*

$\hat{S} = S/\equiv_{\leq}$, *called the set of* connected components *of* $(S, \leq)$, *is the quotient of* S *under the equivalence relation* $\equiv_{\leq} = (\leq \cup \geq)^{+}$. *The order* $\leq$ *and equivalence* $\equiv_{\leq}$ *are extended to sequences of the same length in the usual way, e.g.,* $s'_1 \ldots s'_n \leq s_1 \ldots s_n$ *iff* $s'_i \leq s_i$, $1 \leq i \leq n$.

Σ *is called* sensible *(resp. monotonic) if for any two operators* $f : w \to s$, $f : w' \to s' \in \Sigma$, *with* w *and* w' *of same length, we have* $w \equiv_{\leq} w' \Rightarrow s \equiv_{\leq} s'$. *(resp.* $w \geq w' \Rightarrow s \geq s'$). *Note that a many-sorted signature* Σ *is the special case in which the poset* $(S, \leq)$ *is discrete, i.e.,* $s \leq s'$ *iff* $s = s'$.

For connected components $[s_1], \ldots, [s_n], [s] \in \hat{S}$

$$f^{[s_1] \ldots [s_n]}_{[s]} = \{f : s'_1 \ldots s'_n \to s' \in \Sigma \mid s'_i \in [s_i] \ 1 \leq i \leq n, \ s' \in [s]\}$$

is the family of "subsort polymorphic" operators f *for those components.*   □

We will assume throughout that *each connected component* $[s] \in \hat{S}$ *contains a top element* $\top_{[s]} \in [s]$ *such that for each* $s' \in [s]$, $\top_{[s]} \geq s'$.

Definition 2. *For* $\Sigma = (S, \leq, \Sigma)$ *an OS signature,* A *is an* order-sorted Σ-algebra *iff:*

- A *is a many-sorted* (S, Σ)-algebra A,
- *whenever* $s \leq s'$, *then we have* $A_s \subseteq A_{s'}$, *and*
- *whenever* $f : w \to s$, $f : w' \to s' \in f^{[s_1] \ldots [s_n]}_{[s]}$ *and* $\overline{a} \in A^w \cap A^{w'}$, *then we have* $A_{f:w \to s}(\overline{a}) = A_{f:w' \to s'}(\overline{a})$, *where* $A^{s_1 \ldots s_n} = A_{s_1} \times \ldots \times A_{s_n}$.

An order-sorted Σ-homomorphism $h : A \to B$ *is a many-sorted* (S, Σ)-*homomorphism such that whenever* $[s] = [s']$ *and* $a \in A_s \cap A_{s'}$, *then we have* $h_s(a) = h_{s'}(a)$. h *is injective, resp.* surjective, *resp.* bijective, *iff for each* $s \in S$ h_s *is injective, resp.* surjective, *resp.* bijective. *We call* h *an* isomorphism *if there is another order-sorted* Σ-*homomorphism* $g : B \to A$ *such that for each* $s \in S$, $h_s \circ g_s = 1_{B_s}$, *and* $g_s \circ h_s = 1_{A_s}$, *with* $1_{A_s}, 1_{B_s}$ *the identity functions on* A_s, B_s. *If each* $[s] \in \hat{S}$ *has a top element* $\top_{[s]}$, *one can show that* f *is an isomorphism iff* f *is bijective. Order-sorted* Σ-*algebras and homomorphisms define a category* **OSAlg**$_\Sigma$. □

Theorem 1. [16] *The category* **OSAlg**$_\Sigma$ *has an initial algebra. Furthermore, if* Σ *is sensible, then the term algebra* T_Σ *with:*

- *if* $a : \epsilon \to s$ *then* $a \in T_{\Sigma,s}$, (ϵ *denotes the empty string*),
- *if* $t \in T_{\Sigma,s}$ *and* $s \leq s'$ *then* $t \in T_{\Sigma,s'}$,
- *if* $f : s_1 \ldots s_n \to s$ *and* $t_i \in T_{\Sigma,s_i}$ $1 \leq i \leq n$, *then* $f(t_1, \ldots, t_n) \in T_{\Sigma,s}$,

is initial, i.e., there is a unique Σ-*homomorphism to each* Σ-*algebra.*

Say Σ *has non-empty sorts* iff for each $s \in S$, $T_{\Sigma,s} \neq \emptyset$. To ensure $[\![t]\!] \neq \emptyset$ for any term t *we will assume throughout that* Σ has non-empty sorts.

An S-sorted set $X = \{X_s\}_{s \in S}$ of *variables*, satisfies $s \neq s' \Rightarrow X_s \cap X_{s'} = \emptyset$, and the variables in X are always assumed disjoint from all constants in Σ. The Σ-*term algebra with variables in* X, $T_\Sigma(X)$, is the *initial algebra* for the signature $\Sigma(X)$ obtained by adding to Σ the variables in X *as extra constants*. Since a $\Sigma(X)$-algebra is just a pair (A, α), with A a Σ-algebra, and α an *interpretation of the constants* in X, i.e., an S-sorted function $\alpha \in [X \to A]$, the $\Sigma(X)$-initiality of $T_\Sigma(X)$ can be expressed as the following corollary of Theorem 1:

Theorem 2. *(Freeness Theorem). If* Σ *is sensible, for each* $A \in$ **OSAlg**$_\Sigma$ *and* $\alpha \in [X \to A]$, *there exists a unique* Σ-*homomorphim,* $\_\alpha : T_\Sigma(X) \longrightarrow A$ *extending* α, *i.e., such that for each* $s \in S$ *and* $x \in X_s$ *we have* $x\alpha_s = \alpha_s(x)$.

The first-order language of *equational* Σ-*formulas* is defined in the usual way: its atoms are Σ-*equations* $t = t'$, where $t, t' \in T_\Sigma(X)_{\top_{[s]}}$ for some $[s] \in \hat{S}$ and each X_s is assumed countably infinite. The set $Form(\Sigma)$ of *equational* Σ-*formulas* is then inductively built from atoms by: conjunction ($\wedge$), disjunction ($\vee$) negation ($\neg$), and universal ($\forall x{:}s$) and existential ($\exists x{:}s$) quantification with sorted variables $x{:}s \in X_s$ for some $s \in S$. The literal $\neg(t = t')$ is denoted $t \neq t'$.

The *satisfaction* relation between Σ-algebras and formulas is defined in the usual way: given a Σ-algebra A, a formula $\varphi \in Form(\Sigma)$, and an assignment $\alpha \in [Y \to A]$, with $Y = fvars(\varphi)$ the free variables of φ, we define the satisfaction relation $A, \alpha \models \varphi$ inductively as usual: for atoms, $A, \alpha \models t = t'$ iff $t\alpha = t'\alpha$; for Boolean connectives it is the corresponding Boolean combination of the satisfaction relations for subformulas; and for quantifiers: $A, \alpha \models (\forall x{:}s) \varphi$ (resp. $A, \alpha \models (\exists x{:}s) \varphi$) holds iff for all $a \in A_s$ (resp. some $a \in A_s$) we have $A, \alpha \uplus \{(x{:}s, a)\} \models \varphi$, where the assignment $\alpha \uplus \{(x{:}s, a)\}$ extends α by mapping $x{:}s$ to a. Finally, $A \models \varphi$ holds iff $A, \alpha \models \varphi$ holds for each $\alpha \in [Y \to A]$, where $Y = fvars(\varphi)$. We say that φ is *valid* (or *true*) in A iff $A \models \varphi$.

Definition 3. *A signature morphism* $H : \Sigma \to \Sigma'$ *(called a* view *in Maude [1]) is a monotonic function* $H : (S, \leq) \to (S', \leq')$ *of the underlying posets of sorts, together with a mapping* H *sending each* $f : s_1 \ldots s_n \to s$ *in* Σ *to a term* $H(f) \in T_{\Sigma'}(\{x_1 : H(s_1), \ldots, x_n : H(s_n)\})_{H(s)}$. H *defines a well-typed translation of the syntax of* Σ *into that of* Σ'. *It inductively maps each* Σ-*term* t *to a* Σ'-*term* $H(t)$ *by mapping* $x : s$ *to* $x : H(s)$, *and* $H(f(t_1, \ldots, t_n))$ *to* $H(f)\{x_1 : H(s_1) \mapsto H(t_1), \ldots, x_n : H(s_n) \mapsto H(t_n)\}$, *where* $\{x_1 : H(s_1) \mapsto H(t_1), \ldots, x_n : H(s_n) \mapsto H(t_n)\}$ *denotes the obvious substitution.* H *extends naturally to a translation of equational formulas* $H : Form(\Sigma) \to Form(\Sigma')$ *by mapping atoms according to* H, *respecting Boolean connectives, and mapping each quantifier* $\forall x : s$ *(resp.* $\exists x : s$) *to* $\forall x : H(s)$ *(resp.* $\exists x : H(s)$).

A signature inclusion, denoted $\Sigma \hookrightarrow \Sigma'$, *is a signature morphism that is a poset inclusion* $(S, \leq) \hookrightarrow (S', \leq')$ *on sorts and maps each* $f : s_1 \ldots s_n \to s$ *to itself: more precisely, to the term* $f(x_1:s_1, \ldots, x_n:s_n)$. $\square$

A signature morphism $H : \Sigma \to \Sigma'$ induces a functor in the opposite direction $\_ |_H : \mathbf{OSAlg}_{\Sigma'} \to \mathbf{OSAlg}_\Sigma$, where for each $B \in \mathbf{OSAlg}_{\Sigma'}$, the algebra $B |_H \in \mathbf{OSAlg}_\Sigma$, called its H-*reduct*, is defined using H as follows: (i) for each $s \in S$, $(B |_H)_s = B_{H(s)}$; and (ii) for each $f : s_1 \ldots s_n \to s$ in Σ, $(B |_H)_f$ is the function $\lambda(x_1 \in B_{H(s_1)}, \ldots, x_n \in B_{H(s_n)}). H(f) : B_{H(s_1)} \times \ldots \times B_{H(s_n)} \to B_{H(s)}$ defined by the term $H(f)$ in the Σ'-algebra B.

In Goguen and Burstall's sense, the key point about order-sorted signature morphisms is that they make order-sorted logic an *institution* [8], so that *truth is preserved along translations*. That is, for any $B \in \mathbf{OSAlg}_{\Sigma'}$ and any sentence (i.e., $fvars(\varphi) = \emptyset$) $\varphi \in Form(\Sigma)$ we have the equivalence:

$$(\dagger) \quad B \models H(\varphi) \quad \Leftrightarrow \quad B |_H \models \varphi.$$

This equivalence can be checked in several ways. For example, one can use the embedding of order-sorted logic in membership equational logic, itself embedded in many-sorted first-order logic, as detailed in [16]. This reduces the issue to the same well-known equivalence for many-sorted first-order logic.

An important requirement on a sensible and monotonic signature is *regularity* [9]. It requires for each $f \in \Sigma$ and $u \in S^*$ that, if the set $\{ws \in S^* \mid f : w \to s \in \Sigma \wedge w \geq u\}$ is non-empty, then it has a smallest element. Regularity (or just *preregularity* [1]) ensures that each Σ-term $t \in T_\Sigma(X)$ has a *least sort*, denoted $ls_\Sigma(t)$, with $t \in T_\Sigma(X)_{ls_\Sigma(t)}$. This makes order-sorted automated deduction tasks like term rewriting or unification much easier: the matching of a term t to a variable $x : s$ will succeed iff $ls_\Sigma(t) \leq s$. Without regularity, or preregularity, a costly determination of all possible sorts of t is needed.

3   The $\Sigma \mapsto \Sigma^\#$ Signature Transformation

We define a signature transformation $\Sigma \mapsto \Sigma^\#$ that will give us the key to study validity of equational formulas in initial order-sorted algebras in Sect. 4 and pattern operations in Sect. 5. Σ is a regular order-sorted finite signature with poset of sorts $(S, \leq)$. As first remarked by H. Comon-Lundh in [3], an order-sorted signature Σ is just a Σ^u-tree automaton, with Σ^u the unsorted version of Σ, set of states S, and transitions rules: (i) $f(s_1, \ldots, s_n) \to s$ for each $f : s_1 \ldots s_n \to s$ in Σ, and (ii) ϵ-rules $s \to s'$ for each $s < s'$ in $(S, \leq)$. $T_{\Sigma,s}$ is the language accepted by the accepting state s. This means that the problem of whether $T_{\Sigma,s} = \emptyset$, or whether any Boolean combination of sets $T_{\Sigma,s_1}, \ldots, T_{\Sigma,s_n}$ is empty, are problems decidable by an emptiness check on a regular tree language.

To construct $\Sigma^\#$ we must first define its set $S^\#$ of sorts. Call $s \in S$ *atomic* iff s is a minimal element in the poset $(S, \leq)$. The key idea is to add to S new atomic sorts $s^\#$ characterizing all terms whose least sort is exactly s, where s is non-atomic. But we want $s^\#$ to be non-empty. Let $\downarrow s = \{s' \in S \mid s' < s\}$, and $glbs(s)$ the maximal elements of $\downarrow s$. Call $s \in S$ *redundant* iff $T_{\Sigma,s} - \bigcup_{s' \in glbs(s)} T_{\Sigma,s'} = \emptyset$. We only add $s^\#$ to $S^\#$ if s is non-atomic and irredundant. Since non-emptiness is decidable, we can effectively construct $S^\#$ as the set containing all atomic sorts in S and all new sorts $s^\#$ with $s \in S$ non-atomic and irredundant.

We want a *many-sorted* signature $\Sigma^\#$ on sorts $S^\#$ such that: (i) for s an atomic sort in Σ, we have $T_{\Sigma^\#,s} = T_{\Sigma,s}$, (ii) for each $s^\# \in S^\#$ we have $T_{\Sigma^\#,s^\#} = T_{\Sigma,s} - \bigcup_{s' \in glbs(s)} T_{\Sigma,s'}$; and (iii) if $s, s' \in S^\#$ and $s \neq s'$, then $T_{\Sigma^\#,s} \cap T_{\Sigma^\#,s'} = \emptyset$. Thus, we will be able to represent each sort $s \in S$ as a *disjoint union* of sorts in $S^\#$. That is, define the function $atoms : S \to \mathcal{P}(S^\#)$ inductively as follows: $atoms(s) = $ **if** s is atomic **then** $\{s\}$ **else if** s is irredundant **then** $\{s^\#\} \cup atoms(s_1) \cup \ldots atoms(s_n)$ **else** $atoms(s_1) \cup \ldots atoms(s_n)$ **fi fi**, where $glbs(s) = \{s_1, \ldots, s_n\}$. It then follows from (i)–(iii) above that for any $s \in S$ we will have:

$$T_{\Sigma,s} = \biguplus_{s' \in atoms(s)} T_{\Sigma^\#,s'}$$

This is what we want. We still have to define $\Sigma^\#$. For this, it is useful to decompose Σ as a "telescope" $\Sigma_0 \subset \Sigma_1 \subset \ldots \Sigma_{k-1} \subset \Sigma$. We assume that each constant $a : \epsilon \to s$ in Σ has a single declaration of the specified sort s. To simplify the $\Sigma^\#$ construction we also assume, without real loss of generality, that Σ can have "subsort overloading" but does not have any "ad-hoc overloading;" that is, if $(f : s_1 \ldots s_m \to s), (f : s'_1 \ldots s'_m \to s') \in \Sigma$ then $[s_i] = [s'_i]$ $1 \leq i \leq m$, and $[s] = [s']$. Recall the notation $f_{[s]}^{[s_1] \ldots [s_m]}$ for the set of all subsort-overloaded operators f for these components. Given $(f : s_1 \ldots s_m \to s) \in f_{[s]}^{[s_1] \ldots [s_m]}$ define:

$$(f : s_1 \ldots s_m \to s) \downarrow = \{(f : s'_1 \ldots s'_m \to s') \in f_{[s]}^{[s_1] \ldots [s_m]} \mid s'_1 \ldots s'_m s' < s_1 \ldots s_m s\}.$$

as its set of *strictly smaller typings*. Define: $\Sigma_0 = \{(f : s_1 \ldots s_m \to s) \in \Sigma \mid (f : s_1 \ldots s_m \to s) \downarrow = \emptyset\}$, and, inductively, $\Sigma_{n+1} = \{(f : s_1 \ldots s_m \to s) \in \Sigma \mid (f : s_1 \ldots s_m \to s) \downarrow \subseteq \Sigma_n\}$. Because of the finiteness of Σ, we get a fixpoint $\Sigma_k = \Sigma_{k+1} = \Sigma$, giving us the above-mentioned telescope. Note that regularity of Σ_n, $n \geq 0$, follows easily by construction from the regularity of Σ. Furthermore, for any $t \in T_{\Sigma_n}(X)$ we have $ls_{\Sigma_n}(t) = ls_\Sigma(t)$. For example, for Σ a signature with sorts Nat and Nat (non-zero naturals) with 0, s (successor), and with + subsort overloaded for sorts *Nat* and *NzNat*, its telescope reaches the fixpoint for $\Sigma_1 = \Sigma$, as shown below:

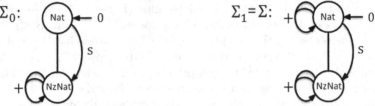

We will define a telescope $\Sigma_0^\# \subseteq \Sigma_1^\# \subseteq \ldots \Sigma_{k-1}^\# \subseteq \Sigma^\#$ that closely mirrors that of Σ. First of all, note that the map $atoms : S \to \mathcal{P}(S^\#)$ naturally extends to a map on strings, $atoms : S^* \to \mathcal{P}((S^\#)^*)$ by defining: $atoms(\epsilon) = \{\epsilon\}$, and $atoms(sw) = \{s'w' \mid s' \in atoms(s) \wedge w' \in atoms(w)\}$. Note also that the mapping $(f : s_1 \ldots s_m \to s) \mapsto s_1 \ldots s_m$ defines a function $arity : \Sigma \to S^*$. Define $\Sigma_0^\# = \{(f : w \to s^\bullet) \mid (f : s_1 \ldots s_m \to s) \in \Sigma_0, \ w \in atoms(s_1 \ldots s_m)\}$, where $s^\bullet$ = **if** s atomic **then** s **else** $s^\#$ **fi**. Then define $\Sigma_{n+1}^\#$ inductively as follows: $\Sigma_{n+1}^\# = \Sigma_n^\# \cup \{(f : w \to s^\#) \mid (f : s_1 \ldots s_m \to s) \in \Sigma_{n+1} - \Sigma_n, \ s \ irredundant, \ w \in atoms(s_1 \ldots s_m) - \{arity(f : w' \to s') \mid (f : w' \to s') \in \Sigma_n^\#\}\}$. If $\Sigma_k = \Sigma$, we define $\Sigma_k^\# = \Sigma^\#$ and obtain a telescope $\Sigma_0^\# \subseteq \Sigma_1^\# \subseteq \ldots \Sigma_{k-1}^\# \subseteq \Sigma^\#$ as claimed. For example, for the above-mentioned signature Σ with 0, s, and + subsort overloaded for sorts *Nat* and *NzNat*, the telescope for its associated $\Sigma^\#$ reaches its fixpoint for $\Sigma_1^\# = \Sigma^\#$ as shown in the figure below:

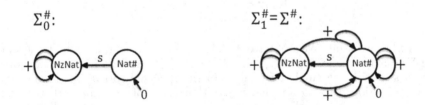

The main properties of the $\Sigma \mapsto \Sigma^\#$ transformation are as follows:

Theorem 3. *Let Σ satisfy the above assumptions. Then:*

1. *$\Sigma^\#$ is sensible*
2. *for $s, s' \in S^\#$, $s \neq s' \Rightarrow T_{\Sigma^\#,s} \cap T_{\Sigma^\#,s'} = \emptyset$*
3. *for each $s \in S$, $T_{\Sigma,s} = \biguplus_{s' \in atoms(s)} T_{\Sigma^\#,s'}$*
4. *$t \in T_\Sigma \wedge ls_\Sigma(t) = s \Leftrightarrow t \in T_{\Sigma^\#} \wedge ls_{\Sigma^\#}(t) = s^\bullet$.*

3.1   Variations on the $\Sigma^\#$ Theme

Several signatures closely related to Σ and $\Sigma^\#$ are also very useful. The most obvious is their union $\Sigma \cup \Sigma^\#$, with set of operators the set-theoretic union $\Sigma \cup \Sigma^\#$ and poset of sorts $(S \cup S^\#, (\leq \cup <^\#)^*)$, with $\leq$ the order in $(S, \leq)$, and

$<^\# = \{(s^\#, s) \mid s$ *nonatomic and irredundant*$\}$. $\Sigma \cup \Sigma^\#$ is even more intuitive than $\Sigma^\#$, because it *refines* Σ into a richer semantics-preserving signature by just adding to it the new atoms $s^\#$, so that now the least sort of any ground term t will always be an atomic sort. This means that we have *sharpened* the typing of any such t as much as possible, which is the reason for the $\Sigma^\#$ notation. For example, for Σ the signature discussed in the Introduction, with subsort inclusion $A < B$, constants a of sort A, b of sort B, and subsort-overloaded unary operator f, $\Sigma \cup \Sigma^\#$ is the signature depicted on the right of the figure in the Introduction.

Note that we have subsignature inclusions $J : \Sigma \hookrightarrow \Sigma \cup \Sigma^\#$ and $J' : \Sigma^\# \hookrightarrow \Sigma \cup \Sigma^\#$. Furthermore, $\Sigma \cup \Sigma^\#$ enjoys very good properties, which make it an initial-semantics-preserving enrichment of both Σ and $\Sigma^\#$:

Lemma 1. $\Sigma \cup \Sigma^\#$ *is regular,* $T_{\Sigma \cup \Sigma^\#} \mid_J = T_\Sigma$, *and* $T_{\Sigma \cup \Sigma^\#} \mid_{J'} = T_{\Sigma^\#}$.

Two other useful signatures are $\Sigma_\top^\#$ and $\Sigma_c^\#$. $\Sigma_\top^\#$ is an order-sorted signature with operations those in $\Sigma^\#$ and with sorts $S_\top \cup S^\#$, where $S_\top = \{\top_{[s]} \mid [s] \in \hat{S}\}$ is the set of top sorts of each connected component in $(S, \leq)$. Its order is defined as the identity relation on $S^\# \cup S_\top$, plus the subsort inclusions $s' \leq \top_{[s]}$ for each $s' \in atoms(\top_{[s]})$. We have a subsignature inclusion $K : \Sigma_\top^\# \hookrightarrow \Sigma \cup \Sigma^\#$. Reasoning as in Lemma 1 it is easy to show that $T_{\Sigma \cup \Sigma^\#} \mid_K = T_{\Sigma_\top^\#}$.

$\Sigma_c^\#$ is a *many-sorted version* of $\Sigma_\top^\#$. Its set of sorts is $S_\top \cup S^\#$, but now $s \leq s'$ iff $s = s'$. The operations of $\Sigma_c^\#$ are those of $\Sigma^\#$ plus the coercion operators $\{c : s' \to \top_{[s]} \mid [s] \in \hat{S},\ s' \in atoms(\top_{[s]}) - \{\top_{[s]}\}\}$, which mimic the subsort inclusions $s' < \top_{[s]}$ in $\Sigma_\top^\#$. We then have a signature morphism $H : \Sigma_c^\# \to \Sigma_\top^\#$ that is the identity on sorts and on the operators in $\Sigma^\#$ and maps each coercion $c : s' \to \top_{[s]}$ to the term $x_1{:}s'$.

For example, for the above-mentioned signature Σ with 0, s, and + subsort overloaded for sorts Nat and $NzNat$, the signatures $\Sigma_\top^\#$ and $\Sigma_c^\#$ are as follows:

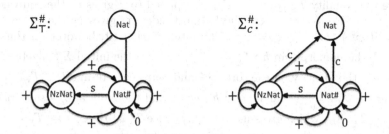

The following diagram summarizes this section:

$$(\ddagger) \quad \Sigma \xhookrightarrow{J} \Sigma \cup \Sigma^\# \xhookleftarrow{K} \Sigma_\top^\# \xleftarrow{H} \Sigma_c^\#$$

4   Equational Formulas in Initial Order-Sorted Algebras

The main goal of this section is to *reduce* the validity of equational first-order formulas in an initial *order-sorted* algebra to the validity of semantically equivalent formulas in an initial *many-sorted* algebra. The main idea of this reduction is to exploit diagram ($\ddagger$) at the end of Sect. 3.1, which begins with an order-sorted signature Σ and ends with a many-sorted signature $\Sigma_c^\#$. Like Alice in Wonderland's Cheshire cat's smile, all order-sorted features vanish in the passage from Σ to $\Sigma_c^\#$. This reduction seems useful for at least two reasons:

1. Its provides a new, very simple proof of the *decidability* of first-order formulas in initial order-sorted algebras. A non-trivial proof of such a decidability result goes back to [3,4], but it requires quite complex formulas and formula transformations involving sort constraints based on quite general sort expressions, whose semantics is defined using tree automata.
2. The reduction-based proof given here provides a useful new *transfer principle*, by which problems with a perhaps unclear solution at the order-sorted level can be reduced to problems having a clear solution at the many-sorted level. For example, as further explained in Sect. 5, the puzzling anomaly about pattern operations in initial order-sorted algebras discussed in the Introduction has a systematic solution thanks to this transfer principle.

   The main idea of the reduction is to assign to each first-order sentence φ in the language of a finite and regular order-sorted signature Σ a corresponding sentence $\varphi_c^\#$ in the language of the many-sorted signature $\Sigma_c^\#$, and then prove that we have an equivalence $T_\Sigma \models \varphi \Leftrightarrow T_{\Sigma_c^\#} \models \varphi_c^\#$. To obtain such an equivalence we make our way from T_Σ and φ to $T_{\Sigma_c^\#}$ and $\varphi_c^\#$ by moving from left to right along the diagram ($\ddagger$). Since some of the steps in this sequence of signature morphisms are easy consequences of the equivalence ($\dagger$) at the end of Sect. 2, we can quickly get such easy equivalences out of the way. Indeed, since J is a subsignature inclusion, it is the identity on formulas, and since by Lemma 1 we have the equality $T_{\Sigma \cup \Sigma^\#}|_J = T_\Sigma$, ($\dagger$) applied to J gives us the equivalence $T_\Sigma \models \varphi \Leftrightarrow T_{\Sigma \cup \Sigma^\#} \models \varphi$. On the leftmost side, ($\dagger$) gives us the equivalence $T_{\Sigma_\top^\#} \models H(\varphi_c^\#) \Leftrightarrow T_{\Sigma_\top^\#}|_H \models \varphi_c^\#$. The interesting twist, however, in that the unique $\Sigma_c^\#$-homomorphism $h : T_{\Sigma_c^\#} \to T_{\Sigma_\top^\#}|_H$ from the initial $\Sigma_c^\#$-algebra $T_{\Sigma_c^\#}$ is obviously the identity on the sorts $S^\#$ and maps each term $c(t) \in T_{\Sigma_c^\#, \top_{[s]}}$ to the term $t \in T_{\Sigma_\top^\#, \top_{[s]}}$. That is, h is *bijective*, and therefore a $\Sigma_c^\#$-*isomorphism* $h : T_{\Sigma_c^\#} \cong T_{\Sigma_\top^\#}|_H$, which gives us the equivalence $T_{\Sigma_\top^\#}|_H \models \varphi_c^\# \Leftrightarrow T_{\Sigma_c^\#} \models \varphi_c^\#$. Therefore, stringing these last two equivalences together, we get the equivalence $T_{\Sigma_\top^\#} \models H(\varphi_c^\#) \Leftrightarrow T_{\Sigma_c^\#} \models \varphi_c^\#$. We will then be done proving our desired equivalence $T_\Sigma \models \varphi \Leftrightarrow T_{\Sigma_c^\#} \models \varphi_c^\#$ if we can define a mapping $\varphi \mapsto \varphi^\#$ such that $H(\varphi_c^\#) = \varphi^\#$ and we show an equivalence $T_{\Sigma \cup \Sigma^\#} \models \varphi \Leftrightarrow T_{\Sigma_\top^\#} \models \varphi^\#$.

   What makes the mapping $\varphi \mapsto \varphi^\#$ not entirely obvious is that $\Sigma_\top^\#$ has considerably fewer sorts than the plentiful $\Sigma \cup \Sigma^\#$. In particular, we have to find a way to express equations and quantifiers involving variables with sorts of $\Sigma \cup \Sigma^\#$ not present in $\Sigma_T^\#$ in the poorer language of $\Sigma_T^\#$. The key idea for this is to observe that every ground $\Sigma \cup \Sigma^\#$-term has an atomic least sort in $S^\#$, and that, by Theorem 3–(3) and Lemma 1, we have the equality $T_{\Sigma \cup \Sigma^\#,s} = \biguplus_{s' \in atoms(s)} T_{\Sigma^\#,s'}$. Therefore, abbreviating $t \in T_{\Sigma \cup \Sigma^\#,s}$ to $t : s$, we have, $t : s \Leftrightarrow \bigvee_{s' \in atoms(s)} t : s'$, which is a property expressible in the language of $\Sigma_T^\#$. Here is now the detailed mapping $\varphi \mapsto \varphi^\#$ using these ideas. Without loss of generality we may assume φ in prenex form, that is, $\varphi = Q\varphi_0$, with Q a sequence of quantifiers and φ_0 quantifier-free. The mapping $\varphi \mapsto \varphi^\#$ decomposes into a mapping $\varphi_0 \mapsto \varphi_0^\#$ for the quantifier-free part and a mapping for the quantifiers.

We first need some notation. $\overline{x{:}s}$ abbreviates a sequence of variables $x_1 : s_1, \ldots, x_n{:}s_n$. We can always decompose the free variables of φ_0 as $fvars(\varphi_0) = \overline{x{:}s}, \overline{y{:}p}$, with $\overline{x{:}s}$ variables having non-atomic sorts, and $\overline{y{:}p}$ variables having atomic sorts. Also, if $\overline{x{:}s} = x_1{:}s_1, \ldots, x_n{:}s_n$, then $\overline{x{:}s}_T$ denotes the variables $\overline{x{:}s}_T = x_1 : \top_{[s_1]}, \ldots, x_n : \top_{[s_n]}$. In the same spirit, $\overline{x{:}s} = \overline{t}$ abbreviates the *conjunction of equations* $x_1{:}s_1 = t_1 \wedge \ldots \wedge x_n{:}s_n = t_n$, and $\{\overline{x{:}s} = \overline{t}\}$ abbreviates the *substitution* $\{x_1{:}s_1 \mapsto t_1, \ldots, x_n{:}s_n \mapsto t_n\}$. Given variables $\overline{x{:}s}$ with sorts in S, let $Spec(\overline{x{:}s}, S^\#)$, called the set of $S^\#$-*specializations* of $\overline{x{:}s}$, be the set $Spec(\overline{x{:}s}, S^\#) = \{\overline{x{:}s} = \overline{z{:}q} \mid |\overline{x{:}s}| = |\overline{z{:}q}| \wedge q_i \in atoms(s_i), 1 \leq i \leq |x{:}s|\}$, where $|\overline{x{:}s}|$ denotes the length of the sequence of variables $\overline{x{:}s}$. To avoid variable capture we will always assume that the variables $\overline{z{:}q}$ are *fresh* variables, different for each $(\overline{x{:}s} = \overline{z{:}q}) \in Spec(\overline{x{:}s}, S^\#)$ and not appearing anywhere else. Viewed as a substitution $\{\overline{x{:}s} = \overline{z{:}q}\}$, each specialization $\overline{x{:}s} = \overline{z{:}q}$ is just a variable mapping lowering the sort s_i of each x_i to a sort $q_i \in atoms(s_i)$ for z_i. We can now define the mapping $\varphi_0 \mapsto \varphi_0^\#$ —where $fvars(\varphi_0) = \overline{x{:}s}, \overline{y{:}p}$, with $\overline{x{:}s}$ variables having non-atomic sorts, and $\overline{y{:}p}$ variables having atomic sorts—as follows:

$$\varphi_0^\# = \bigvee_{(\overline{x{:}s} = \overline{z{:}q}) \in Spec(\overline{x{:}s}, S^\#)} (\exists \overline{z{:}q}) \, (\overline{x{:}s}_T = \overline{z{:}q} \wedge (\varphi_0\{\overline{x{:}s} = \overline{z{:}q}\})).$$

Note that $fvars(\varphi_0^\#) = \overline{x{:}s}_T, \overline{y{:}p}$. For example, for the above-mentioned signature Σ with 0, s, and $+$ subsort overloaded for sorts Nat and $NzNat$, if φ is $x + y = y + x$, with $x, y : Nat$, then, assuming $x_1, y_1 : NzNat$, and $x_2, y_2 : Nat^\#$, $\varphi^\#$ is:

$(\exists x_1, y_1) \, x = x_1 \wedge y = y_1 \wedge x_1 + y_1 = y_1 + x_1 \, \vee$
$(\exists x_2, y_2) \, x = x_2 \wedge y = y_2 \wedge x_2 + y_2 = y_2 + x_2 \, \vee$
$(\exists x_1, y_2) \, x = x_1 \wedge y = y_2 \wedge x_1 + y_2 = y_2 + x_1 \, \vee$
$(\exists x_2, y_1) \, x = x_2 \wedge y = y_1 \wedge x_2 + y_1 = y_1 + x_2.$

The semantic equivalence between φ_0 and $\varphi_0^\#$ can then be expressed as follows:

Lemma 2. *For* φ_0 *as above,* $\alpha \in [\overline{x{:}s}_T, \overline{y{:}p} \rightarrow T_{\Sigma^\#}]$ *satisfies* $T_{\Sigma^\#}, \alpha \models \varphi_0^\#$ *iff there exists* $\beta \in [\overline{x{:}s}, \overline{y{:}p} \rightarrow T_\Sigma]$ *such that* $\alpha = \beta \circ \{\overline{x{:}s}_T = \overline{x{:}s}\}$ *and* $T_\Sigma, \beta \models \varphi_0$.

Since $\varphi = Q\varphi_0$, to define $\varphi^\#$ we still need to deal with the quantifiers Q. This is done inductively for each individual quantifier as follows. If $s \in S \cap (S^\# \cup S_T)$, then $((\forall x{:}s) \, \varphi)^\# = (\forall x{:}s) \, \varphi^\#$, and $((\exists x{:}s) \, \varphi)^\# = (\exists x{:}s) \, \varphi^\#$. Otherwise, let

$atoms(s) = \{q_1, \ldots, q_k\}$, then, $((\forall x{:}s)\ \varphi)^{\#} = (\forall x{:}\top_{[s]})\ (((\exists \overline{z{:}q})\ \bigvee_{i=1}^{k} x{:}\top_{[s]} = z_i{:}$ $q_i) \Rightarrow \varphi^{\#})$, and $((\exists x{:}s)\ \varphi)^{\#} = (\exists x{:}\top_{[s]}, \overline{z{:}q})\ (\bigvee_{i=1}^{k} x{:}\top_{[s]} = z_i{:}q_i) \wedge \varphi^{\#})$.

The key syntactic invariant maintained by this translation is of course that if $fvars(\varphi) = \overline{x{:}s}, \overline{y{:}p}$, then $fvars(\varphi^{\#}) = \overline{x{:}s}_\top, \overline{y{:}p}$. And the key semantic invariant is that for each $\alpha \in [\overline{x{:}s}_\top, \overline{y{:}p}{\to}T_{\Sigma_\top^\#}]$ we have $T_{\Sigma_\top^\#}, \alpha \models \varphi^{\#}$ iff there exists $\beta \in [\overline{x{:}s}, \overline{y{:}p}{\to}T_\Sigma]$ such that $\alpha = \beta \circ \{\overline{x{:}s}_\top = \overline{x{:}s}\}$ and $T_\Sigma, \beta \models \varphi$. For quantifier-free formulas this has already been proved in Lemma 2. That this remains true after each quantification step is easy to check and left to the reader: indeed, the above treatment of quantifiers is analogous to how in set theory we restrict quantifiers ranging over all sets to quantifiers ranging over a given set A by defining $(\forall x \in A)\ \varphi = (\forall x)\ (x \in A \Rightarrow \varphi)$, and $(\exists x \in A)\ \varphi = (\exists x)\ (x \in A \wedge \varphi)$. Our treatment is not just analogous, but in fact a special case: we have just captured $x \in T_{\Sigma,s}$ by the formula $(\exists \overline{z{:}q})\ \bigvee_{i=1}^{k} x = z_i{:}q_i$. Therefore, for any sentence φ (i.e., $fvars(\varphi) = \emptyset$) we get $T_\Sigma \models \varphi \Leftrightarrow T_{\Sigma^\#} \models \varphi^{\#}$.

To close all the proof steps along the Cheshire cat's sequence ($\ddagger$) we need to define the formula $\varphi_c^{\#}$ such that $H(\varphi_c^{\#}) = \varphi^{\#}$. We can get $\varphi_c^{\#}$ from $\varphi^{\#}$ as follows. Since $\Sigma_\top^\#$ and $\Sigma_c^\#$ have the same sorts, the variables and quantifiers in $\varphi^{\#}$ and $\varphi_c^{\#}$ stay the same. We just replace each equation $u = v$ appearing somewhere in $\varphi^{\#}$ by the equation $c(u) = c(v)$ at the same position in $\varphi_c^{\#}$, unless: (i) $\top_{[s]}$ is atomic (then $u = v$ is left unchanged), or (ii) $\top_{[s]}$ is non-atomic and either u or v are variables of sort $\top_{[s]}$, which are then left unchanged. This gives us the desired semantic equivalence $T_\Sigma \models \varphi \Leftrightarrow T_{\Sigma_c^\#} \models \varphi_c^{\#}$.

Since both the technical report version [14] of Maher's paper [13], and the disunification paper by Comon and Lescanne [2] prove that the first-order theory of a *many-sorted* initial algebra T_Ω is *decidable*—i.e., that there is an algorithm to decide for any formula ϕ whether $T_\Omega \models \phi$ holds or not—we then get as a corollary of the above equivalence the following theorem,[1] already known since [3,4], but now obtained in a different way and with a considerably simpler proof:

Theorem 4. *Let Σ be a finite and regular order-sorted signature. For any first-order formula $\varphi \in Form(\Sigma)$ the validity problem $T_\Sigma \models \varphi$ is decidable.*   □

5   Pattern Operations in Initial Order-Sorted Algebras

Given an order-sorted signature Σ, by a *Σ-pattern* we just mean a term $t \in T_\Sigma(X)$, where we assume X_s countably infinite for each sort $s \in S$. We call t a pattern to emphasize that t is a symbolic description of a *language*, namely the set $[\![t]\!] = \{t\sigma \mid \sigma \in [X{\to}T_\Sigma]\}$ of its *ground instances*. Similarly, a finite set of patterns $\{t_1, \ldots, t_n\}$ is a symbolic description of the language $[\![t_1, \ldots, t_n]\!] = [\![t_1]\!] \cup \ldots \cup [\![t_n]\!]$. A language need not be a set of strings. Since strings are just a

---

[1] Theorem 4 holds for Σ finite and regular because any such Σ can be transformed into a semantically equivalent signature with no ad-hoc overloading (by symbol renaming) and with each connected component having a top sort (added when missing).

special case of trees, it can be a *tree language*, that is, a subset $L \subseteq T_\Sigma$ for some Σ. Therefore, $\mathcal{P}(T_\Sigma)$ is the set of all Σ-tree languages, and we have a function

$$[\![-]\!] : \mathcal{P}_{\mathit{fin}}(T_\Sigma(X)) \longrightarrow \mathcal{P}(T_\Sigma) : \{t_1, \ldots, t_n\} \mapsto [\![t_1, \ldots, t_n]\!]$$

sending each finite set of patterns to its associated language. Call a language $L \subseteq T_\Sigma$ a *pattern language* iff $L = [\![t_1, \ldots, t_n]\!]$ for some finite set of patterns $\{t_1, \ldots, t_n\}$. The most obvious question is that of *representability*: which languages $L \subseteq T_\Sigma$ are pattern languages, i.e., can be symbolically *represented* by some $\{t_1, \ldots, t_n\}$? Pattern languages are closed under finite unions by construction. Are they closed under finite intersections? Obviously *yes*, since, by distributivity we can reduce the problem to the intersection of two patterns $[\![u]\!] \cap [\![v]\!]$, and we have $[\![u]\!] \cap [\![v]\!] = [\![u\sigma_1]\!] \cup \ldots \cup [\![u\sigma_n]\!]$, where $\{\sigma_1, \ldots, \sigma_n\} = DUnif_\Sigma(u, v)$, the set of *most general disjoint order-sorted unifiers* of u and v in Σ [17]; that is, before unifying u and v, we rename v if necessary to make its variables disjoint from those of u. Are T_Σ and $\emptyset$ pattern languages? Yes: $\emptyset = [\![\emptyset]\!]$, and $T_\Sigma = [\![x_1 : \top_{[s_1]}, \ldots, x_k : \top_{[s_k]}]\!]$, where $\hat{S} = \{[s_1], \ldots, [s_k]\}$. So, the ony missing Boolean operation is *complement*. But since complement and difference are expressible in terms of each other: $\overline{A} = \top - A$, and $A - B = A \cap \overline{B}$, we can rephrase the question thus: are pattern languages closed under differences? In general they are *not*. For example, for Σ unsorted and having a constant a and a binary f, the language $[\![f(x, y)]\!] - [\![f(z, z)]\!]$ is *not* a pattern language (see Proposition 4.5 in [12]). However, in the unsorted case (see Corollary in pg. 314, [12]) $[\![t_1, \ldots, t_n]\!] - [\![t'_1, \ldots, t'_m]\!]$ *is* a pattern language when the t_i and the t'_j are *linear* terms—have no repeated variables—and more general cases than just sets of linear patterns also yield differences that are pattern languages [6,11,12,18,19].

Since all other Boolean operations are already taken care of, all we need is a way of symbolically defining the *difference* $\{t_1, \ldots, t_n\} - \{t'_1, \ldots, t'_m\}$ of two finite sets of *order-sorted* patterns whenever this represents a pattern language. As illustrated by the example in the Introduction, if we insist on remaining in the given signature Σ this cannot be done, even for sets of linear patterns. However, we can use the $\Sigma \mapsto \Sigma^\#$ transformation and the *transfer principle* from order-sorted problems to many-sorted ones discussed in Sect. 4 to obtain a solution based on the following two simple observations:

1. As *sets* (not as algebras) we have $T_\Sigma = T_{\Sigma^\#}$.
2. For any *order-sorted* pattern $t \in T_\Sigma(X)$ we have the *language equality* $[\![t]\!] = \bigcup_{(\overline{x:s} = \overline{z:q}) \in Spec(\overline{x:s}, S^\#)} [\![t\{\overline{x:s} = \overline{z:q}\}]\!]$, where $\overline{x:s} = fvars(t)$.

where both (1) and (2) are simple corollaries of Theorem 3. This then yields a straightforward way of representing a difference of finite sets of *order-sorted* Σ-patterns $\{t_1, \ldots, t_n\} - \{t'_1, \ldots, t'_m\}$ as a difference of finite sets of *many-sorted* $\Sigma^\#$-patterns: we just replace each t_i (resp t'_j) by the finite set of many-sorted $\Sigma^\#$-patterns $\{t_i\{\overline{x:s} = \overline{z:q}\} \mid (\overline{x:s} = \overline{z:q}) \in Spec(\overline{x:s}, S^\#)\}$, where $\overline{x:s} = fvars(t_i)$. For the example in the Introduction, this method transforms the order-sorted symbolic difference $\{x{:}B\} - \{y{:}A\}$ into the many-sorted symbolic difference: $\{x{:}A, z{:}B^\#\} - \{y{:}A\}$.

Since—with the possible exception of the treatment of *finite sorts* (see below), which warrants an extension of the unsorted algorithms—the unsorted algorithms for computing the symbolic difference of two sets of patterns have a straightforward generalization to the many-sorted case, we can just use the above reduction to the many-sorted case and many-sorted versions of the difference algorithms in [6,12,18,19] to solve the problem of computing when possible the symbolic difference of order-sorted patterns $\{t_1, \ldots, t_n\} - \{t'_1, \ldots, t'_m\}$ as a finite set of (many-sorted) patterns.

But is this the best we can do? There can be some practical limitations, both in performance and at the representational level. For order-sorted signatures with rich type structures a set $atoms(s)$ may have a considerable number of sorts in $S^{\#}$, so that the sets $\{t_i\{\overline{x : s} = \overline{z : q}\} \mid (\overline{x : s} = \overline{z : q}) \in Spec(\overline{x : s}, S^{\#})\}$ for each t_i (resp. t'_i) can become quite big, affecting performance. It also means that the representation of the *solutions* to symbolic difference problems, besides being possibly quite big, may also be more *verbose* than necessary. For example, in the signature of the Introduction, we can compute the *order-sorted* symbolic difference $\{x:B\} - \{b\} = \{f(y:B), a\}$, which is shorter and more intuitive than the equivalent many-sorted representation $\{x:B\} - \{b\} = \{f(z:B^{\#}), f(z':A), a\}$.

We present below an attractive alternative, namely, an *order-sorted* algorithm for computing symbolic differences $\{t_1, \ldots, t_n\} - \{t'_1, \ldots, t'_m\}$ in the extended order-sorted signature $\Sigma \cup \Sigma^{\#}$ that does not require any transformation of the original problem and can significantly overcome the above limitations by yielding simpler and shorter representations and better performance (see [15]).

Let us describe this algorithm. First of all, thanks to the Boolean equation $(A \cup B) - C = (A - C) \cup (B - C)$, we can decompose $\{t_1, \ldots, t_n\} - \{t'_1, \ldots, t'_m\}$ as a union $\{t_1\} - \{t'_1, \ldots, t'_m\} \cup \ldots \cup \{t_n\} - \{t'_1, \ldots, t'_m\}$. Second, thanks to the Boolean equation $A - B = A - (A \cap B)$ we can reduce $\{t\} - \{t_1, \ldots, t_n\}$ to the equivalent symbolic expression $\{t\} - \{t\sigma \mid \sigma \in DUnif_{\Sigma}(t, t_1) \cup \ldots \cup DUnif_{\Sigma}(t, t_n)\}$. Thus, all our symbolic difference problems can be reduced to unions of problems of the form $\{t\} - \{t\sigma_1, \ldots, t\sigma_n\}$ with $\sigma_1, \ldots, \sigma_n$ substitutions instantiating t. Our algorithm gives priority to the easier and frequently occurring cases, using the order-sorted extension of the more general algorithm of Lassez and Marriott [12] only when the simpler algorithms cannot be applied. We also exploit the fact that a sort s may be *finite*—i.e., $T_{\Sigma \cup \Sigma^{\#}, s}$ is a finite set—plus the decidability of sort finiteness to increase the successful difference cases. Specifically:

1. If $t, t\sigma_1, \ldots, t\sigma_n$ are all *linear* terms, we apply the inference rules below.
2. Otherwise, when $\sigma_1, \ldots, \sigma_n$ are all *linear*, i.e., $\sigma_i(x), \sigma_i(y)$ are linear terms not sharing any variables when $x \neq y$, we reduce to case (1) (see [15]).
3. Otherwise, if σ_i is non-linear and $y:s$ occurs more than once either in $\sigma_i(x)$ or in $\sigma_i(x), \sigma_i(z)$, $x \neq z$, with s finite, $T_{\Sigma \cup \Sigma^{\#}, s} = \{u_1, \ldots, u_k\}$, then we replace the problem $\{t\} - \{t\sigma_1, \ldots, t\sigma_n\}$ by the problem $\{t\} - \{t\sigma_1, \ldots, t\sigma_i\{y \mapsto u_1\}, \ldots, t\sigma_i\{y \mapsto u_k\}, \ldots, t\sigma_n\}$ and check again the new problem.
4. Outside cases (1)–(3) above, we invoke the order-sorted version of the algorithm in [12], which is more efficient than those in [6,18,19] and gives a

full answer to difference problems $\{t\} - \{t_1, \ldots, t_n\}$, whereas those in [6,18,19] give a full answer to arbitrary Boolean combinations (see [15]).

In case all terms $t, t\sigma_1, \ldots, t\sigma_n$ are linear, the following rewrite rules are applied:

1. $\{t\} - \{t\sigma_1, \ldots, t\sigma_n\} \rightarrow (\{t\} - \{t\sigma_1\}) \cap \ldots \cap (\{t\} - \{t\sigma_n\})$
2. $\{t\} - \emptyset \rightarrow \{t\}$
3. $\{f(t_1, \ldots, t_n)\} - \{f(t_1\sigma, \ldots, t_n\sigma)\} \rightarrow$
   $\{f(t_1, \ldots, u, \ldots, t_n) \mid u \in (\{t_i\} - \{t_i\sigma\}), 1 \le i \le n\}$,
   where $fvars(u)$ are fresh variables.
4. $\{x{:}s\} - \{y{:}s'\} \rightarrow \{z_1{:}q_1, \ldots, z_k{:}q_k\}$,
   where $\{q_1, \ldots, q_k\} = atoms(s) - atoms(s')$.
5. $\{x{:}s\} - \{f(t_1, \ldots, t_n)\} \rightarrow$
   $\{\overline{z{:}q}\} \cup \bigcup \{\{x_p{:}p\} - \{f(t_1, \ldots, t_n)\{\rho\}\} \mid \rho \in Spec(Y, S^{\#}),$
   $p = ls_{\Sigma^{\#}}(f(t_1, \ldots, t_n)\{\rho\})\} \mid p \in atoms(s) \cap atoms(f(t_1, \ldots, t_n))\}$
   $if \ s \notin S^{\#}$,
   where $Y = fvars(f(t_1, \ldots, t_n))$, $\overline{z{:}q} = z_1{:}q_1, \ldots, z_k{:}q_k$, $\{q_1, \ldots, q_k\} =$
   $atoms(s) - atoms(f(t_1, \ldots, t_n))$, and
   $atoms(f(t_1, \ldots, t_n)) = \{ls_{\Sigma^{\#}}(f(t_1, \ldots, t_n)\{\rho\}) \mid \rho \in Spec(Y, S^{\#})\}$.
6. $\{x{:}s\} - \{f(t_1, \ldots, t_n)\} \rightarrow$
   $\{u \mid u \in Pat(s) - \{f(\overline{x{:}s})\}\} \cup \{f(\overline{x{:}s})\} - \{f(t_1, \ldots, t_n)\}$
   $if \ s = ls_{\Sigma^{\#}}(f(t_1, \ldots, t_n)) \in S^{\#}$,
   where $\overline{x{:}s} = x_1{:}s_1, \ldots, x_n{:}s_n$, $s_i = ls_{\Sigma^{\#}}(t_i)$, and
   $Pat(s) = \{g(x_1{:}s_1, \ldots, x_n{:}s_n) \mid g : s_1 \ldots s_n \rightarrow s \in \Sigma^{\#}\}$.

The correctness of these rules and of the algorithm is proved in [15].

What advantages do we gain from this algorithm? Quite substantial ones to reason effectively about languages. Let $LT_\Sigma(X) \subseteq T_\Sigma(X)$ denote the set of *linear* terms in $T_\Sigma(X)$. Note that if $u \in LT_\Sigma(X)$ then $[\![u]\!]$ is a *regular* tree language. This follows from order-sorted signatures being tree automata, plus the regular expression fact that if $L_1, \ldots, L_n$ are regular languages, then $f(L_1, \ldots, L_n)$ is a regular language. Also, $\mathcal{P}_{fin}(LT_{\Sigma \cup \Sigma^{\#}}(X))$ is closed under symbolic: (i) unions; (ii) intersections, because disjoint unifiers of linear terms are linear; and (iii) differences, since rules (1)–(6) preserve linearity of terms. Furthermore, given $\{t_1, \ldots, t_n\}, \{t'_1, \ldots, t'_m\} \in \mathcal{P}_{fin}(LT_{\Sigma \cup \Sigma^{\#}}(X))$ we can use pattern differences to *decide* whether $[\![\{t_1, \ldots, t_n\}]\!] = [\![\{t'_1, \ldots, t'_m\}]\!]$. Indeed, $[\![\{t_1, \ldots, t_n\}]\!] = [\![\{t'_1, \ldots, t'_m\}]\!] \Leftrightarrow \{t_1, \ldots, t_n\} \equiv \{t'_1, \ldots, t'_m\}$, where the relation $\equiv$ is defined by the equivalence: $\{t_1, \ldots, t_n\} \equiv \{t'_1, \ldots, t'_m\} \Leftrightarrow \{t_1, \ldots, t_n\} - \{t'_1, \ldots, t'_m\} = \emptyset \wedge \{t'_1, \ldots, t'_m\} - \{t_1, \ldots, t_n\} = \emptyset$. By the homomorphism theorem for Boolean algebras, this means that $[\![\_]\!]$ defines an *injective homomorphism of Boolean algebras*

$$[\![\_]\!] : \mathcal{P}_{fin}(LT_{\Sigma \cup \Sigma^{\#}}(X))/{\equiv} \rightarrow \mathcal{P}(T_\Sigma).$$

This is as good as it gets, since $\mathcal{P}_{fin}(LT_{\Sigma \cup \Sigma^{\#}}(X))/{\equiv}$ is a *computable* Boolean algebra, where all operations become effective. This offers an attractive, simpler alternative to tree automata to effectively compute Boolean operations on linear pattern languages in a symbolic way.

6   Implementation and Experiments

The algorithms described in this paper are highly *reflective*. That is, they are *parametric* on signatures Σ and perform meta-level operations on signatures and Σ-terms, such as order-sorted unification, matching, sort comparisons, and so on, to ultimately compute pattern operations. Fortunately, many of these auxiliary meta-level operations are available, or can be easily programmed, in the Maude language through its reflective features using its `META-LEVEL` module [1]. In `META-LEVEL`, a signature Σ is meta-represented as a term $\overline{\Sigma}$ of sort `Module`, and a Σ-term t is meta-represented as a so-called *meta-term* $\overline{t}$ of sort `Term`.

Since: (i) Maude syntax at the meta-level essentially mirrors the syntax at the object level; and (ii) inference rules such as above rules (1)–(6) can be directly expressed as rewrite rules operating on meta-terms, the *representational distance* between the theoretical description of the algorithm in Sect. 5 and its actual meta-level implementation in Maude is relatively short.

We have implemented in Maude the signature transformation $\Sigma \mapsto \Sigma^{\#}$ described in Sect. 3. The implementation essentially coincides with the telescoping procedure described therein. The procedure takes a reflected signature $\overline{\Sigma}$ as an argument and proceeds by non-deterministically selecting an operator f in Σ which has not been processed and whose strictly smaller typings have all been processed. Using the signature transformation procedure we have also implemented the order-sorted pattern operation algorithms described in Sect. 5. The overall algorithm takes as arguments a reflected signature $\overline{\Sigma}$ and a symbolic Boolean expression composed of meta-terms $\overline{t}$ representing Σ-term patterns using a mixfix syntax where _U_ represents union, _&_ represents intersection, and _-_ represents difference. A set of Boolean equations first reduces each Boolean symbolic pattern expression to a *normal form* (essentially pushing conjunctions/differences down the expression tree). A normal form is then solved using an algorithm that, with some additional optimizations and small variations, deals with each symbolic difference problem according to the steps described in Sect. 5: the problem is first classified according to cases (1)–(4), iterating over the finite-sort transformation of case (3) if needed. Then, either the simpler algorithm for case (1) (essentially rules (1)–(6)), or its case (2) extension (see [15]), or the more general order-sorted extension of the Lassez-Marriott algorithm [12] are invoked. Finally, symbolic union and intersection operations are performed to obtain either: (i) a finite set of patterns if the algorithm computed a pattern language, or (ii) a Boolean expression containing some symbolic differences that do not denote pattern languages otherwise.

A user interface has also been constructed in Maude which allows the user to directly enter pattern expressions and theories using the Full Maude [1] syntax, obviating the need to first convert to the slightly more complex meta-term syntax. The user interface is implemented as an extension of Full Maude using Maude's `LOOP-MODE` module [1].

We give below a few examples of how the tool is used. The tool provides to the user two primary commands: `solve-pattern` and `ms-solve-pattern` which are used to solve pattern intersections and differences in an order-sorted (resp.

many-sorted) way. After loading the tool, we can input theories we wish to reason about. An example theory and queries are shown below:

```
(fmod DIAMOND is
    sorts A B C D .                (solve-pattern {a} U {b} .)
    subsorts A < B C < D .         {a} U {b}
    op a : -> A .
    op b : -> B .                  (solve-pattern {f(X:A,Y:B)} &
    op c : -> C .                                 {f(W:B,Z:A)} .)
    op d : -> D .                  {f(A1:A,A2:A)}
    op f : A A -> A .
    op f : B B -> B .              (solve-pattern {f(X:C,b)} -
    op f : C C -> C .                             {f(Y:B,Z:D)} .)
    op f : D D -> D .              {f(C1:C#,b)}
endfm)
```

Note that both theory declarations and commands are enclosed in parentheses () and that commands are ended with a period before the closing parenthesis (.). Also note that, thanks to Maude's mixfix syntax capabilities, pattern syntax at the tool interface level is almost identical to that used internally by the library. The syntax for variables is the usual **name:sort** notation, so that X:B is a variable named X which has sort B. The **ms-solve-pattern** command (not shown above), first reduces the pattern to all of its many-sorted instances and solves each of them using the many-sorted pattern algorithm. The tool and examples can be downloaded online (see: http://maude.cs.illinois.edu/w/index. php?title=Maude_Tools:Order-sorted_Term_Patterns).

Additionally, we conducted experiments comparing our order-sorted pattern operations algorithm to its many-sorted reduction. To ground our discussion, we work in a module COMPLEX-RAT adapted from [9] that defines the complex numbers. We also fix a term set T by randomly selecting operators to generate terms upto depth 2. Then, given $(t_1, t_2) \in T^2$, we generate the pattern operation $[\![t_1]\!] - [\![t_2]\!]$, which we compute by both our order-sorted algorithm and the many-sorted reduction. Our experiments show that, on average, the many-sorted reduction requires about a 1,000 times as many rewrites as the order-sorted algorithm, with the median being 55 times as many rewrites. While not a proof, this presents a strong case that for (non-toy) examples, the order-sorted algorithm is more expressive (no input encoding, shorter output) and performant than its many-sorted cousin. For more details on our experiments, the library source code, and proofs related to the tool implementation see [15].

7   Related Work and Conclusions

On pattern operations the most closely related work is [6,11,12,18,19] and references there. On equational formulas in initial algebras the most closely related work is [2–4,13,14] and references there. The relationships to work in both these areas have been discussed in detail in previous sections (see also [15]).

To conclude, we have shown that the untyped algorithms break down when performing the order-sorted pattern operations needed in current declarative languages, and shown that such operations can be defined using a signature transformation $\Sigma \mapsto \Sigma^{\#}$. We have also shown that this transformation yields new insights and a new, quite simple proof of the known decidability of the first-order theory of an initial order-sorted algebra. The Introduction mentioned many applications of pattern operations. We illustrate a sufficient completeness one in [15], but plan to work on many others and to further advance the current implementation to make it part of the Maude formal tool environment.

Acknowledgements. Partially supported by NSF Grant CNS 13-19109. We thank the referees for their excellent suggestions to improve the paper.

References

1. Clavel, M., Durán, F., Eker, S., Meseguer, J., Lincoln, P., Martí-Oliet, N., Talcott, C.: All About Maude. LNCS, vol. 4350. Springer, Heidelberg (2007)
2. Comon, H., Lescanne, P.: Equational problems and disunification. J. Symbol. Comput. **7**, 371–425 (1989)
3. Comon, H.: Equational formulas in order-sorted algebras. In: Comon, H. (ed.) Automata, Languages and Programming. LNCS, vol. 443, pp. 674–688. Springer, Heidelberg (1990)
4. Comon, H., Delor, C.: Equational formulae with membership constraints. Inf. Comput. **112**(2), 167–216 (1994)
5. Ehrig, H., Mahr, B.: Fundamentals of Algebraic Specification 1, vol. 6. Springer, Heidelberg (1985)
6. Fernández, M.: Negation elimination in empty or permutative theories. J. Symb. Comput. **26**(1), 97–133 (1998)
7. Futatsugi, K., Diaconescu, R.: CafeOBJ Report. World Scientific, Singapore (1998)
8. Goguen, J., Burstall, R.: Institutions: abstract model theory for specification and programming. J. ACM **39**(1), 95–146 (1992)
9. Goguen, J., Meseguer, J.: Order-sorted algebra I: equational deduction for multiple inheritance, overloading, exceptions and partial operations. Theo. Comput. Sci. **105**, 217–273 (1992)
10. Goguen, J., Winkler, T., Meseguer, J., Futatsugi, K., Jouannaud, J.P.: Introducing OBJ. In: Software Engineering with OBJ: Algebraic Specification in Action, pp. 3–167. Kluwer (2000)
11. Lassez, J.-L., Maher, M., Marriott, K.: Elimination of negation in term algebras. In: Tarlecki, Andrzej (ed.) MFCS 1991. LNCS, vol. 520, pp. 1–16. Springer, Heidelberg (1991)
12. Lassez, J.L., Marriott, K.: Explicit representation of terms defined by counter examples. J. Autom. Reason. **3**(3), 301–317 (1987)
13. Maher, M.J.: Complete axiomatizations of the algebras of finite, rational and infinite trees. In: Proceedings of LICS 1988, pp. 348–357. IEEE Computer Society (1988)
14. Maher, M.J.: Complete axiomatizations of the algebras of finite, rational and infinite trees. Technical report, IBM T. J. Watson Research Center (1988)
15. Meseguer, J., Skeirik, S.: Equational formulas and pattern operations in initial order-sorted algebras. Technical Report, University of Illinois at Urbana-Champaign, June 2015. http://hdl.handle.net/2142/78055

16. Meseguer, J.: Membership algebra as a logical framework for equational specification. In: Parisi-Presicce, Francesco (ed.) WADT 1997. LNCS, vol. 1376, pp. 18–61. Springer, Heidelberg (1998)

17. Meseguer, J., Goguen, J., Smolka, G.: Order-sorted unification. J. Symbolic Comput. **8**, 383–413 (1989)

18. Pichler, R.: Explicit versus implicit representations of subsets of the Herbrand universe. Theo. Comput. Sci. **290**(1), 1021–1056 (2003)

19. Tajine, M.: The negation elimination from syntactic equational formula is decidable. In: Kirchner, Claude (ed.) RTA 1993. LNCS, vol. 690, pp. 316–327. Springer, Heidelberg (1993)

Efficient Compilation of Functional Logic Programs

Compiling Collapsing Rules
in Certain Constructor Systems

Sergio Antoy$^{(\boxtimes)}$ and Andy Jost

Computer Science Department, Portland State University, Portland, OR, USA
antoy@cs.pdx.edu, andrew.jost@synopsys.com

Abstract. The implementation of functional logic languages by means of graph rewriting requires a special handling of collapsing rules. Recent advances about the notion of a needed step in some constructor systems offer a new approach to this problem. We present two results: a transformation of a certain class of constructor-based rewrite systems that eliminates collapsing rules, and a rewrite-like relation that takes advantage of the absence of collapsing rules. We formally state and prove the correctness of these results. When used together, these results simplify without any loss of efficiency an implementation of graph rewriting and consequently of functional logic computations.

1   Introduction

Functional logic programming [6,18,19] integrates the best features of the functional and the logic paradigms. For instance, demand-driven evaluation, higher-order functions, and polymorphic typing from functional programming are combined with logic variables, constraint solving, and non-deterministic search from logic programming. Narrowing makes this combination seamless and enables encoding problems into programs in a style elegant, understandable, and easier to reason about [5].

Graph rewriting [9,25,27] is an approach to the implementation of functional and functional logic computations. The objects of a computation are *term graphs*, also referred to as *expressions*, i.e., singly rooted, acyclic graphs. For any graph t, $\mathcal{N}(t)$ is the set of nodes of t. A graph's *node* q has two attributes: a *label*, $\mathcal{L}(q)$, and a sequence of *successors*, $\mathcal{S}(q)$. The label and the successors abstract respectively a symbol of the signature of a rewrite system and the arguments to which the symbol's occurrence is applied in an expression. An implementation represents a node as a dynamic linked data structure holding a label and a sequence of pointers to other nodes. For technical convenience, graphs that differ only for a renaming of nodes are considered equal [15,25].

A graph rewriting system, or *program*, is a set of *rules*, where a rule is a graph with two roots abstracting the left- and right-hand sides of the rule, respectively. Rules are *left linear* [12, Definition 1.4.1], i.e., the left-hand side is a tree. A consequence is that a variable occurs at most once in a left-hand side. A *step* of a computation of a host graph consists of three phases: (1) matching a rule

© Springer International Publishing Switzerland 2015
M. Falaschi (Ed.): LOPSTR 2015, LNCS 9527, pp. 57–72, 2015.
DOI: 10.1007/978-3-319-27436-2_4

left-hand side to a subgraph called the *redex*, (2) constructing the corresponding right-hand side called the redex's *contractum*, and (3) replacing the redex with its contractum. The signature from which the labels of the nodes are drawn is partitioned into *constructors* and *operations*. The left-hand side of a rule is a *pattern*, i.e., a graph rooted (by a node labeled) by an operation and every other node is labeled by either a variable or a constructor. A *constructor form*, or *value*, is a graph whose nodes are all labeled by constructors. A *head constructor form* is a graph rooted by a constructor.

Finding redexes in a graph according to some program is typically an expensive activity. However, this is not our case. For the inductively sequential graph rewriting systems (recalled below), a sound, complete and optimal strategy that finds redexes very efficiently is presented in [14,15]. We consider a slightly more general class [3], that allows a well-behaved form of overlapping. The exact same strategy is applicable to our graphs with the only difference that some redexes have more than one contractum. In this case, in the spirit of functional logic programming, the contractum is chosen non-deterministically.

For example, the following rules, in Curry's syntax, define the function that computes the length of a list, where "[]" represents the empty list and $(x : xs)$ the list with head x and tail xs:

$$
\begin{aligned}
&\texttt{length [] = 0}\\
&\texttt{length (x:xs) = 1 + length xs}
\end{aligned}
\tag{1}
$$

A finite list is denoted $[x_1, \ldots x_n]$, where x_i, for any appropriate i, is an element of the list. The expression $t = length\,[3, 4]$, which is a redex, is pictorially represented in Fig. 1. Conceptually, a rewrite step of t first constructs the contractum of t, $u = 1 + length\,[4]$, which is also shown in Fig. 1, and then redirects to u any reference to t (none occurs in the figure) because "t has become u." The redirection portion of a step [17] is a focus of our work.

Executing steps as described above would be naive and impractical. In fact, t can be a subexpression of a larger expression, called the context of t. The context of t may contain several references to t, i.e., the root of t is a successor of some nodes of its context. All these references should be tracked down and

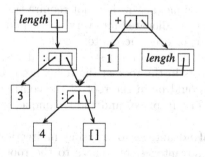

Fig. 1. Graph representation of the expression *length* [3, 4] (left) and its contractum $1 + length\,[4]$ (right). An outer box represents a node. Inside an outer box/node there is the label and a possibly null sequence of boxes representing references to the successors.

changed. This activity is potentially very expensive since a step is no longer a local operation, rather the entire context of t must be traversed. Our work deals with this specific aspect.

In this section, we recalled only the key concepts of graph rewriting needed to understand the problem and present our solution. Some familiarity with this framework is desirable. In Sect. 2 we recall two popular implementation techniques for graph rewriting. Since finding redexes in a host graph is easy and efficient in our framework, we focus only on the low-level details of nodes and pointers manipulation. In Sect. 3 we define the class of programs that we consider and recall recent results about properties of needed redexes in the class. These results are at the core of our technique. In Sect. 4 we define a program transformation that simplifies some aspects of executing those programs by graph rewriting. We state and prove our first correctness claim. In Sect. 5 we define a relation on graphs, called ripping, that produces results similar to rewriting, but is simpler to implement and more efficient to execute. We state and prove our second correctness claim. In Sect. 6 we statically quantify some effects of our technique on the performance of computations. In Sect. 7 we discuss a practical adaptation of our technique to evaluating expressions containing free variables. In Sect. 8 we discuss related work and offer our conclusion.

2   Implementation Techniques

For the sake of efficiency, implementations of graph rewriting are usually "in-place." This means that in a step when the redex is replaced by its contractum, the context of the redex is re-used as the context of the contractum. This in-place rewriting still requires redirecting the pointers of the context pointing to the root of the redex. To avoid the cost of this operation, as discussed in the previous section, implementations of graph rewriting adopt special techniques.

The first technique is based on *indirection pointers* [23, Sect. 8.1]. Every node of an expression has an indirection pointer and is accessed only through this indirection pointer. The replacement of a redex t with its contractum u only needs redirecting to u the indirection pointer of t. Any reference within the context of t to the indirection pointer of t is unaffected. A step is a local operation using this technique, i.e., it does not require traversing the context of t. However, extra memory is allocated for every node of an expression and extra machine cycles are spent for every access to a node.

The second technique is based on destructive updates. In a step, the label and sequence of successors in the root of the redex are overwritten by the corresponding items that would be in the root of the contractum. We call such a step a *rip step* (re-labeling in place) and the technique, which we formalize in Sect. 5, *ripping*.

Ripping has several advantages over using indirection pointers—and one drawback. Among the advantages, references to the root of the redex do not need to be redirected to the root of the contractum; no indirection node is used; no node is allocated for the root of the contractum; and the root of the redex is reused rather than garbage collected. The drawback is that ripping may produce unintended results when a collapsing rule is applied. A *collapsing rule* is a rule whose right-hand side is a variable, which is called the *collapsing variable*. We show the problem on an example. Consider the following expression:

$$t = (\text{id x, id x}) \text{ where x = 0 ? 1} \tag{2}$$

where *id* is the identity function:

$$\text{id x = x} \tag{3}$$

and "?" denotes the *choice* operation defined by the rules:

$$\begin{aligned}\text{x ? y = x}\\\text{x ? y = y}\end{aligned} \tag{4}$$

Contrary to popular functional programming languages, there is no textual order among the rules. Thus, the expression $t\,?\,u$, for any subexpressions t and u, non-deterministically rewrites to t or to u.

The meaning of the *where* clause in (2) is to introduce potentially shared nodes, where "shared" means having multiple predecessors. In the example, x is indeed shared.

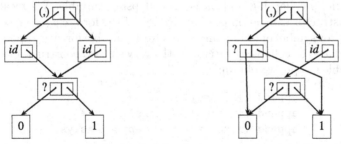

Fig. 2. The expression on the left-hand side has two values, $(0,0)$ and $(1,1)$. The expression on the right-hand side has 4 values, all possible pairs of zeros and ones.

The graph on the left-hand side of Fig. 2 pictorially shows t defined in program (2). This graph has two values, $(0,0)$ and $(1,1)$, resulting from each alternative of the *choice*. The graph on the right-hand side is obtained by a rip step of the redex in the first component of the pair. This graph has four values, all the pairs of zeros and ones. Two of these values, $(0,1)$ and $(1,0)$, are not intended. In a functional, hence deterministic setting, a graph has at most one value, thus, unintended values are not produced. However, the problem of duplicating portions of a computation still occurs and affects the efficiency of a computation rather than its input/output relation.

The problem we just showed is corrected by using a *forward node*. A forward node is a low-level device similar to an indirection pointer, but it is created only by steps applying collapsing rules, as opposed to systematically for every node, and explicitly to avoid the duplication of subexpressions. A program that manipulates graphs, e.g., for printing or evaluating them, must be aware of the possibility of encountering forward nodes and must be able to deal with them. During a computation, there is the danger of creating chains of forward nodes and the opportunity of compacting these chains to avoid the possibility of traversing them over and over.

In this paper, we propose a variation of the second technique, discussed in the previous page, based on destructive updates. Our variation does not require forward nodes. In short, we replace the collapsing rules of a program with non-collapsing rules in a way that does not change the "interesting" computations of the program. The motivation of our work is an implementation with destructive updates. Thus, we also formalize this implementation and discuss its correctness.

3   Detour on Need

Our overall approach to deal with collapsing rules is not to have any in a program. For example, consider the usual operation that concatenates two lists:

$$\begin{aligned} &\texttt{append [] ys = ys} \\ &\texttt{append (x:xs) ys = x : append xs ys} \end{aligned} \tag{5}$$

The first rule is collapsing and ys is its collapsing variable. We recall that a *shallow* constructor expression is an expression of the form $c(x_1, \ldots c_n)$, where c is a constructor symbol of arity n and x_i is a fresh variable for every appropriate i. If we instantiate the collapsing variable with every shallow constructor expression of the variable's type, we obtain:

$$\begin{aligned} &\texttt{append [] [] = []} \\ &\texttt{append [] (x:xs) = (x:xs)} \\ &\texttt{append (x:xs) ys = x : append xs ys} \end{aligned} \tag{6}$$

where there are no collapsing rules. Programs (5) and (6) are similar. Given two lists, t_1 and t_2, if the expression *append* $t_1 \, t_2$ has a value according to (5), then it has the same value according to (6) and vice versa.

However, if *append* $t_1 \, t_2$ has no value according to (5) there is a difference. Consider the following non-terminating nullary operation:

$$\texttt{loop = loop} \tag{7}$$

The expression *append* [] *loop* is a redex according to (5), but it is not and it will never become a redex according to (6). In this section, we show that this difference is irrelevant for the execution of a program.

Our programs are modeled by a class of rewrite systems called overlapping inductively sequential [3]. *Inductive sequentiality* means that operations are defined by cases resembling those of a proof by structural induction. The rules of each operation can be organized in a hierarchical structure, called a *definitional tree* [2], that guides the evaluation strategy. *Overlapping*, in conjunction with the inductive sequentiality, means that if a redex is reduced by distinct rules, these rules have the same left-hand side. The epitome of an overlapping inductively sequential function is the *choice* operation defined in (4).

Every reducible expression t in the overlapping inductively sequential systems has a redex which is reduced by every computation of t to a value, a result that extends to a non-orthogonal class of systems the seminal result of [21]. A strategy that reduces only these redexes is optimal *modulo non-deterministic choices* [3].

A novel notion of need, more appropriate for constructor-based systems, was recently proposed in [7]. This notion depends only on the rules' left-hand side in a way that makes it applicable to the class of the overlapping inductively sequential systems that we just described.

Definition 1. [7] *Let t and u be operation-rooted expressions with u subexpression of t, we say that u is* needed *for t iff in any derivation of t to a head constructor form, u is derived to a head constructor form.*

Observe that u needs neither be a redex nor be a proper subexpression. In fact, u may be irreducible and t is a needed subexpression of itself. We abuse the word "needed" because our notion generalizes the definition of needed redex [21] as follows. The contrapositive formulation is Definition 1 more expressively captures this concept of need: t cannot be derived to a head constructor form, unless u is derived to a head constructor form.

The following statement establishes the connection between the classic formulation of need [21] and our formulation.

Lemma 1. [7] *Let $\mathcal{R}$ be an overlapping inductively sequential system. If u is both a needed (in the sense of [21]) subexpression of t and a redex, then u is a needed (in the sense of our Definition 1) redex of t, i.e., it is reduced to a head constructor form in any derivation of t to a head constructor form.*

From now on, "need" and "needed" will refer to the concept defined in Definition 1. The following immediate consequence of the above lemma is at the core of our technique.

Corollary 1. *Let $\mathcal{R}$ be an overlapping inductively sequential system. If t is a redex according to $\mathcal{R}$ needed for some context $C[]$, u is the contractum of t, and u is (still) operation-rooted, then u is needed for $C[]$ as well.*

This result justifies our claim that programs (5) and (6) are equivalent in practice. Let $t = append\ []\ u$ be a needed expression, where u is an operation-rooted subexpression. Program (6) attempts to evaluate u for matching a rule of *append* to t. Program (5) does not. However, since t is a needed redex, u is its contractum, and u is operation-rooted, by Corollary 1, u is needed as well. Thus, program (5) will eventually attempt to evaluate u to a head constructor form as program (6). In other words, u is equally needed and evaluated by both programs.

4   Transformation

We define below a transformation that takes a rewrite system possibly containing collapsing rules and produces an equivalent rewrite system without collapsing rules. The precise meaning of the equivalence of input and output systems of the transformation is formalized by Theorem 1.

Definition 2. *Let $\mathcal{R}$ be a constructor-based rewrite system. The collapse-free variant of $\mathcal{R}$, denoted $\mathcal{R}_u$, is defined as follows: for each rule R of $\mathcal{R}$, if R is not collapsing, then R is in $\mathcal{R}_u$. Otherwise, for every constructor symbol c of the signature of $\mathcal{R}$, R_c is in $\mathcal{R}_u$, where R_c is the instance of R obtained by instantiating the collapsing variable of R to a shallow constructor expression rooted by c. No other rule is in $\mathcal{R}_u$.*

Of course, in a typed system only well-typed instantiations of the collapsing variable are considered. For example, program (6) is the collapse-free variant of program (5).

Collapsing rules in which the collapsing variable is polymorphic give raise to a potentially large number of instantiations. In modern computers with gigabytes of core memory, the amount of memory for holding these instantiations should hardly be a problem. A rule in these instantiations is selected according to the root symbol of the rule left-hand side argument. This is an efficient operation executed in constant time, i.e., independently of the number of rules. A technique in which the instantiations of collapsing rules are not explicitly generated in the executable code, is discussed later.

Observe that for any program $\mathcal{R}$, $\mathcal{R}$ and its collapse-free variant $\mathcal{R}_u$ have the same signature. A sound, complete, and optimal strategy exists [3] for overlapping inductively sequential term rewriting systems. The same strategy is applicable to overlapping inductively sequential graph rewriting systems. Eventually, we would like to replace a program with its collapse-free variant. Thus, we are pleased to discover that the same strategy exists for the replacement program.

Lemma 2. *Let $\mathcal{R}$ be an overlapping inductively sequential system. Then, the collapse-free variant of $\mathcal{R}$, $\mathcal{R}_u$, is an overlapping inductively sequential system.*

Proof. We prove that every operation of $\mathcal{R}_u$ has a definitional tree, hence $\mathcal{R}_u$ is inductively sequential. The signatures of $\mathcal{R}$ and $\mathcal{R}_u$ are the same. If f is an operation of $\mathcal{R}_u$, then it is an operation of $\mathcal{R}$. Since $\mathcal{R}$ is inductively sequential, f has a definitional tree, say $\mathcal{T}$. If f has a collapsing rule $l \rightarrow r$, there is a *leaf* node L of $\mathcal{T}$ whose pattern π is equal to l modulo a renaming of nodes and variables. Let x be the collapsing variable of $l \rightarrow r$. We replace this *leaf* node of $\mathcal{T}$ with a branch node B that has the same pattern π, and x as the inductive variables. The children of B are leaves whose rules are all and only the rules of f instantiating $l \rightarrow r$ in $\mathcal{T}_u$ according to Definition 2. Hence f has a definitional tree in $\mathcal{R}_u$.   □

The following result precisely states the equivalence between a program $\mathcal{R}$ and its collapse-free variant $\mathcal{R}_u$. The values of an expression e in $\mathcal{R}_u$ are all and only the values of e in $\mathcal{R}$.

Theorem 1. *Let $\mathcal{R}$ be an overlapping inductively sequential system and $\mathcal{R}_u$ its collapse-free variant. For all expressions t and s over the signature of $\mathcal{R}$ (and $\mathcal{R}_u$), with s head constructor form, $t \xrightarrow{*} s$ in $\mathcal{R}$ iff $t \xrightarrow{*} s$ in $\mathcal{R}_u$.*

Proof. The "if" direction is immediate. If $t \to t'$ in $\mathcal{R}_u$, then $t \to t'$ in $\mathcal{R}$, since every rule of $\mathcal{R}_u$ is an instance of a rule of $\mathcal{R}$. Hence, any computation in $\mathcal{R}_u$ is also a computation in $\mathcal{R}$. The "only if" direction is proved by strong induction on the number of collapsing rules applied in $A = t \xrightarrow{*} s$ in $\mathcal{R}$. The base case is immediate, since every non-collapsing rule of $\mathcal{R}$, by construction, is a rule of $\mathcal{R}_u$. For the induction case, consider the first step of A, say a, that applies a collapsing rule. We consider whether the match of the collapsing variable in step a is a head constructor form. Case *true*: the computation in $\mathcal{R}_u$ can make the same step and the claim holds by the induction hypothesis. Case *false*: let w be the match. Corollary 1 proves that w is needed, hence A must derive it to a head constructor form w'. We can re-arrange the steps of A [3, Lemma 20] (as in the Parallel Moves Lemma) so that the derivation of w into w' occurs before step a of A. By the induction hypothesis, $w \to w'$ in $\mathcal{R}_u$. After re-arranging the steps of A, the residual of step a satisfies case *true*, and the claim holds. $\square$

The previous result easily extends from head constructor forms to constructor forms.

Corollary 2. *Let $\mathcal{R}$ be an overlapping inductively sequential system and $\mathcal{R}_u$ its collapse-free variant. For all expressions t and s over the signature of $\mathcal{R}$ (and $\mathcal{R}_u$), with s constructor form, $t \xrightarrow{*} s$ in $\mathcal{R}$ iff $t \xrightarrow{*} s$ in $\mathcal{R}_u$.*

Proof. By induction on the length of a derivation using Theorem 1. $\square$

Curry is a candidate for the application of our results, but some programs that could benefit from our technique cannot be entirely or directly modeled by rewrite systems because of the presence of built-in types. Program (2) makes this point. The collapse-free variant of (2) should contain an instance of the rule of *id* for every integer.

A solution to this problem is to avoid the explicit instantiation of collapsing rules, and instead to compile them slightly differently from non-collapsing rules. When a collapsing rule R is going to be applied to a redex, the match of the collapsing variable is checked. If the match, say t, is rooted by a constructor c, the application proceeds as if R were instantiated by mapping the collapsing variable to a shallow constructor expression rooted by c. Otherwise, t is evaluated in an attempt to obtain a head constructor form t'. If t' is obtained, the rule application proceeds again as described above. Otherwise, it must be that either the evaluation of t does not terminate or terminates in an operation-rooted expression. The latter is a failure of the entire computation, since t is needed. The same outcome, whether non-termination or failure, would be obtained by any implementation, since t must be evaluated to a head constructor form.

Evaluating an expression to obtain a head constructor form is an activity provided by many implementations. Hence, a major task for the adoption of our technique is already available in these implementations. For example, the PAKCS implementation [20] of Curry, which maps Curry source code to Prolog source code, defines a predicate, *hnf*, exactly for this task. The same is true for the Basic Scheme [8], which defines an abstract function, H, for this task and implements it in OCaml.

Some compilers of Curry, e.g. PAKCS [20], use a similar approach to encode polymorphic functions, such as Boolean and constrained equalities. These functions are applicable to instances of every algebraically defined and built-in typed. They could be defined by one rule for every constructor or value. Instead, the availability of a test for head constructor form and a procedure that evaluates an expression to head constructor form avoid the proliferation of rules.

5   Ripping

The proof of correctness of the previous section to some extent completes our work. Given a program $\mathcal{R}$ possibly containing collapsing rules, we transform it into a program $\mathcal{R}_u$ without collapsing rules. This allows us to compile $\mathcal{R}_u$ according to any desired scheme without concerns for collapsing rules. We are guaranteed that the values computed by $\mathcal{R}_u$ are all and only those computed by $\mathcal{R}$ and that they are obtainable with the same strategy and in the same number of steps. Furthermore, the proof of Theorem 1 implicitly shows that a computation to constructor form has the same length in the two systems.

Of course, there is the expectation that the scheme adopted to compile $\mathcal{R}_u$ is correct. The motivation of our work is to compile $\mathcal{R}_u$ for ripping. We are not aware of any proof of its correctness and, indeed, we have not even found a statement of it. In this section we address this issue.

We recall that given two graphs t and s, a (graph) *homomorphism* [15,26] of t into s is a mapping $\sigma : \mathcal{N}(t) \rightarrow \mathcal{N}(s)$ that preserves roots and for nodes not labeled by a variable, labels and successors, i.e.,

1. $\sigma(\mathcal{R}oot(t)) = \mathcal{R}oot(s)$
2. $\mathcal{L}(\sigma(q)) = \mathcal{L}(q)$, for every node $q \in \mathcal{N}(t)$ with $\mathcal{L}(q) \in \Sigma$;
3. $\mathcal{S}(\sigma(q))_i = \sigma(\mathcal{S}(q)_i)$, for every node $q \in \mathcal{N}(t)$ and appropriate index i.

Let t be a graph, $l \rightarrow r$ a rewrite rule, q a node of t and $\sigma : l \rightarrow t|_q$ a homomorphism, i.e., q is the root of a redex of t. We call *ripping*, denoted "$\circ\!\!\rightarrow$" the binary relation on graphs defined as follows: Let p be the root of $\sigma(r)$. $t' = t + \sigma(r)$ except at node q for which, in t', $\mathcal{L}(q) = \mathcal{L}(p)$ and $\mathcal{S}(q) = \mathcal{S}(p)$. In other words, the label and successors of q, in t', are replaced by those of p. This update makes the need of pointer redirection, which occurs during the replacement phase of a rewrite step, unnecessary.

Ripping produces results different from rewriting. Consider again program (2). During the evaluation of t, the rule of *id* is applied to the first component of the pair. Since the rule is collapsing, the argument is evaluated to a head constructor form. The result is non-deterministic, thus let us suppose that 0

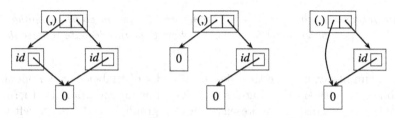

Fig. 3. The second graph is obtained from the first graph with a *rip* step, the technique formalized in this paper. The third graph is obtained from the first graph with a rewrite step.

is produced (if 1 were produced, the reasoning would be identical). The entire expression at this point is pictorially represented in the left-hand side of Fig. 3.

The second graph of Fig. 3 shows the result of a rip step where the redex is the first component of the pair. The result is a graph with two nodes labeled by zero. We remark that no new node is created by this step, rather the root of the redex has been re-labeled with the label of the root of the contractum. The third graph is obtained by applying the same rewrite step to the first graph. We introduce the following concept to precisely characterize the significant differences between these graphs.

Definition 3. *Given two graphs t and s, t is an* adequate representation *of s iff there exists a homomorphism σ of t into s such that, for all distinct nodes p and q of t, if $\sigma(p) = \sigma(q)$, then the label of p (and hence of q) is a constructor symbol. We call such homomorphism an* adequate homomorphism.

For example, the second graph of Fig. 3 is an adequate representation of the third graph.

Fig. 4. Commutative diagram for adequate representation (vertical arrows).

Observe that the match of the left-hand side of a rule to a redex is an adequate homomorphism since rules are left linear and that the composition of adequate homomorphisms is an adequate homomorphism. The diagram in Fig. 4 pictorially represents Lemma 3, where the vertical arrows stand for adequate homomorphisms.

Lemma 3. *Let $\mathcal{R}$ be an overlapping inductively sequential system and $\mathcal{R}_u$ its collapse-free variant. Let t and s be graphs over the signature of $\mathcal{R}$ with t an* *adequate representation of s. Then, $t \circ\!\!\!\rightarrow t'$ in $\mathcal{R}_u$ (a rip step) for some t' iff $s \rightarrow s'$ in $\mathcal{R}_u$ (a rewrite step) for some s', where t' is an adequate representation of s'.*

Proof. Preliminarily, observe that the set of nodes of t labeled by an operation is in a bijection with the set of nodes of s labeled by an operation. Furthermore, if a graph g is an adequate representation of a graph h, and l is the left-hand side of a rewrite rule, then l matches g iff l matches h. Thus, for every step of t there is corresponding step of s, with the same rule, and vice versa.

Assuming we apply the same rule at corresponding nodes of t and s, we constructively prove the existence of an adequate homomorphism of t' into s'. Let's partition the nodes of t' into 3 classes: (1) the root of the redex, (2) the remaining nodes of t' that are also in t, and (3) the nodes created by the step, which originate from the nodes of the rule's right-hand side which are not labeled by a variable. A node in class (2) is also in t, thus it is mapped to make the diagram of Fig. 4 commutative. A node in class (3) is also in s, thus it is mapped to make the diagram of Fig. 4 commutative. The node, say q, in class (1) is mapped from a node in t, that is mapped to the root of the redex in s. Let p the root of the contractum of this redex. Thus, map q to p. This define a homomorphism which is adequate.   □

The following result shows that ripping and rewriting compute the same values of an expression modulo an adequate representation.

Theorem 2. *Let $\mathcal{R}$ be an overlapping inductively sequential system and $\mathcal{R}_u$ its collapse-free variant. Let t and s be graphs over the signature of $\mathcal{R}$ with s a constructor form. If s is a value of t by rewriting in $\mathcal{R}_u$, then there exists an s' that is a value of t by ripping in $\mathcal{R}_u$ and s' is an adequate representation of s. If s' is a value of t by ripping in $\mathcal{R}_u$, then there exists an s that is a value of t by rewriting t in $\mathcal{R}_u$ and s' is an adequate representation of s.*

Proof. By induction on the length of a derivation.   □

The combination of Theorems 1 and 2 shows that the evaluation of an expression by graph rewriting can also be obtained by ripping, in-place rewriting with re-labeling, which appears simpler and more efficient than other alternatives. This technique is simpler and faster when the rule being applied is not a collapsing rule. Our work shows that this is possible for every system in the class that we consider.

A computation in $\mathcal{R}_u$ executed by rewriting has a corresponding computation executed by ripping. We regard these two computation as the same. For every step of one computation, there is a step of the other computation that applies the same rule at a node that we regard as the same because in the hosting graphs there is a bijection between the nodes labeled by operations. The results of the two computations, that have nodes labeled by constructors only, may not be isomorphic graphs. However, they are equal both when printed as (tree) terms, because they are bisimilar [11], and when printed in fully collapsed form[1] [10, 26], because one is an adequate representation of the other.

6 Performance

The major contribution of our work is not a speedup of computations or a reduction of both static and dynamic memory consumption, though they all do occur in some degree, but a simplification of the compiler architecture—forward nodes, and the machinery to handle them, can be entirely eliminated at nearly no cost.

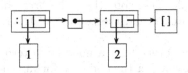

Fig. 5. The evaluation of *append* [1] [2] produces a list containing a forward node represented by the large black dot in the above diagram.

We begin our discussion with an example. Consider a program that concatenates some lists and computes the length of the result. For concreteness, we choose very simple lists, i.e., the program computes *length* (*append* [1, 2]). The rules of *length* and *append* were given in (1) and (5) respectively. The value of *append* [1, 2], say L, computed *without* the use of our technique is shown in Fig. 5. The large black dot represents the forward node created when the first rule of (5) is applied. The same value computed *with* our technique, is equal to L except that the forward node is absent. List L may never be entirely present in memory because operation *length* consumes portions of L as soon as operation *append* constructs portions of L due to the lazy evaluation strategy, but the order of evaluation does not affect our reasoning.

The execution time of each program is too short to be reliably measured with ordinary tools. As far as memory consumption is concerned, our technique saves the allocation and the traversal of the forward node. There is a similar program that instead of constructing a list of two elements separated by a single forward node it constructs a similar list with an arbitrarily long chain of forward nodes. Computing the length of this list takes an arbitrarily long time. More relevant is that the implementation of *length* must be prepared to encounter forward nodes. Hence, extra instructions are executed to check for their presence. When a forward node is encountered, extra instructions are executed to reach the node that the forward node points to. Thus, the object code of *length* is longer, is more complicated, takes longer to execute, and allocates extra dynamic memory.

---

[1] The word "collapse" is overloaded in graph parlance. In this context, its refers to a relation on graphs defined in the cited reference.

For the reasons discussed above, we do not measure execution times to assess the performance of our technique. The average speed up of our technique and the savings in memory consumption depend on the programs used for a benchmark and for many programs these differences are too small to benchmark. As we already pointed out, our contributions are the formalization of our technique and a simplification of the compiler architecture. Our goal is to show that this simplification does not incur execution time or memory consumption penalties, rather the opposite. Below we precisely summarize the effects of our technique on these parameters.

1. Without our technique, every time a collapsing rule is applied, a forward node is allocated and initialized. By contrast, our technique executes the same step with an instantiated rule. Therefore, the node corresponding to the collapsing variable is pattern matched and the content of the root of the redex is reassigned.
2. Without our technique, every time a node is pattern matched, a test must be performed to check whether the node is a forward node. In the affirmative case, the node pointed to by the forward node must be fetched and pattern matched again. The fetched node could be a forward node again. By contrast, our technique avoids the test, and never has to fetch a second node.

7   Narrowing

Functional logic programs compute with unknown information which is abstracted by logic (also called *free*) unbound variables. A free variable is bound during a computation if and when without the binding the computation could not continue. The combination of binding some variables and making a rewrite step is called *narrowing*. Narrowing supports a simple and elegant programming style [5] unique to the functional logic paradigm.

For a contrived example, consider again the rule of (5) and the expression $t = append\ v\ []$, where v is an unbound free variable. No rule can be applied to t. To compute the value of t, v is bound to either $[]$ or $(x:xs)$, non-deterministically, where x and xs are fresh unbound free variables. For example, if v is bound to $[]$, the value of t is $[]$. By contrast, consider the expression $s = append\ []\ v$, where v is again an unbound free variable. In this case, s is rewritten to v, where v is unaffected by the step. Variable v might be bound later depending on the context in which it occurs.

During the execution of a program, we store the bindings of free variables in an array called the *bind-table*. A variable is internally represented as an index in the bind-table array. The k-th entry in the array, holds the binding, if any, of the variable represented by k. A conventional value marks unbound variables. Any node standing for a variable is labeled by the same distinguished symbol, which we denote "*free*". In addition, in a node standing for a variable v, we store the index of v in the bind-table.

Regarding the integration of free variables with our technique, the only relevant question is what happens when, during the application of a collapsing rule, the *collapsing* variable is bound to a *free* variable. The answers is that we simply treat the free variable as if it were a head constructor form. I.e., the step replaces the content of the root of the redex with the content of the root of the replacement, in this case the node representing the free variable.

Graph rewriting stipulates that, for each variable v, in any expression there is at most one node labeled by v [15,25]. Our approach violates this stipulation, but only in appearance. The index k of a node with label *free* is immutable. The binding, if any, indexed by k is in the bind-table. Thus, there is invariably one and only one binding of any variable regardless of the number of nodes standing for that variable. The claims leading to the correctness of our technique, Theorem 2, carry over to narrowing with no significant changes. We only need a minimal extension to the notion of adequate representation. Referring to the notation of Definition 3, if $\sigma(p) = \sigma(q)$, then the label of p (and hence of q) is either a constructor symbol or *free*, and when the label is *free*, the indexes in p and q are the same.

8 Discussion and Related Work

Graph rewriting is a viable mean for the implementation of functional and functional logic languages that has lead to the discovery and development of optimal strategies [4]. Transformations of rewrite systems for compilation purposes are described in [16,22]. The specialization of rules through the instantiations of collapsing variables is typical of partial evaluation [1]. Our goal differs from those of the above techniques. Our transformation is specialized in that its only purpose is to eliminate collapsing rules. Its merit is in the property that, for the class of systems that we consider, which is perfectly suited for functional logic programming, every computation to a value in a system with collapsing rules can be executed, with the same effort, in a system without collapsing rules. An implementation of rewriting without collapsing rules is easier to code and faster to execute. We have not found any work close enough to ours for a direct comparison.

Literature on the implementation of graph rewriting abounds. With respect to our work, papers fall into either of two groups, graph reduction machines [13,24], or some specialized aspects of rewriting [23]. Our implementation of ripping as rewriting is theoretical in that we do not address data structures, register allocation, bit use for tags, and similar. Its merit is to make the pointer redirection phase of a rewrite step effortless in a concrete implementation. We have not found any description of this technique or claim of its correctness.

Acknowledgments. This material is based upon work partially supported by the National Science Foundation under Grant No. CCF-1317249. Michael Hanus provided valuable comments on a preliminary version of this paper.

References

1. Alpuente, M., Falaschi, M., Vidal, G.: Partial evaluation of functional logic programs. ACM Trans. Program. Lang. Syst. **20**(4), 768–844 (1998)

2. Antoy, S.: Definitional trees. In: Kirchner, H., Levi, G. (eds.) ALP 1992. LNCS, vol. 632. Springer, Heidelberg (1992)
3. Antoy, S.: Optimal non-deterministic functional logic computations. In: Hanus, M., Heering, J., Meinke, K. (eds.) ALP 1997 and HOA 1997. LNCS, vol. 1298. Springer, Heidelberg (1997). http://cs.pdx.edu/antoy/homepage/publications/alp97/full.pdf
4. Antoy, S.: Evaluation strategies for functional logic programming. J. Symb. Comput. **40**(1), 875–903 (2005)
5. Antoy, S.: Programming with narrowing. J. Symb. Comput. **45**(5), 501–522 (2010)
6. Antoy, S., Hanus, M.: Functional logic programming. Comm. ACM **53**(4), 74–85 (2010)
7. Antoy, S., Jost, A.: Are needed redexes really needed? In: Proceedings of the 15th Symposium on Principles and Practice of Declarative Programming, PPDP 2013, pp. 61–71. ACM, New York (2013)
8. Antoy, S., Peters, A.: Compiling a functional logic language: *The Basic Scheme*. In: Schrijvers, T., Thiemann, P. (eds.) FLOPS 2012. LNCS, vol. 7294, pp. 17–31. Springer, Heidelberg (2012)
9. Ariola, Z.M., Klop, J.W.: Equational term graph rewriting. Fundam. Inf. **26**, 207–240 (1996)
10. Ariola, Z.M., Klop, J.W., Plump, D.: Bisimilarity in term graph rewriting. Inf. Comput. **156**(12), 2–24 (2000)
11. Barendregt, H.P., van Eekelen, M.C.J.D., Glauert, J.R.W., Kennaway, J.R., Plasmeijer, M.J., Sleep, M.R.: Term graph rewriting. In: de Bakker, J.W., Nijman, A.J., Treleaven, P.C. (eds.) PARLE Parallel Architectures and Languages Europe. Lecture Notes in Computer Science, vol. 259, pp. 141–158. Springer, Heidelberg (1987)
12. Bezem, M., Klop, J.W., de Vrijer, R.: Term Rewriting Systems. Cambridge University Press, Cambridge (2003)
13. Clarke, T.J.W., Gladstone, P.J.S., MacLean, C.D., Norman, A.C.: Skim - the s, k, i reduction machine. In: Proceedings of the 1980 ACM Conference on LISP and Functional Programming, LFP 1980, pp. 128–135. ACM, New York (1980)
14. Echahed, R.: Inductively sequential term-graph rewrite systems. In: Ehrig, H., Heckel, R., Rozenberg, G., Taentzer, G. (eds.) ICGT 2008. LNCS, vol. 5214. Springer, Heidelberg (2008)
15. Echahed, R., Janodet, J.C.: On constructor-based graph rewriting systems. Technical Report 985-I, IMAG (1997). ftp://ftp.imag.fr/pub/labo-LEIBNIZ/OLD-archives/PMP/c-graph-rewriting.ps.gz
16. Fokkink, W., van de Pol, J.: Simulation as a correct transformation of rewrite systems. In: Privara, I., Ružička, P. (eds.) MFCS 1997. LNCS, vol. 1295. Springer, Heidelberg (1997)
17. Glauert, J.R.W., Kennaway, R., Papadopoulos, G.A., Sleep, M.R.: Dactl: an experimental graph rewriting language. J. Prog. Lang. **5**(1), 85–108 (1997)
18. Hanus, M.: The integration of functions into logic programming: from theory to practice. J. Log. Program. **19 & 20**, 583–628 (1994)
19. Hanus, M.: Functional logic programming: from theory to curry. In: Voronkov, A., Weidenbach, C. (eds.) Programming Logics. LNCS, vol. 7797, pp. 123–168. Springer, Heidelberg (2013)
20. Hanus, M.: PAKCS 1.11.4: The Portland Aachen Kiel Curry System (2014). http://www.informatik.uni-kiel.de/pakcs

21. Huet, G., Lévy, J.-J.: Computations in orthogonal term rewriting systems, I. In: Lassez, J.-L., Plotkin, G. (eds.) Computational logic.: Essays Honour Alan Robinson, pp. 395–414. MIT Press, Cambridge (1991)
22. Kamperman, J.F.T., Walters, H.R.: Simulating TRSs by minimal TRSs a simple, efficient, and correct compilation technique. Technical report CS-R9605, CWI (1996)
23. Kennaway, J.R., Klop, J.K., Sleep, M.R., de Vries, F.J.: The adequacy of term graph rewriting for simulating term rewriting. In: Sleep, M.R., Plasmeijer, M.J., van Eekelen, M.C.J.D. (eds.) Term Graph Rewriting Theory and Practice, pp. 157–169. J. Wiley & Sons, Chichester (1993)
24. Landin, P.J.: The mechanical evaluation of expressions. Comput. J. **6**(4), 308–320 (1964)
25. Plump, D.: Term graph rewriting. In: Kreowski, H.-J., Ehrig, H., Engels, G., Rozenberg, G., (eds.) Handbook of Graph Grammars, vol. 2, pp. 3–61. World Scientific (1999)
26. Plump, D.: Essentials of term graph rewriting. Electr. Notes Theor. Comput. Sci. **51**, 277–289 (2001)
27. Sleep, M.R., Plasmeijer, M.J., van Eekelen, M.C.J.D. (eds.): Term Graph Rewriting Theory and Practice. J. Wiley & Sons, Chichester (1993)

From Boolean Equalities to Constraints

Sergio Antoy[1] and Michael Hanus[2]([✉])

[1] Computer Science Department, Portland State University, Portland, OR, USA
antoy@cs.pdx.edu
[2] Institut Für Informatik, CAU Kiel, 24098 Kiel, Germany
mh@informatik.uni-kiel.de

Abstract. Although functional as well as logic languages use equality to discriminate between logically different cases, the operational meaning of equality is different in such languages. Functional languages reduce equational expressions to their Boolean values, `True` or `False`, logic languages use unification to check the validity only and fail otherwise. Consequently, the language Curry, which amalgamates functional and logic programming features, offers two kinds of equational expressions so that the programmer has to distinguish between these uses. We show that this distinction can be avoided by providing an analysis and transformation method that automatically selects the appropriate operation. Without this distinction in source programs, the language design can be simplified and the execution of programs can be optimized. As a consequence, we show that one kind of equational expressions is sufficient and unification is nothing else than an optimization of Boolean equality.

1 Motivation

Functional as well as logic programming languages are based on the common idea to specify computational problems in a high-level and descriptive manner. However, the computational entities and, thus, the programming styles are different. This can be seen in a prominent feature of such languages: the discrimination between logically different cases of a given problem. Functional (as well as imperative) languages use Boolean equations for this purpose, i.e., an equational expression is reduced to either `True` or `False` and, depending on the computed result, a different computation path is selected. A typical example is the factorial function where the base case is distinguished from the recursive case by comparing the argument with 0:[1]

```
fac n = if n==0 then 1
              else n * fac (n-1)
```

On the other hand, logic languages, like Prolog, use separate rules for different cases where (equational) constraints restrict the applicability of the rules. For instance, the following Prolog program defines the concatenation relation

---

This material is based in part upon work supported by the National Science Foundation under Grant No. 1317249.

[1] We use the syntax of Haskell [24] for functional programs.

© Springer International Publishing Switzerland 2015
M. Falaschi (Ed.): LOPSTR 2015, LNCS 9527, pp. 73–88, 2015.
DOI: 10.1007/978-3-319-27436-2_5

between three lists (where we do not use patterns in left-hand sides to make the equational constraints explicit):

```
append(X,Y,Z)  :- X=[], Y=Z.
append(X,Y,Z)  :- X=[E|T], Z=[E|U], append(T,Y,U).
```

The equality symbol "=" used in this program is different from the Boolean equality "==" above. For instance, in the first rule it is not intended to evaluate X=[] to True or False, but this equality must hold to proceed with this rule, i.e., it is a constraint for subsequent evaluation steps. As a consequence, it is not necessary to fully evaluate equational expressions but one can continue a computation even with partial knowledge, as long as the constraint is ensured to hold. For instance, if we want to ensure that a list L ends with the element 0, we can write

```
append(_,[0],L)
```

which is solvable even if the values of the list elements are not known. Thus, if L=[A,B,C] is a list of three variables, then the literal above is solved by binding C to 0 but leaving all other list elements unspecified. Operationally, this is done by unification [27] instead of evaluation to Boolean values.

Functional logic languages attempt to combine the most important features of functional and logic programming in a single language (see [5,18] for recent surveys). In particular, the functional logic language Curry [21] extends Haskell by common features of logic programming, i.e., non-determinism, free variables, and equational constraints. Due to its roots in functional *and* logic programming, Curry provides two kinds of equalities: Boolean equality ("==") as in functional programming and equational constraints ("=:=") as in logic programming. The motivation for this decision is to support nested case distinctions, like in functional programming, as well as rule-oriented programming with partial information, like in logic programming. Although one might argue that it is always possible to guess values for unknowns, so that one kind of equality is sufficient, an important insight of logic programming is that unification can restrict the search space by binding variables instead of guessing values [27]. For instance, if X and Y are Boolean variables, the equational constraint "X=Y" can be solved by simply binding X to Y instead of enumerating appropriate values for X and Y.

Although the distinction between these two kinds of equalities is present in Curry from its early design [15], it also causes some complications. A programmer might not always easily understand which equality should be chosen in a particular situation. Moreover, the distinction between solving and evaluating equalities is also present in the type system, i.e., "==" has the result type Bool whereas "=:=" has the result type Success (indicating the type of constraints). As a consequence, various standard (combinator) functions on Booleans need also be duplicated for the type Success.

In order to improve this situation, we argue in this paper that one kind of equality, namely Boolean equality, is sufficient for the programmer. This will be justified by an automatic method to transform Boolean equalities into constraint equalities, if it is appropriate. Hence, we automatically obtain the nice features of unification, i.e., reduction of the search space. For this purpose, we present a program analysis and transformation method that automatically selects the appropriate kind of equality. This leads to a simpler language design without sacrificing program efficiency.

2   Functional Logic Programming and Curry

We briefly review those elements of functional logic languages and Curry which are necessary to understand the contents of this paper. More details can be found in recent surveys on functional logic programming [5,18] and in the language report [21].

Curry is a declarative multi-paradigm language combining in a seamless way features from functional, logic, and concurrent programming (concurrency is irrelevant as our work goes, hence it is ignored in this paper). The syntax of Curry is close to Haskell [24], i.e., type variables and names of defined operations usually start with lowercase letters and the names of type and data constructors start with an uppercase letter. $\alpha \rightarrow \beta$ denotes the type of all functions mapping elements of type α into elements of type β (where β can also be a functional type, i.e., functional types are "curried"), and the application of an operation f to an argument e is denoted by juxtaposition ("$f\ e$"). In addition to Haskell, Curry allows *free (logic) variables* in conditions and right-hand sides of rules and expressions evaluated by an interpreter.

A *Curry program* consists in the definition of *functions* or *operations* and the *data types* on which the functions operate. Functions are defined by (conditional) equations and are evaluated lazily. Function calls with free variables are evaluated by a possibly non-deterministic instantiation of demanded arguments which corresponds to narrowing [25,28]. Curry narrows with possibly non-most-general unifiers to ensure the optimality of computations [4].

Example 1. We present the above features in a program chosen for its simplicity and brevity, rather than its power. The program defines the data type of Boolean values and polymorphic lists and operations to concatenate two lists and compute the last element of a list:[2]

```
data Bool   = True | False
data List a = []    | a : List a

(++) :: [a]  → [a]  → [a]
[]      ++ ys = ys
(x:xs) ++ ys = x : (xs ++ ys)

last :: [a]  → a
last xs | _ ++ [x] =:= xs   = x
```

The **data** type declarations define **True** and **False** as Boolean values and [] (empty list) and : (non-empty list) as the constructors for polymorphic lists

---

[2] Note that Curry requires the explicit declaration of free variables, as x in the rule of last, to ensure checkable redundancy, but we omit them in this paper for the sake of simplicity.

(a is a type variable ranging over all types and the type "List a" is written as [a] for conformity with Haskell). The (optional) type declaration ("::") of the operation "++" specifies that "++" takes two lists as input and produces an output list, where all list elements are of the same (unspecified) type. Since "++" can be called with free variables in arguments, the equation "_ ++ [x] =:= xs" in the condition of last is solved by instantiating the anonymous free variable _ to the list xs without the last argument, i.e., the only solution to this equation satisfies that x is the last element of xs.

The (optional) condition of a program rule is typically a conjunction of constraints. Each Curry system provides at least *equational constraints* of the form $e_1 =:= e_2$ which are satisfiable if both sides e_1 and e_2 are reducible to unifiable data terms.

In order to use equations to discriminate between different cases, as in the definition of the factorial function fac shown in Sect. 1, Curry also offers a Boolean equality operator "==" which evaluates to True if both arguments can be evaluated to identical data terms, and to False if the arguments evaluate to different data terms. Conceptually, "==" can be considered as defined by rules comparing constructors of the same type, i.e., by the following rules ("&&" is the Boolean conjunction):

```
True  == True  = True      []      == []      = True
False == False = True      (x:xs)  == (y:ys)  = x==y && xs==ys
True  == False = False     []      == (y:ys)  = False
False == True  = False     (x:xs)  == []      = False
```

As already discussed in [6], the presence of the types Success and Bool together with two equality operators, rooted in the history of Curry, might cause confusions and should be avoided in order to obtain a simpler definition of Curry. Hence, [6] proposes to omit the type Success and the operator "=:=" from the definition of Curry, and we follow this proposal in our paper. Note that one can also solve equations by narrowing with the above rules. For instance, [x,x]==[True,y] is solved by instantiating x and y to True while evaluating "==". However, solving equations by narrowing with "==" rules has also a drawback compared to logic programming. If there is an equation between two variables, narrowing enumerates all values for these variables whereas unification (deterministically!) binds one variable to the other. Hence, the expression "xs == ys && xs++ys == [True]" has an infinite search space with solely False results.

This was the motivation for the inclusion of the operator "=:=" in Curry. Conceptually, it can be considered as defined by "positive" rules:

```
True  =:= True  = True      []      =:= []      = True
False =:= False = True      (x:xs)  =:= (y:ys)  = x=:=y && xs=:=ys
```

Thus, "=:=" yields True for identical data terms or fails.[3] Operationally, these rules are not applied by narrowing but combined with the unification principle

---

[3] Note that we omit the type Success, as proposed in [6]. Hence, equational constraints as well as rule conditions are of type Bool rather than Success, in contrast to the current definition of Curry [21].

[27], i.e., if one argument is a free variable, it is bound to the evaluated data term of the other side (if the variable is not contained in this term, see [21] for details). Therefore, the expression "xs =:= ys" evaluates to True by binding xs to ys and the expression "xs =:= ys && xs++ys =:= [True]" has a finite search space without any result.

It would be desirable to automatically replace occurrences of "==" by "=:=" whenever it can be done without losing solutions (see the next section). This would free the programmer from having to select the "right" equality and simplify the language: programmers always use "==" so that the operator "=:=" is just an optimization of "==". This is the motivation for our current work.

Since Curry with all its syntactic sugar (we have only presented a small fragment of it) is a quite rich source language, a simpler intermediate representation of Curry programs has been shown to be useful to describe the operational semantics [1], compile programs [10,19], or implement analyzers [20] and similar tools. Programs of this intermediate language, called FlatCurry, contain a single rule for each function where the pattern matching strategy is represented by case expressions. The basic structure of FlatCurry is defined as follows (where x_i denotes variables, f defined functions, C constructors, and $\overline{o_k}$ a sequence of objects $o_1 \ldots o_k$):

$$
\begin{aligned}
P &::= D_1 \ldots D_m & \text{(program)} \\
D &::= f\ \overline{x_n} = e & \text{(function definition)} \\
p &::= C\ \overline{x_n} & \text{(flat pattern)} \\
e &::= x & \text{(variable)} \\
&\ \ |\ \ C\ \overline{e_n} & \text{(constructor application)} \\
&\ \ |\ \ f\ \overline{e_n} & \text{(function application)} \\
&\ \ |\ \ case\ e_0\ of\ \{\overline{p_k \to e_k}\} & \text{(case distinction)} \\
&\ \ |\ \ e_1\ ?\ e_2 & \text{(non-deterministic choice)}
\end{aligned}
$$

A program P (we omit data type declarations) consists of a sequence of function definitions D with pairwise different variables in the left-hand sides. The right-hand sides are expressions e composed by variables, constructor and function calls, case expressions, and disjunctions. A case expression[4] has the form $case\ e\ of\ \{C_1\ \overline{x_{n_1}} \to e_1, \ldots, C_k\ \overline{x_{n_k}} \to e_k\}$, where e is an expression, $C_1, \ldots, C_k$ are different constructors of the type of e, and $e_1, \ldots, e_k$ are expressions. The *pattern variables* $\overline{x_{n_i}}$ are local variables which occur only in the corresponding subexpression e_i.

By fixing a strategy to match arguments, one can translate Curry programs into FlatCurry programs. The higher-order constructs of Curry are translated into FlatCurry by defunctionalization [26]. Thus, lambda abstractions are transformed into top-level functions and there is a predefined operation *apply* to apply an expression of functional type to an argument (see [18,29] for more details).

---

[4] Since we do not discuss residuation and concurrent computations, we also omit the difference between rigid and flexible case expressions [18].

Conditional rules are not present in FlatCurry since, as shown in [3], they can be transformed into unconditional ones by introducing a "conditional" operator cond defined by

```
cond True x = x
```

For instance, the rule defining last as shown above can be transformed into

```
last xs = cond (_++[x] =:= xs) x
```

The evaluation strategy of Curry is by-need. Hence, the second argument of $cond$ is evaluated only if the first argument is True.

3   Transforming Equalities

In this section we discuss an automatic method to replace occurrences of Boolean equalities of the form $e_1 == e_2$ by an equational constraint $e_1 =:= e_2$. Obviously, such a replacement is not always correct. For instance, consider the following contrived example:

```
isEmpty xs = if xs==[] then True else False
```

If we evaluate the expression "isEmpty xs", where xs is a free variable, we obtain the following two results (e.g., with the Curry system KiCS2 [10]):

```
{xs = []} True
{xs = (_x1:_x2)} False
```

These two results are computed by narrowing the equation xs==[] w.r.t. the rules defining "==" shown in the previous section. However, if we replace the Boolean equality by an equational constraint, as in

```
isEmpty' xs = if xs=:=[] then True else False
```

and evaluate the expression "isEmpty' xs", then we obtain only the single result

```
{xs = []} True
```

since the constraint "xs=:=[]" can only be satisfied, i.e., delivers the value True only.

Thus, in order to avoid losing solutions, a Boolean equation $e_1 == e_2$ can be replaced by the equational constraint $e_1 =:= e_2$ if it is ensured that only the value True is required as the result of this equation. In general, this depends on the context of the equation. Fortunately, there are many situations in functional logic programs where this requirement can be deduced. For instance, consider the following definition of last:

```
last xs | xs == _++[x] = x
```

As discussed above, this rule is transformed into the unconditional rule

```
last xs = cond (xs == _++[x]) x
```

Since the definition of cond requires that the first argument must have the value True in order to evaluate a cond expression, the condition can be replaced by an equational constraint:

```
last' xs = cond (xs =:= _++[x]) x
```

Hence, if we evaluate `last' [x,42]`, where x is a free variable, we obtain the single result

    {x = _x1} 42

On the other hand, we obtain infinitely many answers for the expression `last [x,42]` (where in each answer x is bound to a different integer value). Similarly, we can replace the occurrences of "`==`" by "`=:=`" in the rule

    f xs ys | xs == _++[x] && ys == _++[x]++_ = x

However, in the rule

    g xs ys | xs == _++[x] && not (ys == _++[x]++_) = x

only the first occurrence of "`==`" can be replaced by "`=:=`", since the second occurrence is required to be evaluated to `False` in order to apply the rule.[5]

These examples show that a careful analysis of the kind of values required for a successful evaluation is necessary in order to perform our proposed transformation. Note that such an analysis is different from a strictness analysis in purely functional programming [23]. A strictness analysis provide information about the necessary demand of computation in order to compute any value, whereas we need information about possible values in order to compute other values. For instance, in order to transform the definition of `f` above, it is necessary to know that both arguments of the conjunction operator "`&&`" need to be `True` in order to obtain the overall value `True`. For this purpose, we define in the next section an appropriate analysis for "required" values.

4   Analysis of Required Values

Our goal is to develop a program analysis to infer which kind of values are required at some position in a program in order to compute a result, i.e., some value. To obtain a manageable analysis, we consider only top-level constructors in the analysis so that a *value* is some constructor-rooted expression. In principle, this could be extended to any depth bound k (as used in the abstract diagnosis of functional programs [2] or in the abstraction of term rewriting systems [8,9]), but in practice only a depth $k = 1$ (i.e., top-level constructors) is useful due to the quickly growing size of the abstract domain for $k > 1$. For instance, for lists we distinguish the values `[]` (empty list) and "`:`" (non-empty lists) and for Booleans we distinguish the values `True` and `False`.

Following the framework of abstract interpretation [13], we define for each type τ an *abstract domain* τ^α, i.e., a set of *abstract values*, as follows. If $C_\tau = \{C_1, \ldots, C_k\}$ denotes the set of all constructors of type τ, then $\tau^\alpha = 2^{C_\tau} \cup \{Any\}$, i.e., an abstract value of τ^α is either a subset of the constructors of type τ or the specific constant Any denoting any expression. For instance, the abstract domain for Boolean values is

    Bool^α = { ∅, {True}, {False}, {True,False}, Any }

---

[5] The latter equality could also be improved if disequality constraints [7,22] are available in the target language, but since this is not the case for standard implementations of Curry, we do not consider them in this paper.

Abstract values are ordered by: for all τ_1 and τ_2, $\tau_1 \sqsubseteq Any$, and $\tau_1 \sqsubseteq \tau_2$ if $\tau_1 \subseteq \tau_2$ and both are not Any. Thus, the least upper bound of two abstract values $\tau_1 \neq Any \neq \tau_2$ is their set union, i.e., $\tau_1 \sqcup \tau_2 = \tau_1 \cup \tau_2$.

The meaning of an abstract value a, i.e., the concretization $[\![a]\!]$ of a, is the set of all expressions, if $a = Any$, or, if $a \neq Any$, the set of all values rooted by some constructor of a (where $root(e)$ denotes the symbol at the root of the expression e): $[\![a]\!] = \{e \mid root(e) \in a\}$. We call two abstract values $a, a' \in \tau^\alpha$ *compatible* if $[\![a]\!] \cap [\![a']\!] \neq \varnothing$, i.e., if they have some element in common.

As discussed above, we are interested to deduce required argument values from required result values. For instance, if True is the required value of a conjunction e_1 && e_2, then True is also the required value of both e_1 and e_2. We denote this property by (&&) $::^\alpha \{\text{True}\}, \{\text{True}\} \rightarrow \{\text{True}\}$.

We can read this type as: in order to compute the result True, the argument values are required to be True. Or: unless both arguments are evaluated to True, the result cannot be True.

Definition 1. *A typing $f ::^\alpha a_1, \ldots, a_n \rightarrow a$ of a function f is correct if, for all $e = f\ e_1 \ldots e_n$, the following implication holds: if e evaluates to some value (constructor-rooted term) $t \in [\![a]\!]$, then, for $i = 1, \ldots, n$, e_i evaluates to some $t_i' \in [\![a_i]\!]$.*

The above notion of correctness establishes a condition on the values of the arguments of a function application to produce a certain value as the result of the application. For each function f of (concrete) type $\tau_1, \ldots, \tau_n \rightarrow \tau$, the typing $f ::^\alpha Any, \ldots, Any \rightarrow C_\tau$ (with appropriate numbers of arguments) is correct since any expression is an element of $[\![Any]\!]$. Clearly, defined functions can have more than one correct typing. For instance, the negation operator **not** has the types

>   not $::^\alpha \{\text{True}\} \rightarrow \{\text{False}\}$
>   not $::^\alpha \{\text{False}\} \rightarrow \{\text{True}\}$

and the conjunction operator (&&) has the types

>   (&&) $::^\alpha \{\text{True}\}, \{\text{True}\} \rightarrow \{\text{True}\}$
>   (&&) $::^\alpha Any, Any \rightarrow \{\text{False}\}$

These abstract types can be used as follows. If the condition of a program rule has the form e_1 && e_2, the value True is required as the result of this conjunction. By the first type of "&&", we can deduce that True is also required as the result of both expressions e_1 and e_2, otherwise the conjunction cannot be evaluated to True. However, if a condition has the form **not** $(e_1$ && $e_2)$, we cannot deduce a single value required for e_1 or e_2 (by the second type of "&&"), since this condition yields True if e_1 has the value False or if e_1 has the value True and e_2 has the value False. Note that

>   (&&) $::^\alpha \{\text{False}\}, Any \rightarrow \{\text{False}\}$

is not a correct typing: True $\notin [\![\{\text{False}\}]\!]$ but True && False $\in [\![\{\text{False}\}]\!]$. This is intended: we cannot deduce from the required result value False that the first argument is required to be False.

In order to define well-typed programs, we assume a *type environment* F (for a given program) which contains for each n-ary function symbol f occurring in the program at least one element of the form $f ::^\alpha a_1, \ldots, a_n \to a$. Since we want to know required values of arguments in order to compute some value of an expression, our type analysis also returns a *variable type environment* E containing variable types $x ::^\alpha a$ for variables x occurring in the expression. The least upper bound $E_1 \sqcup E_2$ of two variable type environments E_1 and E_2 is the element-wise least upper bound of the associated types (where absent type information is interpreted as *Any*), e.g., $\{x ::^\alpha \{\text{True}\}, y ::^\alpha \{\text{True}\}\} \sqcup \{x ::^\alpha \{\text{False}\}\} = \{x ::^\alpha \{\text{True}, \text{False}\}, y ::^\alpha Any\}$. Observe that $y ::^\alpha Any$ is in the upper bound because the second environment places no restrictions on y. Similarly, $E_1 \sqcap E_2$ denotes the greatest lower bound of E_1 and E_2.

Var
$$\frac{}{F \vdash x ::^\alpha a \mid \{x ::^\alpha a\}} \quad \text{if } x \text{ is a variable}$$

Con
$$\frac{}{F \vdash C\, e_1 \ldots e_n ::^\alpha a \mid \varnothing} \quad \text{if } \{C\} \text{ and } a \text{ are compatible}$$

Fun
$$\frac{F \vdash e_1 ::^\alpha a_1 \mid E_1 \ldots F \vdash e_n ::^\alpha a_n \mid E_n}{F \vdash f\, e_1 \ldots e_n ::^\alpha a \mid \sqcap\{E_i \mid a_i \neq Any\}} \quad \text{if } f ::^\alpha a_1, \ldots, a_n \to a \in F$$

Or
$$\frac{F \vdash e_1 ::^\alpha a \mid E_1 \quad F \vdash e_2 ::^\alpha a \mid E_2}{F \vdash e_1 \,?\, e_2 ::^\alpha a \mid E_1 \sqcup E_2}$$

$Case$
$$\frac{\begin{array}{c}F \vdash e_0 ::^\alpha a' \mid E_0 \quad F \vdash e_1 ::^\alpha a \mid E_1 \ldots F \vdash e_j ::^\alpha a \mid E_j \\ F \vdash e_{j+1} ::^\alpha a_{j+1} \mid E_{j+1} \ldots F \vdash e_n ::^\alpha a_n \mid E_n\end{array}}{F \vdash case\ e_0\ of\ \{C_1\ \overline{x_{k_1}} \to e_1; \ldots; C_n\ \overline{x_{k_n}} \to e_n\} ::^\alpha a \mid E_0 \sqcap (E_1 \sqcup \ldots \sqcup E_j)}$$
$$\text{if } C_1, \ldots, C_j \in a' \text{ and, for } i = j+1, \ldots, n, a_i \text{ and } a \text{ are not compatible}$$

Fig. 1. Abstract typing rules for FlatCurry expressions

The (abstract) typing rules are shown in Fig. 1. The notation $F \vdash e ::^\alpha a \mid E$ should be read as: "if e is evaluated to some value of type a w.r.t. type environment F, then E are the required values of variables occurring in e." Rule *Var* requires the type of a variable as the type of the expression. Rule *Con* does not put requirements on variables since the term is already a value. Rule *Fun* requires well-typed arguments and an appropriate function typing to apply a function, but joins only the requirements of arguments where a value is required, since other arguments might not be evaluated. Rule *Or* requires that both alternatives of a choice expression must have the same type where the variable type environments are unified from both alternatives. Finally, rule *Case* requires that the constructors of the patterns in the various branches must be contained in the type of the discriminating expression. However, branches with a type that is not compatible with the overall result type are ignored. By this refinement, we can obtain more precise information about required arguments.

Definition 2. *A program P is* well typed *w.r.t. a type environment F for P if, for each rule $f\ x_1 \ldots x_n = e \in P$ and each $f ::^\alpha a_1, \ldots, a_n \to a \in F$, $F \vdash e ::^\alpha a \mid E$ is derivable by the rules in Fig. 1, for some variable type environment E, and, for $i = 1, \ldots, n$, $a_i' \subseteq a_i$ if $x_i ::^\alpha a_i' \in E$, otherwise $a_i = Any$, i.e., the deduced required value is more specific or does not occur.*

We show the usage of this type system by a few examples that are relevant for the application intended with this paper. In these examples, we write T and F for the abstract types {True} and {False}, respectively. The first example is the operator cond introduced in Sect. 2 to transform conditional equations. In FlatCurry, this operator is defined by the rule

```
cond x y = case x of { True   → y }
```

This rule is well-typed w.r.t. cond $::^\alpha$ T, $Any \to Any$ so that we can deduce that the first argument is required to be True in order to compute any value. Note that this rule is also well typed w.r.t. cond $::^\alpha Any, Any \to Any$, but this typing provides less precise information about required arguments.

The second example is the negation operator not defined by

```
not x = case x of { True   → False
                  ; False  → True }
```

It is easy to check that not $::^\alpha$ T $\to$ F is a well-typing of not since the following derivation is valid w.r.t. $F = \{$not $::^\alpha$ T $\to$ F$\}$:

$$\dfrac{\dfrac{}{F \vdash x ::^\alpha T \mid \{x ::^\alpha T\}}\ Var \quad \dfrac{}{F \vdash False ::^\alpha F \mid \varnothing}\ Con \quad \dfrac{}{F \vdash True ::^\alpha T \mid \varnothing}\ Con}{F \vdash \text{case } x \text{ of } \{True \to False; False \to True\} ::^\alpha F \mid \{x ::^\alpha T\}}\ Case$$

Note that the second case branch is ignored in the application of the *Case* rule since its result type T is not compatible with the overall result type F. Similarly, the following types (among others) can be derived to be well typed:

> not $::^\alpha$ F $\to$ T
> not $::^\alpha$ {False, True} $\to Any$

Finally, we consider the conjunction operator (&&) defined by

```
x && y = case x of { True   → y
                   ; False  → False }
```

(&&) $::^\alpha$ T, T $\to$ T is a well-typing since the following derivation holds for the type environment $F = \{$(&&) $::^\alpha$ T, T $\to$ T$\}$:

$$\dfrac{\dfrac{}{F \vdash x ::^\alpha T \mid \{x ::^\alpha T\}}\ Var \quad \dfrac{}{F \vdash y ::^\alpha T \mid \{y ::^\alpha T\}}\ Var \quad \dfrac{}{F \vdash False ::^\alpha F \mid \varnothing}\ Con}{F \vdash \text{case } x \text{ of } \{True \to y; False \to False\} ::^\alpha T \mid \{x ::^\alpha T, y ::^\alpha T\}}\ Case$$

The correctness of our type analysis can be stated by the following theorem:

Theorem 1. *If a program P is well typed w.r.t. a type environment F for P, then each $f ::^\alpha a_1, \ldots, a_n \to a \in F$ is correct.*

We have seen in various examples that there does not exist a meaningful most general type for each function. Although we could type each function f by

$f ::^{\alpha} Any, \ldots, Any \rightarrow Any$, this type does not provide any useful information about required arguments. Thus, the inference of types is more complex than in classical type inference systems [14].

Instead, we use the idea to compute types by a fixpoint analysis [12]. The analysis is started with no information about each function (e.g., $f ::^{\alpha} Any, \ldots$, $Any \rightarrow Any$) and uses the rules in Fig. 1 to compute values for required arguments. If the analysis computes some more precise information about the result of a function, i.e., a result type like $\{C\}$, then the analysis is started again with all constructors (of the corresponding concrete data type): if $C_1, \ldots, C_k$ are all constructors of the data type to which C belongs, we restart the analysis with the environment containing $f ::^{\alpha} Any, \ldots, Any \rightarrow \{C_i\}$ (for $i = 1, \ldots, k$). In this way we obtain more meaningful results without testing all constructors from the beginning, which seems a good compromise between efficiency and precision of the analysis.

5   Implementation

The analysis of required values is a prerequisite to implement the transformation of equalities as discussed in Sect. 3. To implement the analysis, we used the Curry analysis system CASS [20]. CASS is a generic program analysis system which provides an infrastructure to implement new bottom-up analyses. CASS requires only the definition of the abstract domain and the abstract operations to compute the abstract values for each function based on given abstract values for the operations on which the operation to be analyzed depend. Then the reading, parsing, and analysis of modules in their import order and the fixpoint computations are managed by CASS.

The results of the analysis are used to transform Boolean equations as follows. For each function f, we apply the rules in Fig. 1 in order to compute the required values at an occurrence of an expression of the form $e_1 == e_2$ in the right-hand side of the rule of f. If the abstract type is always $\{\texttt{True}\}$, we replace this expression by $e_1 =:= e_2$. This is justified by the fact that the result $\texttt{False}$ is never required when this function must be evaluated.

Hence, our implementation automatically transforms the occurrences of "==" shown in Sect. 3. Since this transformation is performed on FlatCurry programs, it can be easily integrated into the compilation chain for Curry programs. In fact, the transformation is fully integrated into the current releases of the Curry systems PAKCS [19] and KiCS2 [10].

In order to evaluate the usefulness of our transformation, we tested it on some benchmarks. As discussed in Sect. 2, our transformation can reduce infinite search spaces into finite ones. For instance, the expression

```
cond (xs == ys && xs++ys == [True]) True
```

has an infinite search space, whereas the transformed expression

```
cond (xs =:= ys && xs++ys =:= [True]) True
```

has a finite search space. Even in the case of finite search spaces, replacing Boolean equations by equational constraints often has a good impact on the run

Expression	==	=:=
last 10	0.01	0.00
last 15	0.41	0.00
last 20	13.12	0.00
fromPeano (half (toPeano 10000))	31.09	12.98
grep	0.54	0.37
simplify	22.41	16.68
varInExp	0.95	0.42

Fig. 2. Benchmarks: comparing Boolean equations and equational constraints

time since non-deterministic search is transformed to deterministic bindings, as demonstrated by some benchmarks.

We used the Curry implementation KiCS2 [10] for the benchmarks. KiCS2 evaluates the Boolean equality operator by narrowing with the "==" rules shown in Sect. 2 and the equational constraints by managing variable bindings [11]. The benchmarks were executed on a Linux machine (Debian 8.0) with an Intel Core i7-4790 (3.60Ghz) processor and 8GiB of memory. KiCS2 (Version 0.4.0) has been used with the Glasgow Haskell Compiler (GHC 7.6.3, option -O2) as its backend. The timings were performed with the time command measuring the execution time to compute all solutions (in seconds) of a compiled executable for each benchmark as a mean of three runs. The programs used for the benchmarks are last n (compute the last element of a list containing $n - 1$ variables and True at the end), half (compute the half of a Peano number using logic variables), grep (string matching based on a non-deterministic specification of regular expressions [5]), simplify (simplify a symbolic arithmetic expression), and varInExp (non-deterministically return a variable occuring in a symbolic arithmetic expression). Figure 2 shows the execution times to evaluate some expressions without (==) or with (=:=) our transformation. As expected, the creation and traversal of a large search space introduced by "==" is much slower than manipulating variable bindings by "=:=".

6   Practical Evaluation

In this section we discuss some practical experiences we made with our transformation tool.

As mentioned above, the transformation tool is integrated into the compilation chain of the recent releases of the Curry systems PAKCS [19] and KiCS2 [10]. The configuration files of these systems allow the user to set the following usage modes: "off" (do not apply this transformation), "full" (analyze programs as described in Sect. 4 and perform the transformation described in Sect. 5), or "fast" (which is the default: use pre-computed analysis information of standard operations from the prelude to perform the transformation described in Sect. 5). The advantage of the "fast" mode is that it is a reasonable compromise between effectiveness and efficiency. In this mode, the transformation described

Program	#lines	full	fast	=:= (orig.)	== (transf.)
CHR	474	2.65	0.76	11	11
CurryStringClassifier	194	0.82	0.25	21	21
HTML	1316	6.04	2.14	13	13
Parser	49	0.22	0.02	6	6
SetFunctions	90	0.40	0.07	28	28
AddTypes	117	1.46	0.20	4	4
Curry2JS	633	2.85	0.85	6	6
maxtree	17	0.18	0.01	3	3
queens	12	0.19	0.01	5	2

Fig. 3. Benchmarks: transforming Boolean equations into equational constraints

in Sect. 5 does not perform the fixpoint analysis of Sect. 4, but it simply uses the pre-computed abstract types for the most relevant Boolean functions defined in the prelude, like (`&&`) (conjunction), (`||`) (disjunction), `not` (negation), and the conditional operator `cond`. The transformation itself can be efficiently performed by considering only functions that contain occurrences of "`==`". Thus, even large modules are transformed without any perceivable slowdown in the compilation chain.

Although the "fast" mode uses only the results of a few Boolean operations defined in the standard prelude, it is sufficient in practice, as our tests indicate. For these tests, we replaced in various existing Curry programs all equational constraints by Boolean equalities and checked how many of these Boolean equalities are replaced by equational constraints with our transformation tool. The results are shown in Fig. 3. The first group of Curry programs are standard libraries distributed with KiCS2, where HTML is the largest one (supporting programming of dynamic web pages [16]). The next two programs (AddTypes, Curry2JS) are tools contained in the KiCS2 distribution to add type signatures to top-level operations and compiling Curry programs into JavaScript programs (which is used to implement type-safe dynamic web pages [17]), respectively. The last two programs are small examples demonstrating typical functional logic programming techniques.

The first three result columns show the number of lines of code and the transformation times in the "full" and "fast" mode (in seconds, where the same machine as for the benchmarks in the previous section has been used). These numbers clearly indicate the advantage of the "fast" mode. Moreover, there was no difference in the transformation results between these modes. These results are summarized in the last two columns: they show the number of equational constraints ("`=:=`") occurring in the original programs[6] and the number of Boolean equalities ("`==`") that have been transformed back into equational constraints by

---

[6] A logic programmer might wonder about the low number of equational constraints even in larger functional logic programs. This is mainly due to the fact that functional logic programming supports nested expressions (where Prolog programmers have to use auxiliary variables and unification to connect the result from an inner

our transformation tool. The numbers in these columns show that our tool was able to transform almost all of them into constraints. The rare cases where this was not possible (queens) are operations that return constraints to be solved instead of using them in a condition of a program rule. For instance, consider an operation that returns True if its three arguments are pairwise equal:

```
equ3 x y z = x==y && y==z
```

Obviously, our transformation cannot replace the Boolean equalities by equational constraints since this may cause a loss of solutions. For instance, for Boolean values, the expression "not (equ3 x y z)" evaluates to True by binding x to True and y to False (among other solutions). Such solutions would be lost if we replace "==" by "=:=". However, if it is intended that the operation equ3 should only be used for "positive" evaluations, one can easily redefine it by

```
equ3 x y z | x==y && y==z = True
```

With this definition, our transformation tool is able to replace both occurrences of "==" by "=:=".

7 Conclusions

We have presented an automatic method to replace Boolean equalities by equational constraints in functional logic programs. This can be done only if it is ensured that True is required as the result of a Boolean equality, which is the case, e.g., in conditions of rules. To this aim, we developed an analysis for required values. This analysis can be seen as a non-standard type inference where abstract types represent sets of required values. The results of this analysis are then used to drive the actual program transformation.

Our transformation method has the following advantages over the current design of functional logic languages like Curry:

1. The source language becomes simpler. Since equational constraints are considered as an optimization of Boolean equality, the existing type Success can be omitted (as proposed in [6]). This has the consequence that quite similar operations, like inequalities between values ((<=)), do not need to be duplicated for the type of Boolean and constraints, as it is currently the case.
2. It is not necessary to consider the subtle differences between the type Bool and Success and the operators "==" and "=:=". A programmer uses "==" only (where the operator "=:=" must still be provided for the transformation target and in exceptional cases where a programmer wants to write efficient code independent of a program transformation). This also simplifies the teaching of declarative multi-paradigm languages [15].
3. Equational constraints can be considered as an optimized implementation of Boolean equalities. Hence, from a declarative point of view, one has to deal with Boolean equalities only, which are easy to define by standard rewrite rules as shown in Sect. 2.

---

computation to an outer one). Moreover, predicates delivering multiple results can also be expressed as non-deterministic functions.

If the target system also supports disequality constraints, as proposed in early functional logic languages [7,22], one could exploit them in an extension of our transformation tool. For instance, if an expression e_1==e_2 requires always `False` as its result, one could replace it by e_1=/=e_2, where the operator "=/=" represents a disequality constraint. This might be more efficient than guessing values by narrowing with the standard "==" rules but requires a specific implementation of a solver for "=/=".

References

1. Albert, E., Hanus, M., Huch, F., Oliver, J., Vidal, G.: Operational semantics for declarative multi-paradigm languages. J. Symb. Comput. **40**(1), 795–829 (2005)
2. Alpuente, M., Comini, M., Escobar, S., Falaschi, M., Lucas, S.: Abstract diagnosis of functional programs. In: Leuschel, M. (ed.) LOPSTR 2002. LNCS, vol. 2664. Springer, Heidelberg (2003)
3. Antoy, S.: Constructor-based conditional narrowing. In: Procedings of the 3rd International ACM SIGPLAN Conference on Principles and Practice of Declarative Programming (PPDP 2001), pp. 199–206. ACM Press (2001)
4. Antoy, S., Echahed, R., Hanus, M.: A needed narrowing strategy. J. ACM **47**(4), 776–822 (2000)
5. Antoy, S., Hanus, M.: Functional logic programming. Commun. ACM **53**(4), 74–85 (2010)
6. Antoy, S., Hanus, M.: Curry without Success. In: Proceedings of the 23rd International Workshop on Functional and (Constraint) Logic Programming (WFLP 2014), CEUR Workshop Proceedings, vol. 1335, pp. 140–154. CEUR-WS.org, (2014)
7. Arenas-Sánchez, P., Gil-Luezas, A., López-Fraguas, F.J.: Combining lazy narrowing with disequality constraints. In: Penjam, J. (ed.) PLILP 1994. LNCS, vol. 844. Springer, Heidelberg (1994)
8. Bert, D., Echahed, R.: Abstraction of conditional term rewriting systems. In: Proceedings of the 1995 International Logic Programming Symposium, pp. 147–161. MIT Press (1995)
9. Bert, D., Echahed, R., Østvold, M.: Abstract rewriting. In: Cousot, P., Filé, G., Falaschi, M., Rauzy, A. (eds.) WSA 1993. LNCS, vol. 724. Springer, Heidelberg (1993)
10. Braßel, B., Hanus, M., Peemöller, B., Rcck, F.: KiCS2: A new compiler from curry to haskell. In: Kuchen, H. (ed.) WFLP 2011. LNCS, vol. 6816, pp. 1–18. Springer, Heidelberg (2011)
11. Braßel, B., Hanus, M., Peemöller, B., Reck, F.: Implementing equational constraints in a functional language. In: Sagonas, K. (ed.) PADL 2013. LNCS, vol. 7752, pp. 125–140. Springer, Heidelberg (2013)
12. Cousot, P.: Types as abstract interpretations. In: Proceedings of the 24th ACM Symposium on Principles of Programming Languages (Paris), pp. 316–331 (1997)
13. Cousot, P., Cousot, R.: Abstract interpretation: a unified lattice model for static analysis of programs by construction of approximation of fixpoints. In: Proceedings of the 4th ACM Symposium on Principles of Programming Languages, pp. 238–252 (1977)

14. Damas, L., Milner, R.: Principal type-schemes for functional programs. In: Proceedings 9th Annual Symposium on Principles of Programming Languages, pp. 207–212 (1982)

15. Hanus, M.: Teaching functional and logic programming with a single computation model. In: Hartel, P.H., Kuchen, H. (eds.) PLILP 1997. LNCS, vol. 1292. Springer, Heidelberg (1997)

16. Hanus, M.: High-Level Server Side Web Scripting in Curry. In: Ramakrishnan, I.V. (ed.) PADL 2001. LNCS, vol. 1990, pp. 76–92. Springer, Heidelberg (2001)

17. Hanus, M.: Putting declarative programming into the web: translating curry to JavaScript. In: Proceedings of the 9th ACM SIGPLAN International Conference on Principles and Practice of Declarative Programming (PPDP 2007), pp. 155–166. ACM Press (2007)

18. Hanus, M.: Functional logic programming: from theory to curry. In: Voronkov, A., Weidenbach, C. (eds.) Programming Logics. LNCS, vol. 7797, pp. 123–168. Springer, Heidelberg (2013)

19. Hanus, M., Antoy, S., Braßel, B., Engelke, M., Höppner, K., Koj, J., Niederau, P., Sadre, R., Steiner, F.: PAKCS: The portland aachen kiel curry system (2015). http://www.informatik.uni-kiel.de/pakcs/

20. Hanus, M., Skrlac, F.: A modular and generic analysis server system for functional logic programs. In: Proceedings ACM SIGPLAN 2014 Workshop on Partial Evaluation and Program Manipulation (PEPM 2014), pp. 181–188. ACM Press (2014)

21. Hanus, M.: Curry: an integrated functional logic language (vers. 0.8.3). http://www.curry-language.org, 2012

22. Kuchen, H., López-Fraguas, F.J., Moreno-Navarro, J.J., Rodríguez-Artalejo, M.: Implementing a lazy functional logic language with disequality constraints. In: Proceedings of the 1992 Joint International Conference and Symposium on Logic Programming. MIT Press (1992)

23. Mycroft, A.: The theory and practice of transforming call-by-need into call-by-value. In: Robinet, B. (ed.) International Symposium on Programming. Lecture Notes in Computer Science, vol. 83, pp. 269–281. Springer, Heidelberg (1980)

24. Peyton Jones, S.: Haskell 98 Language and Libraries-The Revised Report. Cambridge University Press, Cambridge (2003)

25. Reddy, U.S.: Narrowing as the operational semantics of functional languages. In: Proceedings of IEEE International Symposium on Logic Programming, pp. 138–151, Boston (1985)

26. Reynolds, J.C.: Definitional interpreters for higher-order programming languages. In: Proceedings of the ACM Annual Conference, pp. 717–740. ACM Press (1972)

27. Robinson, J.A.: A machine-oriented logic based on the resolution principle. J. ACM 12(1), 23–41 (1965)

28. Slagle, J.R.: Automated theorem-proving for theories with simplifiers, commutativity, and associativity. J. ACM 21(4), 622–642 (1974)

29. Warren, D.H.D.: Higher-order extensions to Prolog: are they needed? Mach. Intell. 10, 441–454 (1982)

Types and Security

A Type-Theoretic Approach to Resolution

Peng Fu[(✉)] and Ekaterina Komendantskaya

Computer Science, University of Dundee, Dundee, Scotland
{pfu,ekomendantskaya}@dundee.ac.uk

Abstract. We propose a new type-theoretic approach to SLD-resolution and Horn-clause logic programming. It views Horn formulas as types, and derivations for a given query as a construction of the inhabitant (a proof-term) for the type given by the query. We propose a method of program transformation that allows to transform logic programs in such a way that proof evidence is computed alongside SLD-derivations. We discuss two applications of this approach: in recently proposed productivity theory of structural resolution, and in type class inference.

Keywords: Logic programming · Typed lambda calculus · Realizability transformation · Reduction systems · Structural resolution

1 Introduction

Logic Programming (LP) is a programming paradigm based on first-order Horn formulas. Informally, given a logic program Φ and a query A, LP provides a mechanism for automatically inferring whether or not $\Phi \vdash A$ holds, i.e., whether or not Φ logically entails A. The inference mechanism is based on the SLD-resolution, which uses the resolution rule together with first-order unification.

Example 1. Consider the following logic program Φ, consisting of Horn formulas labelled by κ_1, κ_2, κ_3, defining connectivity for a graph with three nodes:

$$\kappa_1 : \forall x. \forall y. \forall z. \mathrm{Connect}(x, y), \mathrm{Connect}(y, z) \Rightarrow \mathrm{Connect}(x, z)$$
$$\kappa_2 : \Rightarrow \mathrm{Connect}(\mathrm{Node}_1, \mathrm{Node}_2)$$
$$\kappa_3 : \Rightarrow \mathrm{Connect}(\mathrm{Node}_2, \mathrm{Node}_3)$$

In the above program, Connect is a predicate, and Node_1 – Node_3 are constants. SLD-derivation for the query $\mathrm{Connect}(x, y)$ can be represented as the following reduction:

$$\Phi \vdash \{\mathrm{Connect}(x, y)\} \leadsto_{\kappa_1, [x/x_1, y/z_1]}$$
$$\{\mathrm{Connect}(x, y_1), \mathrm{Connect}(y_1, y)\} \leadsto_{\kappa_2, [\mathrm{Node}_1/x, \mathrm{Node}_2/y_1, \mathrm{Node}_1/x_1, y/z_1]}$$
$$\{\mathrm{Connect}(\mathrm{Node}_2, y)\} \leadsto_{\kappa_3, [\mathrm{Node}_3/y, \mathrm{Node}_1/x, \mathrm{Node}_2/y_1, \mathrm{Node}_1/x_1, \mathrm{Node}_3/z_1]} \emptyset$$

This work was funded by EPSRC grant EP/K031864/1.

M. Falaschi (Ed.): LOPSTR 2015, LNCS 9527, pp. 91–106, 2015.
DOI: 10.1007/978-3-319-27436-2_6

The first reduction $\leadsto_{\kappa_1,[x/x_1,y/z_1]}$ unifies the query $\mathrm{Connect}(x,y)$ with the head of the rule κ_1 (which is $\mathrm{Connect}(x_1,z_1)$ after renaming) with the substitution $[x/x_1,y/z_1]$ (x_1 is replaced by x and z_1 is replaced by y). So the query is *resolved* with κ_1, producing the next queries: $\mathrm{Connect}(x,y_1)$, $\mathrm{Connect}(y_1,y)$. Note that the substitution in the subscript of $\leadsto$ is a state that will be updated alongside the derivation.

Viewing a program as a collection of Horn clauses, the above derivation first assumed that $\mathrm{Connect}(x,y)$ is false, and then deduced a contradiction (an empty goal) from the assumption. As every SLD-derivation is essentially a proof by contradiction, traditionally the exact content of such proofs plays little role in determining entailment. However, it is desirable to have methods which capture the proof-theoretic content of SLD-derivations. For example, one may wish to reason in a proof-relevant way, and compute not just $\Phi \vdash A$, but $\Phi \vdash p : A$, where p is the proof-witness for the query A and the program Φ. LP and its dialects are used as part of type inference engines underlying functional [6,11] and dependently typed [4] languages. These applications require proof-relevant automated reasoning.

In type class inference (e.g. Haskell), a type class can be seen as an atomic formula and an instance declaration – as a Horn formula. The instance resolution process in type class inference can then be seen as an SLD-derivation, with one additional requirement: this SLD-derivation must compute the evidence for the type class (or construct a dictionary). For example, the following declaration specifies a way to construct equality class instances for datatypes List and Char:

$$\kappa_1 : \forall x.\mathrm{Eq}(x) \Rightarrow \mathrm{Eq}(\mathrm{List}(x))$$

$$\kappa_2 : \Rightarrow \mathrm{Eq}(\mathrm{Char})$$

Here List is a function symbol, Char is a constant and x is a variable; κ_1,κ_2 will be used as primitives for the evidence construction. When we make a comparison of two lists of characters, such as (eq $['a']$ $['b']$), the compiler will insert the evidence d of the type $\mathrm{Eq}(\mathrm{List}(\mathrm{Char}))$ in (eq d $['a']$ $['b']$). The construction of this evidence can be viewed as resolving the query $\mathrm{Eq}(\mathrm{List}(\mathrm{Char}))$, which is witnessed by applying Horn formulas κ_1 and κ_2. Thus, $(\kappa_1\ \kappa_2)$ is the evidence we want for d.

In order to specify the proof-theoretic meaning of derivations, we introduce a type-theoretic approach to recover the notion of proof in LP. It has been noticed by Girard [3], that the resolution rule $\frac{A\vee B \quad \neg B\vee D}{A\vee D}$ can be expressed by means of the cut rule in intuitionistic sequent calculus: $\frac{A\Rightarrow B \quad B\Rightarrow D}{A\Rightarrow D}$. Although the resolution rule is classically equivalent to the cut rule, the cut rule is better suited for performing computation while preserving constructive content. In Sect. 2 we present a type system reflecting this intuition: if p_1 is the proof of $A \Rightarrow B$ and p_2 is the proof of $B \Rightarrow D$, then $\lambda x.p_2\ (p_1\ x)$ is the proof of $A \Rightarrow D$. Thus, proof can be recorded alongside with each cut rule.

We prove that SLD-resolution is sound with respect to the type system (Sect. 2). We give a formulation of SLD-resolution in the form of a reduction rule, called *LP-Unif*. The soundness result shows that, given a logic program Φ and a query A, if A can be LP-Unif reduced to the empty goal with a substitution γ as an answer, then a proof term can be constructed for γA.

In Sect. 3, we introduce a technique called *realizability transformation*, that, given a program Φ, produces a program $F(\Phi)$ in which one extra argument is added to every predicate, in order to record the proof-evidence in derivations. The proof evidence is computed by applying substitution to variables held by this additional argument in the course of running SLD-resolution. Let us revisit the List example. Its transformed version will look as follows:

$$\kappa_1 : \forall x. \forall u_1. \mathrm{Eq}(x, u_1) \Rightarrow \mathrm{Eq}(\mathrm{List}(x), f_{\kappa_1}(u_1))$$
$$\kappa_2 : \Rightarrow \mathrm{Eq}(\mathrm{Char}, c_{\kappa_2})$$

The query $\mathrm{Eq}(\mathrm{List}(\mathrm{Char}))$ of the original program becomes $\mathrm{Eq}(\mathrm{List}(\mathrm{Char}), u)$ after the transformation, where u is a variable. The derivation reaches the empty goal and outputs the substitution $[f_{\kappa_1}(c_{\kappa_2})/u]$, which corresponds to the proof term $(\kappa_1 \ \kappa_2)$ for the query $\mathrm{Eq}(\mathrm{List}(\mathrm{Char}))$.

Realizability transformation bears resemblance to Kleene's [7] method under the same name. We show that realizability transformation preserves the proof-theoretic meaning of the original program and the computational behaviour of LP-Unif reductions. With the help of the transformation, we prove completeness of LP-Unif with repect to the type system.

Together, Sects. 2 and 3 introduce a method of constructing proof evidence in the process of LP derivations. Recently, a variant of resolution for Horn Clauses, called *structural resolution (S-resolution)* has been introduced [5]. S-resolution represents derivations by SLD-resolution as a combination of derivations by term-matching and by substitution. We explain this idea in detail in Sect. 4. The main reason for separating out two components of SLD-resolution in such a way is to make use of structural properties of term-matching that have already been exploited in functional programming and term-rewriting. In particular, S-resolution allowed to define a theory of universal productivity for LP that resembles a similar theory in functional programming [2]: given a potentially infinite derivation by S-resolution, termination of term-matching derivations that comprise it determines *productivity* of the derivation (or in other words, observability of finite fragments of the infinite computation).

We conjecture that the combination of the two ideas – the theory of productivity introduced by S-resolution and the proof-witness construction introduced in this paper bear promise for future development of resolution-based methods. This is why, in Sect. 4 we give a full formal study of how these two methods can be formally combined. We show how S-resolution can be represented by means of *LP-Struct reductions*, combining term-matching reductions and unification. We extend the type-theoretic semantics to S-Resolution. We define conditions which guarantee equivalence of S-Resolution and SLD-resolution, one of which happens to be exactly the property of productivity. We use the realizability transformation as a method for guaranteeing productivity of programs.

Finally, in Sect. 5 we conclude and explain how the combination of S-Resolution and the type-theoretic approach of this paper could be used in non-terminating cases of type class inference. Detailed proofs for lemmas and theorems in this paper may be found in the extended version[1].

2   A Type System for LP: Horn-Formulas as Types

We first formulate a type system to model LP. We show that LP-Unif is sound with respect to the type system.

Definition 1. *Term* $t ::= x \mid f(t_1, ..., t_n)$
 Atomic Formula $A, B, C, D ::= P(t_1, ..., t_n)$
 (Horn) Formula $F ::= [\forall \underline{x}].A_1, ..., A_n \Rightarrow A$
 Proof Term $p, e ::= \kappa \mid a \mid \lambda a.e \mid e\, e'$
 Axioms/LP Programs $\Phi ::= \cdot \mid \kappa : F, \Phi$

Functions of arity zero are called *term constants*, $\mathrm{FV}(t)$ returns all free term variables of t. We use $\underline{A}$ to denote $A_1, ..., A_n$, when the number n is unimportant. If n is zero for $\underline{A} \Rightarrow B$, then we write $\Rightarrow B$. Note that B is an atomic formula, but $\Rightarrow B$ is a formula, we distinguish the notion of atomic formulas from (Horn) formulas. The formula $A_1, ..., A_n \Rightarrow B$ can be informally read as "the conjunction of A_i implies B". We write $\forall \underline{x}.F$ for quantifying over all the free term variables in F; $[\forall x]. F$ denotes F or $\forall x.F$. LP program $B \Leftarrow \underline{A}$ are represented as $\forall \underline{x}.\underline{A} \Rightarrow B$ and query is an atomic formula. Proof terms are lambda terms, where κ denotes a proof term constant and a denotes a proof term variable. We write $A \mapsto_\sigma A'$ (resp. $A \sim_\gamma A'$) to mean A is matchable (resp. unifiable) to A' with substitution σ (resp. γ), i.e. $\sigma A \equiv A'$ (resp. $\gamma A \equiv \gamma A'$).

The following is a new formulation of a type system intended to provide a type theoretic interpretation for LP.

Definition 2 (Horn-Formulas-as-Types System for LP).

$$\frac{e : F}{e : \forall \underline{x}.F}\ gen \qquad \frac{e_1 : \underline{A} \Rightarrow D \quad e_2 : \underline{B}, D \Rightarrow C}{\lambda \underline{a}.\lambda \underline{b}.(e_2\ \underline{b})\ (e_1\ \underline{a}) : \underline{A}, \underline{B} \Rightarrow C}\ cut$$

$$\frac{e : \forall \underline{x}.F}{e : [\underline{t}/\underline{x}]F}\ inst \qquad \frac{(\kappa : \forall \underline{x}.F) \in \Phi}{\kappa : \forall \underline{x}.F}\ axiom$$

Note that the notion of type is identified with Horn formulas (atomic intuitionistic sequent), not atomic formulas. The usual sequent turnstile $\vdash$ is internalized as intuitionistic implication $\Rightarrow$. The rule for first order quantification $\forall$ is placed *outside* of the sequent. The cut rule is the only rule that produces new proof terms. In the *cut* rule, $\lambda \underline{a}.t$ denotes $\lambda a_1....\lambda a_n.t$ and $t\ \underline{b}$ denotes $(...(t\ b_1)...b_n)$. The size of $\underline{a}$ is the same as $\underline{A}$ and the size of $\underline{b}$ is the same as $\underline{B}$, and $\underline{a}, \underline{b}$ are not free in e_1, e_2.

---

[1] Extended version is available at both authors' homepages.

Our formulation is given in the style of typed lambda calculus and sequent calculus, the intention for this formulation is to model LP type-theoretically. It has been observed that the cut rule and proper axioms in intuitionistic sequent calculus can emulate LP [3](§13.4). Here we add a proof term annotation and make use of explicit quantifiers. Our formulation uses Curry-style in the sense that for the *gen* and *inst* rule, we do not modify the structure of the proof terms. Curry-style formulation allows us to focus on the proof terms generated by applying the *cut* rule.

Below is a formulation of SLD-derivation as a reduction system [9].

Definition 3 (LP-Unif Reduction). *Given axioms Φ. We define a reduction relation on the multiset of atomic formulas:*
$\Phi \vdash \{A_1, ..., A_i, ..., A_n\} \leadsto_{\kappa, \gamma \cdot \gamma'} \{\gamma A_1, ..., \gamma B_1, ..., \gamma B_m, ..., \gamma A_n\}$ *for any substitution γ', if there exists $\kappa : \forall \underline{x}.B_1, ..., B_n \Rightarrow C \in \Phi$ such that $C \sim_\gamma A_i$.*

The second subscript in the reduction is intended as a state, it will be updated along with reductions. We assume implicit renaming of all quantified variables each time the above rule is applied. We write $\leadsto$ when we leave the underlining state implicit. We use $\leadsto^*$ to denote the reflexive and transitive closure of $\leadsto$. Notation $\leadsto^*_\gamma$ is used when the final state along the reduction path is γ.

Given a program Φ and a set of queries $\{B_1, ..., B_n\}$, LP-Unif uses only unification reduction to reduce $\{B_1, ..., B_n\}$:

Definition 4 (LP-Unif). *Given a logic program Φ, LP-Unif is given by an abstract reduction system $(\Phi, \leadsto)$.*

Lemma 1. *If $\Phi \vdash \{A_1, ..., A_n\} \leadsto^*_\gamma \emptyset$, then there exist proofs $e_1 : \forall \underline{x}. \Rightarrow \gamma A_1, ..., e_n : \forall \underline{x}. \Rightarrow \gamma A_n$, given axioms Φ.*

Proof. By induction on the length of the reduction.
Base Case. Suppose the length is one, namely, $\Phi \vdash \{A\} \leadsto_{\kappa, \gamma} \emptyset$. It implies that there exists $(\kappa : \forall \underline{x}. \Rightarrow C) \in \Phi$, such that $C \sim_\gamma A$. So we have $\kappa : \Rightarrow \gamma C$ by the *inst* rule. Thus $\kappa : \Rightarrow \gamma A$ by $\gamma C \equiv \gamma A$. Hence $\kappa : \forall \underline{x}. \Rightarrow \gamma A$ by the *gen* rule.

Step Case. Suppose $\Phi \vdash \{A_1, ..., A_i, ..., A_n\} \leadsto_{\kappa, \gamma} \{\gamma A_1, ..., \gamma B_1, ..., \gamma B_m, ..., \gamma A_n\} \leadsto^*_{\gamma'} \emptyset$, where $\kappa : \forall \underline{x}.B_1, ..., B_m \Rightarrow C$ and $C \sim_\gamma A_i$. By inductive hypothesis(IH), we know that there exist proofs $e_1 : \forall \underline{x}. \Rightarrow \gamma' \gamma A_1, ..., p_1 : \forall \underline{x}. \Rightarrow \gamma' \gamma B_1, ..., p_m : \forall \underline{x}. \Rightarrow \gamma' \gamma B_m, ..., e_n : \forall \underline{x}. \Rightarrow \gamma' \gamma A_n$. We can use *inst* rule to instantiate the quantifiers of κ using $\gamma' \cdot \gamma$, so we have $\kappa : \gamma' \gamma B_1, ..., \gamma' \gamma B_m \Rightarrow \gamma' \gamma C$. Since $\gamma' \gamma A_i \equiv \gamma' \gamma C$, we can construct a proof $e_i = \kappa \ p_1 \ ... \ p_m$ with $e_i : \Rightarrow \gamma' \gamma A_i$, by applying the cut rule m times. By *gen*, we have $e_i : \forall \underline{x}. \Rightarrow \gamma' \gamma A_i$. The substitution generated by the unification is idempotent, and γ' is accumulated from γ, i.e. $\gamma' = \gamma'' \cdot \gamma$ for some γ'', so $\gamma' \gamma A_j \equiv \gamma'' \gamma \gamma A_j \equiv \gamma'' \gamma A_j \equiv \gamma' A_j$ for any j. Thus we have $e_j : \forall \underline{x}. \Rightarrow \gamma' A_j$ for any j.

Theorem 1 (Soundness of LP-Unif). *If $\Phi \vdash \{A\} \leadsto^*_\gamma \emptyset$, then there exists a proof $e : \forall \underline{x}. \Rightarrow \gamma A$ given axioms Φ.*

For example, by the soundness theorem above, the derivation in Example 1 yields the proof $(\lambda b.(\kappa_1\ b)\ \kappa_3)\ \kappa_2$ for the formula $\Rightarrow \mathrm{Connect}(\mathrm{node}_1, \mathrm{node}_3)$.

Naturally, we would want to prove the following completeness theorem: If $e : \forall \underline{x}. \Rightarrow A$, then $\Phi \vdash \{A\} \leadsto^*_\gamma \emptyset$ for some γ. It is tempting to prove this theorem by induction on the derivation of $e : \forall \underline{x}. \Rightarrow A$. However, it becomes quite involved. We will discuss a simpler way to prove this theorem at the end of the next section, where we take advantage of the realizability transformation.

3   Realizability Transformation

We define *realizability transformation* in this section. Realizability [7](Sect. 82) is a technique that uses a number representing the proof of a number-theoretic formula. The transformation described here is similar in the sense that we use a first order term to represent the proof of a formula. More specifically, we use a first order term as an extra argument for a formula to represent a proof of that formula. Before we define the transformation, we first state several basic results about the type system in Definition 2.

Theorem 2 (Strong Normalization). *Let beta-reduction on proof terms be the congruence closure of the following relation: $(\lambda a.p)p' \rightarrow_\beta [p'/a]p$. If $e : F$, then e is strongly normalizable with respect to beta-reduction on proof terms.*

The proof of strong normalization (SN) is an adaptation of Tait-Girard's reducibility proof. Since the first order quantification does not impact the proof term, the proof is very similar to the SN proof of simply typed lambda calculus.

Lemma 2. *If $e : [\forall \underline{x}.]\underline{A} \Rightarrow B$ given axioms Φ, then either e is a proof term constant or it is normalizable to the form $\lambda \underline{a}.n$, where n is first order normal proof term.*

Theorem 3. *If $e : [\forall \underline{x}.] \Rightarrow B$, then e is normalizable to a first order proof term.*

Lemma 2 and Theorem 3 show that we can use first order terms to represent normalized proof terms; and thus pave the way to realizability transformation.

Definition 5 (Representing First Order Proof Terms). *Let ϕ be a mapping from proof term variables to first order terms. We define a representation function $[\![\cdot]\!]_\phi$ from first order normal proof terms to first order terms.*
- *$[\![a]\!]_\phi = \phi(a)$.*
- *$[\![\kappa\ p_1...p_n]\!]_\phi = f_\kappa([\![p_1]\!]_\phi, ..., [\![p_n]\!]_\phi)$, where f_κ is a function symbol.*

Definition 6. *Let $A \equiv P(t_1, ..., t_n)$ be an atomic formula, we write $A[t']$, where $(\bigcup_i \mathrm{FV}(t_i)) \cap \mathrm{FV}(t') = \emptyset$, to abbreviate a new atomic formula $P(t_1, ..., t_n, t')$.*

Definition 7 (Realizability Transformation). *We define a transformation F on a formula and its normalized proof term:*

$$- F(\kappa \;\; : \;\; \forall \underline{x}.A_1, ..., A_m \;\; \Rightarrow \;\; B) \;\; = \;\; \kappa \;\; : \;\; \forall \underline{x}.\forall \underline{y}.A_1[y_1], ..., A_m[y_m] \;\; \Rightarrow$$
$$B[f_\kappa(y_1, ..., y_m)], \text{ where } y_1, ..., y_m \text{ are all fresh and distinct.}$$

$$- F(\lambda \underline{a}.n : [\forall \underline{x}].A_1, ..., A_m \Rightarrow B) = \lambda \underline{a}.n : [\forall \underline{x}.\forall \underline{y}].A_1[y_1], ..., A_m[y_m] \Rightarrow$$
$$B[[n]]_{[\underline{y}/\underline{a}]}], \text{ where } y_1, ..., y_m \text{ are all fresh and distinct.}$$

The realizability transformation systematically associates a proof to each predicate, so that the proof can be recorded alongside with reductions.

Example 2. The following logic program is the result of applying realizability transformation on the program in Example 1.

$$\kappa_1 : \forall x.\forall y.\forall u_1.\forall u_2.\text{Connect}(x, y, u_1), \text{Connect}(y, z, u_2) \Rightarrow \text{Connect}(x, z, f_{\kappa_1}(u_1, u_2))$$

$$\kappa_2 : \Rightarrow \text{Connect}(\text{node}_1, \text{node}_2, c_{\kappa_2})$$

$$\kappa_3 : \Rightarrow \text{Connect}(\text{node}_2, \text{node}_3, c_{\kappa_3})$$

Before the realizability transformation, we have the following judgement:

$$\lambda b.(\kappa_1 \; b) \; \kappa_2 : \text{Connect}(\text{node}_2, z) \Rightarrow \text{Connect}(\text{node}_1, z)$$

We can apply the transformation, we get:

$$\lambda b.(\kappa_1 \; b) \; \kappa_2 : \text{Connect}(\text{node}_2, z, u_1) \Rightarrow \text{Connect}(\text{node}_1, z, [[(\kappa_1 \; b) \; \kappa_2]]_{[u_1/b]})$$

which is the same as

$$\lambda b.(\kappa_1 \; b) \; \kappa_2 : \text{Connect}(\text{node}_2, z, u_1) \Rightarrow \text{Connect}(\text{node}_1, z, f_{\kappa_1}(u_1, c_{\kappa_2}))$$

Observe that the transformed formula:
$\text{Connect}(\text{node}_2, z, u_1) \Rightarrow \text{Connect}(\text{node}_1, z, f_{\kappa_1}(u_1, c_{\kappa_2}))$ is provable by $\lambda b.(\kappa_1 \; b) \; \kappa_2$ using the transformed program.

Let $F(\Phi)$ mean applying the realizability transformation to every axiom in Φ. We write $(F(\Phi), \leadsto)$, to mean given axioms $F(\Phi)$, use LP-Unif to reduce a given query. Note that for query A in $(\Phi, \leadsto)$, it becomes query $A[t]$ for some t such that $\text{FV}(A) \cap \text{FV}(t) = \emptyset$ in $(F(\Phi), \leadsto)$.

The following theorem shows that realizability transformation does not change the proof-theoretic meaning of a program. This is important because it means we can apply different resolution strategies to resolve the query on the transformed program without worrying about the change of meaning. Later we will see that the behavior of LP-Struct is different for the original program and the transformed program.

Theorem 4. *Given axioms Φ, if $e : [\forall \underline{x}].\underline{A} \Rightarrow B$ holds with e in normal form, then $F(e : [\forall \underline{x}].\underline{A} \Rightarrow B)$ holds for axioms $F(\Phi)$.*

The other direction for the theorem above is not true if we ignore the transformation F, namely, if $e : \forall \underline{x}. \Rightarrow A[t]$ for axioms Φ, it may not be the case that $e : \forall \underline{x}. \Rightarrow A$, since the axioms Φ may not be set up in a way such that t is a representation of proof e. The following theorem shows that the extra argument is used to record the term representation of the corresponding proof.

Theorem 5. *Suppose* $F(\Phi) \vdash \{A[y]\} \leadsto_\gamma^* \emptyset$. *We have* $p : \forall \underline{x}. \Rightarrow \gamma A[\gamma y]$ *for* $F(\Phi)$, *where* p *is in normal form and* $[\![p]\!]_\emptyset = \gamma y$.

Now we are able to show that realizability transformation will not change the unification reduction behaviour.

Lemma 3. $\Phi \vdash \{A_1, ..., A_n\} \leadsto^* \emptyset$ *iff* $F(\Phi) \vdash \{A_1[y_1], ..., A_n[y_n]\} \leadsto^* \emptyset$.

Proof. For each direction, by induction on the length of the reduction. Each proof will be similar to the proof of Lemma 1.

Theorem 6. $\Phi \vdash \{A\} \leadsto^* \emptyset$ *iff* $F(\Phi) \vdash \{A[y]\} \leadsto^* \emptyset$.

Example 3. Consider the logic program after realizability transformation in Example 2. Realizability transformation does not change the behaviour of LP-Unif, we still have the following successful unification reduction path for query $\text{Connect}(x, y, u)$:

$$F(\Phi) \vdash \{\text{Connect}(x, y, u)\} \leadsto_{\kappa_1, [x/x_1, y/z_1, f_{\kappa_1}(u_3, u_4)/u]}$$
$$\{\text{Connect}(x, y_1, u_3), \text{Connect}(y_1, y, u_4)\}$$

$$\leadsto_{\kappa_2, [c_{\kappa_2}/u_3, \text{node}_1/x, \text{node}_2/y_1, \text{node}_1/x_1, b/z_1, f_{\kappa_1}(c_{\kappa_2}, u_4)/u]}$$
$$\{\text{Connect}(\text{node}_2, y, u_4)\}$$

$$\leadsto_{\kappa_3, [c_{\kappa_3}/u_4, c_{\kappa_2}/u_3, \text{node}_3/y, \text{node}_1/x, \text{node}_2/y_1, \text{node}_1/x_1, \text{node}_3/z_1, f_{\kappa_1}(c_{\kappa_2}, c_{\kappa_3})/u]} \emptyset$$

Now let us come back to the completeness theorem. The following lemma shows that completeness result holds for the transformed program.

Lemma 4. *For* $F(\Phi)$, *if* $n : \Rightarrow A[[\![n]\!]_\emptyset]$ *where* n *is in normal form, then* $F(\Phi) \vdash \{A[[\![n]\!]_\emptyset]\} \leadsto^* \emptyset$.

Proof. By induction on the structure of n.

- **Base Case:** $n = \kappa$. In this case, $[\![n]\!]_\emptyset = f_\kappa$, $\kappa : \forall \underline{x}. \Rightarrow A'[f_\kappa] \in F(\Phi)$ and $\gamma(A'[f_\kappa]) \equiv A[f_\kappa]$ for some substitution γ. Thus $A'[f_\kappa] \sim_\gamma A[f_\kappa]$, which implies $F(\Phi) \vdash \{A[f_\kappa]\} \leadsto_{\kappa, \gamma} \emptyset$.
- **Step Case:** $n = \kappa \, n_1 \, n_2 \, ... \, n_m$. In this case, $[\![n]\!]_\emptyset = f_\kappa([\![n_1]\!]_\emptyset, ..., [\![n_m]\!]_\emptyset)$, $\kappa : \forall \underline{xy}. \; C_1[y_1], ..., C_m[y_m] \Rightarrow B[f_\kappa(y_1, ..., y_m)] \in F(\Phi)$. To obtain $n : \Rightarrow A[[\![n]\!]_\emptyset]$, we have to use $\kappa : \forall \underline{x}. \; C_1[[\![n_1]\!]_\emptyset], ..., C_m[[\![n_m]\!]_\emptyset] \Rightarrow B[f_\kappa([\![n_1]\!]_\emptyset, ..., [\![n_m]\!]_\emptyset)]$ with $\gamma(B[f_\kappa([\![n_1]\!]_\emptyset, ..., [\![n_m]\!]_\emptyset)]) \equiv A[[\![n]\!]_\emptyset]$. By the inst rule, we have $\kappa : \gamma C_1[[\![n_1]\!]_\emptyset], ..., \gamma C_m[[\![n_m]\!]_\emptyset] \Rightarrow \gamma B[f_\kappa([\![n_1]\!]_\emptyset, ..., [\![n_m]\!]_\emptyset)]$. Furthermore, it has to be the case that $n_1 : \Rightarrow \gamma C_1[[\![n_1]\!]_\emptyset], ..., n_m : \Rightarrow \gamma C_m[[\![n_m]\!]_\emptyset]$. Thus we have $F(\Phi) \vdash \{A[[\![n]\!]_\emptyset]\} \leadsto_{\kappa, \gamma} \{\gamma C_1[[\![n_1]\!]_\emptyset], ..., \gamma C_m[[\![n_m]\!]_\emptyset]\}$. By IH, we have $F(\Phi) \vdash \{\gamma C_1[[\![n_1]\!]_\emptyset]\} \leadsto_{\gamma_1}^* \emptyset$. So $F(\Phi) \vdash \{A[[\![n]\!]_\emptyset]\} \leadsto_{\kappa, \gamma} \cdot \leadsto_{\gamma_1}^* \{\gamma_1 \gamma C_2[[\![n_2]\!]_\emptyset], ..., \gamma_1 \gamma C_m[[\![n_m]\!]_\emptyset]\}$. Again, we have $n_2 : \Rightarrow \gamma_1 \gamma C_2[[\![n_2]\!]_\emptyset], ..., n_m : \Rightarrow \gamma_1 \gamma C_m[[\![n_m]\!]_\emptyset]$. By applying IH repeatedly, we obtain $F(\Phi) \vdash \{A[[\![n]\!]_\emptyset]\} \leadsto^* \emptyset$.

Lemma 5. *For* $F(\Phi)$, *if* $F(\Phi) \vdash \{A_1[t_1], ..., A_n[t_n]\} \leadsto^* \emptyset$ *with* $\text{FV}(t_i) = \emptyset$ *for all* i, *then* $\Phi \vdash \{A_1, ..., A_n\} \leadsto^* \emptyset$.

Proof. By induction on the length of $\rightsquigarrow^*$.

- **Base Case:** $F(\Phi) \vdash \{A[f_\kappa]\} \rightsquigarrow^* \emptyset$. We have $\kappa : \Rightarrow A' \in \Phi$ such that $A' \sim_\gamma A$. Thus $\Phi \vdash \{A\} \rightsquigarrow_\kappa \emptyset$.
- **Step Case:** $F(\Phi) \vdash \{A_1[t_1], ..., A_i[t_i], ..., A_n[t_n]\} \rightsquigarrow_{\kappa,\gamma}$
  $\{\gamma A_1[t_1], ..., \gamma B_1[t'_1], ..., \gamma B_l[t'_l], ..., \gamma A_n[t_n]\} \rightsquigarrow^* \emptyset$ with $t_i \equiv f_\kappa(t'_1, ..., t'_l)$ and $\kappa : B_1, ..., B_l \Rightarrow C \in \Phi$ where $C \sim_\gamma A_i$. So by IH, we have $\Phi \vdash \{A_1, ..., A_n\} \rightsquigarrow_{\kappa,\gamma} \{\gamma A_1, ..., \gamma B_1, ..., \gamma B_l, ..., \gamma A_n\} \rightsquigarrow^* \emptyset$.

Now we are ready to prove the completeness result.

Theorem 7 (Completeness). *If* $n : [\forall \underline{x}]. \Rightarrow A$, *where* n *is in normal form, then* $\Phi \vdash \{A\} \rightsquigarrow_\gamma^* \emptyset$.

Proof. By Theorem 4, we have $n : [\forall \underline{x}]. \Rightarrow A[[n]_\emptyset]$ holds in $F(\Phi)$. By Lemma 4, we have $F(\Phi) \vdash \{A[[n]_\emptyset]\} \rightsquigarrow^* \emptyset$. By Lemma 5, we have $\Phi \vdash \{A\} \rightsquigarrow_\gamma^* \emptyset$.

The completeness result relies on realizability transformation to record the proof steps for a query, so the LP-Unif reduction can just follow the proof steps to reduce the query to the empty set. Together with Theorem 1, this proof system gives new semantics for derivations in LP.

4   Structual Resolution

S-resolution [5] is a newly proposed alternative to SLD-resolution that allows a systematic separation of derivations into term-matching and unification steps. A logic program is called *productive* if the term-matching reduction is terminating for any query. For productive programs with coinductive meaning, finite term-rewriting reductions can be seen as measures of observation in an infinite derivation. The ability to handle corecursion in a productive way is an attractive computational feature of S-resolution.

Example 4. The following program defines the predicate Stream:

$$\kappa_1 : \forall x. \forall y. \text{Stream}(y) \Rightarrow \text{Stream}(\text{Cons}(x, y))$$

It will result in infinite LP-Unif reduction:

$$\Phi \vdash \{\text{Stream}(\text{Cons}(x, y))\} \rightsquigarrow_{\kappa_1, [x/x_1, y/y_1]} \{\text{Stream}(y)\} \rightsquigarrow_{\kappa_1, [\text{Cons}(x_2, y_2)/y]}$$
$$\{\text{Stream}(y_2)\} \rightsquigarrow_{\kappa_1, [\text{Cons}(x_3, y_3)/y_2]} \cdots$$

But it will yield finite term-matching reduction since $\text{Stream}(y)$ can not be matched by the head of κ_1 ($\text{Stream}(\text{Cons}(x, y))$):

$$\Phi \vdash \{\text{Stream}(\text{Cons}(x, y))\} \rightarrow_{\kappa_1} \{\text{Stream}(y)\} \nrightarrow$$

In general, term-matching reductions are not complete relative to LP-Unif reductions, but we can combine them with substitutional steps to complete derivations. This is exactly the idea behind S-resolution.

Example 5. The following program defines bits and lists of bits:

$$\kappa_1 : \Rightarrow \text{Bit}(0)$$
$$\kappa_2 : \Rightarrow \text{Bit}(1)$$
$$\kappa_3 : \Rightarrow \text{BList}(\text{Nil})$$
$$\kappa_4 : \forall x.\forall y.\text{BList}(y), \text{Bit}(x) \Rightarrow \text{BList}(\text{Cons}(x,y))$$

LP-Unif would give a complete reduction:

$$\Phi \vdash \{\text{BList}(\text{Cons}(x,y))\} \rightsquigarrow_{\kappa_4,[x/x_1,y/y_1]} \{\text{Bit}(x), \text{BList}(y)\} \rightsquigarrow_{\kappa_1,[0/x,0/x_1,y/y_1]}$$
$$\{\text{BList}(y)\} \rightsquigarrow_{\kappa_3,[\text{Nil}/y,0/x,0/x_1,\text{Nil}/y_1]} \emptyset$$

But term-matching reduction will not be able to compute an answer in this case.

$$\Phi \vdash \{\text{BList}(\text{Cons}(x,y))\} \rightarrow_{\kappa_4} \{\text{Bit}(x), \text{BList}(y)\} \not\rightarrow$$

This is why, S-resolution combines term-matching reductions with additional substitutional steps, in order to compute the same answer:

$$\Phi \vdash \{\text{BList}(\text{Cons}(x,y))\} \rightarrow_{\kappa_4} \{\text{Bit}(x), \text{BList}(y)\} \hookrightarrow_{\kappa_1,[0/x]} \{\text{Bit}(0), \text{BList}(y)\} \rightarrow_{\kappa_1,[0/x]}$$
$$\{\text{BList}(y)\} \hookrightarrow_{\kappa_3,[0/x,\text{Nil}/y]} \{\text{BList}(\text{Nil})\} \rightarrow_{\kappa_3,[0/x,\text{Nil}/y]} \emptyset$$

Completing derivation for Stream in the same way will result in an infinite derivation, in which every term-matching reduction is finite.

In this section, we embed S-resolution into the type theoretic framework we have developed in the previous sections. We first define S-derivations in terms of LP-Struct reductions, in the uniform style with LP-Unif reductions, thereby also defining LP-TM reductions, which resemble reductions in term-rewriting systems [10]. We then prove that LP-Unif and LP-Struct are operationally equivalent subject to two conditions: productivity and non-overlapping. Finally, we show how realizability transformation can be used to guarantee productivity of logic programs in the setting of S-resolution.

4.1   S-Resolution in the Type-Theoretic Setting

Definition 8.   – *Term-matching(LP-TM) reduction:*
$\Phi \vdash \{A_1, ..., A_i, ..., A_n\} \rightarrow_{\kappa,\gamma'} \{A_1, ..., \sigma B_1, ..., \sigma B_m, ..., A_n\}$ *for any substitution* γ', *if there exists* $\kappa : \forall \underline{x}.B_1, ..., B_n \Rightarrow C \in \Phi$ *such that* $C \mapsto_\sigma A_i$.
– *Substitutional reduction:*
$\Phi \vdash \{A_1, ..., A_i, ..., A_n\} \hookrightarrow_{\kappa,\gamma\cdot\gamma'} \{\gamma A_1, ..., \gamma A_i, ..., \gamma A_n\}$ *for any substitution* γ', *if there exists* $\kappa : \forall \underline{x}.B_1, ..., B_n \Rightarrow C \in \Phi$ *such that* $C \sim_\gamma A_i$.

The second subscript of term-matching reduction is used to store the substitutions obtained by unification, it is only used when we combine term-matching reductions with substitutional reductions. The second subscript in the substitutional reduction is intended as a state, it will be updated along with reductions.

Given a program Φ and a set of queries $\{B_1, ..., B_n\}$, LP-TM uses only term-matching reduction to reduce $\{B_1, ..., B_n\}$:

Definition 9 (LP-TM). *Given a logic program Φ, LP-TM is given by an abstract reduction system $(\Phi, \rightarrow)$.*

LP-TM is also sound w.r.t. the type system of Definition 2, which implies that we can obtain a proof for each successful query.

Theorem 8 (Soundness of LP-TM). *If $\Phi \vdash \{A\} \rightarrow^* \emptyset$, then there exists a proof $e : \forall \underline{x}. \Rightarrow A$ given axioms Φ.*

Comparing Theorems 1 and 8, we see that for LP-TM, there is no need to accumulate substitutions, and the resulting formula is proven as stated. This difference is due to the use of term-matching instead of unification for the reduction. The following example shows that the LP-TM is incomplete with respect to the type system.

Example 6. Consider the following program Φ.

$$\kappa_1 : \Rightarrow Q(\mathrm{C})$$
$$\kappa_2 : \forall x. \forall y. Q(x) \Rightarrow P(y)$$

For query $P(\mathrm{C})$, we have $\Phi \vdash \{P(\mathrm{C})\} \rightarrow_{\kappa_2} \{Q(x)\} \not\rightarrow$. However, there exist a proof $(\kappa_2\ \kappa_1) : \Rightarrow P(\mathrm{C})$, by instantiating x, y to C in κ_2.

We use $\rightarrow^\mu$ to denote a reduction path to a $\rightarrow$-normal form. If the $\rightarrow$-normal form does not exist, then $\rightarrow^\mu$ denotes an infinite reduction path. We write $\hookrightarrow^1$ to denote at most one step of $\hookrightarrow$.

We can now formally define S-Resolution within our formal framework. Given a program Φ and a set of queries $\{B_1, \ldots, B_n\}$, LP-Struct first uses term-matching reduction to reduce $\{B_1, \ldots, B_n\}$ to a normal form, then performs one step substitutional reduction, and then repeats this process.

Definition 10 (Structural Resolution (LP-Struct)). *Given a logic program Φ, LP-Struct is given by an abstract reduction system $(\Phi, \rightarrow^\mu \cdot \hookrightarrow^1)$.*

If a finite term-matching reduction path does not exist, then $\rightarrow^\mu \cdot \hookrightarrow^1$ denotes an infinite path. When we write $\Phi \vdash \{\underline{A}\}(\rightarrow^\mu \cdot \hookrightarrow^1)^*\{\underline{C}\}$, it means a nontrivial finite path will be of the shape $\Phi \vdash \{\underline{A}\} \rightarrow^\mu \cdot \hookrightarrow \cdot \ldots \cdot \rightarrow^\mu \cdot \hookrightarrow \cdot \rightarrow^\mu \{\underline{C}\}$.

Now let us see the execution trace of Stream using LP-Struct:

$$\Phi \vdash \{\mathrm{Stream}(\mathrm{Cons}(x, y))\} \rightarrow_{\kappa_1} \{\mathrm{Stream}(y)\} \hookrightarrow_{\kappa_1, [\mathrm{Cons}(x_2, y_2)/y]}$$
$$\{\mathrm{Stream}(\mathrm{Cons}(x_2, y_2))\} \rightarrow_{\kappa_1, [\mathrm{Cons}(x_2, y_2)/y]}$$
$$\{\mathrm{Stream}(y_2)\} \hookrightarrow_{\kappa_1, [\mathrm{Cons}(x_3, y_3)/y_2, \mathrm{Cons}(x_2, \mathrm{Cons}(x_3, y_3))/y]}$$
$$\{\mathrm{Stream}(\mathrm{Cons}(x_3, y_3))\} \rightarrow_{\kappa_1, [\mathrm{Cons}(x_3, y_3)/y_2, \mathrm{Cons}(x_2, \mathrm{Cons}(x_3, y_3))/y]} \{\mathrm{Stream}(y_3)\} \ldots$$

Note that the overall reduction is infinite, but each LP-TM reduction is finite.

4.2   LP-Struct and LP-Unif

The next question one may ask is how LP-Struct compares to LP-Unif. They are not equivalent. Consider the program and the finite LP-Unif derivation of Example 1. LP-Unif has a finite successful derivation for the query $Connect(x, y)$, but we have the following non-terminating reduction by LP-Struct:

$$\Phi \vdash \{Connect(x, y)\} \rightarrow_{\kappa_1} \{Connect(x, y_1), Connect(y_1, y)\}$$
$$\rightarrow_{\kappa_1} \{Connect(x, y_2), Connect(y_2, y_1), Connect(y_1, y)\} \rightarrow_{\kappa_1} \cdots$$

The diverging behavior above is due to the divergence of LP-TM reduction. Therefore, the program of Example 1 is not productive in the sense of [5,8].

Definition 11 (Productivity). *We say a program Φ is productive iff every $\rightarrow$-reduction is finite.*

Perhaps LP-Unif and LP-Struct are operationally equivalent for all productive programs? The following example shows this is not the case.

Example 7.

$$\kappa_1 : \Rightarrow P(C)$$
$$\kappa_2 : \forall x.Q(x) \Rightarrow P(x)$$

Here C is a constant. The program is $\rightarrow$-terminating. However, for query $P(x)$, we have $\Phi \vdash \{P(x)\} \leadsto_{\kappa_1, [C/x]} \emptyset$ with LP-Unif, but $\Phi \vdash \{P(x)\} \rightarrow_{\kappa_2} \{Q(x)\} \not\rightarrow$ for LP-Struct.

Thus, productivity is insufficient for establishing the relation between LP-Struct and LP-Unif. In Example 7, the problem is caused by the overlapping heads $P(C)$ and $P(x)$. Motivated by the notion of non-overlapping rules in term rewriting systems ([1,10]), we introduce the following definition.

Definition 12 (Non-overlapping   Condition). *Axioms Φ are   non-overlapping if for any two formulas $\forall \underline{x}.\underline{B} \Rightarrow C, \forall \underline{x}.\underline{D} \Rightarrow E \in \Phi$, there are no substitution σ, δ such that $\sigma C \equiv \delta E$.*

Theorem 9. *Suppose Φ is non-overlapping. $\Phi \vdash \{A_1, ..., A_n\} \leadsto_\gamma^* \{C_1, ..., C_m\}$ with $\{C_1, ..., C_m\}$ in $\leadsto$-normal form iff $\Phi \vdash \{A_1, ..., A_n\}(\rightarrow^\mu \cdot \hookrightarrow^1)_\gamma^* \{C_1, ..., C_m\}$ with $\{C_1, ..., C_m\}$ in $\rightarrow^\mu \cdot \hookrightarrow^1$-normal form.*

The theorem above still requires the termination of the $\leadsto$ to establish equivalence LP-Unif and LP-Struct. We can weaken this requirement by only requiring termination of the $\rightarrow$-reduction, i.e. by requiring productivity.

Theorem 10 (Equivalence of LP-Struct and LP-Unif). *Suppose Φ is non-overlapping and productive.*

1. *If $\Phi \vdash \{A_1, ..., A_n\} \leadsto \{B_1, ..., B_m\}$, then $\Phi \vdash \{A_1, ..., A_n\}(\rightarrow^\mu \cdot \hookrightarrow^1)^* \{C_1, ..., C_l\}$ and $\Phi \vdash \{B_1, ..., B_m\} \rightarrow^* \{C_1, ..., C_l\}$.*

2. *If* $\Phi \vdash \{A_1, ..., A_n\}(\rightarrow^\mu \cdot \hookrightarrow^1)^*\{B_1, ..., B_m\}$, *then* $\Phi \vdash \{A_1, ..., A_n\} \leadsto^*$
   $\{B_1, ..., B_m\}$.

Note that the above theorem does not rely on termination of LP-Unif reductions and therefore establishes equivalence of LP-Unif and LP-Struct even for coinductive programs like Stream of Example 4, as long as they are productive and non-overlapping. This effect of productivity has not been described in previous work.

4.3   Realizability Transformation and LP-Struct

Even when programs are overlapping and unproductive (as e.g. the program of Example 1), we would still like to obtain a meaningful execution behaviour for LP-Struct, especially if LP-Unif allows successful derivations for the programs. Luckily, we already have a method to achieve that, it is the realizability transformation defined in Sect. 3:

Proposition 1. *For any program* Φ, $F(\Phi)$ *is productive and non-overlapping.*

Proof. First, we need to show $\rightarrow$-reduction is strongly normalizing in $(F(\Phi), \rightarrow)$. By Definition 7, we can establish a decreasing measurement(from right to left, using the strict subterm relation) for each rule in $F(\Phi)$, since the last argument in the head of each rule is strictly larger than the ones in the body. Then, non-overlapping property is due to the fact that all the heads of the rules in $F(\Phi)$ will be *guarded* by the unique function symbol in Definition 7.

Corollary 1. $F(\Phi) \vdash \{A_1, ..., A_n\}(\rightarrow^\mu \cdot \hookrightarrow^1)^*\{B_1, ..., B_m\}$ *iff* $F(\Phi) \vdash \{A_1, ..., A_n\} \leadsto^* \{B_1, ..., B_m\}$.

Proof. By Theorems 1 and 10.

Example 8. For the program in Example 2, the query $\text{Connect}(x, y, u)$ can be reduced by LP-Struct successfully:

$$F(\Phi) \vdash \{\text{Connect}(x, y, u)\} \hookrightarrow_{\kappa_1, [x/x_1, y/z_1, f_{\kappa_1}(u_3, u_4)/u]}$$
$$\{\text{Connect}(x, y, f_{\kappa_1}(u_3, u_4))\} \rightarrow_{\kappa_1} \{\text{Connect}(x, y_1, u_3), \text{Connect}(y_1, y, u_4)\}$$
$$\hookrightarrow_{\kappa_2, [c_{\kappa_2}/u_3, \text{node}_1/x, \text{node}_2/y_1, \text{node}_1/x_1, b/z_1, f_{\kappa_1}(c_{\kappa_2}, u_4)/u]}$$
$$\{\text{Connect}(\text{node}_1, \text{node}_2, c_{\kappa_2}), \text{Connect}(\text{node}_2, y, u_4)\} \rightarrow_{\kappa_2} \{\text{Connect}(\text{node}_2, y, u_4)\}$$
$$\hookrightarrow_{\kappa_3, [c_{\kappa_3}/u_4, c_{\kappa_2}/u_3, \text{node}_3/y, \text{node}_1/x, \text{node}_2/y_1, \text{node}_1/x_1, \text{node}_3/z_1, f_{\kappa_1}(c_{\kappa_2}, c_{\kappa_3})/u]}$$
$$\{\text{Connect}(\text{node}_2, \text{node}_3, c_{\kappa_3})\} \rightarrow_{\kappa_3} \emptyset$$

Note that the answer for u is $f_{\kappa_1}(c_{\kappa_2}, c_{\kappa_3})$, which is the first order term representation of the proof of $\Rightarrow \text{Connect}(\text{node}_1, \text{node}_3)$.

Realizability transformation uses the extra argument as decreasing measurement in the program to achieve termination of $\rightarrow$-reduction. At the same time this extra argument makes the program non-overlapping. Realizability transformation does not modify the proof-theoretic meaning and the execution behaviour of LP-Unif. The next example shows that not every transformation technique for obtaining structurally decreasing LP-TM reductions has such properties:

Example 9. Consider the following program:

$$\kappa_1 : \Rightarrow P(\text{Int})$$
$$\kappa_2 : \forall x.P(x), P(\text{List}(x)) \Rightarrow P(\text{List}(x))$$

It is a folklore method to add a structurally decreasing argument as a measurement to ensure finiteness of $\rightarrow^{\mu}$.

$$\kappa_1 : \Rightarrow P(\text{Int}, 0)$$
$$\kappa_2 : \forall x.\forall y.P(x, y), P(\text{List}(x), y) \Rightarrow P(\text{List}(x), s(y))$$

We denote the above program as Φ'. Indeed with the measurement we add, the term-matching reduction in Φ' will be finite. But the reduction for query $P(\text{List}(\text{Int}), z)$ using unification will fail:

$$\Phi' \vdash \{P(\text{List}(\text{Int}), z)\} \rightsquigarrow_{\kappa_2, [\text{Int}/x, s(y_1)/z]}$$
$$\{P(\text{Int}, y_1), P(\text{List}(\text{Int}), y_1)\} \rightsquigarrow_{\kappa_2, [0/y_1, \text{Int}/x, s(0)/z]} \{P(\text{List}(\text{Int}), 0)\} \nrightarrow$$

However, the query $P(\text{List}(\text{Int}))$ on the original program using unification reduction will diverge. Divergence and failure are operationally different. Thus adding arbitrary measurement may modify the execution behaviour of a program (and hence the meaning of the program). In contrast, by Theorems 4–6, realizability transformation does not modify the execution behaviour of unification reduction.

Example 10. Consider the following non-productive and non-overlapping program and its version after the realizability transformation:

$$Original\,program : \kappa : \forall x.P(x) \Rightarrow P(x)$$
$$After\,transformation : \kappa : \forall x.\forall u.P(x, u) \Rightarrow P(x, f_\kappa(u))$$

Both LP-Struct and LP-Unif will diverge for the queries $P(x), P(x, y)$ in both original and transformed versions. LP-Struct reduction diverges for different reasons in the two cases, one is due to divergence of $\rightarrow$-reduction:
$$\Phi \vdash \{P(x)\} \rightarrow \{P(x)\} \rightarrow \{P(x)\}...$$
The another is due to $\hookrightarrow$-reduction:
$$\Phi \vdash \{P(x, y)\} \hookrightarrow \{P(x, f_k(u))\} \rightarrow \{P(x, u)\} \hookrightarrow \{P(x, f_k(u'))\} \rightarrow \{P(x, u')\}...$$
Note that a single step of LP-Unif reduction for the original program corresponds to infinite steps of term-matching reduction in LP-Struct. For the transformed version, a single step of LP-Unif reduction corresponds to finite steps of LP-Struct reduction, which is exactly the correspondence we were looking for.

5   Conclusions and Future Work

We proposed a type system that gives a proof theoretic interpretation for LP: Horn formulas correspond to the notion of type, and a successful query yields a first order proof term. The type system also provided us with a precise tool to show that realizability transformation preserves both proof-theoretic meaning of the program and the operational behaviour of LP-Unif.

We formulated S-resolution as LP-Struct reduction, which can be seen as a reduction strategy that combines term-matching reduction with substitutional reduction. This formulation allowed us to study the operational relation between LP-Struct and LP-Unif. The operational equivalence of LP-Struct and LP-Unif is by no means obvious. Previous work ([5,8]) only gives soundness and completeness of LP-Struct with respect to the Herbrand models. We identified that productivity and non-overlapping are essential for showing their operational equivalence.

Realizability transformation proposed here ensures that the resulting programs are productive and non-overlapping. It preserves the proof-theoretic meaning of the program, in a formally defined sense of Theorems 4–6. It is general, applies to any logic program, and can be easily mechanised. Finally, it allows to automatically record the proof content in the course of reductions, as Theorem 5 establishes, which helps to prove completeness of LP-Unif (Theorem 7).

With the proof system for LP-reductions we proposed, we are planning to further investigate the interaction of LP-TM/Unif/Struct with typed functional languages. We expect to find a tight connection between our work and the type class inference, cf. [6,11].

In the context of type class inference [6,11], the infinite term-matching behaviour seems pervasive. The example below specifies a possible equality instance declaration for nested datatype such as
`data Bush a = Nil | Con a (Bush (Bush a))`:

$$\kappa_1 : \mathrm{Eq}(x), \mathrm{Eq}(\mathrm{Bush}(\mathrm{Bush}(x))) \Rightarrow \mathrm{Eq}(\mathrm{Bush}(x))$$
$$\kappa_2 : \Rightarrow \mathrm{Eq}(\mathrm{Char})$$

Here Bush is a function symbol, Char is a constant and x is variable. Consider the query $\mathrm{Eq}(\mathrm{Bush}(\mathrm{Char}))$, both LP-Unif and LP-Struct will generate an infinite reduction path by repeatedly applying κ_1. Using the realizability transformation, we can obtain a well-behaved (productive) program:

$$\kappa_1 : \mathrm{Eq}(x, y_1), \mathrm{Eq}(\mathrm{Bush}(\mathrm{Bush}(x)), y_2) \Rightarrow \mathrm{Eq}(\mathrm{Bush}(x), f_{\kappa_1}(y_1, y_2))$$
$$\kappa_2 :\Rightarrow \mathrm{Eq}(\mathrm{Char}, c_{\kappa_2})$$

The substitution for u in the query $\mathrm{Eq}(\mathrm{Bush}(\mathrm{Char}), u)$ will be an infinite term. But we need a finite representation for such infinite term to construct a dictionary. Such coinductive dictionary construction is the subject of our further investigations. We would also like to investigate generalizing the type-theoretic approach from Horn formulas to implicational intuitionistic formulas, the type system in this case will correspond to a version of simply type lambda calculus.

References

1. Baader, F., Nipkow, T.: Term Rewriting and All That. Cambridge University Press, New York (1998)
2. Bertot, Y., Komendantskaya, E.: Inductive and coinductive components of core-cursive functions in coq. Electron. Notes Theor. Comput. Sci. **203**(5), 25–47 (2008)
3. Girard, J.-Y., Taylor, P., Lafont, Y.: Proofs and Types. Cambridge University Press, New York (1989)
4. Gonthier, G., Ziliani, B., Nanevski, A., Dreyer, D.: How to make ad hoc proof automation less ad hoc. ACM SIGPLAN Not. **46**(9), 163–175 (2011)
5. Johann, P., Komendantskaya, E., Komendantskiy, V.: Structural resolution for logic programming. In: Technical Communications, ICLP (2015)
6. Jones, M.P.: Qualified Types: Theory and Practice. Cambridge University Press, Cambridge (2003)
7. Kleene, S.C.: Introduction to Metamathematics. North-Holland Publishing Company, New York (1952). Co-publisher: Wolters-Noordhoff; 8th revised ed.1980
8. Komendantskaya, E., Power, J., Schmidt, M.: Coalgebraic logic programming: from semantics to implementation. J. Log. Comput. (2014)
9. Nilsson, U., Małuszyński, J.: Logic, Programming and Prolog. Wiley, Chichester (1990)
10. Terese.: Term rewriting systems. In: Bezem, M., Willem Klop, J., deVrijer, R. (eds.) Cambridge Tracts in Theoretical Computer Science, vol. 55. pp. xxii + 884. Cambridge University Press (2003)
11. Wadler, P., Blott, S.: How to make ad-hoc polymorphism less ad hoc. In: Symposium on Principles of Programming Languages, pp. 60–76. ACM (1989)

A Typed Language for Events

Sandra Alves[1][(✉)], Sabine Broda[2], and Maribel Fernández[3]

[1] Faculty of Sciences, CRACS/INESCTEC, University of Porto, Porto, Portugal
`sandra@dcc.fc.up.pt`
[2] Faculty of Sciences, CMUP, University of Porto, Porto, Portugal
[3] Department of Informatics, King's College London, London, UK

Abstract. We define a general typed language to deal with the notion of event in the context of access control systems. We distinguish between *generic* events, which represent the kind of actions that can occur in a system, and *specific* events, which represent actual occurrences of those kinds of actions. A relation is given associating specific to generic events, as well as a method for obtaining intervals from a history of events. We describe applications in access control systems with obligations.

Keywords: Event · Event type · Access control · Obligation · Rewriting

1 Introduction

The notion of event, as a particular action or happening taking place in a system, is a pervasive notion in today's computing (and real life) systems. Events can take up many forms, from messages exchanged over a network, to actions performed by users of the system, to occurrences of physical phenomena such as a disk error or a fire alarm.

In the context of access control policies, there are many situations when granting or denying access to certain resources depends on the occurrence of particular events. For example, in a hospital environment, an access control policy may specify that any doctor in the ward should have access to a patient p's medical records, if patient p suffers a cardiac arrest. Several access control models have been designed to deal with policies defined in terms of events (see, for example, [3,8]). From the semantic point of view, the notions of action and event were extensively studied by Davidson [10]. Representation of events inspired by Davidson's work and adapted from Kowalski and Sergot's work on the event calculus [17] have been used in literature [1,5]. In particular, in [1], events were used to define an abstract metamodel for access control and obligations. Obligations differ from permissions in the sense that, although permissions can be issued

S. Alves—Partially funded by FCT, Portuguese Foundation for Science and Technology within project UID/EEA/50014/2013.
S. Broda—Partially supported by CMUP (UID/MAT/00144/2013), which is funded by FCT (Portugal) with national (MEC) and European structural funds through the programs FEDER, under the partnership agreement PT2020.

© Springer International Publishing Switzerland 2015
M. Falaschi (Ed.): LOPSTR 2015, LNCS 9527, pp. 107–123, 2015.
DOI: 10.1007/978-3-319-27436-2_7

but not used, an obligation usually is associated with some mandatory action, which must be performed at a time defined by some temporal constraints or by the occurrence of events. Therefore, effectively dealing with events is a key issue when reasoning about systems with obligations. The model described in [1] is an extension of Barker's category-based metamodel for access control [4] (CBAC): a notion of event adapted from [17] is used to describe a set of core axioms for defining obligations in an abstract way, without making any specific assumptions on the components of the system. In fact, two notions of events are defined in [1]: generic and specific. Generic events are used to represent the kind of events that can occur in a particular system, and specific events correspond to particular occurrences of events in a run of the system. The axiomatisation of the notion of obligation given in [1] relies on an *event typing* relation (associating specific events with generic events), and an *event interval* relation, which defines a link between an event that triggers a specific behaviour, and the event that terminates it. For example, the event associated to a fire alarm going off may start an emergency interval, which will be closed by the event associated to a call to the fire department.

In this paper we provide a general term-based language for events, and formally define the notions of *event typing* and *event interval*, to deal with event classification in a uniform way. Events are presented as typed-terms, built from a user-defined signature, that is, a particular set of typed function symbols that are specific to the system modelled. For each system we also define how to compute the events that close intervals initiated by previous events, based on a system specific function on generic events. This function allows us to extract intervals from a particular history (which is a sequence of events that have occurred in a system). Both event classification and interval computation have applications in access control and obligation management systems.

To summarise, the main contributions of the paper are the following:

- A general typed-language for events, and a typing relation associating specific and generic events;
- A general method for extracting event-intervals from a history;
- An implementation of this general method in Prolog together with methods for dealing with obligations, and an application of these methods in the context of obligation policies.

Overview: In Sect. 2, we recall some basic notions on term rewriting as well as the CBAC metamodel. In Sect. 3 we introduce a typed term language for representing events, and in Sect. 4 we recall the notion of event history and define an algorithm to extract event intervals from a history. Section 5 presents details of an implementation in Prolog of the relation between specific and generic events, the computation of intervals from history, as well as how this can be used in the obligation model. In Sect. 6 we discuss related work and finally, in Sect. 7, conclusions are drawn and further work is suggested.

2   Preliminaries

In this section we recall some basic notions and notations for term rewriting and access control policies involving obligations (see [1,2] for more details).

Term Rewriting. Term rewriting systems can be seen as programming or specification languages, or as formulae manipulating systems. We recall briefly the definition of first-order terms and term rewriting systems [2].

A signature $\mathcal{F}$ is a finite set of function symbols together with their (fixed) arity, where constants a are function symbols of arity zero. $\mathcal{X}$ denotes a denumerable set of variables $X_1, X_2 \ldots$, and $T(\mathcal{F}, \mathcal{X})$ denotes the set of terms built up from $\mathcal{F}$ and $\mathcal{X}$. Terms are identified with finite labeled trees. Positions are strings of positive integers. The subterm of t at position p is denoted by $t|_p$ and the result of replacing $t|_p$ with u at position p in t is denoted by $t[u]_p$. $\mathcal{V}(t)$ denotes the set of variables occurring in t. A term is ground (closed) if $\mathcal{V}(t) = \varnothing$. Substitutions are written $\theta = \{t_1/X_1, \ldots, t_n/X_n\}$ where t_i is assumed to be different from the variable X_i and $\mathsf{dom}(\theta) = \{X_1, \ldots, X_n\}$. We use Greek letters for substitutions and postfix notation for their application.

Given a signature $\mathcal{F}$, a *term rewrite system* on $\mathcal{F}$ is a set of rewrite rules $R = \{l_i \rightarrow r_i\}_{i \in I}$, where $l_i, r_i \in T(\mathcal{F}, \mathcal{X})$, $l_i \notin \mathcal{X}$, and $\mathcal{V}(r_i) \subseteq \mathcal{V}(l_i)$. A term t *rewrites* to a term u at position p with the rule $l \rightarrow r$ and the substitution σ, written $t \rightarrow_p^{l \rightarrow r, \sigma} u$, or simply $t \rightarrow_R u$, if $t|_p = l\sigma$ and $u = t[r\sigma]_p$. Such a term t is called *reducible*. Irreducible terms are said to be in *normal form*. We denote by $\rightarrow_R^+$ (resp. $\rightarrow_R^*$) the transitive (resp. transitive and reflexive) closure of $\rightarrow_R$. The subindex R will be omitted when it is clear from the context.

Access Control and Obligations. The Category-Based Access Control (CBAC) metamodel [4] is an abstract framework for the definition of access control policies, which can be instantiated to derive well-known access control models, such as Role-Based Access Control [11], Bell-La Padula's model [6], and dynamic models [3,8]. The latter permit the definition of access control policies where users' rights depend on their actions, or more generally, on events that happened in the system. In this paper, we present an event language and show how it can be applied in access control and obligation models. More precisely, we consider the extension of the CBAC metamodel that incorporates obligations, which in the following will be referred to as CBACO [1].

The CBAC metamodel is defined using a basic set of primitive, abstract notions: principals (which are the users of the system), resources (which are the objects that should be protected) and actions (which are the operations that users can perform on resources). These entities can be grouped into categories (in access control models we mostly consider categories of users). A category is a class of entities that share some property. Classic types of groupings used in access control, like a role, a security clearance, a discrete measure of trust, etc., are particular instances of the more general notion of category. Permissions, that is, pairs of action and resource, are assigned to categories of users rather than to individual users. Categories can be defined on the basis of e.g., user attributes, geographical constraints, resource attributes. In this way, permissions change in a dynamic and autonomous way (e.g., when a registered user has a birthday), unlike, e.g., role-based access control models, which require the intervention of a security administrator. Then, the axiomatic specification of the model allows us to derive, at any point, the rights of a principal by computing the principal's category and checking the permissions associated to it. In this way, an access request can be evaluated to decide whether it should be granted or denied. In CBACO, in addition to the basic notion of permissions available in CBAC, there are also two abstract notions of obligations, defined as follows.

Definition 1 (Obligation). *A generic obligation is a tuple (a, r, ge_1, ge_2), where a is an action, r a resource, and ge_1, ge_2 two event types (ge_1 triggers the obligation, and ge_2 ends it). If there is no starting event (resp., no ending event) we write $(a, r, \perp, ge)$ (resp., $(a, r, ge, \perp)$), meaning that the action on the resource must be performed at any point before an event of type ge (resp. at any point after an event of type ge).*

Example 1. Assume that in an organisation, the members of the security team must call the fire-department if a fire alarm is activated, and this must be done before they de-activate the alarm. This obligation could be represented by the tuple $(call, firedept, alarmON, alarmOFF)$.

Definition 2 (Duty). *A duty is a tuple (p, a, r, e_1, e_2, h), where p is a principal, a an action, r a resource, e_1, e_2 are two events and h is an event history that includes an interval opened by e_1 and closed by e_2. We replace e_1 (resp. e_2) with $\perp$ if there is no starting (resp. closing) event.*

Unlike access control models, which do not need to check whether the authorised actions are performed or not by the principals, obligation models need to include mechanisms to check whether duties were discharged or not. Specifically, obligation models distinguish four possible states for duties: *invalid* (when the duty is issued after the completion point); *fulfilled* (when carried out within the associated interval); *violated* (when not carried out within the associated interval, although issued with a valid interval) and *pending* (when has not yet been carried, but the interval is still valid). We refer the reader to [1] for the axiomatic and operational semantics of obligation policies in CBACO.

3   Events as Typed Terms

In this section we present a typed term language to represent events. We consider events as particular actions or happenings occurring at a particular time. Types are used to restrict the terms that correspond to events in our language.

In this section, and in the rest of the paper, we will present examples considering a hospital scenario, where several types of events can occur: patients can be triaged, admitted, receive consultation, be discharged, submitted to exams/procedures, etc. Sporadically there can also occur events such as fire alarms that can lead to the hospital evacuation, etc.

3.1   Types and Terms

Let b range over a finite set $\mathbb{B}$ of base types, and l over a finite set $\mathcal{L}$ of labels. The set $\mathbb{B}$ will always contain the types Tm and Ev, which are the types for time and event expressions respectively.

Definition 3 (Type). *The set $\mathbb{T}$ of types is built from $\mathbb{B}$:*

$$\tau \in \mathbb{T} :: = b \mid \{l_1 : \tau_1, \ldots, l_n : \tau_n\} \mid \tau_1 \to \tau_2$$

Record types, of the form $\{l_1 : \tau_1, \ldots, l_n : \tau_n\}$, represent structures labelled with $l_1, \ldots, l_n$, with types $\tau_1, \ldots, \tau_n$ respectively.

We consider a (system specific) function $\mathsf{type} : \mathcal{F} \to \mathbb{T}$, *which assigns a type to each function symbol in $\mathcal{F}$. If f is a function symbol with arity n, then* $\mathsf{type}(f) = \tau_1 \to \cdots \to \tau_n \to \tau$, *for some $\tau_1, \ldots, \tau_n, \tau \in \mathbb{T}$.*

Because terms in our language can contain free variable occurrences, type declarations for variables must be taken into consideration when typing expressions. As usual, an *environment* env, is a set of declarations of the form $X : \tau$ where all the variables X are distinct.

We now present our language to model events. We consider event expressions as terms that can be built from other event expressions, atomic actions or sets of attributes (represented as labelled structures).

Definition 4 (Event Specification). *Consider $X \in \mathcal{X}$, $\tau \in \mathbb{T}$ and $f \in \mathcal{F}$, then* values *and* specifications *are defined in the following way:*

$$\begin{aligned}
\nu \quad &:: = X^\tau \mid f(\nu_1, \ldots, \nu_n), \qquad n \geq 0 \\
\mathsf{spec} &:: = \{l_1 = \nu_1, \ldots, l_n = \nu_n\}, n > 0
\end{aligned}$$

The value X^τ represents a term variable of type τ. An atomic value a is a particular case of a value of the form $f(\nu_1, \ldots, \nu_n)$ where $n = 0$. The event specification $\{l_1 = \nu_1, \ldots, l_n = \nu_n\}$ represents the structure with labels $l_1, \ldots, l_n$ and values $\nu_1, \ldots, \nu_n$ respectively.

Definition 5 (Generic and Specific Events). *A generic event, denoted by $ge(X_1^{\tau_1}, \ldots, X_n^{\tau_n}) \in \mathcal{GE}$, is defined by an equation of the form:*

$$ge(X_1^{\tau_1}, \ldots, X_n^{\tau_n}) = \{\mathsf{spec}_1, \ldots, \mathsf{spec}_m\}^C$$

where the variables $X_1^{\tau_1}, \ldots, X_n^{\tau_n}$ occur in the right-hand side of the equation. The expression $\{\mathsf{spec}_1, \ldots, \mathsf{spec}_m\}^C$ represents the compound generic event, formed from the generic event specifications $\mathsf{spec}_1, \ldots, \mathsf{spec}_m$. If $m = 1$, then just write $ge(X_1^{\tau_1}, \ldots, X_n^{\tau_n}) = \mathsf{spec}$. Compound events represent sets of events that can occur separately in the history, but should be identified as a single event occurrence.

Specific events, denoted by $e \in \mathcal{E}$, are defined in the following way, where $\mathsf{spec}^\varnothing$ denotes ground event specifications (see Definition 4):

$$e :: = \{\mathsf{spec}_1^\varnothing, \ldots, \mathsf{spec}_n^\varnothing\}^C$$

As before, we write $\{\mathsf{spec}^\varnothing\}^C$ as $\mathsf{spec}^\varnothing$.

Example 2. We describe the action of a doctor P_1 reading the medical record of a patient P_2, by the generic event $gen\_read(P_1^{\mathcal{D}}, P_2^{\mathcal{P}})$, where $\mathcal{D}$ and $\mathcal{P}$ represent the types for *doctor* and *patient*, respectively:

$$gen\_read(P_1^{\mathcal{D}}, P_2^{\mathcal{P}}) = \{\texttt{act} = read, \texttt{doc} = P_1^{\mathcal{D}}, \texttt{obj} = rec(P_2^{\mathcal{P}})\}$$

We state that an order to evacuate the neurology ward was issued by Chief Jones, with the specific event:

$$\{\texttt{act} = evacuate, \texttt{ward} = neurology, \texttt{principal} = chief\_jones\}$$

The compound event gen_pregnD represents the events that must occur, before a doctor can make a pregnancy diagnosis.

$$gen\_pregnD(X^{\mathcal{P}}) = \{\ \{\texttt{act} = lab\_test, \texttt{pat} = X^{\mathcal{P}}\},$$
$$\{\texttt{act} = ultrasound, \texttt{obj} = abdomen, \texttt{pat} = X^{\mathcal{P}}\}\ \}^C$$

3.2 Typing Rules

We now assign types to values and event specifications, to ensure that we only deal will well-typed entities (wrt. the type signature specific to each system). We use record types to type labelled structures, with an implicit notion of subtyping (inspired by Ohori's system with polymorphic record types [20]). For the moment we only consider (implicit) subtyping between record types, but this can later be extended to general event types.

Definition 6 (Typing Rules for Values, Specifications and Events). *A typing judgement is a declaration of the form* env $\vdash v\colon \tau$, env $\vdash$ spec$\colon \tau$, *or* env $\vdash ge(X_1^{\tau_1}, \ldots, X_n^{\tau_n}) : $ Ev. *We say that v (resp. spec) has type τ given* env, *and write* env $\vdash v : \tau$ *(resp.* env $\vdash$ spec $: \tau$), *if the judgement can be derived using the following axioms and rules:*

$$\text{env} \vdash X^\tau : \tau, \ \text{if } X : \tau \in \text{env}$$

$$\frac{\text{type}(f) = \tau_1 \to \cdots \to \tau_n \to \tau \quad \text{env} \vdash v_1 : \tau_1 \quad \cdots \quad \text{env} \vdash v_n : \tau_n}{\text{env} \vdash f(v_1, \ldots, v_n) : \tau} \ (n \geq 0)$$

$$\frac{\text{env} \vdash v_1 : \tau_1 \quad \cdots \quad \text{env} \vdash v_{n+k} : \tau_{n+k}}{\text{env} \vdash \{\texttt{l}_1 = v_1, \ldots, \texttt{l}_n = v_n\} \cup \Gamma : \{\texttt{l}_1 : \tau_1, \ldots, \texttt{l}_n : \tau_n\}}$$

where, $\Gamma = \{\texttt{l}_{n+1} = v_{n+1}, \ldots, \texttt{l}_{n+k} = v_{n+k}\}$.

Given the rules above, a generic event expression $ge(X_1^{\tau_1}, \ldots, X_n^{\tau_n})$ *is well-typed given* env, *if* env $\vdash ge(X_1^{\tau_1}, \ldots, X_n^{\tau_n}) : $ Ev *can be derived from:*

$$\frac{ge(X_1^{\tau_1}, \ldots, X_n^{\tau_n}) = \{\text{spec}_1, \ldots, \text{spec}_m\}^C \quad \text{env} \vdash \text{spec}_i : \sigma_i \ (i = 1, \ldots, m)}{\text{env} \vdash ge(X_1^{\tau_1}, \ldots, X_n^{\tau_n}) : \text{Ev}}$$

Given the condition on variables in Definition 5, if env $\vdash ge(X_1^{\tau_1}, \ldots, X_n^{\tau_n})$: Ev then env contains declarations $X_1 : \tau_1, \ldots, X_n : \tau_n$.

Example 3. Using the fact that type$(read) = \mathcal{A}$, type$(dr.\ paul) = \mathcal{D}$, type$(john) = \mathcal{P}$, and type$(rec) = \mathcal{P} \to \mathcal{R}$, we obtain

$$\mathsf{env} \vdash \{\mathsf{act} = read, \mathsf{doc} = P_1^{\mathcal{D}}, \mathsf{obj} = rec(P_2^{\mathcal{P}})\} : \{\mathsf{act} : \mathcal{A}, \mathsf{doc} : \mathcal{D}, \mathsf{obj} : \mathcal{R}\}$$

and, for env $= \{P_1 : \mathcal{D}, P_2 : \mathcal{P}\}$, env $\vdash gen\_read(P_1^{\mathcal{D}}, P_2^{\mathcal{P}})$: Ev using the definition of *gen_read* given in Example 2.

Because we have an implicit subtyping rule for typing records, types are not unique. The event specification in the previous example, spec $= \{\mathsf{act} = read, \mathsf{doc} = P_1^{\mathcal{D}}, \mathsf{obj} = rec(P_2^{\mathcal{P}})\}$, can also be typed with $\{\mathsf{act} : \mathcal{A}, \mathsf{obj} : \mathcal{R}\}$ and $\{\mathsf{obj} : \mathcal{R}\}$. In fact any non-empty subset of $\{\mathsf{act} : \mathcal{A}, \mathsf{doc} : \mathcal{D}, \mathsf{obj} : \mathcal{R}\}$ is a valid type for spec. In this paper we are doing type-checking (and not type-inference), therefore we do not focus on most general types for specifications. But this could be achieved by using the notion of kinds of records as it is done in [20].

We now define an instance relation, associating ground values (denoted $\nu^{\varnothing}$) to values and specific event specifications to generic event specifications, under a substitution.

Definition 7 (Instantiation). *We define the relation* $\vdash_\theta \quad \nu^{\varnothing} :: \nu$ *(resp.* $\vdash_\theta$ *spec*$^{\varnothing} ::$ *spec), where* θ *is a substitution, in the following way:*

$$\frac{\vdash \nu^{\varnothing} : \tau}{\vdash_{\{\nu^{\varnothing}/X\}} \nu^{\varnothing} :: X^{\tau}} \qquad \frac{\vdash_{\theta_1} \nu_1^{\varnothing} :: \nu_1 \quad \cdots \quad \vdash_{\theta_n} \nu_n^{\varnothing} :: \nu_n}{\vdash_{\theta_1 \cup \cdots \cup \theta_n} f(\nu_1^{\varnothing}, \ldots, \nu_n^{\varnothing}) :: f(\nu_1, \ldots, \nu_n)} \quad (n \geq 0)$$

$$\frac{\vdash_{\theta_1} \nu_1^{\varnothing} :: \nu_1 \quad \cdots \quad \vdash_{\theta_n} \nu_n^{\varnothing} :: \nu_n}{\vdash_{\theta_1 \cup \cdots \cup \theta_n} \{\mathsf{l}_1 = \nu_1^{\varnothing}, \ldots, \mathsf{l}_n = \nu_n^{\varnothing}\} \cup \Gamma :: \{\mathsf{l}_1 = \nu_1, \ldots, \mathsf{l}_n = \nu_n\}}$$

where, $\Gamma = \{\mathsf{l}_{n+1} = \nu_{n+1}^{\varnothing}, \ldots, \mathsf{l}_{n+k} = \nu_{n+k}^{\varnothing}\}$. *Whenever we write* $\theta_1 \cup \cdots \cup \theta_n$, *we assume that* $\theta_1, \ldots, \theta_n$ *are compatible substitutions, in the sense that they do not assign different values to the same variable.*

Definition 8 (Event Instance). *The relation* $\vdash_\theta e :: ge(\overrightarrow{X})$, *extends the previous definition to event expressions in the following way:*

$$\frac{ge(\overrightarrow{X}) = \{\mathsf{spec}_1, \ldots, \mathsf{spec}_n\}^C \quad \vdash_{\theta_1} \mathsf{spec}_1^{\varnothing} :: \mathsf{spec}_1 \quad \cdots \quad \vdash_{\theta_n} \mathsf{spec}_n^{\varnothing} :: \mathsf{spec}_n}{\vdash_{\theta_1 \cup \cdots \cup \theta_n} \{\mathsf{spec}_1^{\varnothing}, \ldots, \mathsf{spec}_n^{\varnothing}\}^C :: ge(\overrightarrow{X})}$$

Example 4. Recall $gen\_read(P_1^{\mathcal{D}}, P_2^{\mathcal{P}})$, from Example 2. For the substitution $\theta = \{dr.\ paul/P_1, john/P_2\}$, we can derive

$$\vdash_\theta \{\mathsf{act} = read, \mathsf{doc} = dr.\ paul, \mathsf{obj} = rec(john)\} :: gen\_read(P_1^{\mathcal{D}}, P_2^{\mathcal{P}}).$$

Proposition 1. *If* env $\vdash \nu : \tau$ *(resp.* env $\vdash$ spec $: \tau$) *and* $\vdash_\theta \nu^{\varnothing} :: \nu$ *(resp.* $\vdash_\theta$ spec$^{\varnothing} ::$ spec), *then:*

1. $\theta = \{\nu_1^\varnothing/X_1, \ldots, \nu_n^\varnothing/X_n\}$, where $\{X_1, \ldots, X_n\} = \mathcal{V}(\nu)$ (resp. $\mathcal{V}(\mathsf{spec})$) and $\vdash \nu_i^\varnothing : \tau_i$ where $X_i : \tau_i \in \mathsf{env}$.
2. $\vdash \nu^\varnothing : \tau$ (resp. $\vdash \mathsf{spec}^\varnothing : \tau$).

The instantiation relation defined in this section is syntactic (replacing variables by terms). Depending on the application and the kind of data used to define events, instantiation may require some computation; we call it a semantic instantiation in the latter case. Formally, semantic instantiation is defined in the context of an equational theory. Although we leave a complete study on different equational theories and its appropriateness for future work, in the next section we will deal with semantic instantiation for time expressions.

4   Event History and Intervals

In this section we will define the notions of event history and intervals in history, which are determined by events. We also show how the instance relation defined in the previous section can be used to extract intervals from a history of events that match two given generic events. A history of events corresponds to a specific sequence of events that occur in a particular time frame. To deal with time frames we need a language that appropriately deals with time.

4.1   Time Expressions and Time Constraints

In this subsection we define a language for expressions representing time and use this to encode the approach for dealing with events in [1], in this setting.

Definition 9 (Time Expressions and Constraints). *Let* c *range over a set* S, *partially ordered by* $\leq$ *and closed under* $+$. *We define the set of* time expressions *and* time constraints *denoted* $\mathsf{t} \in \mathcal{T}$ *and* $\mathsf{tc} \in \mathcal{TC}$ *respectively, in the following way:*

$$\mathsf{t} ::= \mathsf{c} \mid X^{\mathsf{Tm}} \mid \mathsf{t} + \mathsf{c} \qquad \mathsf{tc} ::= \mathsf{t} \mid \mathsf{t}^+ \mid \mathsf{t}^{+\mathsf{c}}$$

Time constraints can be seen as intervals $[\mathsf{t}_1, \mathsf{t}_2]$, *where* $\mathsf{t}_1 = \mathsf{t}_2$, *if the time constraint is a time expression;* $\mathsf{t}_2 = \infty$, *if the time constraint is of the form* t_1^+; *and* $\mathsf{t}_2 = \mathsf{t}_1 + \mathsf{c}$, *if the time constraint is of the form* $\mathsf{t}_1^{+\mathsf{c}}$.

Note that a constant time expression represents a specific instant in time, which can be particular to each modelled system. In a time expression of the form $\mathsf{t} + \mathsf{c}$, c can be seen as a duration. In the rest of the paper we will take S to be $\mathbb{N}$, but other constants can be considered (that is, we consider time constants as clock ticks from a fixed point in time).

Definition 10 *Let* $\sigma = \{\mathsf{c}_1/X_1, \ldots, \mathsf{c}_n/X_n\}$ *be a substitution. We define* $[\![\cdot]\!]\sigma$ *for time expressions and time constraints, as follows:*

$$[\![\mathsf{c}]\!]\sigma = \mathsf{c} \qquad [\![\mathsf{t}^+]\!]\sigma = [[\![\mathsf{t}]\!]\sigma, \infty] \qquad [\![\mathsf{t} + \mathsf{c}]\!]\sigma = [\![\mathsf{t}]\!]\sigma + [\![\mathsf{c}]\!]\sigma$$
$$[\![X]\!]\sigma = \sigma(X) \qquad [\![\mathsf{t}^{+\mathsf{c}}]\!]\sigma = [[\![\mathsf{t}]\!]\sigma, [\![\mathsf{t}]\!]\sigma + [\![\mathsf{c}]\!]\sigma]$$

If $\mathcal{V}(\mathsf{t}_1) \cup \mathcal{V}(\mathsf{t}_2) \subseteq \mathsf{dom}(\sigma)$, *then* $\mathsf{t}_1 \preceq_\sigma \mathsf{t}_2$ *iff* $[\![\mathsf{t}_1]\!]\sigma \leq [\![\mathsf{t}_2]\!]\sigma$. *If* $\mathsf{t}_1, \mathsf{t}_2$ *are both ground we simply write* $\mathsf{t}_1 \preceq \mathsf{t}_2$ *instead of* $\mathsf{t}_1 \preceq_\varnothing \mathsf{t}_2$.

Another Representation for Events. Events in [1] are represented by finite sets of arity-2 facts, containing at least two necessary facts, $\mathtt{happens}(e, t)$ and $\mathtt{act}(e, a)$, where e is the identifier of the specific event, and t the time of its happening. Generic events are defined similarly, but can contain variables (as is the case in this paper), and in particular they always contain a variable (E) to be instantiated with the identifier of a specific event. A specific event e is an instance of a generic event ge, if there is a substitution σ such that $ge\sigma \subseteq e$. For example the events

$$e_1 = \{\mathtt{happens}(e_1, 12.25), \mathtt{act}(e_1, \mathit{activate}), \mathtt{obj}(e_1, \mathit{alarm}), \mathtt{subj}(e_1, \mathit{john})\}$$
$$e_2 = \{\mathtt{happens}(e_2, 12.45), \mathtt{act}(e_2, \mathit{deactivate}), \mathtt{obj}(e_2, \mathit{alarm}), \mathtt{subj}(e_2, \mathit{tom})\}$$

are instances, with respective substitutions $\sigma_1 = \{e_1/E, 12.25/T\}$ and $\sigma_2 = \{e_2/E, 12.25/T, \mathit{tom}/X\}$, of the generic events

$\mathtt{alarmON} = \{\mathtt{happens}(E, T), \mathtt{act}(E, \mathit{activate}), \mathtt{obj}(E, \mathit{alarm})\}$
$\mathtt{alarmOFF} = \{\mathtt{happens}(E, T + 20), \mathtt{act}(E, \mathit{deactivate}), \mathtt{obj}(e, \mathit{alarm}), \mathtt{subj}(E, X)\}$

This is an example where the instantiation relation requires some computation (the instantiation of $T + 20$ with the substitution $12.25/T$ will produce 12.45 in σ_2). Note that the function in Definition 10 defines a semantic instantiation for time expressions and time constraints.

The encoding of this event representation is straightforward. Given a particular event, a record $\mathtt{spec}$ is created containing an entry $\mathtt{fact} = exp$ for each fact $\mathtt{fact}(e, exp)$, except for $\mathtt{happens}(e, t)$. The identifier e can, depending on necessity, either be omitted or be included as a particular entry $\mathtt{id} = e$. Finally, the event will be represented by a pair $(\mathtt{spec}, t)$, where t is a time expression (this notion will be formalised in the next section). The encoding of the events above is:

$$(\{\mathtt{act} = \mathit{activate}, \mathtt{obj} = \mathit{alarm}, \mathtt{subj} = \mathit{john}\}, 12.25)$$
$$(\{\mathtt{act} = \mathit{deactivate}, \mathtt{obj} = \mathit{alarm}, \mathtt{subj} = \mathit{tom}\}, 12.45)$$
$$(\mathtt{alarmON} = \{\mathtt{act} = \mathit{activate}, \mathtt{obj} = \mathit{alarm}\}, T)$$
$$(\mathtt{alarmOFF}(X) = \{\mathtt{act} = \mathit{deactivate}, \mathtt{obj} = \mathit{alarm}, \mathtt{subj} = X)\}, T)$$

4.2   History of Events

We will now consider a history of events and define how one can relate events in a history to define appropriate intervals.

Definition 11 (History). *An event history $h \in \mathcal{H}$ is a sequence of distinct specific events in time of the form $[E_1 = (e_1, t_1), \ldots, E_n = (e_n, t_n)]$, where $t_1, \ldots, t_n$ are ground and such that for $i < j$, $t_i \preceq t_j$. A subsequence of h is called an* event interval *and it is represented as $I = (E_i, E_j)$, where E_i, E_j are respectively the first and the last event in the interval. We say that E_i opens the interval and E_j closes it. We use the constant $\perp$ to represent untimed events as a pair $(e, \perp)$.*

Example 5 (History).

$$h = [(\{\text{act} = enterHosp, \text{patient} = john\}, 900),$$
$$(\{\text{act} = consult, \text{patient} = john, \text{doc} = dr.\ paul\}, 1100),$$
$$(\{\text{act} = request, \text{doc} = dr.\ paul, \text{patient} = john, \text{proc} = x\text{-}ray\}, 1110),$$
$$(\{\text{act} = perform, \text{patient} = john, \text{doc} = dr.\ mary, \text{proc} = x\text{-}ray\}, 1125),$$
$$(\{\text{act} = read, \text{doc} = dr.\ paul, \text{ward} = neurology, \text{obj} = rec(john)\}, 1300)$$

To compute intervals we will define a function that describes how events are linked to subsequent events in history. Because we want to consider different scenarios we will also consider different strategies to select intervals. A strategy is a function $\text{strat} : 2^{\mathcal{I}} \rightarrow 2^{\mathcal{I}}$, which will be used to select elements from a set $\mathcal{I}$ of pairs of timed events (intervals). Examples of strategies can be *select-first*, *select-last*, *select-all*, etc. We will use $\mathscr{S}$ to represent the set of available strategies.

Definition 12. *A closing function* $\text{cl} : \mathcal{GE} \times \mathcal{TC} \rightarrow 2^{\mathcal{GE} \times \mathcal{T} \times \mathscr{S}}$, *is a mapping associating to a pair* (ge, tc), *a set of triples of the form* (ge', t', strat), *which are the generic events in time that are closed by the generic event* ge *provided that some constraints on* t' *and* tc *are satisfied, and selected by the function* strat. *In the rest of the paper we will assume a* select-first *strategy, and omit strategies from function* cl.

Example 6. In our hospital scenario, consider:

– $\text{cl}(exitHosp(P, W), T^+)) = \{(triage(P), T), (inWard(P, W), T)\}$
– $\text{cl}(releaseCR(W), T^{+20})) = \{(codeRED(W), T)\}$

For instance $\text{cl}(releaseCR(W), T^{+20})) = \{(codeRED(W), T)\}$ indicates that the specific event $releaseCR(neurology)$ can close the event $codeRED(neurology)$, provided that the former occurs at most 20 instants after the latter.

Closed and open intervals are key notions when dealing with obligations, as they allow us to determine the status of an obligation at a given point. In the next section we will use the notions of closed and open intervals (Definitions 13 and 14 below) in the context of the CBACO metamodel.

Definition 13. *Let* $h \in \mathcal{H}$, $(ge, tc) \in \mathcal{GE} \times \mathcal{TC}$ *and* $(ge', t) \in \mathcal{GE} \times \mathcal{T}$, *then* $closed(ge', ge, h)$ *is the set of event intervals of the form* $((e_i, t_i), (e_j, t_j))$ *such that, for some compatible substitutions* θ_i, θ_j *and a substitution on time variables* σ, *one has:* $(ge', t) \in \text{cl}((ge, tc))$; $\vdash_{\theta_i} e_i :: ge'$ *and* $\vdash_{\theta_j} e_j :: ge$; $t_i = [\![t]\!]\sigma$ *and* $t_j \in [\![tc]\!]\sigma$. *Then the function* $\text{interval}(e_1, e_2, h)$ *is **true** if, for some compatible substitutions* θ_1 *and* θ_2: $\vdash_{\theta_1} e_1 :: ge_1$, $\vdash_{\theta_2} e_2 :: ge_2$ *and* $(e_1, e_2) \in closed(ge_1, ge_2, h)$, *and* **false** *otherwise.*

Definition 14. *Let* $h \in \mathcal{H}$, $(ge, tc) \in \mathcal{GE} \times \mathcal{TC}$ *and* $(ge', t) \in \mathcal{GE} \times \mathcal{T}$, *then* $open(ge', ge, h)$ *is the set of event intervals of the form* $((e_i, t_i), \bot)$ *such that, for some substitution* θ_i *and a substitution on time variables* σ, *one has:* $(ge', t) \in \text{cl}((ge, tc))$; $\vdash_{\theta_i} e_i :: ge'$; $t_i = [\![t]\!]\sigma$, *but there is not an event* (e_j, t_j), *with* $t_i \preceq t_j$, *such that* $\vdash_{\theta_j} e_j :: ge$, *for a substitution* θ_j *compatible with* θ_i *and* $t_j \in [\![tc]\!]\sigma$.

In the defininions above, we do not distinguish between single and compound events, and assume that compound events appear in history. A more detailed and realistic treatment of coumpound events in history is left for future work.

5   A Prolog Implementation

In this section we describe a prototype implementation of the previous definitions in Prolog. Because Prolog programs are expressed in terms of relations, represented as facts and rules, Prolog is an ideal language to implement the notions defined in this paper. Backtracking, unification and logical variables are also useful features for our implementation (although in this prototype implementation we treat substitutions explicitly, a more efficient implementation can make use of Prolog's logic variables and unification to implicitly propagate substitutions and deal with compatibility of substitutions).

5.1   Defining Events, Event Typing and Intervals

For a particular system, the language of events is determined by the set of functors $\mathcal{F}$, its associated types given by function type, and the equations defining generic events, which can be represented as Prolog facts. For example, we can consider the following typed constants and functors for our hospital scenario:

```
type(neurology,ward).
type(dr_paul,doctor).
type(rec,arrow([patient],resource)).
ge(exitHosp,[var(P1,patient)], rec([lab(action,discharge),
                                     lab(patient,var(P1,patient)),
                                     lab(doc,var(P2,doctor))])).
cl((ge(exitHosp,[var(P,doctor),var(W,ward)]),plus(var(T,time))),
   [(ge(triage,[var(P,doctor)]),var(T,time)),
    (ge(inWard,[var(P,doctor),var(W,ward)]),var(T,time))]).
cl((ge(releaseCR,[var(W,ward)]),plus(var(T,time))),
   [(ge(codeRED,[var(W,ward)]),var(T,time))]).
```

Below we present the predicate ty(Theta, E, GE) implementing the relation $\vdash_{\theta} e::ge$ from Definition 8.

```
ty([],A,A):- atomic(A).
ty([(X,Value)],Value,var(X,Type)):- typed([],Value,Type).
ty(Theta,fun(Name,CValues),fun(Name,Values)):- zip(CValues,Values,L),
                                                tyList(Theta,L).
ty(Theta,lab(Name,CValue),lab(Name,Value)):- ty(Theta,CValue,Value).
ty(Theta,rec(CL),rec(L)):- permut(CL,PCL),
                           zip(CL,L,LRec),
                           tyList(Theta,LRec).
ty(Theta,CSpec,ge(Name,Lvar)):- ge(Name,Lvar,Spec), ty(Theta,CSpec,Spec).
ty(Theta,comp(CSpec),ge(Name,Vars)):- ge(Name,Vars,comp(LSpec)),
                                      permut(CSpec,PCSpec),
```

```
                                   zip(PCSpec,LSpec,Specs),
                                   ty(Theta,Specs).
tyList([],[]).
tyList(Theta,[(CValue,Value)|L]):- ty(Theta1,CValue,Value),
                                   tyList(Theta2,L),
                                   compatible(Theta1,Theta2,Theta).
```

Where the predicate typed(Env, Value, Type) implements the typing relation env $\vdash \nu : \tau$ from Definition 6. The predicate compatible(Theta1, Theta2, Theta) verifies if the two substitutions Theta1 and Theta2 are compatible, eliminating duplicated declarations. The predicate zip(L1, L2, L3), succeeds if in the list of pairs L3, each pair contains elements of lists L1 and L2 occurring at the same position (similar to the Haskell zip function). The predicate permut(L1,L2) succeeds if L2 is a permutation of L1.

The functions closed(ge', ge, h) and open(ge', ge, h) from Definition 13 can be computed using the predicates cinterval(GE1,GE2,H,(E1,E2,Sigma)) and ointerval(GE1,GE2,H,(E1,E2,Sigma)), respectively:

```
cinterval((ge(N1,V1),TC),(ge(N2,V2),T),H,((Ei,Ti),(Ej,Tj),Sigma)):-
    cl((ge(N1,V1),TC),LGEs), member((ge(N2,V2),T),LGEs),
    pick((Ei,Ti),H,RH), ty(Theta,Ei,ge(N2,Vs2)),
    pick((Ej,Tj),RH,_), ty(Theta,Ej,ge(N1,Vs2)),
    tsem(T,Sigma,Ti), tsem(TC,Sigma,Int), belongs(Tj,Int).
ointerval((ge(N1,V1),TC),(ge(N2,V2),T),H,((Ei,Ti),bot,Sigma)):-
    cl((ge(N1,V1),TC),LGEs), member((ge(N2,V2),T),LGEs),
    pick((Ei,Ti),H,RH), tsem(T,Sigma,Ti),ty(Theta,Ei,ge(N2,Vs2)),
    not cinterval((ge(N1,V1),TC),(ge(N2,V2),T),H,((Ei,Ti),(_,_),Sigma)).

closed(GE1,GE2,H,L):- findall(E,cinterval(GE1,GE2,H,E),L).
open(GE1,GE2,H,L):- findall(E,ointerval(GE1,GE2,H,E),L).
```

The predicate tsem(T, Sigma, Ti) implements the semantic for time expressions and time constraints. The predicate pick will pick an event from the history and return the rest of the history after that event.

5.2   Application: Obligation Models

In this section, we consider the rewrite-based semantics of CBACO [1], where the status of an obligation (a, r, ge_1, ge_2) (see Definition 1) for principal p in a given history h is computed using the following rewrite rule:

eval-obligation$(p, a, r, ge_1, ge_2, h) \rightarrow if$ opar(p, a, r, ge_1, ge_2) $then$
append(chk-cl*(closed$(ge_1, ge_2, h), p, a, r)$, chk-op*(open$(ge_1, ge_2, h), p, a, r))$
$$else\ [not\text{-}applicable]$$

Here the function opar, specific to the system being modelled, is such that opar(p, a, r, ge_1, ge_2) holds if principal p has the generic obligation (a, r, ge_1, ge_2); append is a standard function that concatenates two lists; closed computes the sublists of h that start and finish with events e_1, e_2, which are respectively instances of ge_1, ge_2 (in this case e_2 closes the interval for this obligation). Similarly open returns the subhistories of h that start with an event e_1 (instance of ge_1) and for which there is no instance of ge_2 in h that closes the interval for this obligation. The function chk-cl with inputs h', p, a, r checks whether in the subhistory h' there is an event where the principal p has performed the action a on the resource r, returning a result fulfilled if that is the case, and violated otherwise. The function chk-op with inputs h', p, a, r checks whether in the subhistory h' there is an event where the principal p has performed the action a on the resource r, returning a result fulfilled if that is the case, and pending otherwise. The functions chk-cl* and chk-op* do the same but for each element of a list of subhistories, returning a list of results. Using the functions and relations defined in the previous sections, we can evaluate obligations according to the above specification. First, we give an alternative, equivalent specification for the evaluation of obligations, which is closer to the logic-programming implementation discussed in the previous subsection. Assuming that ty is the function that implements the instance relation $\vdash_\theta$ on events (that is, $\text{ty}(e) = \{(ge, \theta) \mid \; \vdash_\theta e :: ge\}$), the status of an obligation can be computed using the following rule, where the extra variables in the right hand side are existentially quantified.

$$\begin{aligned}
\text{status}(p, a, r, ge_1, ge_2, h) \rightarrow \; & \textbf{if } \text{opar}(p, a, r, ge_1, ge_2) \\
& \textbf{then if } ((e_1, t_1), (e_2, t_2)) \in \text{closed}(ge_1, ge_2, h) \\
& \quad and \; (e, t) \in h \; and \; (ge, \theta) \in \text{ty}(e) \; and \\
& \quad ge\,\theta = \{principal = p, action = a, resource = r\} \; \textbf{then} \\
& \qquad \textbf{if } t_1 \prec t \prec t_2 \; \textbf{then}\quad \text{fulfilled } \textbf{else } \text{violated} \\
& \quad \textbf{elseif } ((e_1, t_1), \perp) \in \text{open}(ge_1, ge_2, h) \\
& \quad and \; (e, t) \in h \; and \; (ge, \theta) \in \text{ty}(e) \; and \\
& \quad ge\,\theta = \{principal = p, action = a, resource = r\} \; and \\
& \quad t_1 \prec t \; \textbf{then } \text{fulfilled } \textbf{else } \text{pending} \\
& \textbf{else }\quad \textit{not-applicable}
\end{aligned}$$

We are assuming that h contains all the events occurring in the system up to the moment where we wish to check the status of the obligation.

We now give the Prolog implementation for the rule that computes the status of an obligation. To deal with obligations we need Prolog facts/predicates to represent relations in CBACO. In particular, in order to implement assignment of obligations to principals, we have a predicate opar(P,A,R,GE1,GE2). We assume the existence of a generic event ge(par,[var(P,pl),var(A,act),var(R,res)]) with specification rec([lab(principal,var(P,pl)),lab(action,var(A,act)), lab(resource,var(R,res))]), where pl,act,res are the types for principal, actions and resources, respectively.

```
status(P,A,R,ge(N1,V1),ge(N2,V2),H,notapplicable):-
    not opar(P,A,R,GE1,GE2),!.
status(P,A,R,ge(N1,V1),ge(N2,V2),H,S):- opar(P,A,R,GE1,GE2),
    closed(GE1,GE2,H,CI), member((E1,E2),CI), event(P,A,R,H,(E,T)),
    chktime(T,T1,T2,S).
status(P,A,R,ge(N1,V1),ge(N2,V2),H,fulfilled):- opar(P,A,R,GE1,GE2),
    open(GE1,GE2,H,CI), member((E1,bot),CI), event(P,A,R,H,(E,T)),!,
```

```
   tsem(T1,Theta,Time1) tsem(T,Theta,Time), Time>=Time1.
status(P,A,R,ge(N1,V1),ge(N2,V2),H,pending):- opar(P,A,R,GE1,GE2),
   open(GE1,GE2,H,CI), member((E1,bot),CI).
event(P,A,R,H,(E,T)):- member((E,T),H),
   ty(Theta,E,ge(par,[var(P1,pl),var(A1,act),var(R1,res)])),
   member((P1,P),Theta), member((A1,A),Theta), member((R1,R),Theta).
checktime(Time,Time1,Time2,fulfilled):- tsem(Time1,Theta,T1),
                                         tsem(Time2,Theta,T2),
                                         tsem(Time,Theta,T), T>=T1,T2>=T.
checktime(Time,Time1,Time2,invalid):- tsem(Time1,Theta,T1),
                                       tsem(Time2,Theta,T2),
                                       tsem(Time,Theta,T), (T1>=T;T>=T2).
```

The rewrite-based specification of duties (see Definition 2) in CBACO [1] relies on auxiliary functions interval, and type, which are also specific to the system being modelled: interval(e_1, e_2, h) checks whether the event history includes an interval opened by e_1 and closed by e_2, and type(e, h) computes the generic event ge, of which e occurring in h is an instance (and that, in the rule below, is assumed to be unique).

$$\text{duty}(p, a, r, e_1, e_2, h) \rightarrow \text{opar}(p, a, r, \text{type}(e_1, h), \text{type}(e_2, h)) \text{ and } \text{interval}(e_1, e_2, h)$$

In [1] interval and type are assumed to be defined for each specific system, to respectively implement the relations *event interval* and *event typing*. In this paper we give general definitions/implementations of these relations. The implementation of a checker for duties in Prolog is straightforward, using the predicate defined above to compute intervals, which takes into account the type relation between events.

```
duty(P,A,R,E1,E2,H):- opar(P,A,R,GE1,GE2), cinterval(GE1,GE2,H,(E1,E2,_)).
```

6   Related Work

The notion of event has been treated in various settings in the literature, such as logic-based frameworks, algebraic approaches and query languages, amongst others. In the context of access control, Barker et al. [5] have proposed a representation for events as sets of binary predicates, partially motivated by Davidson's view of events as action occurrences [10]. In this formalism, event descriptions are given as finite sets of ground 2-place facts (atoms) that describe an event, uniquely identified by $e_i, i \in \mathbb{N}$, and which includes three necessary facts: $happens(e_i, t_j)$, $act(e_i, a_l)$ and $agent(e_i, u_n)$, and n non-necessary facts. This was later used in [1] to model obligations in the CBAC metamodel, but considering only two necessary facts *happens* and *act*. This representation is claimed to be more flexible than a term-based representation with a fixed set of attributes. In our language, we do not fix necessary facts, although one can define them as part of the set of typed-functors. Furthermore, event specifications are given as records which may contain extra fields, so these sets of predicates can be easily encoded in our language. A less flexible representation was used in [8], in the context of distributed event-based access control. In this work, events are ground terms of the form $\mathsf{event}(e_i, u, a, t)$ where event is a data constructor of arity four, $e_i (i \in \mathbb{N})$ denotes unique event identifiers, u identifies a user, a is an action associated to the event, and t is the time when the event happened. When it is sufficient to know the chronological order of events, then the history can be ordered as to provide that information and the time parameter may be omitted.

The *Obligation Specification Language* (OSL) defined in [13], presents a language for events to monitor and reason about data usage requirements. The notions of obligational formulas/obligations defined in this paper are closely related to the notions of generic/concrete obligations in [1]. Therefore data usage as specified in [13] can be encoded within the CBACO metamodel. The notion of events presented in [13] is also similar to ours in some aspects, but where logical expressions are used to deal with intervals. The paper also presents a relation *refinesEv*, defining an instance relation between events. This relation is based on a subset relation on labels, as the instance relation in [1]. In our setting this instance relation between events is defined for parameterised generic events (i.e. containing variables), by the implicit subtyping on records but more generally using variable instantiation.

Still in the context of access control systems, Bertino et al. [7] proposed the Temporal Role-Based Access Control Model (TRBAC), using events to activate and deactivate roles. This was later used in [22,23], to deal with security analysis in the presence of static temporal role hierarchies in RBAC. The *time models* used in these works also depend on the notion of time interval, but they use a simpler notion of interval that can easily be encoded in our language. The activation and de-activation of roles, as well as dealing with the so-called *safety problem* (i.e., administrative actions that can lead to a policy in which a user can acquire permissions that can compromise the security of the system), is not the purpose of events in our work. Nevertheless, this can achieved, through the assignment of users to categories in CBAC policies, based on some property depending on a temporal constraint.

An important notion in the above formalisms, and in the language described in this paper, is the notion of interval, which provides means to reason about assignment of status in [5] and status of obligations in [1]. Intervals as sequences that are initiated and terminated by events, and during which certain facts hold, are also a key aspect in the event calculus [17,18]. The initial motivation of the event calculus was to deal with database updating, but it has been applied in a variety of settings [9,12,15,16]. Like in the event calculus, we also consider intervals as being initiated and closed by events, however we do not reason (in general) about facts that hold at a certain point.

Time intervals and time constraints have also been used to appropriately deal with obligations in access control models [1,14,19,21]. In most of these models time intervals are not defined by events, but as fixed points in time, which are easily represented in our language. Time constraints in [19] consider sequences of time intervals, to enforce systematic repetition of obligations. We do not consider this type of constraints, as repetition of obligations can be enforced through the definition of the categories for obligations, but our representation of time constraints could easily be adapted to consider sequences of intervals.

7 Conclusions

We have defined a language to represent events as typed-terms, built from a user-defined signature, to formally deal with the notions of event typing and event intervals in a uniform way, in the context of the CBACO metamodel. In a given system, intervals can be automatically extracted from a history of events by means of a relation that determines how events are closed in the system. This approach allows us to adequately define general functions to implement event typing and to compute event intervals, without having to know the exact type of events that we are dealing with. As future work we would like to extend this formalism to deal with notions such as conflicting events, and automatically generated events. Furthermore, we believe that a type-system for events could be useful in identifying patterns of events in history, which could lead to interesting applications in the context of event processing. We believe that the notions of intervals defined here could be useful in other contexts. In particular, it could be used to infer intervals where a particular status is valid, which can be applied in status-based access control models.

References

1. Alves, S., Degtyarev, A., Fernández, M.: Access control and obligations in the category-based metamodel: a rewrite-based semantics. In: Proietti, M., Seki, H. (eds.) LOPSTR 2014. LNCS, vol. 8981, pp. 148–163. Springer, Heidelberg (2015)
2. Baader, F., Nipkow, T.: Term rewriting and all that. Cambridge University Press, Great Britain (1998)
3. Barker, S.: Action-status access control. In: Proceedings of SACMAT 2007, pp. 195–204. ACM (2007)
4. Barker, S.: The next 700 access control models or a unifying meta-model? In: Proceedings of SACMAT 2009, pp. 187–196. ACM (2009)
5. Barker, S., Sergot, M.J., Wijesekera, D.: Status-based access control. ACM Trans. Inform. Syst. Secur. 12(1), 1:1–1:47 (2008)
6. Bell, D.E., Lapadula, L.J.: Secure Computer System: Unified Exposition and MULTICS Interpretation. Technical report ESD-TR-75-306, The MITRE Corporation (1976)
7. Bertino, E., Bonatti, P.A., Ferrari, E.: TRBAC: A temporal role-based access control model. ACM Trans. Inform. Syst. Secur. 4(3), 191–233 (2001)
8. Bertolissi, C., Fernández, M., Barker, S.: Dynamic event-based access control as term rewriting. In: Barker, S., Ahn, G.-J. (eds.) Data and Applications Security 2007. LNCS, vol. 4602, pp. 195–210. Springer, Heidelberg (2007)
9. Craven, R., Lobo, J., Ma, J., Russo, A., Lupu, E., Bandara, A.: Expressive policy analysis with enhanced system dynamicity. In: Proceedings of ASIACCS 2009, pp. 239–250. ACM (2009)
10. Davidson, D.: Essays on Actions and Events. Oxford University Press (2001)

11. Ferraiolo, D., Kuhn, R., Chandramouli, R.: Role-Based Access Control. Artech House, Norwood (2003)
12. Gelfond, M., Lobo, J.: Authorization and obligation policies in dynamic systems. In: Garcia de la Banda, M., Pontelli, E. (eds.) ICLP 2008. LNCS, vol. 5366, pp. 22–36. Springer, Heidelberg (2008)
13. Hilty, M., Pretschner, A., Basin, D., Schaefer, C., Walter, T.: A policy language for distributed usage control. In: Biskup, J., López, J. (eds.) ESORICS 2007. LNCS, vol. 4734, pp. 531–546. Springer, Heidelberg (2007)
14. Irwin, K., Yu, T., Winsborough, W.H.: On the modeling and analysis of obligations. In: Proceedings of CCS 2006, pp. 134–143. ACM (2006)
15. Kowalski, R.: Database updates in the event calculus. J. Logic Program. **12**(1–2), 121–146 (1992)
16. Kowalski, R., Sadri, F.: A logic-based framework for reactive systems. In: Bikakis, A., Giurca, A. (eds.) RuleML 2012. LNCS, vol. 7438, pp. 1–15. Springer, Heidelberg (2012)
17. Kowalski, R., Sergot, M.: A Logic-based Calculus of Events. New Gen. Comput. **4**(1), 67–95 (1986)
18. Miller, R., Shanahan, M.: The Event calculus in classical logic - alternative axiomatisations. Electron. Trans. Artif. Intell. **3**(A), 77–105 (1999)
19. Ni, Q., Bertino, E., Lobo, J.: An obligation model bridging access control policies and privacy policies. In: Proceedings of SACMAT 2008, pp. 133–142. ACM (2008)
20. Ohori, A.: A polymorphic record calculus and its compilation. ACM Trans. Program. Lang. Syst. **17**(6), 844–895 (1995)
21. Pontual, M., Chowdhury, O., Winsborough, W.H., Yu, T., Irwin, K.: On the management of user obligations. In: Proceedings of SACMAT 2011, pp. 175–184. ACM (2011)
22. Ranise, S., Truong, A.T., Armando, A.: Scalable and precise automated analysis of administrative temporal role-based access control. In: Proceedings of SACMAT 2014, pp. 103–114 (2014)
23. Ranise, S., Truong, A.T., Viganò, L.: Automated analysis of RBAC policies with temporal constraints and static role hierarchies. In: Proceedings of SAC 2015, pp. 2177–2184 (2015)

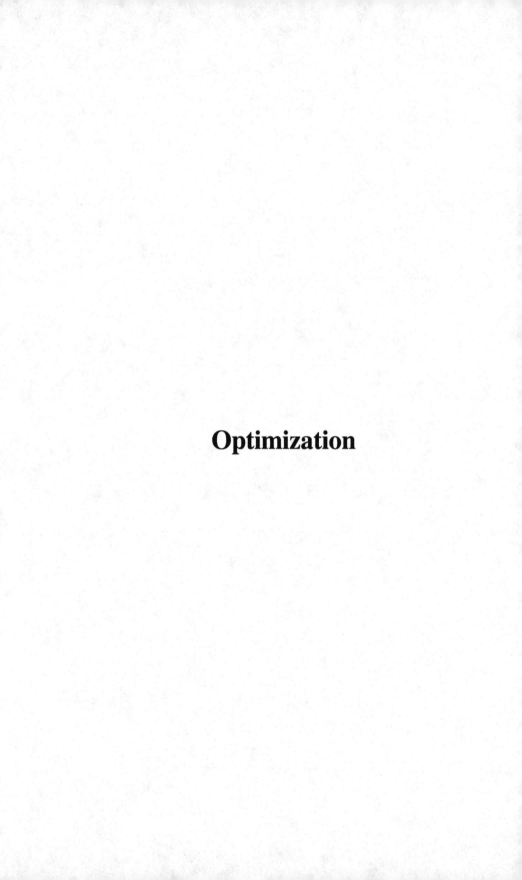

Optimization

Applying Sorting Networks to Synthesize Optimized Sorting Libraries

Michael Codish[1], Luís Cruz-Filipe[2(✉)], Markus Nebel[2],
and Peter Schneider-Kamp[2]

[1] Department of Computer Science,
Ben-Gurion University of the Negev, Beer-Sheva, Israel
mcodish@cs.bgu.ac.il
[2] Department Mathematics and Computer Science,
University of Southern Denmark, Odense, Denmark
{nebel,petersk}@imada.sdu.dk, lcfilipe@gmail.com

Abstract. This paper presents an application of the theory of sorting networks to facilitate the synthesis of optimized general-purpose sorting libraries. Standard sorting libraries are often based on combinations of the classic Quicksort algorithm with insertion sort applied as base case for small, fixed, numbers of inputs. Unrolling the code for the base case by ignoring loop conditions eliminates branching, resulting in code equivalent to a sorting network. This enables further program transformations based on sorting network optimizations, and eventually the synthesis of code from sorting networks. We show that, if considering the number of comparisons and swaps, the theory predicts no real advantage of this approach. However, significant speed-ups are obtained when taking advantage of instruction level parallelism and non-branching conditional assignment instructions, both of which are common in modern CPU architectures. We provide empirical evidence that using code synthesized from efficient sorting networks as the base case for Quicksort libraries results in significant real-world speed-ups.

1   Introduction

General-purpose sorting algorithms are based on comparing, and possibly exchanging, pairs of inputs. If the order of these comparisons is predetermined by the number of inputs to sort and does not depend on their concrete values, then the algorithm is said to be data-oblivious. Such algorithms are well suited for e.g. parallel sorting or secure multi-party computations.

Sorting functions in state-of-the-art programming language libraries (such as the GNU C Library) are typically based on a variant of Quicksort, where the base cases of the recursion apply insertion sort: once the subsequence to sort considered by Quicksort falls under a certain length M, it is sorted using insertion sort. The reasons for using such base cases is that, both theoretically

---

Supported by the Israel Science Foundation, grant 182/13 and by the Danish Council for Independent Research, Natural Sciences.

M. Falaschi (Ed.): LOPSTR 2015, LNCS 9527, pp. 127–142, 2015.
DOI: 10.1007/978-3-319-27436-2_8

and empirically, insertion sort is faster than Quicksort for sorting small numbers of elements. Typical values of M are 4 (e.g. in the GNU C library) or 8.

Generalizing this construction, we can take any sorting algorithm based on the divide-and-conquer approach (e.g. Quicksort, merge sort), and use another sorting method once the number of elements to sort in one partition does not exceed a pre-defined limit M. The guiding idea here is that, by supplying optimized code for sorting up to M inputs, the overall performance of the sorting algorithm can be improved. One obvious way to supply optimized code for sorting up to M inputs is to provide a unique optimized implementation of sorting m elements, for each $m \leq M$.

This approach leads directly to the following problem: *For a given fixed number M, how can we obtain an efficient way to sort M elements on a modern CPU?* Similar questions have been asked since the 1950s, though obviously with a different notion of what constitutes a modern CPU.

Sorting networks are a classical model of comparison-based sorting that provides a framework for addressing such questions. In a sorting network, n inputs are fed into n channels, connected pairwise by comparators. Each comparator compares the two inputs from its two channels, and outputs them sorted back to the same two channels. Consecutive comparators can be viewed as a "parallel layer" if no two touch the same channel. Sorting networks are data-oblivious algorithms, as the sequence of comparisons performed is independent of the actual input. For this reason, they are typically viewed as hardware-oriented algorithms, where data-obliviousness is a requirement and a fixed number of inputs is given.

In this paper, we examine how the theory of sorting networks can improve the performance of general-purpose software sorting algorithms. We show that replacing the insertion sort base case of a Quicksort implementation as found in standard C libraries by optimized code synthesized from logical descriptions of sorting networks leads to significant improvements in execution times.

The idea of using sorting networks to guide the synthesis of optimized code for base cases of sorting algorithms may seem rather obvious, and, indeed, has been pursued earlier. A straightforward attempt, described in [10], has not resulted in significant improvements, though. In this paper we show that this is not unexpected, providing theoretical and empirical insight into the reasons for these rather discouraging results. In a nutshell, we provide an average case analysis of the complexity w.r.t. measures such as number of comparisons and number of swaps. From the complexity point of view, code synthesized from sorting networks can be expected to perform slightly worse than unrolled insertion sort. Fortunately, for small numbers (asymptotic) complexity arguments are not always a good predictor of real-world performance.

The approach taken in [7] matches the advantages of sorting networks with the vectorization instruction sets available in some modern CPU architectures. The authors obtain significant speedups by implementing parallel comparators as vector operations, but they require a complex heuristic algorithm to generate sequences of bit shuffling code that needs to be executed between comparators. Their approach is also not fully general, as they target a particular architecture.

In this paper, we combine the best of both these attempts by providing a straightforward implementation of sorting networks that still takes advantage of the features of modern CPU architectures, while keeping generality. We obtain speedups comparable to [7], but our requirements to the instruction set are satisfied by virtually all modern CPUs, including those without vector operations. The success of our approach is based on two observations.

- Sorting networks are data-oblivious and the order of comparisons is fully determined at compile time, i.e., they are free of any control-flow branching. Comparators can also be implemented without branching, and on modern CPU architectures even efficiently so.
- Sorting networks are inherently parallel, i.e., comparators at the same level can be performed in parallel. Conveniently, this maps directly to implicit *instruction level parallelism* (ILP) common in modern CPU architectures. This feature allows parallel execution of several instructions on a single thread of a single core, as long as they are working on disjoint sets of registers.

Avoiding branching and exploiting ILP are tasks also performed through program transformations by the optimization stages of modern C compilers, e.g., by unrolling loops and reordering instructions to minimize data-dependence between neighbouring instructions. They are though both restricted by the data-dependencies of the algorithms being compiled and, consequently, of only limited use for data-dependent sorting algorithms, like insertion sort.

Throughout this paper, for empirical evaluations we run all code on an Intel Core i7, measuring runtime in CPU cycles using the time stamp counter register using the RDTSC instruction. As a compiler for all benchmarks, we used LLVM 6.1.0 with clang-602.0.49 as frontend on Max OS X 10.10.2. We also tried GCC 4.8.2 on Ubuntu with Linux kernel 3.13.0-36, yielding comparable results.

The remainder of the paper is organized as follows. Section 2 provides background information and formal definitions for both sorting algorithms and hardware features. In Sect. 3, we theoretically compare Quicksort and the best known sorting networks w.r.t. numbers of comparisons and swaps. We aggressively unroll insertion sort until we obtain a sorting network in Sect. 4, and in Sect. 5 we show how to implement individual comparators efficiently. We empirically evaluate our contribution as a base case of Quicksort in Sect. 6, before concluding and giving an outlook on future work in Sect. 7.

2 Background

2.1 Quicksort with Insertion Sort for Base Case

For decades, Quicksort has been used in practice, due to its efficiency in the average case. Since its first publication by Hoare [8], several modifications were suggested to improve it further. Examples are the clever choice of the pivot, or the use of a different sorting algorithm, e.g., insertion sort, for small subproblem sizes. Most such suggestions have in common that the empirically observed

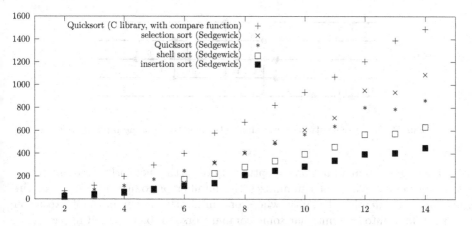

Fig. 1. Comparison of different sorting algorithms for small numbers of inputs.

efficiency can be explained on theoretical grounds by analyzing the expected number of comparisons, swaps, and partitioning stages (see [13] for details).

Figure 1 presents a comparison of the common spectrum of data-dependent sorting algorithms for small numbers of inputs, depicting the number of inputs (x-axis) together with the number of cycles required to sort them (y-axis), averaged over 100 million random executions. The upper curve in the figure is obtained from the standard Quicksort implementation in the C library (which is at some disadvantage, as it requires a general compare function as an argument). The remaining curves are derived from applying standard sorting algorithms, as detailed by Sedgewick [14]; the code was taken directly from the book's web page, http://algs4.cs.princeton.edu/home/. Insertion sort is the clear winner.

2.2 Sorting Networks

A *comparator network* on n channels is a finite sequence $C = c_1, \ldots, c_k$ of *comparators*, where each comparator c_ℓ is a pair (i_ℓ, j_ℓ) with $1 \leq i_\ell < j_\ell \leq n$. The *size* of C is the number k of comparators it contains. Given an input $x \in D^n$, where D is any totally ordered domain, the *output* of C on x is the sequence $C(x) = x^n$, where x^ℓ is defined inductively as follows: $x^0 = x$, and x^ℓ is obtained from $x^{\ell-1}$ by swapping the elements in positions i_ℓ and j_ℓ, in case $x_{i_\ell} < x_{j_\ell}$. C is a *sorting network* if $C(x)$ is sorted for all $C \in D^n$. It is well known (see e.g. [9]) that this property is independent of the concrete domain D.

Comparators may act in parallel. A comparator network C has *depth* d if C is the concatenation of $L_1, \ldots, L_d$, where each L_i is a *layer*: a comparator network with the property that no two of its comparators act on a common channel.

Figure 2 depicts a sorting network on 5 channels in the graphical notation we will use throughout this paper. Comparators are depicted as vertical lines, and layers are separated by a dashed line. The numbers illustrate how the input $10101 \in \{0,1\}^5$ propagates through the network. This network has 6 layers and 9 comparators.

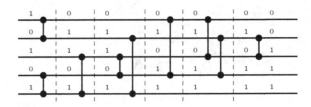

Fig. 2. A sorting network on 5 channels operating on the input 10101.

There are two main notions of optimality of sorting networks in common use: *size* optimality, where one minimizes the number of comparators used in the network; and *depth* optimality, where one minimizes the number of execution steps, taking into account that some comparators can be executed in parallel.

Given n inputs, finding the minimal size s_n and depth t_n of a sorting network is an extremely hard problem that has seen significant progress in recent years. The table below details the best currently known bounds. The values for $n \leq 8$ are already listed in [9]; the values of t_9 and t_{10} were proven exact by Parberry [11], those of t_{11}–t_{16} by Bundala and Závodný [1], and t_{17} was recently computed by Ehlers and Müller [5] using results from [3,4]. Finally, the values of s_9 and s_{10} were first given in [2].

n	1	2	3	4	5	6	7	8	9	10	11	12	13	14	15	16	17
s_n	0	1	3	5	9	12	16	19	25	29	35	39	45	51	56	60	73
											33	37	41	45	49	53	58
t_n	0	1	3	3	5	5	6	6	7	7	8	8	9	9	9	9	10

Oblivious versions of classic sorting algorithms can also be implemented as sorting networks, as described in [9]. Figure 3(a) shows an oblivious version of insertion-sort. The vertical dashed lines highlight the 4 iterations of "insertion" required to sort 5 elements. Figure 3(b) shows the same network, with comparators arranged in parallel layers. Bubble-sort can also be implemented as a sorting network as illustrated in Fig. 3(c), where the vertical dashed lines illustrate the 4 iterations of the classic bubble-sort algorithm. When ordered according to layers, this network becomes identical to the one in Fig. 3(b).

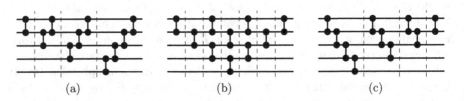

(a)   (b)   (c)

Fig. 3. Sorting networks for insertion sort (a) and bubble-sort (c) on 5 inputs, dashed lines separating iterations. When parallelized, both networks become the same (b).

2.3   Modern CPU Architectures

Modern CPU architectures allow multiple instructions to be performed in parallel on a single thread. This ability is called *instruction-level parallelism* (ILP), and is built on three modern micro-architectural techniques[1]:

- *superscalar instruction pipelines*, i.e., pipelines with the ability to hold and execute multiple instructions at the same time
- *dynamic out-of-order execution*, i.e., dynamic reordering of instructions respecting data dependencies
- *redundant execution units*, i.e., multiple Arithmetic Logic Units per core

Together, these features allow execution of instructions in an order that minimizes data dependencies, so that multiple redundant execution units can be used at the same time. This is often termed *implicit* ILP, in contrast to the explicit ILP found in vector operations.

Example 1. Consider the C expression (x+y)*(z+u). Assume the variables x, y, z, and u are loaded in registers eax, ebx, ecx, and edx. Then the evaluation of the above expression is compiled to three machine instructions: ADD eax,ebx; ADD ecx,edx; MUL eax,ecx, with the result in ecx. Here, the first two instructions are data-independent and can be executed in parallel, while the last one depends on the results of those, and is executed in another CPU cycle.

Conditional branching instructions are the most expensive instructions on pipelined CPUs, as they require flushing and refilling the pipeline. In order to minimize their cost, modern CPU architectures employ dynamic branch prediction. By keeping the pipeline filled with the instructions of the predicted branch, the cost of branching is severely alleviated. Unfortunately, branch prediction cannot be perfect, and when the wrong branch is predicted, the pipeline needs to be flushed and refilled – an operation taking many CPU cycles.

In order to avoid branching instructions for "small" decisions, e.g., deciding whether to assign a value or not, modern CPU architectures also feature conditional instructions. Depending on flags set by e.g. a comparison, either an assignment of a value of a register will be performed, or the instruction will be ignored. In both cases, the pipeline is filled with the subsequent instructions, and the cost of the operation is smaller than a possible branch prediction failure.

Example 2. Consider the C statement if (x == 42) x = 23; with variable x loaded in eax. Without conditional move instructions, this is compiled to code with a conditional branching instruction, i.e. CMP eax,42; JNZ after; MOV eax, 23, where after is the address of the instruction following the MOV instruction. Alternatively, using conditional instructions, we obtain CMP eax, 42; CMOVZ eax, 23. This code not only saves one machine code instruction, but most importantly avoids the huge performance impact of a mispredicted branch.

---

[1] For details on these features of modern microarchitectures see e.g. [6,15].

Table 1. Average number of comparisons and swaps when executing optimal sorting networks with at most $M = 14$ inputs.

n	1	2	3	4	5	6	7	8	9	10	11	12	13	14
comparisons	0	1	3	5	9	12	16	19	25	29	35	39	45	51
swaps	0.0	0.5	1.5	2.7	4.8	6.6	8.6	10.6	13.0	11.1	19.4	22.4	20.0	26.5

3 Quicksort with Sorting Networks for Base Case

The general theme of this paper is to derive, from sorting networks, optimized code to sort small numbers of inputs, and then to apply this code as the base case in a Quicksort algorithm. In this section, we compare precise average case results for the number of comparisons and swaps performed by a classic Quicksort algorithm and by a modification that uses sorting networks on subproblems of size at most 14. We choose 14 for this analysis, as it is the largest value n for which we could conveniently measure the number of comparisons and swaps for all $n!$ permutations. We used the best-known (w.r.t. size) sorting networks (optimal for up to 10 inputs) in order to obtain the most favorable comparison numbers for sorting networks. To this end, we assume the algorithm to act on random permutations of size n, each being the input with equal probability.

Let C_n (resp. S_n) denote the expected number of comparisons (resp. swaps) performed by classic Quicksort on (random) inputs of size n. Let furthermore $\hat{C}_n$ and $\hat{S}_n$ denote the corresponding quantities for Quicksort using sorting networks for inputs smaller than 15. It is standard to set up recurrence relations for those quantities which typically obey a pattern such as:

$$T_n(a,b) = \begin{cases} a \cdot n + b + \frac{1}{n} \sum_{1 \leq j \leq n} T_{j-1}(a,b) + T_{n-j}(a,b) & \text{if } n > M, \\ g(n) & \text{otherwise.} \end{cases}$$

Here, a and b have to be chosen properly to reflect the parameter's (comparisons, swaps) behavior, M determines the maximum subproblem size for which a different algorithm (insertion sort, sorting networks) is used, and g accounts for the costs of that algorithm. In order to analyze classic Quicksort as proposed by Hoare, we have to choose $a = 1$, $b = -1$ (resp. $a = \frac{1}{6}$, $b = \frac{2}{3}$) for comparisons (resp. swaps), together with $M = 0$ and $g(0) = 0$. For the analysis of our proposed modification using sorting networks for subproblems of small sizes, we set $M = 14$ together with the values for g as given in Table 1. Using standard algebraic manipulations, it is possible to solve this recurrence explicitly to obtain a formula for $T_n(a,b)$ in terms of n, M, a and b. Defining $t_n = a \cdot n + b$ and $\nabla t_n = t_n - t_{n-1}$, one finds (see [12] for details) that, for $n > M$,

$$T_n(a,b) = 2(n+1) \sum_{M+2 \leq k \leq n} \frac{\nabla t_k}{k+1} + \frac{n+1}{M+2} (t_{M+1} + T_{M+1}(a,b)) - t_n.$$

Computing the closed form expressions for $\sum_{M+2\leq k\leq n}\frac{\nabla t_k}{k+1}$ for the different choices of t_n, we finally get

$$C_n = 2n\ln(n+1) - 2.84557n + o(n) \qquad S_n = \frac{1}{3}n\ln(n+1) + 0.359072n + o(n)$$

$$\hat{C}_n = 2n\ln(n+1) - 2.44869n + o(n) \qquad \hat{S}_n = \frac{1}{3}n\ln(n+1) + 0.524887n + o(n)$$

We see that, when increasing n, both parameters get worse by our modification of classic Quicksort. Even for small n and optimal size sorting networks, there is no advantage w.r.t. the numbers of comparisons or swaps. In conclusion, we cannot hope to get a faster sorting algorithm simply by switching to sorting networks for small subproblems – at least not on grounds of our theoretical investigations. And, by transitivity, replacing insertion sort by sorting networks in the base case should result in an even worse behavior w.r.t. both parameters.

4   Unrolling the Base Case

In this section, we show how to unroll an implementation of insertion sort, step by step, until we finally obtain code equivalent to a sorting network. We take the basic insertion sort code from Sedgewick [14], and, for illustration, assume that the fixed number of inputs is $n = 5$. We experimented also with optimized variants (e.g. making use of sentinels to avoid the j>0 check), but did not find any of them to be faster for small inputs given a modern C compiler.

```
#define SWAP(x,y) {int tmp = a[x]; a[x] = a[y]; a[y] = tmp;}
static inline void sort5(int *a, int n) {
  n=5
  for (int i = 1; i < n; i++)
    for (int j = i; j > 0 && a[j] < a[j-1]; j--)
      SWAP(j-1, j)
}
```

Applying partial evaluation and (outer) loop unrolling results in:

```
static inline void sort5_unrolled(int *a) {
  for (int j = 1; j > 0 && a[j] < a[j-1]; j--)
    SWAP(j-1, j)
  for (int j = 2; j > 0 && a[j] < a[j-1]; j--)
    SWAP(j-1, j)
  for (int j = 3; j > 0 && a[j] < a[j-1]; j--)
    SWAP(j-1, j)
  for (int j = 4; j > 0 && a[j] < a[j-1]; j--)
    SWAP(j-1, j)
}
```

The condition in the inner loop is data-dependent, hence no sound and complete program transformation can be applied to unroll them. To address this, we move the data-dependent part of the loop condition to the statement in the body of the loop, while always iterating the variable j down to 1.

```
static inline void sort5_oblivious(int *a) {
  for (int j = 1; j > 0; j--)
    if (a[j] < a[j-1]) SWAP(j-1, j)
  for (int j = 2; j > 0; j--)
    if (a[j] < a[j-1]) SWAP(j-1, j)
  for (int j = 3; j > 0; j--)
    if (a[j] < a[j-1]) SWAP(j-1, j)
  for (int j = 4; j > 0; j--)
    if (a[j] < a[j-1]) SWAP(j-1, j)
}
```

Now we can now apply (inner) loop unrolling and obtain:

```
static inline void sort5_oblivous_unrolled(int *a) {
  if (a[1] < a[0]) SWAP(0, 1)
  if (a[2] < a[1]) SWAP(1, 2)
  if (a[1] < a[0]) SWAP(0, 1)
  if (a[3] < a[2]) SWAP(2, 3)
  if (a[2] < a[1]) SWAP(1, 2)
  if (a[1] < a[0]) SWAP(0, 1)
  if (a[4] < a[3]) SWAP(3, 4)
  if (a[3] < a[2]) SWAP(2, 3)
  if (a[2] < a[1]) SWAP(1, 2)
  if (a[1] < a[0]) SWAP(0, 1)
}
```

All the statements in the body of sort5_oblivous_unrolled are now conditional swaps. For readability, we move the condition into the macro. COMPs on the same line indicate that they originate from the same iteration of insertion sort:

```
#define COMP(x,y) { if (a[y] < a[x]) SWAP(x,y) }
static inline void sort5_fig3a(int *a) {
  COMP(0, 1)
  COMP(1, 2)  COMP(0, 1)
  COMP(2, 3)  COMP(1, 2)  COMP(0, 1)
  COMP(3, 4)  COMP(2, 3)  COMP(1, 2)  COMP(0, 1)
}
```

This sequence is equivalent to the sorting network in Fig. 3(a). Thus, we can apply the reordering of comparators that resulted in Fig. 3(b) to obtain the following implementation, where we reduce the number of layers to 7 (here, COMPs on the same line indicate a layer in the sorting network):

```
static inline void sort5_fig3b(int *a) {
  COMP(0, 1)
  COMP(1, 2)
  COMP(0, 1)  COMP(2, 3)
  COMP(1, 2)  COMP(3, 4)
  COMP(0, 1)  COMP(2, 3)
  COMP(1, 2)
  COMP(0, 1)
}
```

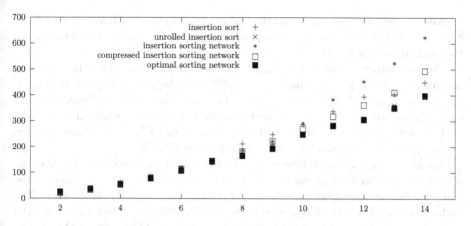

Fig. 4. Comparison of insertion sort with (unrolled) comparator based code for small numbers of inputs.

Figure 4 presents a comparison of a standard insertion sort (code from [14]) with the several optimized versions, depicting the number of inputs (x-axis) together with the number of cycles required to sort them (y-axis), averaged over 100 million random executions. The curve labeled "insertion sort" portrays the same data as the corresponding curve in Fig. 1. The curve labeled "unrolled insertion sort" corresponds to the unrolled version of insertion sort (in the style of function sort5_unrolled). The other three curves correspond to code derived from different types of sorting networks: the "insertion sorting network" from Fig. 3(a) and function sort5_fig3a; the "compressed insertion sorting network" from Fig. 3(b) and function sort5_fig3b; and the "optimal sorting network", corresponding to the use of a best (smallest) known sorting network.

From the figure, it is clear that standard sorting network optimizations such as reordering of independent comparators [9] give a slight performance boost. But there is another clear message: even going beyond standard program transformations by breaking data-dependence and obtaining a sequence of conditional swaps (i.e., a sorting network), we do not manage to make any significant improvements of the performance of sorting implementations for small numbers of inputs. Furthermore, even when using size-optimal sorting networks, we obtain no real benefit over compiler-optimized insertion sort. This is in line with the theoretical results on average case complexity discussed in the previous section.

5 Implementing Sorting Networks Efficiently

The results in the previous two sections explained the rather discouraging results obtained by a naive attempt to use sorting networks as the base case of a divide-and-conquer sorting algorithm: they are simply not faster than e.g. insertion sort – at least when implemented naively. In this section we show how to exploit two main properties of sorting networks, together with features of modern CPU architectures, and obtain speed-ups of a factor higher than 3 compared to unrolled insertion sort.

We first observe that, as sorting networks are data-oblivious, the order of comparisons is fully determined at compile time, i.e., their implementation is free of any control-flow branching. Unfortunately, the naive implementation of each comparator involves branching to decide whether to perform a swap. The path taken depends entirely on the specific inputs to be sorted, and as such branch prediction necessarily does not perform very well.

Luckily, we can also implement comparators without branching. To this end, we use a conditional assignment (defined by the macro COND below), which can be compiled to the conditional move (CMOV) instruction available on modern CPU architectures. This approach proved to be very fruitful. For illustration, from the optimal-size sorting network for 5 inputs portrayed in Fig. 2, we synthesize the following C function sort5_best, where each row in the code corresponds to a layer in the sorting network:

```
#define COND(c,x,y) { x = (c) ? y : x; }
#define COMP(x,y) { int ax = a[x]; COND(a[y]<ax,a[x],a[y]); \
                                   COND(a[y]<ax,a[y],ax  ); }

static inline void sort5_best(int *a) {
  COMP(0, 1)  COMP(3, 4)
  COMP(2, 4)
  COMP(2, 3)  COMP(1, 4)
  COMP(0, 3)
  COMP(0, 2)  COMP(1, 3)
  COMP(1, 2)
}
```

The comparator macro that compares and conditionally swaps the values at indices x and y works as follows:

1. Keep a copy of the value at index x.
2. Compare (once) the value at index y with the stored value from x.
3. If the value was greater, copy the value at index y to index x. Otherwise, do nothing.
4. If the value was greater, write the old copied value from x to index y. Otherwise, do nothing.

Correctness follows directly by case analysis. If the value at index y was not greater than the value at index x, the two conditional assignments do not change anything, and all we did was an unnecessary copy of the valued at index x. If the value at index y was greater than the value at index y, we essentially perform a classic swap using ax as the temporary variable.

Given a sufficient optimization level (-O2 and above), the above code is compiled by the LLVM (or GNU) C compiler to use two conditional move (CMOV) instructions, resulting in a totally branching free code for sort5_best. As can be expected, the other two instructions are a move (MOV) and a compare (CMP) instruction. In other words, each comparator is implemented by exactly four non-branching machine code instructions.

Alternatively, we could implement the comparator applying the folklore idea of swapping values using XORs to eliminate one conditional assignment:[2]

```
#define COND(c,x,y) { x = (c) ? y : x; }
#define COMP(x,y) { int ax = a[x]; COND(a[y]<ax,a[x],a[y]); \
                    a[y] ^= ax ^ a[x]; }
```

This alternative comparator performs a conditional swap as follows:

1. Keep a copy of the value at index x.
2. If the value at index y is greater than the value at index x, copy the value at index y to index x.
3. Bitwise XOR the value at index y with the copied old and the new value at index x.

Step 3 works because, if the condition holds, then ax and the value at index x cancel out, leaving the value at index y unchanged, while otherwise the value at y and ax cancel out, effectively assigning the original value from index x to index y.

We also implemented this variant, and observed that it compiles down to five instructions (MOV, CMP, CMOV, and two XORs). We benchmarked the two variants and observed that they are indistinguishable in practice, with differences well within the margin of measurement error. Thus, we decided to continue with this second version, as the XOR instructions are more "basic" and can therefore be expected to behave better w.r.t. e.g. instruction level parallelism.

A third approach would be to define branching-free minimum and maximum operations,[3] and use them to assign the minimum to the upper channel and the maximum to the lower channel of the comparator. We tested this approach, but found that it did not compile to branching-free code. Even if it did, the number of instructions involved would be rather large, eliminating any chance of competing with the two previous variants.

The reader might wonder whether a different SWAP macro could similarly speed up the working of standard insertion sort. The answer is a clear no, as the standard swapping operation is implemented by only three operations. Tricks like using XORs only increase the number of instructions to execute, while not reducing branching in the code. We implemented and benchmarked several alternative SWAP macros, finding only detrimental effects on measured performance.

Figure 5 compares three sorting algorithms for small numbers of inputs: (1) the unrolled insertion sort (also plotted in Fig. 4); (2) code derived from a standard insertion sorting network (also plotted in Fig. 4); (3) the same insertion sorting network but with a non-branching version of the COMP macro. We compare the number of branches encountered and mispredicted (averaged over 100 million random executions). From the figure it is clear that the number of branches encountered

---

[2] See https://graphics.stanford.edu/~seander/bithacks.html#SwappingValuesXOR.
[3] See https://graphics.stanford.edu/~seander/bithacks.html#IntegerMinOrMax.

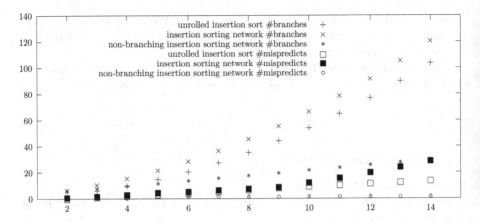

Fig. 5. Comparing the number of branches, encountered and mispredicted, in optimized sorting algorithms for small numbers of inputs.

(and mispredicted) is larger for both unrolled insertion sort and a naive implementation of sorting networks. In contrast, the branching-free implementation exhibits a nearly constant level of branches encountered and mispredicted. These branches actually originate from the surrounding test code (filling an array with random numbers, computing random numbers, and checking that the result is actually sorted).

Our second observation is that sorting networks are inherently parallel, i.e., comparators at the same level can be performed simultaneously. This parallelism can be mapped directly to instruction level parallelism (ILP). The ability to make use of ILP has further performance potential. In order to demonstrate this potential, we constructed artificial test cases with varying levels of data dependency. Given a natural number m, we construct a comparator network of size 1000 consisting of subsequences of m parallel comparators. We would expect that, as m grows, we would see more use of ILP.

In Fig. 6, the values for m are represented on the x-axis, while the y-axis (as usual) indicates the averaged number of CPU cycles. Indeed, we see significant

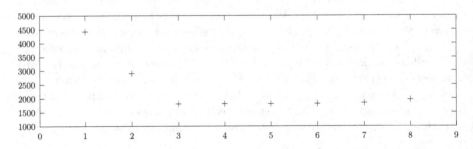

Fig. 6. ILP on comparator networks of length 1000 with differing levels of parallelism.

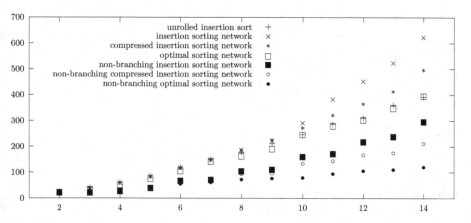

Fig. 7. Comparison of sorting networks for small numbers of inputs: non-branching sorting networks are fastest.

performance gains when going from $m = 1$ to $m = 2$ and $m = 3$. From this value onwards, performance stays unchanged. This is the result of each comparator being compiled to 5 assembler instructions when using optimization level -O3. Then we obtain slightly under 2 CPU cycles per comparator.

Combining the gains from ILP with the absence of branching, we obtain large speed-ups for small inputs when comparing to both insertion sort and naive implementations of sorting networks. In Fig. 7, we show the magnitude of the improvements obtained. Once again we plot the number of inputs on the x-axis against the number of cycles required to sort then on the y-axis, averaged over 100 million random executions. We consider the unrolled insertion sort, the three sorting networks from Fig. 4 (insertion sorting network, compressed insertion sorting network, and optimal sorting network), and these same three sorting networks using non-branching comparators (non-branching insertion sorting network, non-branching compressed insertion sorting network, and non-branching optimal sorting network). The figure shows that using the best known (optimal) sorting networks in their non-branching forms results in a speed-up by a factor of more than 3.

6   Quicksort with Sorting Network Base Case

We now demonstrate that optimizing the code in the base case of a Quicksort algorithm translates to real-world savings when applying the sorting function. To this end, we use as base cases (1) the (empirically) best variant of insertion sort unrolled by applying program transformations to the algorithm from [14], and (2) the fastest non-branching code derived from optimal (size) sorting networks.

In Fig. 8 we depict the results of sorting lists of 10,000 elements. The y-axis measures the number of cycles (averaged over one million random runs), and the x-axis specifies the limit at which Quicksort reverts to a base case. For example,

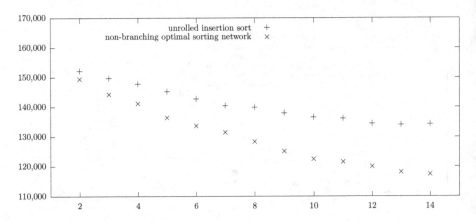

Fig. 8. Quicksort: comparing insertion sort at the base case with non-branching optimal sorting networks at the base case. Plotting base case size (x axis) and number of cycles (averaged over one million random runs).

the value 8 indicates that the algorithm uses a base case whenever it is required to sort a sequence of length *at most* 8 elements. The value 2 corresponds to the case where the base case has no impact. To quantify the impact of the choice of base case, we compare to the case for value 2 (on the x-axis). For insertion sort we see a 2–12 % reduction in runtime depending on the limit, and for non-branching sorting networks we achieve instead 7–23 % reduction in runtime.

7 Conclusion

In this paper, we showed, both theoretically and empirically, that using code derived naively from sorting networks is not advantageous to sort small numbers of inputs, compared to the use of standard data-dependent sorting algorithms like insertion sort. Furthermore, we showed that program transformations are of only limited utility for improving insertion sort on small numbers of inputs.

By contrast, we showed how to synthesize simple yet efficient implementations of sorting networks, and gave insight into the microarchitectural features that enable this implementation. We demonstrated that we do obtain significant speed-ups compared to naive implementations such as [10]. A further empirical comparison between our implementation and the one described in [7] (not detailed in this paper) shows similar performance and scaling behavior. However, our approach allows the exploitation of instruction-level parallelism without the need for a complex instruction set-specific algorithm, as required by [7]. We also provided further evidence that efficient sorting networks are useful as a base case in divide-and-conquer sorting algorithms such as, e.g., Quicksort.

Our results also show that using different sorting networks has measurable impact on the efficiency of the synthesized C code. While previous research on finding optimal sorting networks has focused on optimal depth or optimal size, in the future we plan to identify criteria that will lead to optimal performance in this context. What are the parameters that determine real-world efficiency of the synthesized code, and how can we find sorting networks that optimize these parameters? We also plan to explore other target architectures, such as GPUs, and to benchmark our approach as base case for other sorting algorithms, such as merge sort.

References

1. Bundala, D., Závodný, J.: Optimal sorting networks. In: Dediu, A.-H., Martín-Vide, C., Sierra-Rodríguez, J.-L., Truthe, B. (eds.) LATA 2014. LNCS, vol. 8370, pp. 236–247. Springer, Heidelberg (2014)
2. Codish, M., Cruz-Filipe, L., Frank, M., Schneider-Kamp, P.: Twenty-five comparators is optimal when sorting nine inputs (and twenty-nine for ten). In: ICTAI 2014, pp. 186–193. IEEE, December 2014
3. Codish, M., Cruz-Filipe, L., Schneider-Kamp, P.: The quest for optimal sorting networks: efficient generation of two-layer prefixes. In: Winkler, F., Negru, V., Ida, T., Jebelan, T., Petcu, D., Watt, S.M., Zaharie, D., (eds.) SYNASC 2014, pp. 359–366. IEEE (2015)
4. Codish, M., Cruz-Filipe, L., Schneider-Kamp, P.: Sorting networks: the end game. In: Dediu, A.-H., Formenti, E., Martín-Vide, C., Truthe, B. (eds.) LATA 2015. LNCS, vol. 8977, pp. 664–675. Springer, Heidelberg (2015)
5. Ehlers, T., Müller, M.: New bounds on optimal sorting networks. In: Beckmann, A., Mitrana, V., Soskova, M. (eds.) CiE 2015. LNCS, vol. 9136, pp. 167–176. Springer, Heidelberg (2015)
6. Fisher, J.A., Faraboschi, P., Young, C.: Embedded Computing: A VLIW Approach to Architecture, Compilers, and Tools. Morgan Kaufman, San Francisco (2005)
7. Furtak, T., Amaral, J.N., Niewiadomski, R.: Using SIMD registers and instructions to enable instruction-level parallelism in sorting algorithms. In: SPAA 2007, pp. 348–357. ACM (2007)
8. Hoare, C.A.R.: Quicksort. Comput. J. **5**(1), 10–15 (1962)
9. Knuth, D.E.: The Art of Computer Programming, Volume III: Sorting and Searching. Addison-Wesley, New York (1973)
10. Lopez, B., Cruz-Cortes, N.: On the usage of sorting networks to big data. In: Arabnia, H.R., Yang, M.Q., Jandieri, G., Park, J.J., Solo, A.M.G., Tinetti, F.G., (eds.) Advances in Big Data Analytics: The 2014 WorldComp International Conference Proceedings. Mercury Learning and Information (2014)
11. Parberry, I.: A computer-assisted optimal depth lower bound for nine-input sorting networks. Math. Syst. Theor. **24**(2), 101–116 (1991)
12. Sedgewick, R.: The analysis of quicksort programs. Acta Inf. **7**, 327–355 (1977)
13. Sedgewick, R., Flajolet, P.: An Introduction to the Analysis of Algorithms. Addison-Wesley-Longman, New York (1996)
14. Sedgewick, R., Wayne, K.: Algorithms, 4th edn. Addison-Wesley, Reading (2011)
15. Silc, J., Robic, B., Ungerer, T.: Processor Architecture: From Dataflow to Superscalar and Beyond. Springer, New York (1999)

Impact of Accuracy Optimization on the Convergence of Numerical Iterative Methods

Nasrine Damouche[1]([✉]), Matthieu Martel[1], and Alexandre Chapoutot[2]

[1] University of Perpignan Via Domitia, LAMPS, France
{nasrine.damouche,matthieu.martel}@univ-perp.fr
[2] ENSTA ParisTech, Palaiseau, France
alexandre.chapoutot@ensta-paristech.fr

Abstract. Among other objectives, rewriting programs serves as a useful technique to improve numerical accuracy. However, this optimization is not intuitive and this is why we switch to automatic transformation techniques. We are interested in the optimization of numerical programs relying on the IEEE754 floating-point arithmetic. In this article, our main contribution is to study the impact of optimizing the numerical accuracy of programs on the time required by numerical iterative methods to converge. To emphasize the usefulness of our tool, we make it optimize several examples of numerical methods such as Jacobi's method, Newton-Raphson's method, etc. We show that significant speedups are obtained in terms of number of iterations, time and flops.

Keywords: Program transformation · Floating-point numbers · IEEE754 Standard · Numerical analysis · Convergence acceleration

1 Introduction

A few decades ago, program transformation techniques have been successfully applied to specialize programs by partial evaluation [16]. For example, the performances of Knuth-Morris-Pratt's algorithm were reached by specialized versions of the naive, quadratic, pattern matching algorithm [4]. Other killer applications of partial evaluation were ranging from ray-tracing [16] to communication protocol optimization [22]. In this context, partial evaluation was used to optimize the execution time of programs. Our current work seeks another grail, namely the optimization of the numerical accuracy of computations carried out in the IEEE754 floating-point arithmetic [2,23]. As for partial evaluation, we perform source to source transformations guided by partial information on the data used at run-time [14]. In our case, we need ranges for the input variables of the programs, obtained by abstract interpretation of their codes [5]. In former articles, we have shown how our techniques make it possible to improve

This work was supported by the ANR Project ANR-12-INSE-0007 "CAFEIN".

M. Falaschi (Ed.): LOPSTR 2015, LNCS 9527, pp. 143–160, 2015.
DOI: 10.1007/978-3-319-27436-2_9

the accuracy of various algorithms coming from control theory or from numerical analysis [8,9,15]. In this article, we study the effect of our transformation on the convergence of well-known iterative numerical methods such as Newton-Raphson's or Jacobi's method [17]. We show that more accurate implementations obtained by automatic program transformation converge much faster than the original ones. In other words, less iterations are needed to reach a result with a given accuracy. As a consequence, improving the accuracy significantly improves the performances of this class of programs, bringing us back to the concerns of partial evaluation.

In former work we have shown how to optimize automatically intra-procedural programs [8]. To optimize programs, we use static analysis by abstract interpretation [5,11] to over-approximate the roundoff errors as well as a set of rewriting rules for the transformation itself, applied to programs that are written in SSA Form [7]. We have experimented our tool to improve the numerical accuracy of small control command programs (e.g. PID and lead-lag controllers) and numerical procedures (trapeze rule and Runge-Kutta methods [8]). We have also demonstrated the efficiency of our tool to optimize slightly larger codes like a rocket trajectory simulation code of about $\mathcal{O}(100)$ lines of code [9].

Our main contribution in this article is to show that our technique improves the execution time of programs by increasing their numerical accuracy. By optimizing programs to be more accurate, we accelerate their convergence speed. In order to demonstrate the impact of the accuracy on the convergence time, we have chosen a set of four representative iterative methods which are Jacobi's and Newton-Raphson's method, a method to compute the largest Eigenvalue and Gram-Schmidt's method. Significant speedups are obtained in terms of number of iterations, time and total number of floating-point operations (flops).

In Sect. 2, we discuss how we compute the error on the numerical accuracy as well as the basic techniques used to rewrite programs. In Sect. 3, we detail the programs that we want to optimize. We give the programs before and after optimization together with experimental results. We conclude in Sect. 4.

2 Program Transformation for Numerical Accuracy

In this section, we first introduce the method that we use to compute the errors on the numerical accuracy. In Sects. 2.2 and 2.3, we also recall the transformation techniques used to optimize the numerical accuracy of expressions and programs and which are detailed in [8,15]. All the material introduced in this section is used in the tool that we use to optimize the programs of Sect. 3.

2.1 Floating-Point Arithmetic and Error Bound Computation

The floating-point arithmetic is defined by the IEEE754 Standard [2,23]. Floating-point numbers are used to encode real numbers. However, because they are a finite representation of their mathematical cousins, roundoff errors arise during computations. A floating-point number x is defined by

$$x \approx s_x \cdot (x_0.x_1 \ldots x_{p-1}) \cdot b^{e_x} = s_x \cdot m_x \cdot b^{e_x} \qquad (1)$$

where $s_x \in \{-1, 1\}$ is the sign, $m_x = x_0 x_1 \ldots x_{p-1}$ is the mantissa with digits $0 \leq x_i < b$, $0 \leq i \leq p-1$, p is the precision and e_x is the exponent, $e_{min} \leq e_x \leq e_{max}$. The IEEE754 Standard specifies several formats for the floating-point numbers by providing specific values for p, b, e_{min} and e_{max}. It also defines some rounding modes, towards $+\infty$, $-\infty$, 0 and to the nearest. Our transformation technique, introduced in Sects. 2.2 and 2.3, is independent of the selected rounding mode and, in this article, we assume that all the floating-point computations are done by using the rounding mode to the nearest. Let us write $\uparrow_{+\infty}$, $\uparrow_{-\infty}$, $\uparrow_0$ and $\uparrow_\sim$ the rounding functions, the IEEE754 Standard defines the semantics of the elementary operations by:

$$x \circledast_r y = \uparrow_r (x * y) \tag{2}$$

where $\circledast_r \in \{+, -, \times, \div\}$ is computed by using the rounding mode r and $* \in \{+, -, \times, \div\}$ denotes an exact operation. Because of the roundoff errors, the results of the computations are not exact. For example, let us consider two functions f and g which are mathematically equivalent. We have $f(x) = x^2 - 2.0 \times x + 1.0$ and $g(x) = (x - 1.0) \times (x - 1.0)$. If we compute $f(0.999)$ we get $1.00000000002875566e^{-6}$ and if we compute g of the same value, we obtain $1.00000000000000186e^{-6}$. On small computations, we have obtained already different results.

We present now the computation of errors on the numerical accuracy of arithmetic expressions [19]. These errors are stored in an abstract value [5] using a pair of intervals. The first interval contains the range of the floating-point values of the program, and the second one contains the range of the errors obtained by subtracting the floating-point values from the exact ones. In the abstract value denoted by $(x^\sharp, \mu^\sharp) \in E^\sharp$, we have $x^\sharp$ the interval corresponding to the range of the values and $\mu^\sharp$ the interval of errors on $x^\sharp$. This value $x^\sharp$ abstracts a set of concrete values $\{(x, \mu) : x \in x^\sharp \text{ and } \mu \in \mu^\sharp\}$ by intervals in a component-wise way. We introduce now the semantics of arithmetic expressions on $E^\sharp$. We approximate an interval $x^\sharp$ with real bounds by an interval based on floating-point bounds, denoted by $\uparrow_\sim^\sharp (x^\sharp)$. Here bounds are rounded to the nearest (see Eq. (3)).

$$\uparrow_\sim^\sharp [(\underline{x}, \overline{x})] = [\uparrow_\sim (\underline{x}), \uparrow_\sim (\overline{x})]. \tag{3}$$

In the other direction, we have the function $\downarrow_\sim^\sharp$ that abstracts the concrete function $\downarrow_\sim$. It computes the exact value of the error $\downarrow_\sim (x) = x - \uparrow_\sim (x)$. Every error associated to $x \in [\underline{x}, \overline{x}]$ is included in $\downarrow_\sim^\sharp [(\underline{x}, \overline{x})]$. We have

$$\downarrow_\sim^\sharp [(\underline{x}, \overline{x})] = [-y, y] \quad \text{with} \quad y = \frac{1}{2} \text{ulp}(\max(|\underline{x}|, |\overline{x}|)). \tag{4}$$

Formally, the *unit in the last place*, denoted by ulp(x), consists of the weight of the least significant digit of the floating-point number x. Equations (5) and (6) give the semantics of the addition and multiplication over $E^\sharp$. If we sum two floating-point numbers, we may add the errors generated by the operands to the error produced by the roundoff of the result. When multiplying two floating-point numbers, the semantics is given by the development of $(x_1^\sharp + \mu_1^\sharp) \times (x_2^\sharp + \mu_2^\sharp)$.

$$(x_1^\sharp, \mu_1^\sharp) + (x_2^\sharp, \mu_2^\sharp) = (\uparrow_\sim^\sharp (x_1^\sharp + x_2^\sharp), \mu_1^\sharp + \mu_2^\sharp + \downarrow_\sim^\sharp (x_1^\sharp + x_2^\sharp)), \tag{5}$$

$$(x_1^\sharp, \mu_1^\sharp) \times (x_2^\sharp, \mu_2^{,\sharp}) = (\uparrow_\sim^\sharp (x_1^\sharp \times x_2^\sharp), x_2^\sharp \times \mu_1^\sharp + x_1^\sharp \times \mu_2^\sharp + \mu_1^\sharp \times \mu_2^\sharp + \downarrow_\sim^\sharp (x_1^\sharp \times x_2^\sharp)). \tag{6}$$

This semantics is used to select the most accurate expression, in the sense that it minimizes $\mu^\sharp$ during the transformation of expressions introduced in the next section. To analyze statically a full program, we use a standard abstract interpretation of commands [5,19] with the abstract domain $E^\sharp$ of values. Note that dynamical analyses have also been proposed recently [3].

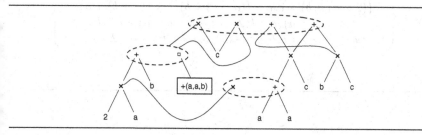

Fig. 1. APEG for the expression $e = ((a + a) + c) \times c$.

2.2 Transformation of Expressions

To introduce the transformation of arithmetic expressions, we consider variables $id \in \mathcal{V}$ with $\mathcal{V}$ a finite set, constants $cst \in \mathcal{F}$ with $\mathcal{F}$ the set of floating-point numbers and the operators $+$, $-$, $\times$ and $\div$. The syntax is

$$\text{Expr} \ni e :: = id \mid cst \mid e + e \mid e - e \mid e \times e \mid e \div e. \tag{7}$$

Here, we briefly present former work [15,20,25] to semantically transform [6] arithmetic expressions using Abstract Program Expression Graph (APEG). This data structure remains in polynomial size while dealing with an exponential number of equivalent expressions [21]. An APEG is defined inductively as follows: (1) A value v or a variable x is an APEG, (2) An expression $p_1 * p_2$ is an APEG, where p_1 and p_2 are APEGs and $*$ is a binary operator, (3) A box $\boxed{*(p_1, \ldots, p_n)}$ is an APEG, where $*$ is a commutative and associative operator and the p_i, $1 \leq i \leq n$, are APEGs and (4) A non-empty set $\{p_1, \ldots, p_n\}$, called equivalence class, of APEGs is an APEG where p_i, $1 \leq i \leq n$, is not a set of APEGs itself.

An example of APEG is given in Fig. 1. When an equivalence class (denoted by a dotted ellipse in Fig. 1) contains many *APEGs* $p_1, \ldots, p_n$ then one of the p_i $1 \leq i \leq n$ may be selected in order to build an expression. A box $\boxed{*(p_1, \ldots, p_n)}$ represents any parsing of the expression $p_1 * \ldots * p_n$. From an implementation point of view, when several equivalent expressions share a common subexpression, the latter is represented only once in the APEG. Then APEGs provide a compact representation of a set of equivalent expressions and make it possible to represent in an unique structure many equivalent expressions of very different shapes. For readability reasons, in Fig. 1, the leafs corresponding to the variables a, b and c are duplicated while, it practice, they are defined only once in the structure. The set $\mathcal{A}(p)$ of expressions contained inside an APEG p is defined inductively as follows: (1) If p is a value v or a variable x then $\mathcal{A}(p) = \{v\}$ or $\mathcal{A}(p) = \{x\}$, (2) If p is an expression $p_1 * p_2$ then $\mathcal{A}(p) = \bigcup_{e_1 \in \mathcal{A}(p_1),\ e_2 \in \mathcal{A}(p_2)} e_1 * e_2$,

(3) If p is a box $\boxed{*(p_1, \ldots, p_n)}$ then $\mathcal{A}(p)$ contains all the parsings of $e_1 * \ldots * e_n$ for all $e_1 \in \mathcal{A}(p_1), \ldots, e_n \in \mathcal{A}(p_n)$ and (4) If p is an equivalence class $\{p_1, \ldots, p_n\}$ then $\mathcal{A}(p) = \bigcup_{1 \leq i \leq n} \mathcal{A}(p_i)$.

For instance, the APEG p of Fig. 1 represents all the following expressions:

$$\mathcal{A}(p) = \left\{ \begin{array}{l} \big((a+a)+b\big) \times c, \big((a+b)+a\big) \times c, \big((b+a)+a\big) \times c, \\ \big((2 \times a)+b\big) \times c, c \times \big((a+a)+b\big), c \times \big((a+b)+a\big), \\ c \times \big((b+a)+a\big), c \times \big((2 \times a)+b\big), (a+a) \times c + b \times c, \\ (2 \times a) \times c + b \times c, b \times c + (a+a) \times c, b \times c + (2 \times a) \times c \end{array} \right\} \qquad (8)$$

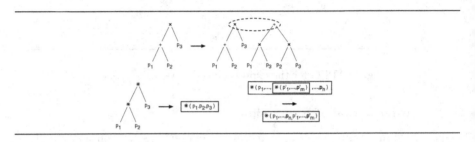

Fig. 2. Some rules for APEG construction by pattern matching.

In their article on EPEGs, R. Tate *et al.* use rewriting rules to extend the structure up to saturation [24, 25]. In our context, such rules would consist of performing some pattern matching in an existing APEG p and then adding new nodes in p, once a pattern has been recognized. For example, the rules corresponding to distributivity and box construction are given in Fig. 2. An alternative technique for APEG construction is to use dedicated algorithms. Such algorithms, working in polynomial time, have been proposed in [15].

2.3   Transformation of Commands

In this section, we focus on the transformation of commands which is done using a set of rewriting rules. Our language is made of assignments, conditionals, loops and sequences of commands. The syntax is

$$\text{Com} \quad \ni \quad c ::= id = e \mid c_1 \;;\; c_2 \mid \text{if}_\Phi \; e \; \text{then} \; c_1 \; \text{else} \; c_2 \mid \text{while}_\Phi \; e \; \text{do} \; c \mid \text{nop.} \qquad (9)$$

The transformation relies on several hypotheses. First of all, programs are assumed to be in static single assignment form (SSA form) [7]. The principle of this intermediary representation is that every variable may be assigned only once in the source code and must be initialized before its use. To understand this intermediary representation, let us consider the example of Fig. 3. In the original program, x is assigned several times. In the program in SSA form, a new variable x_1, x_2, etc. is used for each assignment and at the junction of control

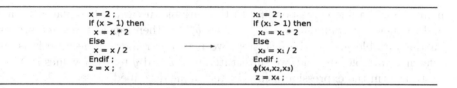

Fig. 3. Original program (left) and its SSA Form (right).

paths (in conditionals or loops), a Φ-node $\Phi(x_1, x_2, x_3)$ indicates that we assign to x_1 the value of x_2 or x_3 depending on where we are coming from.

The second hypothesis is that we optimize a reference variable defined by the user. Our transformation is defined by rules using states $\langle c, \delta, C, \nu, \beta \rangle$ where:

- c is a command, as defined in Eq. (9),
- δ is an environment $\delta : \mathcal{V} \to$ Expr which maps variables to expressions. Intuitively, this environment records the expressions assigned to variables in order to inline them later on in larger expressions,
- $C \in$ Ctx is a single hole context [12]. It records the program enclosing the current expression to be transformed,
- $\nu \in \mathcal{V}$ denotes the reference variable that we aim at optimizing,
- $\beta \subseteq \mathcal{V}$ is a list of assigned variables that should not be removed from the code. Initially, $\beta = \{\nu\}$, i.e., the target variable ν must not be removed.

The environment δ is used to discard assignments from programs and to re-insert the expressions when the variables are read, in order to build larger expressions.

Let us consider first assignments. If (i) the variable v of some assignment $v = e$ does not exist in the domain of δ and (ii) $v \notin \beta$ and (iii) $v \neq \nu$ then we memorize e in δ and we remove the assignment from the program. Otherwise, if one of the conditions (i), (ii) or (iii) is not satisfied then we rewrite this assignment by inlining the variables saved in δ in the concerned expression. Note that, when transforming programs by inlining expressions in variables, we get larger formulas. The basic idea, in our implementation, when dealing with too large expressions, is to create intermediary variables and to assign to them the sub-expressions obtained by slicing the global expression at a given level of the syntactic tree. The last step consists of re-inserting these intermediary variables into the main program.

For example, let us consider the program below in which three variables x, y and z are assigned. We assume that z is the variable that we aim at optimizing and $a = 0.1$, $b = 0.01$, $c = 0.001$ and $d = 0.0001$ are constants.

$$
\begin{aligned}
& \langle x = a+b \ ; \ y = c+d \ ; \ z = x+y \ , \ \delta, \ [], \ \nu = z, \ [z] \rangle \\
\longrightarrow \ & \langle \text{nop} \ ; \ y = c+d \ ; \ z = x+y, \delta' = \delta[x \mapsto a+b], \ [], \ \nu = z, \ [z] \rangle \\
\longrightarrow \ & \langle \text{nop} \ ; \ \text{nop} \ ; \ z = x+y, \delta'' = \delta'[y \mapsto c+d], \ [], \ \nu = z, \ [z] \rangle \\
\longrightarrow \ & \langle \text{nop} \ ; \ \text{nop} \ ; \ z = ((d+c)+b)+a, \ \delta'', \ [], \ \nu = z, \ [z] \rangle
\end{aligned}
\tag{10}
$$

In Eq. (10), the environment δ and the context C are initially empty and the list β contains the reference variable z. We remove the variable x and memorize it in δ. So, the line corresponding to the variable discarded is replaced by nop and the new environment is $\delta = [x \mapsto a + b]$. We then repeat the same process on the variable y. For the last step, we may not remove z because it is the reference variable. Instead, we substitute, in z, x and y by their values in δ and we transform the expression using the technique described in Sect. 2.2.

Our tool also transforms conditionals. If a certain condition is always true or false, then we keep only the right branch, otherwise, we transform both branches of the conditional. When it is necessary, we re-inject variables that have been discarded from the main program. Let us take another example to explain how we transform conditionals.

$$x_1 = 0; \; \text{if}_{\Phi(y_3, y_1, y_2)} \; x_1 > 1 \; \text{then} \; y_1 = x_1 + 2; \; \text{else} \; y_2 = x_1 - 1; \nu = y_3. \quad (11)$$

First of all, x_1 is stored in δ. Then, we transform recursively the new program

$$\text{if}_{\Phi(y_3, y_1, y_2)} \; x_1 > 1 \; \text{then} \; y_1 = x_1 + 2; \; \text{else} \; y_2 = x_1 - 1; \nu = y_3. \quad (12)$$

This program is semantically incorrect since the test is undefined. So we re-inject the statement $x_1 = 0$ in the program and add x_1 to the list β in order to avoid an infinite loop in the transformation.

For a sequence $c1; \; c2$, the first command c_1 is transformed into c_1' in the current environment δ, C, ν and β and a new context C' is built which inserts c_1' inside C. Then c_2 is transformed into c_2' using the context $C[c_1'; []]$, the formal environments δ' and the list β' resulting from the transformation of c_1. Finally, the state $\langle c_1' \; ; \; c_2', \delta'', \beta'' \rangle$ is returned.

Other transformations have been defined for while loops. A first rule makes it possible to transform the body of the loop assuming that the variables of the condition have not been stored in δ. In this case, the body is optimized in the context $C[\text{while}_\Phi \, e \, \text{do} \, []]$ where C is the context of the loop. A second rule builds the list $V = Var(e) \cup Var(\Phi)$ where $Var(\Phi)$ is the list of variables read and written in the Φ nodes of the loop. The set V is used to achieve two tasks: firstly, it is used to build a new command c' corresponding to the sequence of assignments that must be re-inserted. Secondly, the variables of V are removed from the domain of δ and added to β. The resulting command is obtained by transforming $c'; \text{while}_\Phi \, e \, \text{do} \, c$ with δ' and $\beta \cup V$.

3 Case Studies

In this section, we consider four iterative programs performing numerical computations: Jacobi's Method, Newton-Raphson's Method, an Iterated Power Method used to compute the largest eigenvalue of a matrix and an iterative orthogonalization algorithm more stable than Gram-Schmidt Method. We demonstrate the efficiency of our techniques in accelerating the convergence of these algorithms by measuring the number of iterations before and after rewriting. We present speedups in terms of execution time and number of floating-point operations needed to achieve the computation.

We have implemented the original and optimized numerical iterative methods in the C programming language, compiled with GCC 4.2.1, and made them run on an Intel Core i5 with 4 Go memory in IEEE754 single precision in order to emphasize the effect of the finite precision. Programs are compiled with the default optimization level $-O2$. We have tried other levels of optimization without observing significant changes in our results.

3.1   Linear Systems of Equations

We start with a first case study concerning Jacobi's method [17] which consists of an iterative computation that solves linear systems of the form $\mathbf{A}\mathbf{x} = \mathbf{b}$. From this equation, we build a sequence of vectors $(\mathbf{x}^{(0)}, \mathbf{x}^{(1)}, \dots, \mathbf{x}^{(k)}, \mathbf{x}^{(k+1)}, \dots)$ that converges towards the solution $\mathbf{x}^{(k)}$ of the system of linear equations.

To build the algorithm corresponding to this method, we decompose the initial matrix $\mathbf{A}$ into three matrices. The first one $\mathbf{D}$ contains the diagonal terms a_{ii} of the matrix. The second $\mathbf{U}$ contains the terms of the matrix which are above the main diagonal of $\mathbf{A}$ (a_{ij} with $j > i$) and the last one $\mathbf{L}$ contains the remaining terms of $\mathbf{A}$, i.e., the terms that are below the main diagonal (a_{ij} with $j < i$).

So, after transforming the matrix $\mathbf{A}$, we have the following equation to solve $\mathbf{D}\mathbf{x} = \mathbf{b} - (\mathbf{L} + \mathbf{U})\mathbf{x}$.

To compute $\mathbf{x}^{(k+1)}$, we use:

$$x_i^{(k+1)} = \frac{b_i - \sum_{j=1, j \neq i}^{n} a_{ij} x_j^{(k)}}{a_{ii}} \quad \text{where } \mathbf{x}^{(k)} \text{ is known.} \tag{13}$$

The method iterates until $|x_i^{(k+1)} - x_i| < \epsilon$ for the desired x_i, $1 \leq i \leq n$. A sufficient condition for the stability of Jacobi's method is that

$$|a_{ii}| > \sum_{j=1, j \neq i}^{n} |a_{ij}|. \tag{14}$$

Let us now examine how we can improve the convergence of Jacobi's method on the example given in Eq. (15). This system is stable with respect to the sufficient condition of Eq. (14) but it is close to be unstable in the sense that $\forall i, 1 \leq i \leq 4, |a_{ii}| \approx \sum_{j=1, j \neq i}^{j=4} |a_{ij}|$.

$$\begin{pmatrix} 0.62 & 0.1 & 0.2 & -0.3 \\ 0.3 & 0.602 & -0.1 & 0.2 \\ 0.2 & -0.3 & 0.6006 & 0.1 \\ -0.1 & 0.2 & 0.3 & 0.601 \end{pmatrix} \cdot \begin{pmatrix} x_1 \\ x_2 \\ x_3 \\ x_4 \end{pmatrix} = \begin{pmatrix} 1.0/2.0 \\ 1.0/3.0 \\ 1.0/4.0 \\ 1.0/5.0 \end{pmatrix}. \tag{15}$$

We describe this system using the notations of Eq. (13). To solve Eq. (15) by Jacobi's method, we use the algorithm presented in Fig. 4. This program is transformed with our tool by using the set of transformation rules described in Sect. 3. Note that, in the version of this program given to our tool, we have unfolded the body of the while loop twice. This makes it possible to rewrite more drastically the code by mixing the computations of both iterations. In this example, without unfolding, we win very few iterations and, obviously, if we unfold the body of the loop more than twice, our tool improves even more the accuracy at the price of a longer code. Note that in the examples of the next sections, we do not perform such an unfolding because our tool already optimizes significantly the original codes (results would be even better with unfolding).

```
eps = 10e-16; a11 = 0.61; a22 = 0.602; a33 = 0.6006; a44 = 0.601;
b1 = 0.5; b2 = 1.0/3.0; b3 = 0.25; b4 = 1.0/5.0;
while (e > eps) {
    x_n1 = (b1/a11) - (0.1/a11) * x2 - (0.2/a11) * x3 + (0.3/a11) * x4;
    x_n2 = (b2/a22) - (0.3/a22) * x1 + (0.1/a22) * x3 - (0.2/a22) * x4;
    x_n3 = (b3/a33) - (0.2/a33) * x1 + (0.3/a33) * x2 - (0.1/a33) * x4;
    x_n4 = (b4/a44) + (0.1/a44) * x1 - (0.2/a44) * x2 - (0.3/a44) * x3;
    e = x_n1 - x1; x1 = x_n1; x2 = x_n2; x3 = x_n3; x4 = x_n4; }
```

Fig. 4. Listing of the initial program of Jacobi's method.

The program corresponding to Jacobi's method after optimization is shown in Fig. 5. Note that this code is rather not intuitive and could very difficultly be written by hand. Concerning the accuracy of the variables, our tool states that the percentage of the optimization computed by the abstract semantics of Sect. 2 is up to 44.5 %. This means that the bound on the numerical error of the computed values of x_i, $1 \leq i \leq 4$ at any iteration is reduced by 44.5 %.

In Fig. 6, one can see the difference between the original and the transformed programs in term of the number of iterations needed to compute x_1, x_2, x_3 and x_4. Roughly speaking, about 15 % less iterations are needed with the optimized code. Obviously, the fact that the body of the loop is unfolded twice, in the optimized code is taken into account in the computation of the number of iterations needed to converge.

```
eps = 10e-16 ;
while (e > eps)  {
  TMP_1 = (0.553709856035437 - (x1 * 0.498338870431894)) ;
  TMP_2 = (0.166112956810631 * x3) ;
  TMP_6 = (0.333000333000333 * x1) ;
  x_n1 = (((0.819672131147541 - (0.163934426229508 * ((TMP_1 + TMP_2) - (0.332225913621263
        * x4)))) - (0.327868852459016 * (((0.416250416250416 - TMP_6) + (0.4995004995005 * x2))
        - (0.166500166500167 * x4)))) + (0.491803278688525 * (((0.332778702163062
        + (0.166389351081531 * x1)) - (0.332778702163062 * x2)) - (0.499168053244592 * x3)))) ;
  x_n2 = (((0.553709856035437 - (0.498338870431894 * x_n1)) + (0.166112956810631
        * (((0.416250416250416 - TMP_6) + (0.4995004995005 * x2)) - (0.166500166500167 * x4))))
        - (0.332225913621263 * (((0.332778702163062 + (0.166389351081531 * x1))
        - (0.332778702163062 * x2)) - (0.499168053244592 * x3)))) ;
  x_n3 = (((0.416250416250416 - (0.333000333000333 * x_n1)) + (0.4995004995005 * x_n2))
        - (0.166500166500167 * (((0.332778702163062 + (0.166389351081531 * x1))
        - (0.332778702163062 * x2)) - (0.499168053244592 * x3)))) ;
  x_n4 = (((0.332778702163062 + (0.166389351081531 * x_n1)) - (0.332778702163062 * x_n2))
        - (0.499168053244592 * x_n3)) ;
  e = (x_n4 - x4) ;   x1 = x_n1 ;   x2 = x_n2 ;   x3 = x_n3 ;   x4 = x_n4 ;  }
```

Fig. 5. Listing of the optimized program of Jacobi's method.

x_i	Initial Num of iteration	Iterations Num after optimization	Difference	Percentage
x_1	1891	1628	263	14.0
x_2	2068	1702	366	17.3
x_3	2019	1702	317	15.7
x_4	1953	1628	325	16.7

Fig. 6. Number of iterations of Jacobi's method before and after optimization to compute x_i, $1 \leq i \leq 4$.

3.2   Zero Finding

Newton-Raphson's Method [17] is a numerical method used to compute the successive approximations of the zeros of a real-valued function. In order to understand how this method works, let us start with the derivative $f'(x)$ of the function f which may be used to find the slope, and thus the equation of the tangent to the curve at a specified point. The method starts in an interval, for the equation $f(x) = 0$, in which there exists only one solution, the root a.

```
eps = 0.0005  ;   e = 1.0 ;  x = 0.0 ;
while (e > eps){
    f   = (x*x*x*x*x) - (10.0*x*x*x*x) + (40.0*x*x*x) - (80.0*x*x) + (80.0*x) - (32.0) ;
    ff  = (5.0*x*x*x*x) - (40.0*x*x*x) + (120.0*x*x) - (160.0*x) + (80.0) ;
    x_n = x - (f / ff) ;
    e = (x - x_n) ;   x = x_n ;
    if (e < 0.0) { e = (e * (-1.0)) ; } else { e = (e * 1.0) ; } ;    }
```

Fig. 7. Listing of the initial Newton-Raphson's program.

We choose a value u_0 close enough to a and then we build a sequence $(u_n)_{n \in \mathbb{N}}$ where u_{n+1} is obtained from u_n, as the abscissa of the meet point of the x-axis and the tangent at point $(u_n, f(u_n))$ to the function f. The final formula is given in Eq. (16). Note that the computation stops when $|u_{n-1} - u_n| < \epsilon$.

$$u_{n+1} = u_n - \frac{f(u_n)}{f'(u_n)}. \tag{16}$$

In general, Newton-Raphson's converges very quickly (quadratic convergence) but it may be slower if the computation of f or f' is inaccurate. For our case study, we have chosen functions which are difficult to evaluate in the IEEE754 floating-point arithmetic. Let us consider the function $f(x) = (x - 2)^5$. The developed formula of f and its derivative f' are:

$$f(x) = x^5 - 10x^4 + 40x^3 - 80x^2 + 80x - 32, \tag{17}$$

$$f'(x) = 5x^4 - 40x^3 + 120x^2 - 160x + 80. \tag{18}$$

It is well-known from floating-point arithmetic experts that evaluating the developed form of a polynomial close to a multiple root may be quite inaccurate [18]. Consequently, this example presents some numerical difficulties for Newton-Raphson's method which converges slowly in this case.

```
eps = 0.0005 ; e   = 1.0 ; x   = 0.0 ; x_n = 1.0 ;
while (e > eps){
  TMP_1 = (((((x * x) * x) * x) * x) - ((((10.0 * x) * x) * x) * x)) ;
  TMP_2 = ((x * x) * (40.0 * x)) ;
  TMP_3 = (80.0 * x) ;
  TMP_5 = (((5.0 * x) * x) * (x * x)) ;
  TMP_6 = ((x * x) * (40.0 * x)) ;
  TMP_7 = (120.0 * x) ;
  x_n = (x - ((((TMP_1 + TMP_2) - (TMP_3 * x)) + TMP_3) - 32.0)
        / (((TMP_5 - TMP_6) + (TMP_7 * x)) - (160.0 * x)) + 80.0))) ;
  e = (x - x_n) ;        x = x_n ;
  if (e < 0.0) { e = (e * (-1.0)) ; } else {  e = (e * 1.0) ;  } ; }
```

Fig. 8. Listing of the optimized Newton-Raphson's program.

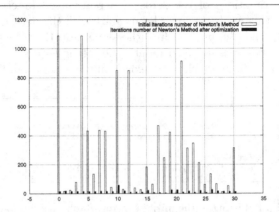

Fig. 9. Number of iterations of the Newton-Raphson's Method before and after optimization for initial values ranging from 0 to 3 (30 runs with a step of 0.1).

The algorithm corresponding to Eq. (16) is given in Fig. 7. We recognize the computation of $f(x)$ and its derivative $f'(x)$ called ff. When optimizing this program with our tool, we get the program of Fig. 8. The accuracy of the x_i's is improved up to 1.53 % following the semantics of Sect. 2.

The results given in Fig. 9 show how much our tool optimizes the number of iterations needed to converge. Obviously, this number of iterations needed to converge to the solution with a given precision depends on the initial value x_0. We have experimented several initial values. We make x_0 go from 0 to 3 with a step of 0.1. The 30 results are presented in Fig. 9. Due to the numerical inaccuracies, the number of iterations ranges from 10 to 1200, approximatively. It is always close to 10 with the transformed program.

3.3   Eigenvalue Computation

The Iterated Power Method is a method used to compute the largest eigenvalue of a matrix and the related eigenvector [10]. We start by fixing an arbitrary initial vector $\mathbf{x}^{(0)}$ containing a single non-zero element. Next, we build an intermediary vector $\mathbf{y}^{(1)}$ such that $\mathbf{A}\mathbf{x}^{(0)} = \mathbf{y}^{(1)}$. Then, we build $\mathbf{x}^{(1)}$ by re-normalizing $\mathbf{y}^{(1)}$ so that the selected component is again equal to 1. This process is repeated up to convergence. Optionally, we may change the reference vector if it converges to 0. Note that the renormalization factor converges to the largest eigenvalue and $\mathbf{x}$ converges to the related eigenvector, under the conditions that the largest eigenvalue is unique and that all eigenvectors are independent. The convergence speed is proportional to the ratio between the two largest eigenvalues (in absolute value). For our experiments, let us take a square matrix $\mathbf{A}$ of dimension 4 with the eigenvector $(0.0 \quad 0.0 \quad 0.0 \quad 1.0)^T$ given on the Eq. (15):

$$\mathbf{A} = \begin{pmatrix} d & 0.01 & 0.01 & 0.01 \\ 0.01 & d & 0.01 & 0.01 \\ 0.01 & 0.01 & d & 0.01 \\ 0.01 & 0.01 & 0.01 & d \end{pmatrix} \text{ with } d \in [175.0, 200.0]. \tag{19}$$

By applying the Iterated Power Method, the first intermediary vector is

$$\mathbf{A}\mathbf{x}^0 = \mathbf{y}^1, \quad \mathbf{A}^{y^1}/y_4^1 = \mathbf{y}^2, \quad \mathbf{A}^{y^2}/y_4^2 = \mathbf{y}^3, \quad \ldots \tag{20}$$

To re-normalize this intermediary vector, we divide it by the last value d, manner to have $y_4^{(1)}$ equal to 1.0. The new vector is: $(0.01/d \quad 0.01/d \quad 0.01/d \quad 1.0)^T$. We keep iterating with the new intermediary vector. We have: We repeat the former operation on this new intermediary vector in order to re-normalize it. By repeating this process several times, the series converges to the eigenvector $(1.0 \quad 1.0 \quad 1.0 \quad 1.0)^T$.

```
eps = 0.0005 ; d = 175.0 ; v1 = 0.0 ; v2 = 0.0 ; v3 = 0.0 ; v4 = 1.0 ; a41 = 0.01 ; a44 = d ;
a11 = d ; a12 = 0.01 ; a13 = 0.01 ; a14 = 0.01 ; a21 = 0.01 ; a22 = d ; a42 = 0.01 ; e = 1.0 ;
a23 = 0.01 ; a24 = 0.01 ; a31 = 0.01 ; a32 =  0.01 ; a33 = d ; a34 = 0.01 ; a43 = 0.01 ;
while (e > eps) {
    vx = a11 * v1 + a12 * v2 + a13 * v3 + a14 * v4 ;
    vy = a21 * v1 + a22 * v2 + a23 * v3 + a24 * v4 ;
    vz = a31 * v1 + a32 * v2 + a33 * v3 + a34 * v4 ;
    vw = a41 * v1 + a42 * v2 + a43 * v3 + a44 * v4 ;
    v1 = vx / vw ;  v2 = vy / vw ;  v3 = vz / vw ;  v4 =  1.0 ;  e = 1.0 - v1;
    if (v1 < 1.0) { e = 1.0 - v1 ;} else { e = v1 - 1.0 ;} }
```

Fig. 10. Listing of the Initial iterated power method.

Our tool has improved the error bounds computed by the semantics of Sect. 2.1 of up to 25.76 %. The optimized code is given in Fig. 11.

When running this program, we observe significant improved results. In other words, the transformed implementation succeeds to reduce the numbers of iterations needed to converge and accelerates the convergence speed of the iterative power method. The experimental results are summarized in Fig. 12.

3.4   Iterative Gram-Schmidt Method

The Gram-Schmidt method is used to orthogonalize a set of non-zero vectors in a Euclidean or Hermitian space $\mathcal{R}^n$. This method takes as input a linear

```
eps = 0.0005 ; d = 175.0 ; v1  =  0.0 ;  v2 =    0.0 ;   v3 =  0.0 ;  v4 =  1.0 ;  e = 1.0 ;
while (e > eps) {
   vx = ((((0.01 * v4) + (0.01 * v2)) + (0.01 * v3)) + (d * v1)) ;
   vy = ((((0.01 * v1) + (0.01 * v4)) + (0.01 * v3)) + (d * v2)) ;
   vz = ((((0.01 * v4) + (0.01 * v2)) + (0.01 * v1)) + (d * v3)) ;
   vw = ((((0.01 * v2) + (0.01 * v1)) + (0.01 * v3)) + (d * v4)) ;
   v1 = (vx / vw) ;  v2 = (vy / vw) ;  v3 = (vz / vw) ;   v4 = 1.0 ;   e = (1.0 - v1) ;
   if (v1 < 1.0) { e = 1.0 - v1 ;} else { e = v1 - 1.0 ;}      }
```

Fig. 11. Listing of the optimized iterated power method.

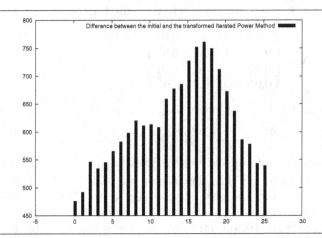

Fig. 12. Difference between numbers of iterations of initial and optimized Iterated Power Method (tests done for $d \in [175, 200]$ with a step of 1).

independent set of vectors $\mathbf{Q} = \{\mathbf{q}_1, \mathbf{q}_2, \ldots, \mathbf{q}_j\}$. The output is the orthogonal set of vectors $\mathbf{Q'} = \{\mathbf{q'}_1, \mathbf{q'}_2, \ldots, \mathbf{q'}_j\}$, with $1 \leq j \leq n$ [1, 10, 13]. The process followed by Gram-Schmidt method starts by defining the projection:

$$proj_{\mathbf{q'}}(\mathbf{q}) = \frac{\langle \mathbf{q}, \mathbf{q'} \rangle}{\langle \mathbf{q'}, \mathbf{q'} \rangle} \mathbf{q'}. \tag{21}$$

In Eq. (21), $\langle \mathbf{q}, \mathbf{q'} \rangle$ is the dot product of the vectors $\mathbf{q}$ and $\mathbf{q'}$. It means that the vector $\mathbf{q}$ is projected orthogonally onto the line spanned by the vector $\mathbf{q'}$. The normalized vectors are $\mathbf{e}_j = \frac{\mathbf{q'}_j}{||\mathbf{q'}_j||}$ where $||\mathbf{q'}_j||$ consists of the norm of the vector $\mathbf{q'}_j$. Explicitly, Gram-Schmidt process can be written as:

$$\mathbf{q'}_1 = \mathbf{q}_1,$$
$$\mathbf{q'}_2 = \mathbf{q}_2 - proj_{\mathbf{q'}_1}(\mathbf{q}_2),$$
$$\vdots$$
$$\mathbf{q'}_j = \mathbf{q}_j - \sum_{j=1}^{j-1} proj_{\mathbf{q'}_j}(\mathbf{q}_j).$$

In general, Gram-Schmidt method is numerically stable and it is not necessary to use an iterative algorithm. However, important numerical errors may arise when the vectors become more and more linearly dependent. In this case iterative algorithms yield better results, as for example the algorithm of Fig. 13 which repeats the orthogonalization step until the ratio $\frac{\|q'_j\|_2}{\|q_j\|_2}$ becomes large enough [13]. First, it starts by computing the orthogonal projection of $span(\{q_1, q_2, q_3\})$. Then, it substracts this projection from the original vector and then normalizes the result to obtain q_3, i.e., $span(\{q_1, q_2, q_3\})$ = $span(\{x_1, x_2, x_3\})$ and q_3 is orthogonal to q_1, q_2. We assume that $r_{jj} > 0$.

To understand how this algorithm works, let us take for example a set of vectors in R^3 that we aim at orthogonalizing.

$$Q_n = \left\{ q_1 = \begin{pmatrix} 1/7n \\ 0 \\ 0 \end{pmatrix}, q_2 = \begin{pmatrix} 0 \\ 1/25n \\ 0 \end{pmatrix}, q_3 = \begin{pmatrix} 1/2592 \\ 1/2601 \\ 1/2583 \end{pmatrix} \right\}. \quad (22)$$

For our experiments, we have chosen the values of n ranging from 1 to 10.

```
Q11 = 1.0 / 7n ; Q12 = 0.0  ; Q13 = 0.0  ;  Q21 = 0.0  ; Q22 = 1.0 / 25n ; Q23 = 0.0  ;
Q31 = 1.0 / 2592.0 ; Q32 = 1.0 / 2601.0 ;  Q33 = 1.0 / 2583.0 ; eps = 0.000005 ;
qj1 = Q31; qj2 = Q32;  qj3 = Q33;   r1 = 0.0; r2 = 0.0; r3 = 0.0;  e = 10.0 ;
r = qj1 * qj1 + qj2 * qj2 + qj3 * qj3 ; rold = sqrt(r) ;
while( e > eps ) {
    h1 = Q11 * qj1 + Q21 * qj2 + Q31 * qj3 ;
    h2 = Q12 * qj1 + Q22 * qj2 + Q32 * qj3 ;
    h3 = Q13 * qj1 + Q23 * qj2 + Q33 * qj3 ;
    qj1 = qj1 - (Q11 * h1 + Q12 * h2 + Q13 * h3) ;
    qj2 = qj2 - (Q21 * h1 + Q22 * h2 + Q23 * h3) ;
    qj3 = qj3 - (Q31 * h1 + Q32 * h2 + Q33 * h3) ;
    r1 = r1 + h1 ;  r2 = r2 + h2 ;  r3 = r3 + h3 ;
    r = qj1 * qj1 + qj2 * qj2 + qj3 * qj3 ;
    rjj = sqrt(r);
    e =  1.0 - (rjj / rold) ;
    if (e < 0.0) { e = -e ; };
    rold = rjj ;   }
```

Fig. 13. Listing of the initial iterative Gram-Schmidt program.

In Fig. 14, we give the transformed iterative Gram-Schmidt algorithm generated by our tool. By applying our techniques to the iterative Gram-Schmidt algorithm presented previously in Fig. 13, we show in Fig. 15 that the transformed algorithm converges faster than the initial one by up to 14.5 %.

3.5 Performance Analysis

We have shown in the former sections that we optimize the number of iterations of our four iterative numerical algorithms. In this section, we provide complementary benchmarks concerning speedups and the number of floating-point operations. Our objective is to check that the gains in the number of iterations

```
Q11 = 1.0 / 7n ; Q12 = 0.0  ; Q13 = 0.0  ; Q21 = 0.0  ; Q22 = 1.0 / 25n ;  Q23 = 0.0  ;
Q31 = 1.0 / 2592.0 ; Q32 = 1.0 / 2601.0 ;  Q33 = 1.0 / 2583.0 ; eps = 0.000005 ;
qj1 = Q31; qj2 = Q32;  qj3 = Q33;  r1 = 0.0; r2 = 0.0; r3 = 0.0; e = 10.0 ;
r = qj1 * qj1 + qj2 * qj2 + qj3 * qj3 ; rold = sqrt(r) ;
while ( e > eps)  {
  TMP_6 = (qj1 * qj3) ;
  TMP_14 = (qj2 * qj3) ;
  qj1 = (qj1 - ((0.14285714285 *(((qj1 * qj3)) + (0.14285714285 * qj1))));
  qj2 = (qj2 - ((0.04 * (((0.0 * qj1) + (qj2 * qj3)) + (0.04 * qj2))))) ;
  qj3 = (qj3 - (((qj2 * ((TMP_14) + (0.04 * qj2))) + (qj3 + (qj3 * qj3))))
        + (qj1 * (((qj1 * qj3)) + (0.14285714285 * qj1))))) ;
  r1 = (r1 + ((TMP_6) + (0.14285714285 * qj1))) ;
  r2 = (r2 + ((TMP_14) + (0.04 * qj2))) ;
  r3 = (r3 + ((qj3 * qj3))) ;
  r = qj1 * qj1 + qj2 * qj2 + qj3 * qj3 ;
  rjj = sqrt(r);
  e = 1.0 - (rjj / rold) ;
  if (e < 0.0) { e = -e ; };
  rold = rjj ;    }
```

Fig. 14. Listing of the optimized iterative Gram-Schmidt program.

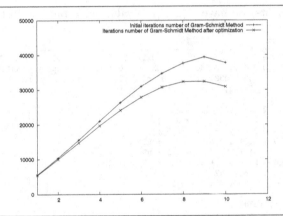

Fig. 15. Iterations number of initial and optimized iterative Gram-Schmidt Method for the family $(Q_n)_n$ of vectors, $1 \leq n \leq 10$.

are not annealed by overheads in the execution time of a single iteration or by other side effects for example due to the compiler.

We have chosen to observe the speedups of x_4 for Jacobi's method, and $x_0 = 3$ for Newton-Raphson's method. We have taken $d = 200$ for the iterated power method and $\mathbf{q}_{11} = \frac{1}{63}$ and $\mathbf{q}_{22} = \frac{1}{225}$ for iterative Gram-Schmidt method.

If we focus on measuring the execution time of the four programs before and after optimization, we observe that the percentage of improvement is rather important. If we take for example Jacobi's method, we remark that we reduce its execution time by 74.5 %. We give in Fig. 16 the speedups results obtained for the four methods. These results are very interesting to emphasize the usefulness of our tool and its ability to improve accuracy and execution time simultaneously.

	Original Code Execution Time in s	Optimized Code Execution Time in s	Percentage Improvement	Mean on n Runs
Jacobi	$1.49 \cdot 10^{-4}$	$0.38 \cdot 10^{-4}$	74.5%	10^4
Newton	$1.34 \cdot 10^{-3}$	$0.02 \cdot 10^{-3}$	98.4%	10^4
Eigenvalue	$4.50 \cdot 10^{-2}$	$3.07 \cdot 10^{-2}$	31.6%	10^3
Gram-Schmidt	$1.99 \cdot 10^{-1}$	$1.70 \cdot 10^{-1}$	14.5%	10^2

Fig. 16. Execution time measurements of programs of Sect. 3.

Method	♯ of ± per it Original Code	♯ of ± per it Optimized Code	Total ♯ of ± Original Code	Total ♯ of ± opt Optimized Code	Percentage of Improvement
Jacobi	13	15	25389	24420	3.81
Newton-Raphson	11	11	3465	132	96.19
Eigenvalue	15	15	694080	685995	1.16
Gram-Schmidt	21	19	791364	715996	9.52

Method	♯ of × per it Original Code	♯ of × per it Optimized Code	Total ♯ of × Original Code	Total ♯ of × opt Optimized Code	Percentage of Improvement
Jacobi	28	14	54684	22792	58.32
Newton-Raphson	27	26	8505	312	96.33
Eigenvalue	19	19	879168	868927	1.16
Gram-Schmidt	22	20	712316	647560	9.09

Fig. 17. Floating-point operations needed by programs of Sect. 3 to converge.

We have also counted the number of floating-point operations (flops) in the original and optimized codes. The numbers are given in Fig. 17. For each method, we count the number of additions and subtractions as well as the number of products and divisions for a single iteration and for the total number of iterations required in each case to converge. These results are coherent with the observed execution times.

4   Conclusion

This article focuses on the impact of automatic transformation of programs in order to improve their numerical accuracy on the convergence time of numerical iterative algorithms. Our experiments show the usefulness of our approach on the time required by numerical iterative methods to converge. We have experimented several representative numerical iterative methods by giving them to our tool and we have shown that the transformed programs converge more quickly than the original ones without loss of accuracy. We have extended this study with complementary results concerning the execution time and the total number of floating-point operations.

What remains to be done is to have a more complete tool implementing other programming language patterns like functions and pointers. In addition, it would be interesting to extend the current work with a case study concerning a real size numerical application. The study described in [9] is a first step in this direction. Another future work would consist in studying the impact of accuracy optimization on the convergence time of distributed numerical algorithms like the ones used usually for high performance computing. In addition, still about distributed systems, an important issue concerns the reproducibility of the results: different runs of the same application yield different results due to the variations in the order of evaluation of the mathematical expression. We would like to study how our technique could improve reproducibility.

References

1. Abdelmalek, N.: Roundoff error analysis for gram-schmidt method and solution of linear least squares problem. BIT **11**, 345–368 (1971)
2. ANSI/IEEE. IEEE Standard for Binary Floating-point Arithmetic, std 754-2008 (second edn) (2008)
3. Benz, F., Hildebrandt, A., Hack, S.: A dynamic program analysis to find floating-point accuracy problems. In: PLDI 2012, pp. 453–462. ACM (2012)
4. Consel, C., Danvy, O.: Partial evaluation of pattern matching in strings. Inf. Proc. Lett. **30**(2), 79–86 (1989)
5. Cousot, P., Cousot, R.: Abstract interpretation: a unified lattice model for static analysis of programs by construction of approximations of fixed points. In: POPL 1977, pp. 238–252. ACM (1977)
6. Cousot, P., Cousot, R.: Systematic design of program transformation frameworks by abstract interpretation. In: POPL 2002, pp. 178–190. ACM (2002)
7. Cytron, R., Gershbein, R.: Efficient accomodation of may-alias information in SSA form. In: PLDI 1993, pp. 36–45. ACM (1993)
8. Damouche, N., Martel, M., Chapoutot, A.: Intra-procedural optimization of the numerical accuracy of programs. In: Núñez, M., Güdemann, M. (eds.) Formal Methods for Industrial Critical Systems. LNCS, vol. 9128, pp. 31–46. Springer, Heidelberg (2015)
9. Damouche, N., Martel, M., Chaoutot, A.: Optimizing the accuracy of a rocket trajectory simulation by program transformation. In: Computing Frontiers, pp. 40:1–40:2. ACM (2015)
10. Golub, G.H., van Loan, C.F.: Matrix Computations, 3rd edn. Johns Hopkins University Press, Baltimore (1996)
11. Goubault, E., Putot, S.: Static analysis of finite precision computations. In: Jhala, R., Schmidt, D. (eds.) VMCAI 2011. LNCS, vol. 6538, pp. 232–247. Springer, Heidelberg (2011)
12. Hankin, E.: Lambda Calculi A Guide For Computer Scientists. Clarendon Press, Oxford (1994)
13. Hernandez, V., Roman, J.E., Tomas, A., Vidal, V.: Orthogonalization routine in SLEPc Technical Report STR-1. In: Polytechnic University of Valencia. STR1 (2007)
14. Hunt, S., Sands, D.: Binding time analysis: a new perspective. In: PEPM 1991, pp. 154–165 (1991)
15. Ioualalen, A., Martel, M.: A new abstract domain for the representation of mathematically equivalent expressions. In: Miné, A., Schmidt, D. (eds.) SAS 2012. LNCS, vol. 7460, pp. 75–93. Springer, Heidelberg (2012)
16. Jones, N.D., Gomard, C.K., Sestoft, P.: Partial Evaluation and Automatic Program Generation. Prentice Hall International, Englewood Cliffs (1993). ISBN 0-13-020249-5

17. Kendall, A.: An Introduction to Numerical Analysis. John Wiley & Sons, New York (1989)
18. Langlois, Ph., Louvet, N.: How to ensure a faithful polynomial evaluation with the compensated horner algorithm. In: ARITH-18, pp. 141–149. IEEE Computer Society (2007)
19. Martel, M.: Semantics of roundoff error propagation in finite precision calculations. Higher-Order Symbolic Comput. **19**(1), 7–30 (2006)
20. Martel, M.: Accurate evaluation of arithmetic expressions (invited talk). Electr. Notes Theor. Comput. Sci. **287**, 3–16 (2012)
21. Mouilleron, C.: Efficient computation with structured matrices and arithmetic expressions. Ph.D. thesis, Université de Lyon-ENS de Lyon, November 2011
22. Muller, G., Volanschi, E.-N., Marlet, R.: Scaling up partial evaluation for optimizing the sun commercial RPC protocol. In: PEPM 1997, pp. 116–126. ACM (1997)
23. Muller, J.-M., Brisebarre, N., de Dinechin, F., Jeannerod, C.-P., Lefèvre, V., Melquiond, G., Revol, N., Stehlé, D., Torres, S.: Handbook of Floating-Point Arithmetic, Birkhäuser Boston (2010)
24. Tate, R., Stepp, M., Tatlock, Z., Lerner, S.: Equality saturation: a new approach to optimization. In: POPL 2009, pp. 264–276. ACM (2009)
25. Tate, R., Stepp, M., Tatlock, Z., Lerner, S.: Equality saturation: a new approach to optimization. Log. Meth. Comput. Sci. **7**(1), 1–37 (2011)

Static Analysis

Abstract Analysis of Universal Properties for *tccp*

Marco Comini[1], María del Mar Gallardo[2], Laura Titolo[2(✉)],
and Alicia Villanueva[3]

[1] DIMI, Università Degli Studi di Udine, Udine, Italy
marco.comini@uniud.it
[2] LCC, Universidad de Málaga, Málaga, Spain
{gallardo,laura.titolo}@lcc.uma.es
[3] DSIC, Universitat Politècnica de València, Valencia, Spain
villanue@dsic.upv.es

Abstract. The Timed Concurrent Constraint Language (*tccp*) is a time extension of the concurrent constraint paradigm of Saraswat. *tccp* was defined to model reactive systems, where infinite behaviors arise naturally. In previous works, a semantic framework and abstract diagnosis method for the language has been defined.

On the basis of that semantic framework, this paper proposes an abstract semantics that, together with a widening operator, is suitable for the definition of different analyses for *tccp* programs. The abstract semantics is correct and can be represented as a finite graph where each node represents a hypothetical computational step of the program containing approximated information for the variables. The widening operator allows us to guarantee the convergence of the abstract fixpoint computation.

Keywords: Concurrent constraint paradigm · Abstract analysis

1 Introduction

The Concurrent Constraint Paradigm (*ccp*, [10]) is a simple, logic model which is different from other (concurrent) programming paradigms mainly due to the notion of store-as-constraint that replaces the classical store-as-valuation model. It is based on an underlying constraint system that handles constraints on variables and deals with partial information. Within this family, [6] introduced the *Timed Concurrent Constraint Language* (*tccp*) by adding to the original *ccp* model the notion of time and the ability to capture the absence of information. With these features, one can specify behaviors typical of reactive systems such as *timeouts* or *preemption* actions.

This work has been supported by the Andalusian Excellence Project P11-TIC7659. This work has been partially supported by the EU (FEDER) and the Spanish MINECO under grant TIN 2013-45732-C4-1-P (DAMAS) and by Generalitat Valenciana PROMETEOII/2015/013.

M. Falaschi (Ed.): LOPSTR 2015, LNCS 9527, pp. 163–178, 2015.
DOI: 10.1007/978-3-319-27436-2_10

It is well-known that modeling and analyzing concurrent systems by hand can be an extremely hard task. Thus, the development of automatic formal methods is essential. The particular characteristics of *ccp* languages make such task even harder, since we have to deal with technical issues due to the infinite computations (natural to reactive systems), use of negative information (particular for *ccp* languages) and non-determinism.

One well established technique to develop semantic-based program analysis is abstract interpretation [5], which relies on the definition of a specific approximated abstract semantics that captures the information needed to perform the analysis. Typically, one defines an over-approximation of the concrete semantics that includes all possible traces of the system, possibly introducing inexistent ones. This allows one to develop (correct) analysis of *universal* properties. It does not allow to analyze *existential* properties, for instance to verify that there exists a suspension trace. In our proposal, we follow such approach starting from the concrete semantics for *tccp* defined in [4]. This semantics addresses (with the minimal amount of information) all thorniest difficulties of *tccp* (i.e., infinite computations, use of negative information and non-determinism). To the best of our knowledge, this is the only bottom-up and condensed semantics which is fully abstract w.r.t. the *full tccp* language. Therefore, such semantics is particularly well-suited as the base to apply abstract interpretation techniques, which take great advantage from a bottom-up and condensed definition. The fully-abstract denotational semantics of [6] captures just finite computations and has a top-down definition thus it is not well-suited for our purposes.

We define a framework of over-approximated abstract semantics parametric w.r.t. an abstract constraint system. This allows us to recycle the work done for developing abstract domains for logic programs (such as groundness analysis). More interestingly, we can also make new analyses for reactive systems such as non-suspension analysis and universal (safety and liveness) properties. Since we need to preserve the notion of time—to be able to express properties of interest like safety or time-depending properties—the abstract semantics domains are not Noetherian (even if we use finite abstract constraint systems). Thus, in order to have an effective approach we use the widening approach of [2,5] to ensure finiteness of the analysis. Applicability of our approach is illustrated by showing different analyses over our guiding example, a lift/passenger system. More specifically, we show how properties such as *the lift direction and floor are consistently updated* or *the lift/passenger never suspends* can be analyzed.

2   The *tccp* language

The *tccp* language [6] is particularly suitable to specify both reactive and time critical systems. As the other languages of the *ccp* paradigm [10], it is parametric w.r.t. a *cylindric constraint system* which handles the data information of the program in terms of constraints. The computation progresses as the concurrent and asynchronous activity of several agents that can accumulate information in a *store*, or query information from it. A cylindric constraint system[1] is an algebraic structure $\mathbf{C} = \langle \mathcal{C}, \preceq, \otimes, \oplus, \mathit{false}, \mathit{true}, \mathit{Var}, \exists \rangle$ composed of a set of constraints $\mathcal{C}$ such that $(\mathcal{C}, \preceq)$ is a complete algebraic lattice where $\otimes$ is the *lub* operator, $\oplus$ is the *glb* operator and *false* and *true* are respectively the greatest and the least element of $\mathcal{C}$; *Var* is a denumerable set of variables and $\exists$ existentially quantifies variables over constraints. The *entailment* $\vdash$ is the inverse of $\preceq$.

Given a cylindric constraint system $\mathbf{C}$ and a set of process symbols Π, the syntax of agents is given by the grammar $A ::= \mathsf{skip} \mid \mathsf{tell}(c) \mid A \parallel A \mid \exists x\, A \mid \sum_{i=1}^{n} \mathsf{ask}(c_i) \to A \mid \mathsf{now}\ c\ \mathsf{then}\ A\ \mathsf{else}\ A \mid p(\vec{x})$ where $c, c_1, \ldots, c_n$ are finite constraints in $\mathbf{C}$; $p_{/m} \in \Pi$, and $\vec{x}$ denotes a generic tuple of m variables. A *tccp* program is an object of the form $D.A$, where A is an agent, called *initial agent*, and D is a set of *process declarations* of the form $p(\vec{x}){:}-A$ (for some agent A). The notion of time is introduced by defining a discrete and global clock.

The *operational semantics* of *tccp*, defined in [6], is formally described by a transition system $T = (\mathit{Conf}, \longrightarrow)$. Informally, the $\mathsf{tell}(c)$ agent adds the constraint c to the store in the next time instant and then stops. The choice agent $\sum_{i=1}^{n} \mathsf{ask}(c_i) \to A_i$ consults the store and non-deterministically executes (at the following time instant) one of the agents A_i whose corresponding guard c_i is entailed by the current store; otherwise, if no guard is entailed by the store, the agent suspends. The conditional agent now c then A else B behaves in the current time instant like A (resp. B) if c is (resp. is not) entailed by the store. $A \parallel B$ models the parallel composition of A and B in terms of maximal parallelism. The agent $\exists x\, A$ makes variable x local to A. To this end, it uses the $\exists$ operator of the constraint system. Finally, the agent $p(\vec{x})$ takes from D a declaration of the form $p(\vec{x}){:}-A$ and then executes A at the following time instant.

Example 1. The following code shows a possible *tccp* implementation of a simple lift/passenger system. We assume that the lift is located at a building with $N+1$ floors numbered as $0, 1, \cdots, N$. The lift process uses variables CF and Dir to store the current floor where the lift is placed and the movement direction (*up/down*), respectively. At each time instant, the lift moves, if possible, to the following floor, according to the current movement direction. When the lift reaches floors 0 or N, then it changes the movement direction. Process *pssngr* models the behavior of a client that wants to take the lift to go from origin floor O to destination floor D. This process makes use of variable St to store its state: *wait*, when it is waiting for the lift, *in*, when it is inside the lift and *out*, when the passenger has arrived at the destination floor. We use a simple constraint system composed of the atoms $\{up, down, in, out, wait\}$ and with arithmetic operations over the numbers $\{0, \ldots, N\}$. Due to the monotonicity of the store, streams (written in a list-fashion way) are used to model *imperative-style* variables [6].

$$main(N, O, D) : - \exists CF, Dir, St\, (lift(N, CF, Dir) \parallel pssngr(CF, O, D, St) \parallel$$
$$\mathsf{tell}(CF = [0 \mid \_]) \parallel \mathsf{tell}(Dir = [up \mid \_]) \parallel \mathsf{tell}(St = [wait \mid \_]))$$

$$lift(N, CF, Dir) : - \exists CF^l, Dir^l, F\, \big(\, \mathsf{now}(Dir = [up \mid \_] \wedge CF = [N \mid \_])$$

---

[1] See [6,10] for more details on cylindric constraint systems, where traditionally, the *glb* $\oplus$ is not explicitly defined.

$$\text{then } (\text{tell}\,(Dir = [up \mid Dir^l]) \parallel \text{tell}\,(Dir^l = [down \mid \_]) \parallel lift(N, CF, Dir^l))$$
$$\text{else now}\,(Dir = [up \mid \_]) \text{ then } (\text{tell}(CF = [F \mid CF^l]) \parallel$$
$$\text{ask}(true) \rightarrow (\text{tell}(CF^l = [F + 1 \mid \_]) \parallel lift(N, CF^l, Dir)))$$
$$\text{else now}\,(Dir = [down \mid \_] \wedge CF = [0 \mid \_])$$
$$\text{then } (\text{tell}(Dir = [down \mid Dir^l]) \parallel \text{tell}(Dir^l = [up \mid \_]) \parallel lift(N, CF, Dir^l))$$
$$\text{else } (\text{tell}(CF = [F \mid CF^l]) \parallel$$
$$\text{ask}(true) \rightarrow (\text{tell}(CF^l = [F - 1 \mid \_]) \parallel lift(N, CF^l, Dir))))$$

$$pssngr(CF, O, D, St) : -\,\exists St'\,\big($$
$$\text{ask}(CF = [D \mid \_] \wedge St = [in|\_]) \rightarrow (\text{tell}(St = [in \mid St']) \parallel \text{tell}(St' = [out \mid \_]))$$
$$+\,\text{ask}(CF = [O \mid \_] \wedge St = [wait|\_]) \rightarrow (\text{tell}(St = [wait|St']) \parallel \text{tell}(St' = [in \mid \_]) \parallel$$
$$\text{tell}(CF = [\_ \mid CF']) \parallel pssngr(CF', O, D, St'))$$
$$+\,\text{ask}((CF \neq [O \mid \_] \wedge CF \neq [D \mid \_]) \vee (CF = [D \mid \_] \wedge St \neq [in|\_])$$
$$\vee\,(CF = [O \mid \_] \wedge St \neq [wait|\_])) \rightarrow (\text{tell}(CF = [\_ \mid CF']) \parallel pssngr(CF', O, D, St')))$$

2.1 The Concrete Denotational Semantics

In this section, we briefly recall the concrete denotational domain and semantics of [4], which is fully-abstract (correct and complete) w.r.t. the small-step operational behavior of *tccp*. The denotational semantics of a *tccp* program [4] consists of a set of *conditional (timed) traces* that represent, in a compact way, all the possible behaviors that the program can manifest when fed with an *input* (initial store). Conditional traces can be seen as hypothetical computations in which, for each time instant, we have a condition representing the information that the global store has to satisfy in order to proceed to the next time instant.

Briefly, a conditional trace is a (possibly infinite) sequence $t_1 \cdots t_n \cdots$ of *conditional states*, which can be of three forms:

conditional store: a pair $\eta \rightarrowtail c$, where η is a *condition* and $c \in \mathbf{C}$ a store;
stuttering: the construct $stutt(C)$, with $C \subseteq \mathbf{C} \setminus \{true\}$;
end of a process: the construct $\boxtimes$.

Intuitively, the conditional store $\eta \rightarrowtail c$ means that, provided condition η is satisfied by the current store, the computation proceeds so that in the following time instant, the store is c. A *condition* η is a pair $\eta = (\eta^+, \eta^-)$ where $\eta^+ \in \mathbf{C}$ and $\eta^- \in \wp(\mathbf{C})$ are called positive and negative condition, respectively. The positive/negative condition represents information that a given store must/must not entail, thus they have to be consistent in the sense that $\forall c^- \in \eta^-\ \eta^+ \nvdash c^-$. The stuttering construct models the suspension of the computation when none of the guards in a non-deterministic agent is satisfied. C is the set of guards in the non-deterministic agent.

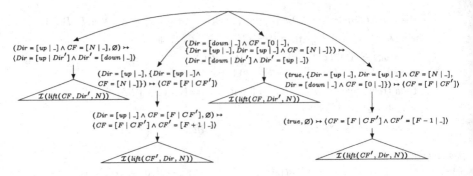

Fig. 1. Graph representation of the semantics of the *lift* process.

Conditional traces are monotone (i.e., for each $t_i = \eta_i \rightarrowtail c_i$ and $t_j = \eta_j \rightarrowtail c_j$ such that $j \geq i$, $c_j \vdash c_i$) and consistent (i.e., each store in a trace does not entail the negative conditions of the following conditional state). $\top$ is the set of all maximal conditional traces, i.e., infinite traces or finite traces ending with $\boxtimes$. With $C \sqsupseteq C' \iff \forall c \in C.\exists c' \in C'.c \vdash c'$ we order $\top$ as:

$$(\eta_1^+, \eta_1^-) \rightarrowtail c \cdot r_1 \sqsubseteq (\eta_2^+, \eta_2^-) \rightarrowtail c \cdot r_2 \iff \eta_1^+ \vdash \eta_2^+ \wedge \eta_2^- \sqsupseteq \eta_1^- \wedge r_1 \sqsubseteq r_2$$

$$stutt(\eta_1^-) \cdot r_1 \sqsubseteq stutt(\eta_2^-) \cdot r_2 \iff \eta_2^- \sqsupseteq \eta_1^- \wedge r_1 \sqsubseteq r_2$$

Intuitively, a trace r is smaller than another trace r' iff the conditions of r are more (or equally) restrictive than those of r'. We denote the domain of *maximal conditional trace sets* as $\mathbb{C}$. $(\mathbb{C}, \sqsubseteq, \bigsqcup, \bigsqcap, \top, \bot)$ is a complete lattice, where $M_1 \sqsubseteq M_2 \Leftrightarrow \forall r_1 \in M_1 \exists r_2 \in M_2.r_1 \sqsubseteq r_2$.

The concrete denotational semantics is based on a semantics evaluation function $\mathcal{A}[\![A]\!]_{\mathcal{I}}$ which, given an agent A and an interpretation $\mathcal{I}$, builds the conditional traces associated to A (defined in [4]). The interpretation $\mathcal{I}$ is a function which associates to each process symbol a set of maximal conditional traces "modulo variance".

Definition 1 (Interpretations). *Let* $\mathbb{PC} := \{p(\vec{x}) \mid \vec{x}$ *are distinct variables and* p *is a process symbol*$\}$. *An* interpretation *is a function* $\mathcal{I} \colon \mathbb{PC} \to \mathbb{C}$ *modulo variance*[2]. *The semantic domain* $\mathbb{I}$ *is the set of all interpretations ordered by the pointwise extension of* $\sqsubseteq$.

The semantics for a set of process declarations D is the fixpoint $\mathcal{F}[\![D]\!] := \text{lfp}(\mathcal{D}[\![D]\!])$ of the continuous operator $\mathcal{D}[\![D]\!]_{\mathcal{I}}(p(\vec{x})) := \bigsqcup_{p(\vec{x}):-A \in D} \mathcal{A}[\![A]\!]_{\mathcal{I}}$. Proof of full abstraction w.r.t. the operational behavior of *tccp* is given in [4].

Example 2 (Semantics of our guiding example). Consider the lift process defined in Example 1. We show in Fig. 1 its concrete semantics. Each branch of the tree corresponds to one of the branches of the nested now agents. The first branch

---

[2] Two functions $I, J \colon \mathbb{PC} \to \mathbb{C}$ are *variants*, denoted by $I \cong J$, if for each $\pi \in \mathbb{PC}$ there exists a variable renaming ρ such that $(I(\pi))\rho = J(\pi\rho)$.

(left-to-right order) represents the case in which the direction of the lift is *up* and the current floor is the last one (N). The second branch is taken when the direction is *up* but the current floor is not N (see the negative condition). In that case, the current floor changes from F to $F+1$. The third branch represents the case when the direction of the lift is *down* and the current floor is 0, thus the direction is changed to *up* by adding the constraint $Dir' = [up \mid \_]$. Finally, the fourth branch is taken when all the guards are not entailed (see the negative condition, composed by all the guards of the nested now agents). In that case, the lift moves to the lower floor $F-1$. In all the aforementioned cases, a recursive call is invoked appropriately. These calls are represented in Fig. 1 by the triangles labeled with the interpretation of the process *lift*.

3   The (finite) Abstract Semantics for *tccp*

In this section, we define our over-approximated abstract semantics for *tccp*. Our abstract semantics is parametric w.r.t. an approximation of the underlying constraint system.

The problem of abstracting constraint systems in the *ccp* paradigm was studied in [7,11], where abstraction meant loss of completeness but not of correctness. However, for the *tccp* case, due to the strong synchronization of parallel processes, over-approximation of stores could lead to lose correctness [1].

In our semantic domain, constraints are used in three components: in the positive part of the condition, in the negative part and in the store. Since these three components represent different information of a trace, we need to approximate them differently. Similarly to [1,3], we use both an over- and an under-approximation of the constraint system. The intuitive idea is that we approximate positive information (positive condition and store) with the over-approximation, whereas we approximate negative information with the under-approximation. This allows us to guarantee that we do not loose concrete behaviors when we abstract the semantics, i.e., it is ensured completeness of the abstract semantics. The over-approximating function $\tau^+ : \mathcal{C} \to \hat{\mathcal{C}}$ abstracts the constraint system $\mathcal{C}$ into an abstract one $\hat{\mathbf{C}} = \langle \hat{\mathcal{C}}, \hat{\preceq}, \hat{\otimes}, \hat{\oplus}, false, true, Var, \hat{\exists} \rangle$ where $true$ and $false$ are the smallest and the greatest abstract constraint, respectively. We often use the inverse relation $\hat{\vdash}$ of $\hat{\preceq}$. The under-approximating function $\tau^- : \wp(\mathcal{C}) \to \check{\mathcal{C}}$ abstracts the constraint system into another abstract constraint system $\check{\mathbf{C}} = \langle \check{\mathcal{C}}, \check{\preceq}, \check{\otimes}, \check{\oplus}, false, true, Var, \check{\exists} \rangle$. We have two "external" operations $\hat{\times} : \mathcal{C} \times \hat{\mathcal{C}} \to \hat{\mathcal{C}}$ and $\check{\times} : \mathcal{C} \times \check{\mathcal{C}} \to \check{\mathcal{C}}$ that update an abstract constraint with a concrete constraint (coming from the program).

Over and under-abstract constraints must satisfy the following properties.

$$c \,\hat{\times}\, \tau^+(a) = \tau^+(c \otimes a)$$
$$\tau^+(a \otimes b) = \tau^+(a) \,\hat{\otimes}\, \tau^+(b)$$
$$a \vdash b \Longrightarrow \tau^+(a) \,\hat{\vdash}\, \tau^+(b)$$
$$\tau^+(\exists_x a) = \hat{\exists}_x \tau^+(a)$$

$$c \,\check{\times}\, \tau^-(C) = \tau^-(\{c\} \cup C)$$
$$\tau^-(C \cup C') = \tau^-(C) \,\check{\oplus}\, \tau^-(C')$$
$$\tau^-(\{a\}) \,\check{\vdash}\, \tau^-(C) \Longrightarrow \exists c \in C. \, a \vdash c$$
$$\tau^-(\{\exists_x c \mid c \in C\}) = \check{\exists}_x \tau^-(C)$$

Moreover, they must be consistent, which means that the "bridge" relation $\tilde{\vdash} \in \hat{\mathcal{C}} \times \check{\mathcal{C}}$ must hold: $\forall c \in C.\ a \not\vdash c \Longrightarrow \tau^+(a) \tilde{\not\vdash} \tau^-(C)$.

Example 3 (Sign abstraction). Given the standard constraint system with inequalities over integer numbers, we abstract it to the abstract constraint system that contains only information about the sign of the system variables. We define the "positive" abstract constraint system as $\hat{\mathbf{S}} := \langle \mathcal{S}, \Leftarrow, \wedge, \vee, \mathit{false}, \mathit{true}, \mathit{Var}, \exists \rangle$ where $\mathcal{S}$ is the set of finite conjunctions of $\{\mathrm{pos}_x, \mathrm{neg}_x, \mathrm{zero}_x \mid x \in \mathit{Var}\} \cup \{\mathit{false},\ \mathit{true}\}$.

The abstract approximation τ^+ is defined by cases as follows:

$$\tau^+(\mathit{true}) = \mathit{true} \qquad \tau^+(x \leq a) = \begin{cases} \mathrm{neg}_x & \text{if } a \leq 0 \\ \mathit{true} & \text{if } a > 0 \end{cases} \qquad \tau^+(x \geq a) = \begin{cases} \mathrm{pos}_x & \text{if } a \geq 0 \\ \mathit{true} & \text{if } a < 0 \end{cases}$$

$$\tau^+(\mathit{false}) = \mathit{false} \qquad \tau^+(x = a) = \begin{cases} \mathrm{pos}_x & \text{if } a > 0 \\ \mathrm{neg}_x & \text{if } a < 0 \\ \mathrm{zero}_x & \text{if } a = 0 \end{cases}$$

Dually, we define $\check{\mathbf{S}} := \langle \mathcal{S}, \Rightarrow, \vee, \wedge, \mathit{true}, \mathit{false}, \mathit{Var}, \exists \rangle$, the "negative" abstract constraint system. The τ^- function is defined as $\tau^-(C) := \bigwedge_{c \in C} \tau'(c)$, where

$$\tau'(\mathit{true}) = \mathit{true} \qquad \tau'(x \leq a) = \begin{cases} \mathrm{neg}_x & \text{if } a \leq 0 \\ \mathit{false} & \text{if } a > 0 \end{cases} \qquad \tau'(x \geq a) = \begin{cases} \mathrm{pos}_x & \text{if } a \geq 0 \\ \mathit{false} & \text{if } a < 0 \end{cases}$$

$$\tau'(\mathit{false}) = \mathit{false} \qquad \tau'(x = a) = \begin{cases} \mathrm{pos}_x & \text{if } a > 0 \\ \mathrm{neg}_x & \text{if } a < 0 \\ \mathrm{zero}_x & \text{if } a = 0 \end{cases}$$

The abstract denotational model $\mathbb{A}$ is formed by *abstract conditional traces*, which are conditional traces where conditions and stores are formed by approximated constraints. An abstract conditional trace is said to be *valid* when all its abstract conditions are consistent. An abstract condition (c^+, c^-) is not consistent when $\tau^+(c^+) \tilde{\vdash} \tau^-(c^-)$.

It is worth noting that $(\mathbb{A}, \sqsubseteq, \bigsqcup, \bigsqcap, \mathbf{A}, \bot)$ is a complete lattice.

3.1   The Abstract Semantics

Now we are ready to define our abstraction approach which works in two steps: the first one abstracts information, and the second one *folds* suspending traces. Formally, concrete and abstract domains are related by the following functions:

$$(\mathbb{C}, \sqsubseteq) \xleftrightarrow[\alpha^{\mathcal{C}}]{\gamma^{\mathcal{C}}} (\mathbb{A}, \sqsubseteq) \xleftrightarrow[\mathit{fold}]{\mathit{unfold}} (\mathbb{A}, \sqsubseteq).$$

The abstraction function $\alpha^{\mathcal{C}}$ applies τ^+ to each positive condition and store and τ^- to each negative condition occurring in the considered trace. $\alpha^{\mathcal{C}}$ is parametric to the abstraction of the constraint system. The adjoint of $\alpha^{\mathcal{C}}$ is the concretization function $\gamma^{\mathcal{C}}$ that, given a set of abstract traces, produces all the concrete traces that can be approximated with it. For example, given a trace of the form $r = stutt\{X > 5\} \cdot (X > 6, \{Y < 0\}) \rightarrowtail X > 9$ the sign abstraction results in the abstract trace $\alpha^{\mathcal{C}}(r) = stutt(\{pos_X\}) \cdot (pos_X, \{neg_Y\}) \rightarrowtail pos_X$.

The second step of our abstract scheme, the *fold* abstraction, just collapses together all the consecutive identical $stutt(C)$ states. For example, consider $r = stutt(\{pos_X\}) \cdot stutt(\{pos_X\}) \cdot stutt(\{pos_X\})$, then $fold(r) = stutt(\{pos_X\})$. The adjoint of this abstraction function is the concretization $unfold$ which expands each state $stutt(C)$ into a sequence of repetitions of $stutt(C)$ of arbitrary length. Note that this second step does not guarantee termination of analysis since in *tccp* infinite behaviors are not only due to stuttering computations.

Lemma 1. $(\alpha^{\mathcal{C}}, \gamma^{\mathcal{C}})$, $(fold, unfold)$ *and their composition* $\alpha = \alpha^{\mathcal{C}} \circ fold$ *and* $\gamma = unfold \circ \gamma^{\mathcal{C}}$ *are Galois Insertions.*

Proof (Sketch). It is easy to see that both α and γ are monotonic functions, for all $M \in \mathbb{C}$, $(\gamma \circ \alpha)(M) \sqsupseteq M$, and $\alpha \circ \gamma$ is the identity for $\mathbb{A}$. $\qquad\square$

The Galois insertion defined before can be naturally lifted to the domain of interpretations. We denote as $\mathbb{I} := [\mathbb{PC} \to \mathbb{A}]$ the abstract counterpart of $\mathbb{I}$.

The abstract semantics for a *tccp* program is based on the evaluation function for *tccp* agents defined below. In order to improve readability, we have lighten the definition by omitting some technical details. However, we still need some auxiliary operators and properties, intuitively described below. Their formal definitions are similar to those in [4].

Given a trace r and a constraint c, $r\downarrow_c$ denotes the propagation of c in the positive conditions occurring in r. The *hiding operator* $\bar{\exists}: Var \times \mathbf{A} \to \mathbf{A}$ hides the information regarding a given variable in a trace. It uses $\hat{\exists}$ and $\check{\exists}$ to hide the information in the positive and in the negative conditions and stores. The $\bar{\|}$ operator composes two traces by consistently merging their conditions and stores. A trace r is said to be *self-sufficient* if the first condition is $(true, \emptyset)$ and each store satisfies the successive condition. Moreover, r is *x-self-sufficient* if $\bar{\exists}_{Var\setminus\{x\}} r$ is self-sufficient. In other words, for self-sufficient conditional traces, no additional information (from other agents) is needed in order to *complete* the computation.

Definition 2 (Abstract Semantics Evaluation Function for Agents).
Given a tccp agent A *and an (abstract) interpretation* $\mathcal{I}^{\alpha} \in \mathbb{I}_{\mathbb{A}}$, *we define the semantics evaluation* $\mathcal{A}^{\alpha}[\![A]\!]_{\mathcal{I}^{\alpha}} \in \mathbb{A}$ *by structural induction as follows.*

$$\mathcal{A}^{\alpha}[\![\mathsf{skip}]\!]_{\mathcal{I}^{\alpha}} := \{\boxtimes\} \tag{3.1}$$

$$\mathcal{A}^{\alpha}[\![\mathsf{tell}(c)]\!]_{\mathcal{I}^{\alpha}} := \{(\hat{true}, \check{false}) \rightarrowtail \tau^+(c) \cdot \boxtimes\} \tag{3.2}$$

$$\mathcal{A}^{\alpha}[\![A \parallel B]\!]_{\mathcal{I}^{\alpha}} := \bigsqcup \{r_A \bar{\|} r_B \mid r_A \in \mathcal{A}^{\alpha}[\![A]\!]_{\mathcal{I}^{\alpha}}, r_B \in \mathcal{A}^{\alpha}[\![B]\!]_{\mathcal{I}^{\alpha}}\} \tag{3.3}$$

$$\mathcal{A}^{\alpha}[\![\exists x\, A]\!]_{\mathcal{I}^{\alpha}} := \bigsqcup \{\bar{\exists}_x r \mid r \in \mathcal{A}^{\alpha}[\![A]\!]_{\mathcal{I}^{\alpha}}, r \text{ is x-self-sufficient}\} \tag{3.4}$$

$$\mathcal{A}^\alpha[\![p(\vec{x})]\!]_{\mathcal{I}^\alpha} := \{(tr\hat{u}e, fa\check{l}se) \rightarrowtail tr\hat{u}e \cdot r | r \in \mathcal{I}^\alpha(p(\vec{x}))\} \tag{3.5}$$

$$\mathcal{A}^\alpha[\![\sum_{i=1}^{n} \mathsf{ask}(c_i) \rightarrow A_i]\!]_{\mathcal{I}^\alpha} := M \sqcup \{stutt(\tau^-(\{c_1, \ldots, c_n\})) \cdot r | r \in M\} \sqcup \tag{3.6}$$
$$\{stutt(\tau^-(\{c_1, \ldots, c_n\}))\}$$

$$where\ M = \bigsqcup \{(\tau^+(c_i), fa\check{l}se) \rightarrowtail tr\hat{u}e \cdot (r\!\downarrow_{\tau^+(c_i)}) | 1 \le i \le n, r \in \mathcal{A}^\alpha[\![A_i]\!]_{\mathcal{I}^\alpha}\}$$

$\mathcal{A}^\alpha[\![\mathsf{now}\ c\ \mathsf{then}\ A\ \mathsf{else}\ B]\!]_{\mathcal{I}^\alpha} :=$

$$\{(\tau^+(c), fa\check{l}se) \rightarrowtail tr\hat{u}e \cdot \boxtimes | \boxtimes \in \mathcal{A}[\![A]\!]_{\mathcal{I}^\alpha}\} \sqcup \tag{3.7a}$$

$$\bigsqcup \{(c \hat{\times} \hat{\eta}, \check{\eta}) \rightarrowtail \hat{d} \cdot (r\!\downarrow_{\tau^+(c)}) | (\hat{\eta}, \check{\eta}) \rightarrowtail \hat{d} \cdot r \in \mathcal{A}^\alpha[\![A]\!]_{\mathcal{I}^\alpha}, c \hat{\times} \hat{\eta} \check{\nvdash} \check{\eta}\} \sqcup \tag{3.7b}$$

$$\bigsqcup \{(\tau^+(c), \check{\eta}) \rightarrowtail tr\hat{u}e \cdot r\!\downarrow_{\tau^+(c)}) | stutt(\check{\eta}) \cdot r \in \mathcal{A}^\alpha[\![A]\!]_{\mathcal{I}^\alpha}, c \check{\nvdash} \check{\eta}\} \sqcup \tag{3.7c}$$

$$\bigsqcup \{(tr\hat{u}e, \tau^-(\{c\})) \rightarrowtail tr\hat{u}e \cdot \boxtimes | \boxtimes \in \mathcal{A}^\alpha[\![B]\!]_{\mathcal{I}^\alpha}\} \sqcup \tag{3.7d}$$

$$\bigsqcup \{(\hat{\eta}, \{c\} \check{\times} \check{\eta}) \rightarrowtail \hat{d} \cdot r | (\hat{\eta}, \check{\eta}) \rightarrowtail \hat{d} \cdot r \in \mathcal{A}^\alpha[\![B]\!]_{\mathcal{I}^\alpha}, \hat{\eta} \hat{\nvdash} \tau^-(\{c\})\} \sqcup \tag{3.7e}$$

$$\bigsqcup \{(tr\hat{u}e, \{c\} \check{\times} \check{\eta}) \rightarrowtail tr\hat{u}e \cdot r | stutt(\check{\eta}) \cdot r \in \mathcal{A}^\alpha[\![B]\!]_{\mathcal{I}^\alpha}\} \tag{3.7f}$$

Note that all the operations regarding the positive part of conditions and the stores are abstracted with the τ^+ abstraction, whereas all the definitions for the negative condition use the τ^- abstraction.

We explain in more detail some significant cases. The semantics of the tell (c) agent (3.1) has a trace with two conditional states: the first one with condition $(tr\hat{u}e, fa\check{l}se)$, which is the less demanding condition since c must be added to the store in any case (in the next time instant). Next, the computation terminates with the end-of-process symbol $\boxtimes$. The parallel, hiding and process call cases are defined like in the concrete semantics. The semantics for the non-deterministic choice (3.6) collects, for each guard c_i, a conditional trace of the form $(\tau^+(c_i), fa\check{l}se) \rightarrowtail tr\hat{u}e \cdot (r\!\downarrow_{\tau^+(c_i)})$. This trace requires that $\tau^+(c_i)$ has to be satisfied by the current store. Then, the constraint $\tau^+(c_i)$ is propagated to the conditions in trace r (the continuation of the computation, which belongs to the semantics of A_i). Furthermore, we collect the stuttering traces, which correspond to the case when the computation suspends. These traces are of the form $stutt(\tau^-(\{c_1, \ldots, c_n\})) \cdot r$ where r is one of the traces above.

The semantics for a set of process declarations D is the fixpoint $\mathcal{F}^\alpha[\![D]\!] := \mathsf{lfp}$ $(\mathcal{D}^\alpha[\![D]\!])$ of the continuous operator $\mathcal{D}^\alpha[\![D]\!]_{\mathcal{I}^\alpha}(p(\vec{x})) := \bigsqcup_{p(\vec{x}): - A \in D} \mathcal{A}^\alpha[\![A]\!]_{\mathcal{I}^\alpha}$. It can be shown that $\mathcal{A}^\alpha$ and $\mathcal{D}^\alpha$ are, the optimal abstractions of $\mathcal{A}$ and $\mathcal{D}$, i.e., $\mathcal{A}^\alpha[\![A]\!] = \alpha \circ \mathcal{A}[\![A]\!] \circ \gamma$ and $\mathcal{D}^\alpha[\![D]\!] = \alpha \circ \mathcal{D}[\![D]\!] \circ \gamma$. Hence, abstract interpretation theory ensures that $\mathcal{F}^\alpha[\![D]\!]$ is the best correct approximation of $\mathcal{F}[\![D]\!]$.

3.2   From Infinite to Finite Semantics

Since the domain of abstract conditional traces is not Noetherian (i.e., it admits infinite increasing chains), the abstract least fixpoint does not necessarily converge in finite time. Our solution is to use a *widening operator* [2,5] that ensures the convergence of the abstract fixpoint in a finite number of steps.

In the following, we use a representation of sets of abstract conditional traces in terms of *conditional graphs*. These graphs are enriched with the information about the process calls, which is necessary to identify the part of the graph corresponding to each iteration of $\mathcal{D}^\alpha[\![D]\!]$ at the moment of applying the widening operator.

Definition 3. *A conditional graph G is a triple $(Init, Nodes, Edges)$ where*

- *Init is the set of initial nodes, each one labeled with a (unique) process symbol, denoted by $\mathrm{init}(G)$*
- *Nodes is a set of nodes, each one containing a conditional step, and*
- *Edges is a set of edges between nodes that can be of two kinds: either simple edges $n \to n'$, or edges of the form $n \overset{\rho}{\underset{p}{\Rightarrow}} n'$ representing a call to process p with variable renaming ρ. Edges represent the passage of one time unit.*

G *denotes the set of all conditional graphs. Moreover, $n \nrightarrow$ denotes a node n that has no outgoing edges.*

We define the function $paths : \mathbf{G} \to \mathbf{A}$ which, given a conditional graph, returns the set of all paths of the graph. When an arc of the form $\overset{\rho}{\underset{p}{\Rightarrow}}$ is traversed, a variant with fresh variables in the co-domain of the renaming ρ is applied to the nodes that follow in the path and the information of the store is propagated to the positive conditions, similarly to what happens when a call is done. The order relation over graphs $\leq$ is defined as $G_1 \leq G_2 \iff paths(G_1) \sqsubseteq paths(G_2)$. We denote as $(\mathbb{G}, \leq, \bigvee, \bigwedge, \mathbf{G}, \perp_\mathbb{G})$ the complete lattice where $\bigvee$ is the least upper bound operator that joins a set of graphs by combining all the sequences that have a prefix in common in the same path, $\bigwedge$ is the greatest lower bound operator that returns the common parts of a set of graphs and $\perp_\mathbb{G}$ is the graph composed only of an empty initial node.

The semantics of a *tccp* process $p(\vec{x})$ can be seen as a conditional graph G with the initial node labeled with p and such that $paths(G) = \mathcal{F}^\alpha[\![D]\!](p(\vec{x}))$. The graph for the process $p(\vec{x})$ is built by linking the initial node of p to the nodes corresponding to the first conditional states of the semantics of an agent A such that $p(\vec{x}) : -A \in D$. The rest of the graph is built following the denotational semantics of Definition 2: each conditional state becomes a node in the graph and it is connected to the following one by a simple edge. When a call to a process $q(\vec{y})$ is found and the declaration $q(\vec{z}) : -A'$ is in D, an arrow $\overset{[\vec{z}/\vec{y}]}{\underset{q}{\Longrightarrow}}$ is added, thus linking the current node to the graph labeled with q by using the variable renaming $[\vec{z}/\vec{y}]$.

Now we are ready to define our widening operator. Widening operators provide a simple solution to the convergence problem by over-approximating infinite increasing chains in a finite number of steps. A *widening operator* [2,5] on the lattice $(\mathbb{L}, \leq)$ is a partial function $\nabla : \mathbb{L} \times \mathbb{L} \to \mathbb{L}$ satisfying: **(covering)** for all $x, y \in \mathbb{L}$ such that $x \leq y$, $x \nabla y$ exists and $y \leq x \nabla y$; and **(termination)** for each increasing chain $x_0 \leq x_1 \leq \dots$ the increasing chain defined as $y_0 = x_0$ and $y_{i+1} = y_i \nabla x_{i+1}$ is not strictly increasing.

We propose a widening operator[3] $\triangledown$ that looks for repeated patterns in consecutive iterations of $\mathcal{D}^\alpha[\![D]\!]$ and converges, in a finite number of steps, in a conditional graph that represents an over-approximation of the abstract fixpoint $\mathcal{F}^\alpha$. In the sequel, we abuse in notation and write $t \overset{\rho_2}{\underset{p}{\Rightarrow}} t'_1 \to \cdots \to t'_n$ to denote the set of the edges occurring in this path, i.e., $\{t \overset{\rho_2}{\underset{p}{\Rightarrow}} t'_1, t'_1 \to t'_2 \ldots, t'_{n-1} \to t'_n\}$.

Definition 4 (Graph Widening). *Let $G_1, G_2 \in \mathbb{G}$ such that $G_1 \leq G_2$. The graph widening of G_1 w.r.t. G_2 is defined as $G_1 \triangledown G_2 := G_1 \vee (I, N, E)$ where $I :=$ init(G_2), N is the set of nodes that occur in the set of edges E, and*

$$E := \bigcup \{t \overset{\rho_2}{\underset{p}{\Rightarrow}} t_1 \mid \textit{it exists a subpath in } G_2 \textit{ of the form } t \overset{\rho_2}{\underset{p}{\Rightarrow}} t'_1 \ldots t'_n \not\to \textit{ s.t. an edge}$$

$$\Rightarrow \textit{ labeled with p does not occur in } t'_1 \ldots t'_n \textit{ and it exists a subpath}$$

$$\textit{in } G_1 \textit{ of the form } \overset{\rho_1}{\underset{p}{\Rightarrow}} t_1 \ldots t_n \overset{\rho'_1}{\underset{p}{\Rightarrow}}, \textit{ s.t. an edge } \Rightarrow \textit{ labeled with p}$$

$$\textit{does not occur in } t_1 \ldots t_n \textit{ and } \forall 1 \leq i \leq n \; \rho_1(t_i) = \rho_2(t'_i)\} \cup$$

$$\bigcup \{t \overset{\rho_2}{\underset{p}{\Rightarrow}} t'_1 \to \cdots \to t'_n \mid \textit{it exists a subpath in } G_2 \textit{ on the form } t \overset{\rho_2}{\underset{p}{\Rightarrow}} t'_1 \ldots t'_n \not\to$$

$$\textit{s.t. in } t'_1 \ldots t'_n \textit{ it does not occur an edge } \Rightarrow \textit{ labeled with p and it does}$$

$$\textit{not exist a a subpath in } G_1 \textit{ of the form } \overset{\rho'_1}{\underset{p}{\Rightarrow}} t_1 \ldots t_n \overset{\rho'_1}{\underset{p}{\Rightarrow}}, \textit{ s.t. in } t_1 \ldots t_n \textit{ it}$$

$$\textit{does not occur an edge } \Rightarrow \textit{ labeled with p and } \forall 1 \leq i \leq n \; \rho_1(t_i) = \rho_2(t'_i)\}$$

At each iteration, the widening checks if a suffix r of a path b in the graph of a process p (which corresponds to the trace produced at the last iteration of p) has already appeared in a previous iteration of p (modulo variables renaming). In this case, it adds an edge, labeled with the necessary variable renaming ρ_2, from the node t precedent to the pattern r to the first node of the equivalent pattern found in the previous widening iteration (first case of Definition 4). Otherwise, if no equivalent (modulo variable renaming) pattern is found, the path b is added to the graph (second case of Definition 4).

Lemma 2. *If the underlying abstract Cylindric Constraint Systems are finite, then the operator $\triangledown$ is a widening operator on $\mathbb{G}$.*

Proof. (Sketch). The covering property is a consequence of the fact that the branches of G_2 that are not included by the widening are already present in G_1 modulo variable renaming; that is the reason why a direct edge is added from the last node before the repetition to the equivalent branch detected in G_1.

Termination of the widening is a consequence of the properties of the abstract constraint systems and of the finiteness of the program syntax. By definition, just a finite number of conditional steps can be computed, thus iteration's length is finite. Furthermore, when a repeated pattern is detected, that (possibly cyclic) branch is not further expandable. □

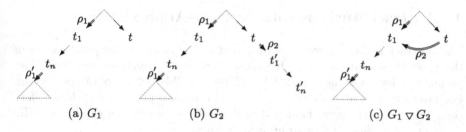

(a) G_1 (b) G_2 (c) $G_1 \triangledown G_2$

Fig. 2. The graph widening behavior.

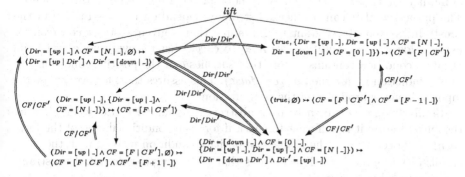

Fig. 3. Graph representation of the abstract semantics of the *lift* process.

Figure 2 shows a graphical representation of the graph widening behavior. To improve readability, in the figure we assume that all process calls involve the same process, thus we just include the renaming for variables in the edges.

Given a *tccp* set of declaration D, we can guarantee ([2]) that the chain

$$I_0 = \bot \qquad I_{i+1} = \begin{cases} I_i & \text{if } \mathcal{D}^\alpha[\![D]\!]_{I_i} \sqsubseteq I_i \\ I_i \triangledown (I_i \sqcup \mathcal{D}^\alpha[\![D]\!]_{I_i}) & \text{otherwise} \end{cases}$$

converges to a graph which is a correct approximation of the abstract semantics in a finite number of steps. That graph contains an initial node for each process declaration such that the subgraph reachable from the initial node represents the corresponding process and subgraphs are linked by edges with renamings.

Example 4. Fig. 3 shows the conditional graph corresponding to the abstract semantics of the *lift* process. We abstract streams of the concrete Constraint System by posing a depth limit for streams, i.e., we keep the first k values of a stream, and then we have the top of the domain. All other constraints are abstracted to themselves. The resulting abstract Constraint System is thus finite.

Due to the application of the widening operator it can be noted how the recursive calls (represented as triangles in Fig. 1) are replaced in Fig. 3 with the (set of) arcs pointing to the possible continuations of the computation.

---

[3] In defining our widening operator, we follow the approach of [2] instead of [5].

4   Abstract Analysis with an Over-Approximation

The abstract semantics we have proposed so far is an over-approximation of the concrete semantics. Thus, it allows us to check *universal properties*, i.e., properties that must be satisfied by all the possible behaviors of the system. For instance, it is possible to analyze some temporal properties such as safety (i.e., something bad never happens) or liveness (i.e., something good eventually happens) or to check if a program never suspends.

In order to check whether some invariant property is satisfied by our program, it is necessary to check if every node of the graph respects this property. The properties that can be checked strongly depend on the abstraction of the constraint system. If we want to guarantee that a given abstract constraint c never holds in a computation, we need to check that for every node, either its negative condition contains a store that satisfies c or the positive condition $\hat{\eta}$ is in contradiction with c (i.e., $\hat{\eta} \hat{\otimes} c = false$). This assures that, for every possible input, c is never produced in the computation.

Similarly, in order to check if an abstract constraint c is always entailed by the current store it is sufficient to check if for each conditional step of the form $(\hat{\eta}, \breve{\eta}) \rightarrowtail d$ occurring in the graph, the positive condition merged with the store entails c (i.e., $\hat{\eta} \hat{\otimes} d \hat{\vdash} c$). This ensures that for every possible initial constraint d, c is entailed by the store.

Example 5. We may be interested in proving several invariant properties on the *lift* process in Example 1. For instance, we can try to verify that "the current floor stream CF never gets a negative number". To this end, we check all the conditions in the graph in Fig. 3, and since we find (at least) a node that does not contradict that CF is negative (see the first node of the right branch), we conclude that it cannot be assured that the *lift* process respects this safety property. As a matter of fact, provided we start the computation with an initial state where CF is initialized to a negative number, then the last else branch of the program can be taken, and CF would keep negative in the subsequent trace.

Consider now the invariant property "each time the direction of the lift is updated, also its floor is updated". In this case, it can be noticed that all the conditional steps in Fig. 3 satisfy this property, since whenever the positive condition in the step merged with the store entails that Dir has a value, then it is also entailed that CF is instantiated.

Verifying liveness properties is harder since it involves analyzing unknown length sequences of steps. For instance, given a process $p(\vec{x})$, assume that we want to check that "every time an abstract constraint c holds, then it exists a future state where another abstract constraint d holds".

Given the conditional graph for $p(\vec{x})$, this property would hold if for each node labeled with a conditional step whose positive condition and store entails c then all paths starting from such node contain a conditional step whose positive condition and store entails d.

Example 6. Observe that *lift* process in Example 1 satisfies the property "every time the current floor is 0 and the direction is *down*, the direction will be *up* eventually". In fact, the first node of the third branch from the left in Fig. 3 is the sole step that contains in its positive condition $CF = [0 \mid \_]$ and $Dir = [down \mid \_]$. Furthermore, for each possible path from this node we find a conditional step where $Dir = [up \mid \_]$ appears in the positive condition or in the store.

Other interesting liveness property that can be analyzed on the *lift* process is "whenever the current floor is 0 it exists a future state when this value changes", i.e., we do not stay indefinitely in floor 0.

Since the number of nodes in the graph is finite, the aforementioned analysis terminates in a finite number of steps.

Let us now analyze non-suspension. Non-suspension analysis consists in assuring that no execution of a *tccp* program suspends. In conditional graphs, in order to check whether $p(\vec{x})$ never suspends, it is sufficient to check that there is no node N in G labeled with a *stutt* construct with an outgoing arc pointing to N itself. Inversely, if the graph contains a stuttering node, we can not guarantee suspension, due to over-approximation of the semantics.

Example 7. Consider the semantics of the *lift* process in Fig. 3. It is worth noting that the graph does not contain any node labeled with *stutt*. Therefore, we can assure that the *lift* process never suspends.

5   Related Work

To the best of our knowledge, this is the first attempt to propose an abstract interpretation framework for a concurrent constraint language adhering to the characteristics of *tccp* (negative information, non-determinism and infinite behaviors). In [8], a framework for dataflow analysis of *tcc* and *utcc* programs is presented. The two main differences between these two languages and *tccp* are the notion of time (*tcc* and *utcc* use dedicated timing constructs) and determinism (*vs.* non-determinism of *tccp*). Moreover, in the case studies, [8] uses a $depth(k)$ abstraction to ensure convergence, which consists in a non-selective cut at some point in time. In [9], it was defined a model checking algorithm for *tccp* which allowed us to verify timed-depending properties. Their algorithm was based on the exploration of a graph representation of the program behavior which resembles the graph representation of the semantics defined in this paper. Thus we could as well employ our graph representation to perform (an efficient) model checking. Note however that the abstract semantics that we propose now is not limited to the verification of temporal properties.

Finally, [1] proposes an abstract semantic framework for *tccp* that, differently from our approach, was based on source-to-source transformations. The two approaches are completely different: [1] aimed at using the concrete semantics to execute the transformed (abstract) program. This could be done thanks to a non-trivial transformation of the program (an analysis on the structure of the program was necessary as a preprocess of the transformation). Our approach aims at defining an abstract semantics that, thanks to the characteristics of the concrete denotational semantics, is guaranteed to be correct and we argue that is precise enough to allow the definition of interesting analyses.

6   Conclusions and Future Work

We have proposed an abstract semantics that, together with a widening operator, is suitable for the definition of different analyses for *full tccp* programs. This is a difficult task because of the presence of infinite computations, use of negative information and non-determinism. However, it is essential since these are the features that make *tccp* well-suited to model reactive systems.

The abstract semantics is an over-approximation, which makes possible to define analysis tools for universal properties. To the best of our knowledge, this is the first proposal that defines an analysis which adaptively ensures termination depending on the program (by means of widening). This should give better results than the non-selective approaches.

This is a first step towards our final goal of defining a rich abstract semantic framework for the analysis of *tccp* programs. We plan to implement the abstract semantics so that we can produce some experimental results. We will need also to implement and develop suitable and useful abstractions for the constraint system, corresponding to the analyses to be performed. We are also interested in defining an under-approximated semantics for *tccp*. Under-approximations produce correct semantics, which means that not all the behaviors are captured, but no spurious behaviors are included. These kind of abstractions allow one to analyze *existential* properties, for instance that there exists a suspension trace.

References

1. Alpuente, M., Gallardo, M.M., Pimentel, E., Villanueva, A.: A semantic framework for the abstract model checking of tccp programs. Theor. Comput. Sci. **346**(1), 58–95 (2005)
2. Bagnara, R., Hill, P.M., Ricci, E., Zaffanella, E.: Precise widening operators for convex polyhedra. Sci Comput. Program. **58**(1–2), 28–56 (2005)
3. Comini, M., Titolo, L., Villanueva, A.: Abstract diagnosis for timed concurrent constraint programs. Theor. Pract. Log. Program. **11**(4–5), 487–502 (2011)
4. Comini, M., Titolo, L., Villanueva, A.: A condensed goal-independent bottom-up fixpoint modeling the behavior of tccp. Technical report, DSIC, Universitat Politècnica de València (2013). http://riunet.upv.es/handle/10251/34328
5. Cousot, P., Cousot, R.: Abstract Interpretation: a unified lattice model for static analysis of programs by construction or approximation of fixpoints. In: Proceedings of the 4th ACM SIGACT-SIGPLAN Symposium on Principles of Programming Languages, Los Angeles, California, January 17–19, pp. 238–252. ACM Press, New York (1977)
6. de Boer, F.S., Gabbrielli, M., Meo, M.C.: A timed concurrent constraint language. Inf. Comput. **161**(1), 45–83 (2000)

7. Falaschi, M. Gabbrielli, M., Marriott, K., Palamidessi, C.: Compositional analysis for concurrent constraint programming. In: Proceedings of the Eighth Annual IEEE Symposium on Logic in Computer Science, pp. 210–221. IEEE Computer Society Press, Los Alamitos (1993)
8. Falaschi, M., Olarte, C., Palamidessi, C.: Abstract interpretation of temporal concurrent constraint programs. Theor. Pract. Log. Program. (TPLP) 15(3), 312–357 (2015)
9. Falaschi, M., Villanueva, A.: Automatic verification of timed concurrent constraint programs. Theor. Pract. Log. Program. 6(3), 265–300 (2006)
10. Saraswat, V.A.: Concurrent Constraint Programming. The MIT Press, Cambridge (1993)
11. Zaffanella, E., Giacobazzi, R., Levi, G.: Abstracting synchronization in concurrent constraint programming. J. Funct. Log. Program. 6, 1997 (1997)

A Global Occurrence Counting Analysis for Brane Calculi

Chiara Bodei[1][(✉)], Linda Brodo[2], Roberta Gori[1], Diana Hermith[3],
and Francesca Levi[1]

[1] Dipartimento di Informatica, Università di Pisa, Pisa, Italy
chiara.bodei@unipi.it
[2] Dipartimento di Scienze Politiche, Scienze della Comunicazione e Ingegneria
dell'Informazione, Università di Sassari, Sassari, Italy
[3] Dipartimento di Ingegneria dell'informazione e Scienze Matematiche,
Università di Siena, Siena, Italy

Abstract. We propose a polynomial static analysis for Brane Calculi [6], based on *Abstract Interpretation* [8] techniques. The analysis provides a description of the possible hierarchical structure of membranes and of the processes possibly associated to each membrane, together with global occurrence counting information. Our analysis can be applied in the biological setting to investigate systems in which the information on the number of membranes occurring in the system plays a crucial role.

1 Introduction

One of the challenges of systems biology is to understand the complex behaviour of biological systems and networks that interact in diverse ways. There is a large amount of data concerning single components and functions. The main difficulty is establishing their relationships and inferring the overall emerging behaviour of a system. Consider, for instance, the crucial problem of identifying biological pathways and reconstructing their inter-connections

Computational frameworks and *in silico* investigation have been recently exploited to support *in vitro* or *in vivo* experiments. Often, these frameworks come with the drawback of high computational cost, due to the expensive inspection of the models that capture dynamic behaviour. Static analysis provides techniques able to reduce the computational cost, at the price of loosing precision. In particular, they provide safe over-approximations of the dynamic behaviour: all the events that the analysis predicts *may* happen, while all the non predicted events will *never* happen.

In this paper we propose a *static analysis* for Brane Calculi [6], based on *Abstract Interpretation* (AI) [8] techniques. Brane calculi have been introduced to more closely model the behaviour of membrane-enclosed compartments. Therefore, these calculi are useful for modelling and reasoning about a large class of biological systems. Our analysis is based on the idea of enriching the standard information on control flow analysis (as the one in [2]) with global occurrence

© Springer International Publishing Switzerland 2015
M. Falaschi (Ed.): LOPSTR 2015, LNCS 9527, pp. 179–200, 2015.
DOI: 10.1007/978-3-319-27436-2_11

counting information. More specifically, we compute an abstract state describing the possible structure of all the derivatives of the analysed system. The abstract state provides information on the possible membrane hierarchy and on the processes that may be associated to each membrane, together with their multiplicity. The global occurrence counting information refers to the number of occurrences of membranes and processes, independently of their location. The analysis can be efficiently computed in polynomial time, thanks to the properties of the abstract semantics.

Our analysis can be applied to investigate systems in which the information on the number of membranes in the system plays a crucial role in biological terms. One could, for instance, be interested in distinguishing between the case of one healthy cell that can let many molecules pass through the cell membrane, and the case in which an ill cell can let at most one molecule to pass. We illustrate our analysis by considering examples of communication via *mobile vesicles*. A mobile vesicle containing the substance to be transmitted springs from a membrane *Source* and eventually fuses with a membrane *Target* releasing its content X inside of it. We consider two different variants of this kind of communication by expanding the encoding proposed in [28], where there are two target membranes $Target_1$ and $Target_2$. In the first version, the membrane *Source* can communicate X either to $Target_1$ or $Target_2$, while in the second one, one instance of X is communicated to $Target_1$ and the other one to $Target_2$. Then, we show that our analysis faithfully predicts that X cannot be simultaneously present in both membranes in the first case, while it may occur in both in the second case, thanks to the occurrence counting information.

Related Work. Static analysis techniques have been applied to many biologically-oriented calculi (see, e.g. the survey in [17]). In particular, Control Flow Analysis (CFA) has been applied to Beta-binders [26] in [1], to Brane Calculi [2], and to BioAmbients [27] in [21,23–25]; while Abstract Interpretation to BioAmbients [14–16], and to Brane Calculi [3,4]. Many of these works are inspired by the application of static analysis techniques [10,18,20] to Mobile Ambients (MA) [7], from which, many bio-inspired calculi derive.

Part of the above mentioned static analyses [1,2,21,23] abstract, with different precision, the behaviour of the investigated system by providing an approximate description of structure of all derivatives. As a consequence, these approaches can be applied to establish invariant properties showing that certain events will not happen in any derivatives of the analysed system. These techniques have polynomial complexity but they provide a less precise description of the possible topological structure of derivatives given that they do not maintain any information on occurrence counting. The richer contextual CFA in [24], and the causality-oriented AI-based analysis in [3,4] improve the prediction accuracy, but, still, they are not able to observe the multiplicity.

Instead, in [15], the authors present a counting analysis for BioAmbients that is able to express that an ambient can reside in alternative locations.

This analysis has exponential complexity and provides accurate information about the number of occurrences of ambients, by counting the local number inside any ambient rather than their global number.

There are several static analysis frameworks that include occurrence counting information applied to MA and to π-calculus [19]. In [20,22] analyses for MA are introduced that are rather expensive from a computational point of view. The authors propose in [20] an exponential analysis for counting the global number of occurrences of ambients. The approach based on CFA substantially differs from our analysis, which is computed by calculating an abstract semantics. At the expense of a higher complexity, the shape analysis in [22] uses context-dependent counts for inferring a more accurate description of the internal structure of an ambient, by taking care of the local multiplicity of ambients.

In [9,11,12] the author proposes a framework based on AI, applied to the π-calculus that, differently from the previous proposals, is non-uniform, i.e. the analysis can distinguish among recursive instances of agents. In this approach, the occurrence number of instances of agents is approximated by using a relational abstraction. The analysis is quite precise and efficient: its complexity is polynomial. For instance, the analysis described in [11] has a worst time cost of n^4, where n is the number of processes in the initial configuration. This approach is adequate to capture mutual exclusion and other security properties of complex mobile systems, formalised in π-calculus. In [10] the author proposes an adaptation of these techniques to MA, with a focus on security properties such as non-interference or confinement. In [13], the author proposes polynomial and precise analyses for MA and BioAmbients, based on both global and local counting. Specifically, the global analysis is the same as the one proposed in [11] for the π-calculus. This approach can handle mass preservation like invariants, which are ubiquitous in biological systems. In particular, it is able to preserve precision when dealing with continuations of replicated prefixes.

The static analyses proposed in [14,16,25] rely on a different approach since that they compute an abstract transition system to approximate the system behaviour, by still exploiting occurrence counting information. These techniques obviously provide useful information to verify temporal properties but at the price of a high complexity.

Structure of the Paper. In Sect. 2, we recall the semantics of Brane calculi. In Sect. 3, we introduce our running examples, based on hypothetical scenarios of communication via mobile vesicles in the style of [28]. In Sect. 4, we present the analysis and in Sect. 5 we apply it to our examples. Conclusions are drawn in Sect. 6.

For lack of space, we present the formal definitions only for the fragment of the calculus that includes the Phago/Exo/Pino (PEP) and Bud interactions. This fragment is sufficiently expressive to formalise our running examples. It is worth pointing out that our analysis can be easily extended to the full calculus.

2   An Overview on Brane Calculus

The Brane Calculi [6] are a family of calculi based on a set of primitives inspired by biological membrane interactions. As already mentioned, we focus here on the fragment of the calculus that includes the Phago/Exo/Pino (PEP) and Bud interactions.

The Phago/Exo/Pino(PEP) actions represent the biological processes of endocytosis and exocytosis. The first indicates the process of incorporating external material into a cell, by engulfing it with the cell membrane, while the second one indicates the reverse process. Endocytosys is rendered by two more basic operations: phagocytosis (*phago*), which consists in engulfing just one external membrane, and pinocytosis (*pino*), which consists in engulfing zero external membranes. Exocytosis is instead denoted by (*exo*). We further use the action (*bud*) to model the membrane splitting process that consists in the splitting off exactly one internal membrane. It is worth mentioning that the bud action can be encoded with a sequence of PEP actions [6]. However, from the analysis point of view it is more convenient to use the bud action as primitive.

We introduce the syntax and the semantics for the calculus, considering a *labelled version* of the calculus. As usual in static analysis, labels are exploited to support the analysis (presented in Sect. 4) and *do not affect* the dynamic semantics of the calculus.

A membrane system consists of nested membranes, where each membrane has associated a membrane process. The syntax of the labelled calculus is described in Table 1, where n is taken from a countable set $\mathcal{N}$ of names, and where we write $P \in \mathsf{Sys}$ for *systems*, $\sigma \in \mathsf{Proc}$ for *membrane processes*, and $a \in \mathsf{Act}$ for *actions*. Each membrane is annotated with a *membrane label* $\Gamma \in \widehat{\mathsf{Lab}}_{\mathcal{M}}$ and each action is annotated with a *process label* $\lambda \in \mathsf{Lab}_{\mathcal{P}}$.

We therefore need two distinct sets of labels. We have the set of *process labels* $\mathsf{Lab}_{\mathcal{P}}$, ranged over by $\alpha, \beta, \gamma \ldots$. Moreover, given a set of basic membrane labels $\mathsf{Lab}_{\mathcal{M}}$, we have the associated set of *membrane labels* $\widehat{\mathsf{Lab}}_{\mathcal{M}}$, ranged over by $\Delta, \Gamma, \Psi \ldots$. The set $\widehat{\mathsf{Lab}}_{\mathcal{M}}$ is inductively defined as follows: (i) $\mathsf{Lab}_{\mathcal{M}} \subseteq \widehat{\mathsf{Lab}}_{\mathcal{M}}$; (ii) if $\Gamma, \Delta \in \widehat{\mathsf{Lab}}_{\mathcal{M}}$ and $\lambda, \mu \in \mathsf{Lab}_{\mathcal{P}}$, then $\mathrm{phago}(\Gamma, \Delta, \lambda, \mu)$, $\mathrm{pino}(\Delta, \lambda)$ and $\mathrm{bud}(\Gamma, \Delta, \lambda, \mu) \in \widehat{\mathsf{Lab}}_{\mathcal{M}}$.

The system $\sigma(\!|P|\!)^{\Gamma}$ describes a *membrane*, decorated by label Γ^1 that contains the system P and that performs the *membrane process* σ, describing its interaction capabilities. The construct $a^{\lambda}.\sigma$ defines a sequential process that executes an action a, decorated by label λ, and then behaves as the process σ. We adopt standard syntactical abbreviations: a^{λ} stands for $a^{\lambda}.0$, $(\!|P|\!)^{\Gamma}$ stands for $0(\!|P|\!)^{\Gamma}$, and $\sigma(\!|\,|\!)^{\Gamma}$ is a shorthand for $\sigma(\!|\diamond|\!)^{\Gamma}$.

The semantics of the calculus is given by the reduction rules in Table 1, modulo the structural congruence rules, here omitted because standard (see [6]). The labelled transition relation is $\overset{l}{\longrightarrow}$, where $P \overset{l}{\longrightarrow} Q$ denotes that the system P

---

[1] For brevity, from now on, we will usually write membrane Γ, instead of membrane labelled by Γ.

Table 1. Syntax and reduction rules for (labelled) Brane.

$$
\begin{aligned}
P, Q &::= \diamond \mid P \circ Q \mid\ !P \mid \sigma(\!|P|\!)^\Gamma && \text{systems } \mathbf{Sys} \\
\sigma, \tau &::= 0 \mid \sigma|\tau \mid\ !\sigma \mid a^\lambda.\sigma && \text{membrane processes } \mathbf{Proc} \\
a, b &::= \mathbf{phago}_n \mid \overline{\mathbf{phago}}_n(\rho) \mid \mathbf{exo}_n \mid \overline{\mathbf{exo}}_n \mid \mathbf{pino}(\rho) \mid \mathbf{bud}_n \mid \overline{\mathbf{bud}}_n(\sigma) && \text{actions } \mathbf{Act}
\end{aligned}
$$

$$(\textsc{Par}) \frac{P \xrightarrow{l} Q}{P \circ R \xrightarrow{l} Q \circ R} \qquad (\textsc{Brane}) \frac{P \xrightarrow{l} Q}{\sigma(\!|P|\!)^\Gamma \xrightarrow{l} \sigma(\!|Q|\!)^\Gamma} \qquad (\textsc{Struct}) \frac{P \equiv P' \wedge P' \xrightarrow{l} Q' \wedge Q' \equiv Q}{P \xrightarrow{l} Q}$$

(Phago) $\mathbf{phago}_n^\lambda.\sigma|\sigma_0(\!|P|\!)^\Delta \circ \overline{\mathbf{phago}}_n^\mu(\rho).\tau|\tau_0(\!|Q|\!)^\Gamma \xrightarrow{\mathbf{phago}_1(\Delta,\Gamma,\lambda,\mu)} \tau|\tau_0(\!|\rho(\!|\sigma|\sigma_0(\!|P|\!)^\Delta|\!)^{\mathbf{phago}(\Delta,\Gamma,\lambda,\mu)} \circ Q|\!)^\Gamma$

(Exo) $\overline{\mathbf{exo}}_n^\mu.\tau|\tau_0(\!|\mathbf{exo}_n^\lambda.\sigma|\sigma_0(\!|P|\!)^\Delta \circ Q|\!)^\Gamma \xrightarrow{\mathbf{exo}_1(\Delta,\Gamma,\lambda,\mu)} P \circ \sigma|\sigma_0|\tau|\tau_0(\!|Q|\!)^\Gamma$

(Pino) $\mathbf{pino}^\lambda(\rho).\sigma|\tau(\!|P|\!)^\Delta \xrightarrow{\mathbf{pino}_1(\Delta,\lambda)} \sigma|\tau(\!|\rho(\!||\!)^{\mathbf{pino}(\Delta,\lambda)} \circ P|\!)^\Delta$

(Bud) $\overline{\mathbf{bud}}_n^\mu(\rho).\tau|\tau_0(\!|\mathbf{bud}_n^\lambda.\sigma|\sigma_0(\!|P|\!)^\Delta \circ Q|\!)^\Gamma \xrightarrow{\mathbf{bud}_1(\Delta,\Gamma,\lambda,\mu)} \rho(\!|\sigma|\sigma_0(\!|P|\!)^\Delta|\!)^{\mathbf{bud}(\Delta,\Gamma,\lambda,\mu)} \circ \tau|\tau_0(\!|Q|\!)^\Gamma$

evolves into the system Q performing a reaction described by the *transition label* $l \in \mathbf{Lab}_T$. The set of transition labels $\mathbf{Lab}_T$ (ranged over by $l_1, l_2 \ldots$) is defined as follows:

$$\mathbf{Lab}_T = \{\mathbf{pino}_1(\Delta,\lambda), \mathbf{phago}_1(\Gamma,\delta,\lambda,\mu), \mathbf{exo}_1(\Gamma,\Delta,\lambda,\mu), \mathbf{bud}_1(\Gamma,\Delta,\lambda,\mu) \mid \Gamma, \Delta \in \widehat{\mathbf{Lab}}_M, \lambda, \mu \in \mathbf{Lab}_P\}$$

Besides the standard reduction rule for congruence (STRUCT), and the contextual rules to propagate reductions across parallel composition (PAR) and membrane nesting (BRANE), there are the axioms specific of the membrane actions.

Rule (PHAGO) models the inclusion of an external membrane, labelled by Δ, inside a membrane, labelled by Γ. The two membranes Δ and Γ exercise the actions $\mathbf{phago}_n^\lambda$ and $\overline{\mathbf{phago}}_n^\mu(\rho)$, respectively. Once engulfed, the membrane Δ is enclosed inside a new membrane with label $\mathbf{phago}(\Delta,\Gamma,\lambda,\mu)$, which has associated the process ρ. The corresponding transition label is $\mathbf{phago}_1(\Delta,\Gamma,\lambda,\mu)$. Rule (EXO) models the expulsion of the membrane Δ, outside the external membrane Γ, triggered by the actions $\mathbf{exo}_n^\lambda$ and $\overline{\mathbf{exo}}_n^\mu$, respectively. The corresponding transition label is $\mathbf{exo}_1(\Delta,\Gamma,\lambda,\mu)$. In the rule (PINO), the membrane Δ, creates a new empty membrane, labelled by $\mathbf{pino}(\Delta,\lambda)$, inside itself. The action $\mathbf{pino}^\lambda(\rho)$ is equipped with a process ρ that will be associated to the new membrane. The corresponding transition label is $\mathbf{pino}_1(\Delta,\lambda)$. Finally, in the rule (BUD), the membrane Γ expels the child membrane Δ, performing the actions $\overline{\mathbf{bud}}_n^\mu(\rho)$ and $\mathbf{bud}_n^\lambda$, respectively. The membrane Δ is wrapped inside a new membrane with label $\mathbf{bud}(\Delta,\Gamma,\lambda,\mu)$ and has associated the process ρ. The corresponding transition label is $\mathbf{bud}_1(\Delta,\Gamma,\lambda,\mu)$.

The semantics of a system is defined as a *Labelled Transition System* (LTS). Given $P \in \mathbf{Sys}$, we use $LTS(P) = (X, \xrightarrow{l}, P)^2$ to denote the LTS obtained from

---

[2] Here, $X \subseteq \mathbf{Sys}$ stands for the set of systems that are reachable from system P.

the initial system P by applying the rules and axioms in Table 1. Moreover, to ensure the correctness of our analysis, we assume that the system P is *well labelled*, i.e. that the process labels occurring in P are all distinct. The analysis is not correct in general for every labelling of the initial systems.

Remark 1. It is worth briefly discussing the role of labels in our calculus. Process and membrane labels will be exploited in the analysis to maintain the information on the topological structure of systems. To decorate the new membranes introduced by reactions of the calculus, we adopt composite membrane labels that record the labels of the membrane and of the actions that interact. This labelling technique may introduce arbitrarily nested membrane labels (e.g. $\mathtt{bud}(\Delta, \mathtt{pino}(\Gamma, \delta), \lambda, \mu)$) and even an infinite number of membrane labels.

Furthermore, we decorate the reduction steps with transition labels giving information on the labels of the actions and on the membranes involved in the interaction. This information will be exploited in the analysis to establish a correspondence between reduction steps and abstract transitions.

3   Communication via Mobile Vesicles

To illustrate our analysis, we consider hypothetical scenarios of communication via mobile vesicles. A mobile vesicle containing the substance to be transmitted springs from a membrane *Source* and eventually fuses with a membrane *Target* releasing its content inside of it. In eucaryotic cells, a large variety of proteins is targeted to its final destination via mobile transport vesicles, i.e. small membrane-enclosed sacs separated from the cytosol by a lipid bilayer. Proteins can be contained in the vesicles (i.e. secretory proteins) or embedded in their membrane (i.e. transmembrane proteins). Through vesicular trafficking, proteins follow routes involving intracellular locations (e.g. endoplasmic reticulum, Golgi apparatus or lysosomes) as well as the plasma membrane, in the case of endo- and exocytosis. Since vesicular transport is essential in the organisation of eukaryotic cells, understanding the mechanisms that control vesicle budding and fusion is an active research topic in cell biology.

Roughly speaking, we are focussing on modelling the communication (through a vesicle) of a molecule X embedded in a *Source* membrane to specified *Target* compartments. More specifically we consider the encoding in Brane calculi presented in [28], where communication based upon the natural budding of mobile vesicle in a cell was modelled as follows. A vesicle containing (embedded in its membrane) a molecule that needs to be shuttled between two compartments, buds from a membrane *Source*. Then, it is engulfed by another compartment (the *Target* membrane) through phagocytosis (creating a coat membrane containing the vesicle) and, finally, the coat of the mobile vesicle is decomposed within the membrane *Target*, releasing the transported molecule in it. Here, we introduce a more general situation, by introducing two substantially different scenarios of communication via mobile vesicles. In both cases *Source* can communicate the molecule X, via a mobile vesicle, to different membranes $Target_1$ and $Target_2$.

In the first version, the membrane *Source* can communicate the molecule X *either* to $Target_1$ *or* to $Target_2$, while in the second one, one instance of molecule X is communicated to membrane $Target_1$ and the other one to $Target_2$. Both the previous systems can be rendered along the lines of [28].

Example 1. To model the first scenario, we slightly extend the original model in [28]. In this case we have two vesicles and two target membranes, and each vesicle can be phagocytosed by one of the target membranes. Note that, in the biological setting, this particular kind of non-deterministic behaviour is of great interest because it may arise, e.g. in the extracellular environment of cells, where extracellular vesicles, which are small vesicles released by donor cells, can be taken up by any recipient target cell. This transport mechanism plays an important role in cell-to-cell communication.

The encoding of this scenario is reported in the upper part of Table 2, where the processes σ_S, σ_{target_i}, τ_X and τ_i' (with $i = 1, 2$) stand for membranes processes (not specified as not relevant at this level of abstraction), and where we decorate actions and membranes with basic membrane labels in $\mathsf{Lab}_\mathcal{M}$ [3]. The molecule X to be transmitted is enclosed inside a membrane labelled by Γ. Such membrane triggers the communication process, exercising the action bud_n^γ and leading to the gemmation of one of the two vesicles that will transport the molecule into one of the two targets.

Table 2. First scenario: Encoding (upper part) and evolution of the first vesicle (lower part).

$$S \overset{def}{=} \sigma_S (\!| Source \circ Target_1 \circ Target_2 |\!)^{skin} \quad Target_i \overset{def}{=} \overline{!phago}_{n_i}^{\delta_i} (\!|\overline{exo}_{n_i}^{\beta_i}|\!)|\sigma_{target_i} (\!||\!)^{target_i}$$

$$Vesicle_i \overset{def}{=} phago_{n_i}^{\mu_i}.exo_{n_i}^{\nu_i}|\tau_i' \quad Source \overset{def}{=} \sigma_{Source}(\!|\sigma_X(\!|X|\!)^{\Gamma}|\!)^{source}$$

$$\sigma_X \overset{def}{=} \mathsf{bud}_n^\gamma|\tau_X \quad \sigma_{source} \overset{def}{=} !\overline{\mathsf{bud}}_n^{\lambda_1} (\!|Vesicle_1|\!)|!\overline{\mathsf{bud}}_n^{\lambda_2} (\!|Vesicle_2|\!)$$

$$S = \sigma_S (\!| Source \circ Target_1 \circ Target_2 |\!)^{skin} \equiv$$

$$\sigma_S (\!|\overline{!\mathsf{bud}}_n^{\lambda_1} (\!|Vesicle_1|\!)|!\overline{\mathsf{bud}}_n^{\lambda_2} (\!|Vesicle_2|\!)(\!|\underline{\mathsf{bud}}_n^\gamma|\tau_X(\!|X|\!)^{\Gamma}|\!)^{source} \circ Target_1 \circ Target_2 |\!)^{skin} \overset{l_1}{\longrightarrow}$$

$$\sigma_S (\!|\overline{!\mathsf{bud}}_n^{\lambda_1} (\!|Vesicle_1|\!)|!\overline{\mathsf{bud}}_n^{\lambda_2} (\!|Vesicle_2|\!)(\!||\!)^{source} \circ Vesicle_1(\!|\tau_X(\!|X|\!)^{\Gamma}|\!)^{\Pi_1} \circ Target_1 \circ Target_2 |\!)^{skin} \equiv$$

$$\sigma_S (\!|\sigma_{source}(\!||\!)^{source} \circ \underline{phago}_{n_1}^{\mu_1}.exo_{n_1}^{\nu_1}|\tau_1'(\!|\tau_X(\!|X|\!)^{\Gamma}|\!)^{\Pi_1} \circ \overline{!phago}_{n_1}^{\delta_1} (\!|\overline{exo}_{n_1}^{\beta_1}|\!)|\sigma_{target_1} (\!||\!)^{target_1} \circ Target_2 |\!)^{skin} \overset{l_{11}}{\longrightarrow}$$

$$\sigma_S (\!|\sigma_{source}(\!||\!)^{source} \circ \overline{!phago}_{n_1}^{\delta_1} (\!|\underline{\overline{exo}}_{n_1}^{\beta_1}(\!|\underline{exo}_{n_1}^{\nu_1}|\tau_1'(\!|\tau_X(\!|X|\!)^{\Gamma}|\!)^{\Pi_1}|\!)^{\Pi_{11}}|\!)|\sigma_{target_1} (\!||\!)^{target_1} \circ Target_2 |\!)^{skin} \overset{l_{12}}{\longrightarrow}$$

$$\sigma_S (\!|\sigma_{source}(\!||\!)^{source} \circ \overline{!phago}_{n_1}^{\delta_1} (\!|\overline{exo}_{n_1}^{\beta_1}|\!)|\sigma_{target_1} (\!|\tau_X(\!|X|\!)^{\Gamma} \circ \tau_1'(\!||\!)^{\Pi_{11}}|\!)^{target_1} \circ Target_2 |\!)^{skin}$$

where

$$l_1 = \mathsf{bud}_1(\Gamma, source, \gamma, \lambda_1) \quad l_{11} = \mathsf{phago}_1(\Pi_1, target_1, \mu_1, \delta_1) \quad l_{12} = \mathsf{exo}_1(\Pi_1, \Pi_{11}, \nu_1, \beta_1)$$

$$\Pi_1 = \mathsf{bud}(\Gamma, source, \gamma, \lambda_1) \quad \Pi_{11} = \mathsf{phago}(\Pi_1, target_1, \mu_1, \delta_1)$$

For simplicity, we only illustrate the dynamic evolution of the first vesicle in the lower part of Table 2, where we underline the prefixes involved in each

---

[3] We also assume the system S to be well labelled.

transition. The membrane Γ buds from the membrane *source* thus creating the first vesicle. Then the vesicle fuses with the corresponding membrane *target*$_1$, by means of actions $phago_{n_1}^{\mu_1}$ and $exo_{n_1}^{\nu_1}$. Note that the membrane created by the bud reaction is decorated with the label Π_1, while the one created by the phago reaction with the label Π_{11}. Furthermore, the bud, phago and exo reactions are decorated with the transition labels l_1, l_{11} and l_{12}, respectively. The dynamic evolution of the second vesicle is analogous.

It should be clear that in this case the molecule X cannot be simultaneously present in membranes target$_1$ and target$_2$: either the molecule is transmitted to the first target or, alternatively, it is transmitted to the second one.

Example 2. To model the second scenario, we again extend the original model in [28]. In this case there are two membranes *Source*, each containing the molecule X that has to be transmitted. One instance of substance X is communicated via a mobile vesicle to membrane $Target_1$ while the other one membrane is communicated via a mobile vesicle to membrane $Target_2$. Also this case may arise in the biological setting, when different types of vesicles are simultaneously present, and each type of cell can selectively interact only with the correct target membrane, i.e. each kind of cargo is transported to the specific programmed location via membrane fusion. This scenario may occur in intracellular vesicles in eukaryotic cells, where membrane-enveloped vesicles travel in between organelles in the cytoplasm.

The encoding of this second scenario is reported in Table 3, using process and membrane labels similar to the ones introduced in Table 2. Each membrane *Source* can communicate its molecule X to the corresponding *Target*. The dynamic evolution is similar to the previous one and it is not reported.

In this case, differently from the previous one, the molecules X will eventually end up in both membranes target$_1$ and target$_2$: one occurrence in target$_1$ and the other one in target$_2$.

Table 3. Second scenario: encoding.

$$S' \stackrel{def}{=} \sigma_S (\!| Source_1 \circ Source_2 \circ Target_1 \circ Target_2 |\!)^{skin}$$
$$Target_i \stackrel{def}{=} !\overline{phago}_{n_i}^{\delta_i}.(\overline{exo}_{n_i}^{\beta_i}) \,|\, \sigma_{target_i} (\!|\, |\!)^{target_i}$$
$$Vesicle_i \stackrel{def}{=} phago_{n_i}^{\mu_i}.exo_{n_i}^{\nu_i} \,|\, \tau_i'$$
$$Source_i \stackrel{def}{=} \sigma_{Source_i} (\!| \sigma_{X_i} (\!|X|\!)^{\Gamma} |\!)^{source}$$
$$\sigma_{X_i} \stackrel{def}{=} bud_n^{\gamma_i} \,|\, \tau_{X_i}$$
$$\sigma_{source_i} \stackrel{def}{=} !\overline{bud}_n^{\lambda_i}.(Vesicle_i)$$

4   The Abstraction

The analysis computes a description of the possible structure of all the derivatives of the system under investigation. Following the AI approach, the analysis result is calculated by collecting all reachable abstract states representing approximate information on the evolution of the system. More specifically, *abstract states* provide information on the possible hierarchical structure of membranes, and on the processes that may be associated to each membrane, together with information about the possible number of occurrences of membrane and process labels. We prove that the analysis is a *safe over-approximation* of the concrete behaviour. Furthermore, we show that the properties of the abstract semantics help us in computing our analysis in polynomial time (see Theorem 1).

Abstract Membrane Labels. To guarantee that the analysis can be computed in a finite number of steps, we need an *abstraction* of *membrane labels*. In the abstract setting, the basic membrane labels are defined as $\mathsf{Lab}^\circ_{\mathcal{M}} = \mathsf{Lab}_{\mathcal{M}} \cup \{@\}$, where the special symbol @ represents the outermost membrane. Then, we derive the corresponding set of *abstract membrane labels* $\widehat{\mathsf{Lab}}^\circ_{\mathcal{M}}$, ranged over by $\Gamma^\circ, \Delta^\circ, ...,$ defined as the least set s.t.: (i) $\mathsf{Lab}^\circ_{\mathcal{M}} \subseteq \widehat{\mathsf{Lab}}^\circ_{\mathcal{M}}$; and (ii) if $\Gamma^\circ, \Delta^\circ \in \widehat{\mathsf{Lab}}^\circ_{\mathcal{M}}$ and $\lambda, \mu \in \mathsf{Lab}_{\mathcal{P}}$ then $(\Gamma^\circ, \Delta^\circ, \lambda, \mu)$ and $(\Gamma^\circ, \lambda) \in \widehat{\mathsf{Lab}}^\circ_{\mathcal{M}}$.

Note that in the previously introduced abstraction of membrane labels, arbitrarily nested membrane labels can still arise (e.g. $(\Gamma^\circ, (\Delta^\circ, \Theta^\circ, \nu, \pi), \lambda, \mu)$). As a consequence, we introduce further approximations to guarantee that the abstract membrane labels generated in the analysis are finite. We then consider the set of *abstract membrane labels parametric w.r.t.* the level of nesting depth $d \in \mathbb{N}^+$ defined as:

$$\widehat{\mathsf{Lab}}^d_{\mathcal{M}} = \{\Delta^\circ | \Delta^\circ \in \widehat{\mathsf{Lab}}^\circ_{\mathcal{M}} \text{ and } depth(\Delta^\circ) \le d\} \cup \{(\top, \top, \lambda, \mu), (\top, \lambda) \mid \lambda, \mu \in \mathsf{Lab}_{\mathcal{P}}\}$$

where $depth(\Delta^\circ)$ gives the maximal number of nesting parenthesis levels occurring in Δ°. Intuitively, all the abstract membrane labels with depth greater than d are approximated with the following new special membrane labels: $(\top, \top, \lambda, \mu)$ and $(\top, \lambda)$.

This is formalised by introducing an *abstraction function* that maps a membrane label into an abstract membrane label with respect to a given parameter d.

Definition 1. *Let $d \in \mathbb{N}^+$ and $\Delta \in \widehat{\mathsf{Lab}}_{\mathcal{M}}$. The* abstract version *of Δ, denoted by $\Delta^\bullet \in \widehat{\mathsf{Lab}}^d_{\mathcal{M}}$[4], is inductively defined as follows,*

1. $\Delta \in \mathsf{Lab}_{\mathcal{M}} \Rightarrow \Delta^\bullet = \Delta$;
2. $\Delta = \#(\Gamma, \Psi, \lambda, \mu)$ with $\# \in \{\text{bud}, \text{phago}\} \Rightarrow \Delta^\bullet = \begin{cases} (\Gamma^\bullet, \Psi^\bullet, \lambda, \mu) & \text{if } depth((\Gamma^\bullet, \Psi^\bullet, \lambda, \mu)) \le d \\ (\top, \top, \lambda, \mu) & \text{otherwise} \end{cases}$
3. $\Delta = \#(\Gamma, \lambda)$ with $\# \in \{\text{pino}\} \Rightarrow \Delta^\bullet = \begin{cases} (\Gamma^\bullet, \lambda) & \text{if } depth((\Gamma^\bullet, \lambda)) \le d \\ (\top, \lambda) & \text{otherwise} \end{cases}$

By summarising, Δ° denotes a generic abstract membrane label, while $\Delta^\bullet$ exactly denotes the abstract membrane label that is the abstract version of the membrane label Δ.

Abstract States. An *abstract state* reports information on the parent-child relation between membranes and a description of the processes possibly associated to each membrane. Furthermore, it reports information about the possible number of occurrences of membrane and process labels. The occurrence counting information approximates the *global* number of membrane and process labels that may appear in any system.

---

[4] For simplicity, we omit the explicit indication of the parameter d (assume fixed once for all).

To describe the structure of systems, we adopt an *abstract representation*, formally represented by a set of pairs, i.e. by a relation, that, for any abstract membrane label Δ°, gives: (i) the abstract membrane labels that *may* be child membranes of Δ°; and (ii) the *sequential processes* that *may* be associated to membrane Δ°.

Definition 2 (Abstract Representation). *An* abstract representation R° *is a relation* $R^\circ \subseteq \widehat{Lab}^d_\mathcal{M} \times (\widehat{Lab}^d_\mathcal{M} \cup SProc)$, *where* $SProc = \{a^\lambda.\sigma \mid a^\lambda.\sigma \in Proc\}$ *denotes the subset of sequential processes. We use* $\mathcal{R}^\circ$ *to denote the set of abstract representations.*

Given R°, if the pair $(\Delta^\circ, \Gamma^\circ) \in R^\circ$, then the abstract membrane Γ° *may* be a child membrane of the membrane Δ°. Similarly, if the pair $(\Delta^\circ, a^\lambda.\sigma) \in R^\circ$, then the sequential process $a^\lambda.\sigma$ *may* be associated with membrane Δ°.

To describe *occurrence counting* information, we adopt the set Mul $=\{1, \omega\}$ where each $x \in$ Mul denotes a *multiplicity* with the expected interpretation: 1 indicates at most one occurrence, while ω indicates any number of occurrences. The set of multiplicities Mul comes equipped with the standard order $1 \leq \omega$ and with the binary addition operator $+^\circ$, that, for any $x_1, x_2 \in$ Mul, gives ω as a result.

Definition 3 (Occurrence Counting). *An* occurrence counting function *is a partial function* $O^\circ : \widehat{Lab}^d_\mathcal{M} \cup Lab_P \to$ Mul. *We use* $\mathcal{O}^\circ$ *for the set of occurrence counting functions.*

By using a standard notation, an occurrence counting function O° can be alternatively represented by a set of pairs: $\{(\ell, x) \mid \ell \in dom(O^\circ) \wedge O^\circ(\ell) = x\}$.

We rely on some auxiliary operators on occurrence counting functions. First, we introduce the substitution operator $O^\circ[x/\ell]$ that, applied to the occurrence counting function O°, returns the function where the multiplicity of $\ell \in \widehat{Lab}^d_\mathcal{M} \cup Lab_P$ is replaced by $x \in$ Mul.

Moreover, we define an operator $\cup^+$ that computes the addition of two functions $O^\circ_1, O^\circ_2 \in \mathcal{O}^\circ$, the occurrence counting function $O^\circ_1 \cup^+ O^\circ_2$ is defined as follows, where $\ell \in \widehat{Lab}^d_\mathcal{M} \cup Lab_P$,

$$O^\circ_1 \cup^+ O^\circ_2(\ell) = \begin{cases} O^\circ_1(\ell) +^\circ O^\circ_2(\ell) & \text{if } \ell \in dom(O^\circ_1) \cap dom(O^\circ_2) \\ O^\circ_1(\ell) & \text{if } \ell \in dom(O^\circ_1), \ell \notin dom(O^\circ_2) \\ O^\circ_2(\ell) & \text{if } \ell \in dom(O^\circ_2), \ell \notin dom(O^\circ_1) \end{cases}$$

We now have all the ingredients to define abstract states.

Definition 4 (Abstract State). *An* abstract state *is a pair* $S^\circ = (R^\circ, O^\circ)$, *where* $R^\circ \in \mathcal{R}^\circ$ *is an abstract representation and* $O^\circ \in \mathcal{O}^\circ$ *is an occurrence counting function. We use* $\mathcal{S}^\circ$ *for the set of abstract states.*

In standard AI style, the abstract states come equipped with an approximation order (denoted by $\sqsubseteq^\circ$) that allows us to compare two approximations in terms of precision.

Definition 5 (Approximation Orders).

- *Given* $O_1^\circ, O_2^\circ \in \mathcal{O}^\circ$, *we say that* $O_1^\circ \sqsubseteq_O O_2^\circ$ *iff for each* $\ell \in \widehat{\mathrm{Lab}}_\mathcal{M}^d \cup \mathrm{Lab}_\mathcal{P}$ *such that* $\ell \in dom(O_1)$, *we have* $O_1(\ell) = x_1$ *and* $O_2(\ell) = x_2$ *with* $x_1 \leq x_2$.
- *Given* $S_1^\circ, S_2^\circ \in \mathcal{S}^\circ$, *we say that* $S_1^\circ \sqsubseteq^\circ S_2^\circ$ *iff* $S_1^\circ = (R_1^\circ, O_1^\circ)$ *and* $S_2^\circ = (R_2^\circ, O_2^\circ)$, $R_1^\circ \subseteq R_2^\circ$ *and* $O_1^\circ \sqsubseteq_O O_2^\circ$.

Given the previous orders, the corresponding least upper bounds (l.u.b.), $\sqcup_O$ over occurrence counting functions and $\sqcup^\circ$ over abstract states, are defined as expected.

To formally relate systems and abstract states, we introduce a *translation function* t° that maps systems into abstract states. The function $t^\circ : \widehat{\mathrm{Lab}}_\mathcal{M}^d \times$ Sys $\rightarrow \mathcal{S}^\circ$, presented in Table 4, returns an abstract state, describing the system, with respect to an abstract membrane that represents the enclosing membrane. The definition relies, in turn, on a corresponding translation function for processes $t^\circ : \widehat{\mathrm{Lab}}_\mathcal{M}^d \times$ Proc $\rightarrow \mathcal{S}^{\circ 5}$.

Table 4. Translation function for systems and processes.

$$
\begin{array}{ll}
t^\circ(\Delta^\circ, \diamond) & = (\emptyset, \emptyset) \\
t^\circ(\Delta^\circ, P \circ Q) & = (R_1^\circ \cup R_2^\circ, O_1^\circ \cup^+ O_2^\circ) \text{ where } t^\circ(\Delta^\circ, P) = (R_1^\circ, O_1^\circ), t^\circ(\Delta^\circ, Q) = (R_2^\circ, O_2^\circ) \\
t^\circ(\Delta^\circ, !P) & = (R^\circ, O^\circ[\omega/\ell]_{\ell \in dom(O^\circ)})) \text{ where } t^\circ(\Delta^\circ, P) = (R^\circ, O^\circ) \\
t^\circ(\Delta^\circ, \sigma(\!|P|\!)^\Gamma) & = \begin{cases} (R_1^\circ \cup R_2^\circ \cup \{(\Delta^\circ, \Gamma^\bullet)\}, O_1^\circ \cup^+ O_2^\circ \cup^+ \{(\Gamma^\bullet, 1)\}) & \text{if } \sigma \not\equiv 0 \vee P \not\equiv \diamond \\ (\emptyset, \emptyset) & \text{otherwise} \end{cases} \\
& \quad \text{where } t^\circ(\Gamma^\bullet, P) = (R_1^\circ, O_1^\circ) \text{ and } t^\circ(\Gamma^\bullet, \sigma) = (R_2^\circ, O_2^\circ) \\[4pt]
t^\circ(\Delta^\circ, 0) & = (\emptyset, \emptyset) \\
t^\circ(\Delta^\circ, \sigma|\tau) & = (R_1^\circ \cup R_2^\circ, O_1^\circ \cup^+ O_2^\circ) \text{ where } t^\circ(\Delta^\circ, \sigma) = (R_1^\circ, O_1^\circ) \text{ and } t^\circ(\Delta^\circ, \tau) = (R_2^\circ, O_2^\circ) \\
t^\circ(\Delta^\circ, !\sigma) & = (R^\circ, O^\circ[\omega/\ell]_{\ell \in dom(O^\circ)}) \text{ where } t^\circ(\Delta^\circ, \sigma) = (R^\circ, O^\circ) \\
t^\circ(\Delta^\circ, a^\lambda.\sigma) & = (\{(\Delta^\circ, a^\lambda.\sigma)\}, \{(\lambda, 1)\})
\end{array}
$$

Based on the above defined translation function, it is immediate to derive a corresponding *abstraction function* that, given a system, returns the abstract state that is its *best approximation*. Intuitively, the best approximation is the *most precise* (with respect to the order $\sqsubseteq^\circ$) abstract state that *safely represents* the information contained in the system.

Definition 6 (Abstraction Function). *We define* $\alpha_{\mathsf{Sys}} : \mathsf{Sys} \rightarrow \mathcal{S}^\circ$ *such that, given* $P \in \mathsf{Sys}$, $\alpha_{\mathsf{Sys}}(P) = (R^\circ, O^\circ \cup^+ \{(@, 1)\})$, *where* $t^\circ(@, P) = (R^\circ, O^\circ)$.

The best approximation of a system is obtained by applying the translation function t° w.r.t. the abstract membrane label @ representing the outermost membrane. Note that the previously introduced notions can be used to express the fundamental notion of safe approximation between abstract states and systems: an abstract state S° *safely approximates* a system P if and only if $\alpha_{\mathsf{Sys}}(P) \sqsubseteq^\circ S^\circ$. Moreover, the abstraction function is exploited to compute the initial abstract state in the abstract semantics.

---

[5] For simplicity, we use t° for both abstract systems and processes.

Table 5. The abstract state $\alpha_{\mathsf{Sys}}(S) = S_0^\circ = (R_0^\circ, O_0^\circ)$, where $i = 1, 2$.

membrane	children	processes
@	skin	
skin	$source, target_i$	σ_S
source	Γ	$\overline{\mathrm{bud}}_n^{\lambda_i}(Vesicle_i)$
$target_i$		$\mathrm{phago}_{n_i}^{\delta_i}(\overline{\mathrm{exo}}_{n_i}^{\beta_i}), \sigma_{target_i}$
Γ	X	$\overline{\mathrm{bud}}_n^{\gamma}, \tau_X$

membrane/process	multiplicity
@	1
$skin, source$	1
$target_1, target_2$	1
Γ	1
γ	1
$\lambda_1, \lambda_2, \delta_1, \delta_2$	ω

Example 3. Let us consider the system S introduced in Example 1 (see Table 2). Assuming the parameter for the depth of abstract membrane labels $d = 3$, the best approximation of S is given by the abstract state $\alpha_{\mathsf{Sys}}(S) = S_0^\circ = (R_0^\circ, O_0^\circ)$, illustrated in Table 5. For convenience, both the abstract representation R_0° and the occurrence counting function O_0° are described by means of tables, on the left and on the right, respectively.

The table on the left contains one row for each abstract membrane label Δ° in the domain of R_0°. For each Δ° the corresponding row reports in the second column, the set of abstract membrane labels that *may* be children of Δ°, and in the third column the set of sequential processes that *may* be associated to membrane Δ°. More formally, $children(\Delta^\circ) = \{\Theta^\circ \mid (\Delta^\circ, \Theta^\circ) \in R_0^\circ\}$ and $processes(\Delta^\circ) = \{\sigma \mid (\Delta^\circ, \sigma) \in R_0^\circ\}$. Hence, the third line can be read as the membrane *source* may include the membrane Γ, and it may have associated the processes $\overline{\mathrm{bud}}_n^{\lambda_1}(Vesicle_1)$ and $\overline{\mathrm{bud}}_n^{\lambda_2}(Vesicle_2)$.

The table on the right reports the multiplicities for each abstract membrane and process label in the domain of O_0°. For instance, the membrane labels *skin* and *source* have multiplicity 1, while the process labels λ_i and δ_i (with $i = 1, 2$) come with multiplicity ω. The corresponding prefixes occur indeed under the scope of a replication (see the rules in Table 4).

Abstract Transitions. The abstract semantics is given in terms of the abstract transition relation $\xrightarrow{l^\circ}_\circ$ among abstract states, where $l^\circ \in \mathsf{Lab}_{\mathcal{T}}^\circ$ is the *abstract transition label* describing the reaction. The abstract transitions are obtained by introducing inference rules for abstract states that model the abstract counterpart of the membrane interactions possible in the concrete system.

The set of *abstract transition labels* $\mathsf{Lab}_{\mathcal{T}}^\circ$ [6] (ranged over by $l_1^\circ, l_2^\circ, \ldots$) is defined as in the concrete case, by replacing membrane labels with abstract membrane labels. Thus, we have:

$$\mathsf{Lab}_{\mathcal{T}}^\circ = \{\mathrm{pino}_1(\Delta^\circ, \lambda), \mathrm{bud}_1(\Gamma^\circ, \Delta^\circ, \lambda, \mu), \mathrm{exo}_1(\Gamma^\circ, \Delta^\circ, \lambda, \mu), \mathrm{phago}_1(\Gamma^\circ, \Delta^\circ, \lambda, \mu) \mid$$
$$\Gamma^\circ, \Delta^\circ \in \widehat{\mathsf{Lab}}_{\mathcal{M}}^d, \lambda, \mu \in \mathsf{Lab}_{\mathcal{P}}\}$$

Due to the lack of space, we comment here only on the abstract inference rule, given in Table 6, corresponding to the (Bud) interaction. The abstract inference rules (PHAGO$^\circ$), (EXO$^\circ$), (PINO$^\circ$) in Table 7, corresponding to (PHAGO),

---

[6] For simplicity, we omit the explicit indication of the parameter d when is clear from the context.

(Exo), (Pino) reactions can be derived in similar way from their concrete versions.

Rule (Bud°) uses an auxiliary operator co to modify an occurrence counting function $O°$ according to a given multiplicity x.

$$co(O°, x) = \begin{cases} O°[\omega/\ell]_{\ell \in dom(O°)} & \text{if } x = \omega \\ O° & \text{otherwise} \end{cases}$$

Rule (Bud°) simulates the concrete (Bud) rule, by modelling the gemmation of a membrane $\Delta°$ from another membrane $\Gamma°$ that may synchronise on actions $\overline{\mathrm{bud}}_n^\mu(\rho)$ and bud_n^λ. This requires that: (i) the abstract membrane $\Delta°$ is reported as a possible child of the membrane $\Gamma°$ (i.e. $(\Gamma°, \Delta°) \in R°$); (ii) according to the abstract representation $R°$, the actions cobud and bud may be associated to membranes $\Gamma°$ and $\Delta°$, respectively. Furthermore, it must be the case that the multiplicities of the process labels μ and λ associated to the actions are defined.

The abstract transition label $l°$ is derived, as in the concrete case, by combining the labels of the membranes and of the actions involved. The resulting abstract state is obtained by enriching the abstract state $(R°, O°)$ with information reporting the effects of the possible movement of the membrane $\Delta°$ out from the membrane $\Gamma°$. This requires to update both the abstract representation and the occurrence counting function. Note that the membrane introduced by the bud reaction is described by the abstract membrane label $\Pi°$, obtained by approximating the membrane $(\Delta°, \Gamma°, \lambda, \mu)$ according to its depth.

The abstract representation is extended by introducing the abstract membrane $\Pi°$ as a possible child of the membrane $\Phi°$ (in turn, parent of $\Gamma°$), and $\Delta°$ as a possible child of membrane $\Pi°$. Moreover, we have to introduce information on the membrane processes that may be associated to membranes $\Gamma°$, $\Delta°$ and $\Pi°$. In the case of membrane $\Pi°$, this requires to add $R°_2$ obtained by applying the translation function to process ρ related to cobud. Similarly, in the case of the membranes $\Gamma°$ and $\Delta°$ the related abstract representations $R°_3$ and $R°_4$ are obtained by applying the continuations of the two coactions (σ and τ), respectively.

Finally, the occurrence counting function is updated by adding one occurrence of membrane $\Pi°$ introduced by the bud reaction and the occurrence counting functions $O°_2$, $O°_3$, and $O°_4$, obtained by the translations the process ρ and of the continuations of the coactions. Note that co operator allows us to propagate the ω multiplicity, in the case of the continuations of prefixes under replication.

The Analysis. The analysis of a system P provides an *abstract state* describing the possible topological structure of all the derivatives of P together with occurrence counting information on membrane and process labels. We aim at calculating such abstract state by collecting all the abstract states that can be reached from the initial one $\alpha_{\mathrm{Sys}}(P)$, by applying the abstract inference rules.

Nevertheless, the application of the abstract inference rules without a strategy would lead us to have a correct, but very coarse approximation, especially

Table 6. Abstract inference rule for (BUD).

$$(\text{BUD}°)$$

$$(\Phi°, \Gamma°) \in R°, (\Gamma°, \Delta°) \in R°,$$
$$(\Gamma°, \overline{\text{bud}}_n^\mu(\rho).\tau) \in R°, (\Delta°, \text{bud}_n^\lambda.\sigma) \in R°, \; O°(\lambda) = x, \; O°(\mu) = y$$

$$(R°, O°) \xrightarrow{l°}{}_\circ (R° \cup R_1° \cup R_2° \cup R_3° \cup R_4°, O° \cup^+ \{(\Pi°, 1)\} \cup^+ O_2° \cup^+ co(O_3°, x) \cup^+ co(O_4°, y))$$
$$R_1° = \{(\Phi°, \Pi°), (\Pi°, \Delta°)\},$$
$$t°(\Pi°, \rho) = (R_2°, O_2°), t°(\Delta°, \sigma) = (R_3°, O_3°), t°(\Gamma°, \tau) = (R_4°, O_4°)$$

where $l° = \text{bud}_1(\Delta°, \Gamma°, \lambda, \mu)$ and $\Pi° = \begin{cases} (\Delta°, \Gamma°, \lambda, \mu) & \text{if } (\Delta°, \Gamma°\lambda, \mu) \in \widehat{\text{Lab}}_\mathcal{M}^d, \\ (\top, \top, \lambda, \mu) & \text{otherwise} \end{cases}$

as far as the counting information is concerned. The reason is that, in principle, any enabled reaction would be applied several times. As a consequence, infinite copies of the corresponding membranes and processes are introduced, even though there are cases in which this behaviour cannot occur in the dynamic evolution of the system. Our strategy for overcoming this problem consists in exploiting occurrence counting information to determine which abstract transitions apply to an abstract state. This allows us to more faithfully model the concrete behaviour and therefore to gain precision in our analysis.

Table 7. Abstract inference rules for (PHAGO), (EXO), and (PINO).

$$(\text{PHAGO}°)$$

$$(\Phi°, \Delta°) \in R°, (\Phi°, \Gamma°) \in R°,$$
$$(\Delta°, \text{phago}_n^\lambda.\sigma) \in R°, (\Gamma°, \overline{\text{phago}}_n^\mu(\rho).\tau) \in R,° \; O°(\lambda) = x, \; O°(\mu) = y$$

$$(R°, O°), \xrightarrow{l°}{}_\circ (R° \cup R_1° \cup R_2° \cup R_3° \cup R_4°, O° \cup^+ \{(\Pi°, 1)\} \cup^+ O_2° \cup^+ co(O_3°, x) \cup^+ co(O_4°, y))$$
$$R_1° = \{(\Gamma°, \Pi°), (\Pi°, \Delta°)\}$$
$$t°(\Pi°, \rho) = (R_2°, O_2°), t°(\Delta°, \sigma) = (R_3°, O_3°), t°(\Gamma°, \tau) = (R_4°, O_4°)$$

where $l° = \text{phago}_1(\Delta°, \Gamma°, \lambda, \mu)$ and $\Pi° = \begin{cases} (\Delta°, \Gamma°, \lambda, \mu) & \text{if } (\Delta°, \Gamma°\lambda, \mu) \in \widehat{\text{Lab}}_\mathcal{M}^d, \\ (\top, \top, \lambda, \mu) & \text{otherwise} \end{cases}$

$$(\text{EXO}°)$$

$$(\Phi°, \Gamma°) \in R°, (\Gamma°, \Delta°) \in R°,$$
$$(\Gamma°, \overline{\text{exo}}_n^\mu.\tau) \in R°, (\Delta°, \text{exo}_n^\lambda.\sigma) \in R°, \; O°(\lambda) = x, \; O°(\mu) = y$$

$$(R°, O°) \xrightarrow{l°}{}_\circ (R° \cup R_1° \cup R_2° \cup R_3° \cup R_4°, O° \cup^+ co(O_3°, x) \cup^+ co(O_4°, y)),$$
$$R_1° = \{(\Phi°, \Theta°) \mid (\Delta°, \Theta°) \in R°\}, \; R_2° = \{(\Gamma°, \tau') \mid \tau' \in sub(C, x, \text{exo}_n^\lambda.\sigma)\},$$
$$C = \{\sigma' \mid (\Delta°, \sigma') \in R°\} \; t°(\Gamma°, \sigma) = (R_3°, O_3°), t°(\Gamma°, \tau) = (R_4°, O_4°)$$

where $l° = \text{exo}_1(\Delta°, \Gamma°, \lambda, \mu)$ and $sub(C, x, a^\lambda.\sigma) = \begin{cases} C \setminus \{a^\lambda.\sigma\} & \text{if } x = 1 \\ C & \text{otherwise} \end{cases}$

$$(\text{PINO}°)$$

$$(\Delta°, \text{pino}^\lambda(\rho).\sigma) \in R°$$

$$(R°, O°) \xrightarrow{l°}{}_\circ (R° \cup R_1° \cup R_2° \cup R_3°, O° \cup^+ \{(\Pi°, 1)\} \cup^+ O_2° \cup^+ co(O_3°, O°(\lambda)))$$
$$R_1° = \{(\Delta°, \Pi°)\}, \; t°(\Pi°, \rho) = (R_2°, O_2°), \; t°(\Delta°, \sigma) = (R_3°, O_3°)$$

where $l° = \text{pino}_1(\Delta°, \lambda)$ and $\Pi° = \begin{cases} (\Delta°, \lambda) & \text{if } (\Delta°, \lambda) \in \widehat{\text{Lab}}_\mathcal{M}^d, \\ (\top, \lambda), & \text{otherwise} \end{cases}$

We start by analysing the information given by *abstract transition labels*, and by distinguishing: (i) the abstract transitions that require updating the occurrence counting information more than once, from (ii) those that require to do it just once. Intuitively, the former model concrete transitions that may occur more than once in a concrete derivation, while the latter model concrete transitions that occur at most once.

To this aim, we need to define the *multiplicity* of abstract transition labels in a given abstract state, computed by the function $mul : S^\circ \times \mathtt{Lab}^\circ_T \to \mathtt{Mul}$ defined as follows:

$$mul((R^\circ, O^\circ), l^\circ) = \begin{cases} \omega & \text{if } l^\circ = \mathtt{pino}_1(\Delta^\circ, \lambda) \wedge O^\circ(\lambda) = \omega, \\ \omega & \text{if } l^\circ = a_1(\Gamma^\circ, \Theta^\circ, \lambda, \mu), a_1 \in \{\mathtt{bud}_1, \mathtt{exo}_1, \mathtt{phago}_1\} \wedge O^\circ(\lambda) = O^\circ(\mu) = \omega, \\ 1 & \text{otherwise} \end{cases}$$

Note that the multiplicity assigned to an abstract transition label entirely depends on the multiplicity of labels associated to the actions that participate in the reaction. For any kind of reaction if all the involved actions have multiplicity ω, then also the associated abstract transition label has multiplicity ω. In this case indeed the reaction may be applied *more than once* in the corresponding concrete derivations. On the contrary, if at least one of the actions involved in the reaction has multiplicity 1, then the corresponding reaction may be executed *no more than once* in any derivation of the concrete system.

The multiplicity of transition labels is indeed exploited to compute the abstract semantics, where abstract states are enriched with information on the involved abstract transitions labels. More precisely, we have *configurations* in the form $T^\circ \triangleright S^\circ$, where $S^\circ \in S^\circ$ is an abstract state and $T^\circ \subseteq \mathtt{Lab}^\circ_T$ is a set of abstract transition labels representing the reactions that have been already exercised. We use C° to denote the set of configurations.

To describe the evolution of configurations we introduce two meta-inference rules that encode our strategy for the application of abstract rules. These rules allow us to define the evolution of a configuration $T^\circ \triangleright (R^\circ_1, O^\circ_1)$ into another configuration, whenever there exists an abstract reaction $(R^\circ_1, O^\circ_1) \xrightarrow{l^\circ}_\circ (R^\circ_2, O^\circ_2)$. The choice of the meta-inference rule depends on the multiplicity of the abstract transition label l° associated to the reaction, i.e. $mul((R^\circ_1, O^\circ_1), l^\circ)$.

$$\frac{(R^\circ_1, O^\circ_1) \xrightarrow{l^\circ}_\circ (R^\circ_2, O^\circ_2) \wedge (l^\circ \notin T^\circ \vee (l^\circ \in T^\circ \wedge mul((R^\circ_1, O^\circ_1), l^\circ) = \omega)}{T^\circ \triangleright (R^\circ_1, O^\circ_1) \xrightarrow{l^\circ}_\triangleright T^\circ \cup \{l^\circ\} \triangleright (R^\circ_2, O^\circ_2)} \quad (1)$$

$$\frac{(R^\circ_1, O^\circ_1) \xrightarrow{l^\circ}_\circ (R^\circ_2, O^\circ_2) \wedge (l^\circ \in T^\circ \wedge mul((R^\circ_1, O^\circ_1), l^\circ) = 1)}{T^\circ \triangleright (R^\circ_1, O^\circ_1) \xrightarrow{l^\circ}_\triangleright T^\circ \triangleright (R^\circ_1, O^\circ_1)} \quad (2)$$

We can apply the first rule (1), provided that either the reaction l° has never been applied before ($l^\circ \notin T^\circ$) or its multiplicity is ω. Thus, either the reaction associated to l° can be applied only once and it has not been realised or it can be realised any number of times. In both cases, the resulting configuration is obtained by recording that the reaction l° has now been performed, and by updating both the abstract representation and the occurrence counting information.

We can apply the second rule (2), if the reaction related to l° has multiplicity 1 and has already been applied ($l^\circ \in T^\circ$). In this case, it may indicate that the concrete reaction approximated by l° can be performed in another context, different from the one considered before. This requires updating the abstract representation reporting the effects of the move, while the occurrence counting information does not have to be modified since it already reports the correct multiplicities of the membrane and process labels involved in the move.

Example 4. To illustrate the application of meta-inference rule (1) let us consider the abstract state $\alpha_{\mathsf{Sys}}(S) = S_0^\circ = (R_0^\circ, O_0^\circ)$ of Example 3 (see Table 5) describing the best approximation of the system S, presented in Example 1. Note that we can apply the abstract rule (BUD°) to $\alpha_{\mathsf{Sys}}(S)$, because its premises are fulfilled: (i) $(skin, source), (source, \Gamma) \in R^\circ{}_0$ and (ii) $(\Gamma, \mathsf{bud}_n^\gamma), (source, \overline{\mathsf{bud}}_n^{\lambda_1}(Vesicle_1)) \in R_0^\circ$. Furthermore, $O_0^\circ(\lambda_1)$ and $O_0^\circ(\gamma)$ are defined. As a consequence, we have $\alpha_{\mathsf{Sys}}(S) \xrightarrow{l_1^\circ}{}_\circ S_1^\circ$, where $l_1^\circ = \mathsf{bud}_1(\Gamma, source, \gamma, \lambda_1)$ and the state $S_1^\circ = (R_1^\circ, O_1^\circ)$ is the one depicted in Table 8. Hence, considering the configuration $\emptyset \triangleright \alpha_{\mathsf{Sys}}(S) = S_0^\circ$, we can apply meta-inference rule (1), since $l_1^\circ \notin \emptyset$, obtaining $\emptyset \triangleright \alpha_{\mathsf{Sys}}(S) \xrightarrow{l_1^\circ}{}_\triangleright \{l_1^\circ\} \triangleright S_1^\circ$. Note that, since $O_0^\circ(\gamma) = 1$ (while $O_0^\circ(\lambda_1) = \omega$) we have that $mul((R_0^\circ, O_0^\circ), l_1^\circ) = 1$.

The analysis of a system P provides an abstract state that is obtained by collecting (taking the l.u.b.) all the abstract states that can be reached from the initial configuration $\emptyset \triangleright \alpha_{\mathsf{Sys}}(P)$, by applying the meta-inference rules (1) and (2).

Definition 7 (The Analysis). *We define a function $\mathcal{A}^\circ : \mathsf{Sys} \to \mathcal{S}^\circ$ such that for $P \in \mathsf{Sys}$ we have $\mathcal{A}^\circ(P) = \bigsqcup_{\{S^\circ | T^\circ \triangleright S^\circ \in X_P^\circ\}}^\circ S^\circ$, where $X_P^\circ = lfp(\mathcal{F}^\circ(\{\emptyset \triangleright \alpha_{\mathsf{Sys}}(P)\}))^7$ and the function $\mathcal{F}^\circ : \wp(\mathcal{C}^\circ) \to \wp(\mathcal{C}^\circ)$ defined as $\mathcal{F}^\circ(X_1^\circ) = X_1^\circ \cup \{C_2^\circ \mid C_1^\circ \xrightarrow{l^\circ}{}_\triangleright C_2^\circ, C_1^\circ \in X_1^\circ\}$.*

Despite the fact that the analysis involves a fixed point over a power domain, which seems to admit exponentially long increasing paths, the analysis can be

Table 8. The abstract state $S_1^\circ = (R_1^\circ, O_1^\circ)$, where $i = 1, 2$.

membrane	children	processes	membrane/process	multipl.
@	skin		@	1
skin	$source, target_i, \Pi_1^\circ$	σs	$skin, source,$	1
source	Γ	$\overline{\mathsf{bud}}_n^{\lambda_i}(Vesicle_i),$	$target_1, target_2,$	1
$target_i$		$\overline{\mathsf{phago}}_{n_i}^{\delta_i}(\mathsf{exo}_{n_i}^{\beta_i}), \sigma_{target_i}$	Γ, Π_1°	1
Γ	X	$\mathsf{bud}_n^\gamma, \tau_X$	γ, μ_1	1
$\Pi_1^\circ = (\Gamma, source, \gamma, \lambda_1)$	Γ	$\mathsf{phago}_{n_1}^{\mu_1}.\mathsf{exo}_{n_1}^{\nu_1}, \tau_1'$	$\lambda_1, \lambda_2, \delta_1, \delta_2$	ω

---

^{7} where *lfp* is the least fixed point.

computed in polynomial time. For its computation there is no need to deal with power sets: crafting a single maximal path is enough to compute the fixed point, which is a singleton set. This allows us to obtain a polynomial bound. It can be shown indeed that the analysis can be effectively computed by building a single maximal path starting from the initial configuration and ending into a final configuration, i.e. a configuration that cannot further evolve according to the meta-inference rules (1) and (2).

This property relies on the following result.

Theorem 1. *Let $P \in \mathsf{Sys}$ be well labelled and let $C_1^\circ, C_2^\circ \in \mathcal{C}^\circ$ be two configurations such that $C_1^\circ, C_2^\circ \in \mathit{lfp}(\mathcal{F}^\circ(\{\emptyset \triangleright \alpha_{\mathsf{Sys}}(P)\}))$, as defined in Definition 7. If C_1° and C_2° are final configurations then $C_1^\circ = C_2^\circ$.*

The previous property allows us to calculate the analysis of a system without computing all the configurations that can be reached from the initial one. Indeed, $\mathcal{A}^\circ(P)$ can be computed by building a single path

$$T_0^\circ \triangleright S_0^\circ, \quad T_1^\circ \triangleright S_1^\circ, \quad \ldots, \quad T_m^\circ \triangleright S_m^\circ$$

where (i) $T_0^\circ = \emptyset$, $S_0^\circ = \alpha_{\mathsf{Sys}}(P)$; (ii) $T_m^\circ \triangleright S_m^\circ$ is a final configuration; and, (iii) for each $i \in [1, m]$ the corresponding configuration is obtained by applying the meta-inference rules (1) and (2) to the previous configuration $T_{i-1}^\circ \triangleright S_{i-1}^\circ$. Note that this path is an ascending chain since, for each $i \in [0, m-1]$, either $T_i^\circ \subset T_{i+1}^\circ$ and $S_i^\circ \sqsubseteq^\circ S_{i+1}^\circ$ or $T_i^\circ \subseteq T_{i+1}^\circ$ and $S_i^\circ \sqsubset^\circ S_{i+1}^\circ$. Hence, we have that the analysis of system P precisely coincides with the final state, i.e. $\mathcal{A}^\circ(P) = S_m^\circ$.

The above reasoning guarantees that the analysis can be computed in *polynomial time*, observing that the number of abstract membranes and transition labels arising in the computation of the analysis is polynomial, when fixing the maximum depth d to a constant value.

Finally, we present the main theorem that shows that the analysis of a system *safely approximates* its concrete behaviour, described by the concrete LTS. This means that each derivative P' of P is over-approximated by the abstract state calculated by the analysis of P.

Theorem 2 (Safety). *Let $P \in \mathsf{Sys}$ be a well labelled system and let $LTS(P) = (X, \xrightarrow{l}, P)$. We have that $\left(\bigsqcup_{P' \in X}^\circ \alpha_{\mathsf{Sys}}(P')\right) \sqsubseteq^\circ \mathcal{A}^\circ(P)$.*

5 Our Analysis at Work

We now apply our analysis to the systems presented in Sect. 3 (assuming again $d = 3$ analogously as in Sect. 4). We illustrate in more the details the analysis of the system S of Example 1, whose first steps have been introduced in Example 3 and 4. Since the analysis of the system S' described in Example 2 is similarly obtained, we only comment its results.

Example 5. The analysis of the system S shown in Table 2 of Example 1 is computed starting from the initial configuration $\emptyset \triangleright \alpha_{\mathsf{Sys}}(S)$, where $\alpha_{\mathsf{Sys}}(S) = S^\circ_0 = (R^\circ_0, O^\circ_0)$ is the abstract state of Table 5 (commented in Example 3).

Table 9. The abstract state $S^\circ = (R^\circ, O^\circ)$, where $i = 1, 2$.

membrane	children	processes		membrane/process	multipl.
@	skin			@	1
skin	source, target$_i$, Π_i°	σS		skin, source,	1
source	Γ	$\overline{\text{bud}}_n^{\lambda_i}(Vesicle_i)$,		target$_1$, target$_2$,	1
target$_i$	Π_{i1}°, Γ	$\overline{\text{phago}}_{n_i}^{\delta_i}(\overline{\text{exo}}_{n_i}^{\beta_i}), \sigma_{target_i}$		$\Gamma, \Pi_1^\circ, \Pi_2^\circ$	1
Γ	X	$\text{bud}_n^\gamma, \tau X$		$\Pi_{11}^\circ, \Pi_{21}^\circ$	1
$\Pi_i^\circ = (\Gamma, source, \gamma, \lambda_i)$	Γ	$\underbrace{\text{phago}_{n_i}^{\mu_i} . \text{exo}_{n_i}^{\nu_i},}_{Vesicle_i}$ $\text{exo}_{n_i}^{\nu_i}, \tau_i'$		$\gamma, \mu_1, \mu_2, \nu_1, \nu_2$	1
				β_1, β_2	1
$\Pi_{i1}^\circ = (\Pi_i^\circ, target_i, \delta_i, \mu_i)$	Π_i°	$\overline{\text{exo}}_{n_i}^{\beta_i}, Vesicle_i, \tau_i'$		$\lambda_1, \lambda_2, \delta_1, \delta_2$	ω

The final configuration is given by $\{l^\circ{}_1, l^\circ{}_{11}, l^\circ{}_{12}, l^\circ{}_2, l^\circ{}_{21}, l^\circ{}_{22}\} \triangleright S^\circ$, where $S^\circ = (R^\circ, O^\circ)$ is the abstract state in Table 9, and the abstract transition labels for $i = 1, 2$ are:

$$l^\circ{}_i = \text{bud}_1(\Gamma, source, \gamma, \lambda_i), l^\circ{}_{i1} = \text{phago}_1(\Pi^\circ{}_i, target_i, \delta_i, \mu_i), l^\circ{}_{i2} = \text{exo}_1(\Pi^\circ{}_i, \Pi^\circ{}_{i1}, \nu_i, \beta_i).$$

Here, the abstract transition labels $l^\circ{}_1$, $l^\circ{}_{11}$ and $l^\circ{}_{12}$ are the abstract versions of the transition labels l_{11} and l_{12} in Table 2. They represent the (bud), (phago) and (exo) reactions performed by the first vesicle, respectively. Analogously, the abstract transition labels $l^\circ{}_2$, $l^\circ{}_{21}$ and $l^\circ{}_{22}$ represent the labels introduced by the similar evolution of the second vesicle. Note that in this case all the abstract transition labels have multiplicity 1, and consequently only the meta-inference rule (2) can be applied to the final configuration. As a consequence, at this point, no matter which reaction is applied, the final configuration cannot further evolve, and, in particular, the information counting information cannot be updated anymore.

Hence, we can conclude that $\mathcal{A}^\circ(S) = S^\circ = (R^\circ, O^\circ)$. For clarity, the membrane hierarchy described by abstract representation R° is shown in the tree in Fig. 1, where the nodes represent the abstract membrane labels and the edges represent the parent-child relation. It is worth noting that the information provided by R° predicts that the membrane Γ, which encloses the molecule X, *may* end up in membrane $target_1$, as well as in membrane $target_2$. However, the occurrence counting information expressed by O° guarantees that the membrane Γ will never reside at the same time inside the membranes $target_1$ and $target_2$. To point this out the two alternative inclusions of membrane Γ inside the membranes $target_1$ and $target_2$ the lines are displayed with dotted edges (blue in the pdf) in Fig. 1. Note that without applying the meta-inference rules (1) and (2), and by repeatedly updating the occurrence counting information, while applying abstract inference rules, we would obtain that $O^\circ(\Gamma) = \omega$, thus losing the information necessary to determine the alternative presence of membrane Γ in the two target membranes.

Example 6. The analysis of the system S' described in Table 3 of Example 2 is given be the abstract state $S'^\circ = (R'^\circ, O'^\circ)$ illustrated in Table 10. The analysis

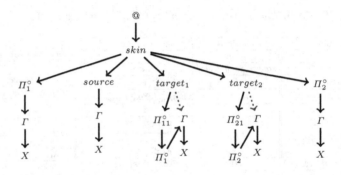

Fig. 1. The membrane hierarchy tree described by R°.

result is obtained as in Example 5 starting from the corresponding initial configuration. Note that the abstract representation R'° roughly describes the same information reported in Example 5, while the occurrence counting information O'° makes a difference. Given that the multiplicity of membrane Γ is in this case ω, the analysis reveals that Γ *may* be enclosed at the same time inside membranes $target_1$ and $target_2$[8].

We can then conclude that our analysis, thanks to the occurrence counting information, allows us to observe that the two biological systems introduced in Sect. 3 exhibit a different dynamical behaviour. In both cases, the analysis predicts that molecule X may end up both in $target_1$ and in $target_2$, but only in the first scenario the two inclusions are alternative.

Note that, in general, information on the possible presence/absence of a component in a membrane could also be exploited when developing a biological model, to detect errors in the model specification.

Table 10. The abstract state $S'^\circ = (R'^\circ, O'^\circ)$, where $i, j = 1, 2$ and $i \neq j$.

membrane	children	processes
@	skin	
skin	$source, target_i,$ $\Pi_i^\circ, \Pi_{Dij}^\circ$	σ_S
source	Γ	$\overline{bud}_n^{\lambda_i}(Vesicle_i),$
$target_i$	$\Pi_{i1}^\circ, \Pi_{Dij1}^\circ, \Gamma$	$\overline{phago}_{n_i}^{\delta_i}(\overline{exo}_{n_i}^{\beta_i}), \sigma_{target_i}$
Γ	X	$bud_n^{\gamma_i}, \tau X$
$\Pi_i^\circ = (\Gamma, source, \gamma_i, \lambda_i)$	Γ	$\underbrace{phago_{n_i}^{\mu_i}.exo_{n_i}^{\nu_i},}_{Vesicle_i}$ $exo_{n_i}^{\nu_i}, \tau_i'$
$\Pi_{Dij}^\circ = (\Gamma, source, \gamma_i, \lambda_j)$	Γ	$exo_{n_j}^{\nu_j}, Vesicle_j, \tau_j'$
$\Pi_{i1}^\circ = (\Pi_i^\circ, target_i, \delta_i, \mu_i)$	Π_i°	$\overline{exo}_{n_i}^{\beta_i}, Vesicle_i, \tau_i'$
$\Pi_{Dij1}^\circ = (\Pi_{Dij}^\circ, target_j, \delta_j, \mu_j)$	Π_{Dij}°	$\overline{exo}_{n_j}^{\beta_j}, Vesicle_j, \tau_j'$

membrane/process	multipl.
@	1
$skin,$	1
$source, \Gamma$	ω
γ_1, γ_2	1
$target_1, target_2,$	1
Π_1°, Π_2°	1
$\Pi_{11}^\circ, \Pi_{21}^\circ$	1
$\Pi_{D12}^\circ, \Pi_{D21}^\circ$	1
$\Pi_{D121}^\circ, \Pi_{D211}^\circ$	1
$\mu_1, \mu_2, \nu_1, \nu_2$	1
β_1, β_2	1
$\lambda_1, \lambda_2, \delta_1, \delta_2,$	ω

---

[8] Note that giving the two membrane sources the same label does not influence the result.

6 Conclusions

We presented an analysis based on *Abstract Interpretation* techniques for approx-imating the behaviour of biological systems described in Brane Calculi [6]. The analysis consists in two components. The first component, which over-approximates the possible membrane hierarchy, is obtained by adapting static analysis techniques used for process algebras handling biological compartments (see e.g. [1–4, 21]). The less standard occurrence counting component of the analysis is used to predict whether some component may occur at most once in any system reachable from the initial one. The two components influence each other. In particular, the prediction on the possible membrane hierarchy is refined with global occurrence counting information, thus allowing us to increase the precision with respect to the previous static approaches for Brane Calculi [2–4]. Note that the analyses providing occurrence counting information proposed for BioAmbients [14–16, 25], the sibling bio-inspired calculus, cannot be straight-forwardly adapted to Brane Calculi. A careful labelling technique for membranes is required indeed because of bitonality, i.e. the fact that brane interactions pos-sibly introduce new membranes, in order not to mix what is inside a membrane with what is outside (entities can be shuttled inside or outside, only if wrapped by another membrane).

To validate the applicability of our analysis in the biological setting, we applied it to two minimal examples of communication via mobile vesicles that exhibit different dynamical behaviours with respect to the presence of a molecule X inside two target membranes: simultaneous in the second case, but not in the first one. Despite its simplicity, without the occurrence counting, the analysis failed to detect differences between the two cases, thus not predicting that, in the first case, X cannot be simultaneously present in the two target membranes, while in the second case, it may occur in both.

As future work, we would like to improve the accuracy of our approach. In particular, we would like to better analyse the systems that contain different instances of the same membrane or of the same process, and to better handle replication. One possible direction would be to refine our analysis by providing local occurrence counting information in the style of [15, 16]. In this way, we could apply our approach to more complex biological case studies, such as the one modelled in [5], for investigating the relationships occurring among events. Note that the formalisation of the LDL Cholesterol Degradation pathway in Brane Calculi presented in [5] requires a version of the calculus with recursive definitions in place of replication. Recursive definitions have been shown useful to provide a more intuitive modelling of real systems with an infinite behaviour.

References

1. Bodei, C.: Control flow analysis for beta-binders with and without static compart-ments. Theor. Comput. Sci. **410**(33–34), 3110–3127 (2009)
2. Bodei, C., Brodo., L.: Brane calculi systems: A static preview of their possible behaviour. CoRR abs/1108.3429 (2011)

3. Bodei, C., Gori, R., Levi, F.: An analysis for causal properties of membrane inter-actions. Electr. Notes Theor. Comput. Sci. **299**, 15–31 (2013)
4. Bodei, C., Gori, R., Levi, F.: Causal static analysis for brane calculi. Theor. Comput. Sci. **587**, 73–103 (2015)
5. Busi, N., Zandron, C.: Modeling and analysis of biological processes by mem(brane) calculi and systems. In: Proceedings of the Winter Simulation Conference WSC 2006, pp. 1646–1655. WSC (2006)
6. Cardelli, L.: Brane calculi. In: Danos, V., Schachter, V. (eds.) CMSB 2004. LNCS (LNBI), vol. 3082, pp. 257–278. Springer, Heidelberg (2005)
7. Cardelli, L., Gordon, A.: Mobile ambients. Theor. Comput. Sci. **240**(1), 177–213 (2000)
8. Cousot, P., Cousot, R.: Static determination of dynamic properties of generalized type unions. In: Language Design for Reliable Software, pp. 77–94 (1977)
9. Feret, J.: Confidentiality analysis of mobile systems. In: Palsberg, J. (ed.) Static Analysis. LNCS, vol. 1824, pp. 135–154. Springer, Heidelberg (2000)
10. Feret, J.: Abstract interpretation-based static analysis of mobile ambients. In: Cousot, P. (ed.) SAS 2001. LNCS, vol. 2126, p. 412. Springer, Heidelberg (2001)
11. Feret, J.: Occurrence counting analysis for the pi-calculus. Electr. Notes Theor. Comput. Sci. **39**(2), 1–18 (2001)
12. Feret, J.: Abstract interpretation of mobile systems. J. Log. Algebr. Program. **63**(1), 59–130 (2005)
13. Feret, J.: Analysis of mobile systems by abstract interpretation. Ph.D. thesis, École Polytechnique, Paris, France (2005)
14. Gori, R., Levi, F.: An analysis for proving temporal properties of biological systems. In: Kobayashi, N. (ed.) APLAS 2006. LNCS, vol. 4279, pp. 234–252. Springer, Heidelberg (2006)
15. Gori, R., Levi, F.: A new occurrence counting analysis for bioambients. In: Yi, K. (ed.) APLAS 2005. LNCS, vol. 3780, pp. 381–400. Springer, Heidelberg (2005)
16. Gori, R., Levi, F.: Abstract interpretation based verification of temporal properties for bioambients. Inf. Comput. **208**(8), 869–921 (2010)
17. Guerriero, M., Prandi, D., Priami, C., Quaglia, P.: Process calculi abstractions for biology. In: Condon, A., Harel, D., Kok, J.N., Salomaa, A., Winfree, E. (eds.) Algorithmic Bioprocesses Natural Computing Series, pp. 463–486. Springer, Heidelberg (2009)
18. Levi, F., Maffeis, S.: On abstract interpretation of mobile ambients. Inf. Comput. **188**(2), 179–240 (2004)
19. Milner, R.: Communicating and mobile systems: the π-calculus. Cambridge University Press, Cambridge (1999)
20. Nielson, F., Hansen, R., Nielson, H.R.: Abstract interpretation of mobile ambients. Sci. Comput. Program. **47**(2–3), 145–175 (2003)
21. Nielson, F., Nielson, H.R., Priami, C., da Rosa, D.S.: Control flow analysis for bioambients. ENTCS **180**(3), 65–79 (2007)
22. Nielson, H.R., Nielson, F.: Shape analysis for mobile ambients. In: Proceedings of the 27th ACM SIGPLAN-SIGACT Symposium on Principles of Programming Languages (POPL 2000), pp. 142–154. ACM Press (2000)
23. Nielson, H.R., Nielson, F., Pilegaard, H.: Spatial analysis of bioambients. In: Giacobazzi, R. (ed.) SAS 2004. LNCS, vol. 3148, pp. 69–83. Springer, Heidelberg (2004)
24. Pilegaard, H., Nielson, H.R., Nielson, F.: Context dependent analysis of bioambients. In: Simulation and Verification of Dynamic Systems. Dagstuhl Seminar Proceedings, vol. 06161 (2006)

25. Pilegaard, H., Nielson, H.R., Nielson, F.: Pathway analysis for bioambients. J. Log. Algebr. Program. **77**(1–2), 92–130 (2008)
26. Priami, C., Quaglia, P.: Beta binders for biological interactions. In: Danos, V., Schachter, V. (eds.) CMSB 2004. LNCS (LNBI), vol. 3082, pp. 20–33. Springer, Heidelberg (2005)
27. Regev, A., Panina, E., Silverman, W., Cardelli, L., Shapiro, E.: Bioambients: An abstraction for biological compartements. Theor. Comput. Sci. **325**(1), 141–167 (2004)
28. Vitale, A., Mauri, G.: Communication via mobile vesicles in brane calculi. ENTCS **171**(2), 187–196 (2007)

Using Dynamic Pushdown Networks to Automate a Modular Information-Flow Analysis

Heiko Mantel[1], Markus Müller-Olm[2],
Matthias Perner[1][✉], and Alexander Wenner[2]

[1] Computer Science Department, TU Darmstadt, Darmstadt, Germany
{mantel,perner}@cs.tu-darmstadt.de
[2] Institut Für Informatik, Westfälische Wilhelms-Universität Münster,
Münster, Germany
{markus.mueller-olm,alexander.wenner}@wwu.de

Abstract. In this article, we propose a static information-flow analysis for multi-threaded programs with shared memory communication and synchronization via locks. In contrast to many prior analyses, our analysis does not only prevent information leaks due to synchronization, but can also benefit from synchronization for its precision. Our analysis is a novel combination of type systems and a reachability analysis based on dynamic pushdown networks. The security type system supports flow-sensitive tracking of security levels for shared variables in the analysis of one thread by exploiting assumptions about variable accesses by other threads. The reachability analysis based on dynamic pushdown networks verifies that these assumptions are sound using the result of an automatic guarantee inference. The combined analysis is the first automatic static analysis that supports flow-sensitive tracking of security levels while being sound with respect to termination-sensitive noninterference.

Keywords: Information-flow security · Concurrency · Static analysis

1 Introduction

Before giving a multi-threaded program access to sensitive information, one might want to know whether the program keeps this information secret. Static information-flow analyses are a solution for checking whether a program keeps sensitive information secret before running the program.

Information-flow security for sequential programs received a lot of attention in research and mature solutions exist, e.g. [2,5,7,12]. Analyzing information-flow security for concurrent programs is conceptually more difficult. In particular, analyses for sequential programs are not sufficient for analyzing concurrent programs [17], because further information leaks can occur. Consider, for instance, the program $o1:=s1; s1:=s2; s2:=o1; o1:=0$, which swaps the values stored in $s1$ and $s2$ via the variable $o1$. Assume the values of $s1$ and $s2$ shall be kept

© Springer International Publishing Switzerland 2015
M. Falaschi (Ed.): LOPSTR 2015, LNCS 9527, pp. 201–217, 2015.
DOI: 10.1007/978-3-319-27436-2_12

Fig. 1. Work flow of the proposed analysis

secret from an attacker who can only observe the variable *o1* after the program run. While the program does not leak the values of *s1* and *s2* if run in isolation, it might leak the value of *s1* to the attacker if the program *o2:=o1; o1:=o2* is run concurrently. Synchronization adds further complexity to this problem, because it can introduce additional information leaks [14].

For verifying that multi-threaded programs have secure information flow, several security type systems were proposed and proven sound wrt. noninterference-like security properties (e.g., [16,17]). While some of this work addresses the danger of information leakage via synchronization (e.g., [14,19,20]), the potential positive effects of synchronization primitives for information-flow security have been neglected for some time. However, programmers use synchronization frequently to limit the possible interferences between threads. In particular, synchronization can be employed to prevent information leakage.

Mantel, Sands, and Sudbrock propose a framework for verifying information-flow security in a modular fashion such that the positive effects of synchronization can be exploited [10]. They present a flow-sensitive security type system that is suitable for rely-guarantee-style reasoning about information-flow security based on code annotations that capture a programmer's intentions and expectations by so called modes. A mode is either an assumption about a given thread's environment that the programmer expects to hold when the thread reaches some program point, or it is a guarantee that the programmer intends to provide to the thread's environment. In [10], the security type system is proven sound under the precondition that all assumptions made by a thread are justified by corresponding guarantees of other threads and that all such guarantees are, indeed, provided. In [3], this approach is adapted to a hybrid information-flow analysis, where monitors enforce the soundness of rely-guarantee-style reasoning by forcing threads to provide all guarantees that are needed to justify the assumptions made by other threads.

In this article, we propose a particular combination of security type systems with dynamic pushdown networks [9] (brief: DPNs). The purpose of this combination is to obtain a solution for rely-guarantee-style reasoning where DPNs are used to effectively check that all assumptions are justified. In addition, we present an inference that soundly computes the guarantees that are provided at each program point. That is, our solution statically ensures that modes are used soundly and our soundness result is unconditional, unlike in [10] where a sound use of modes is assumed. In contrast to [3], we present a solution for a static analysis, i.e. one only needs to verify the information flow security of a program once and no run-time overhead is imposed on the program. Another novelty of this article in comparison to [3,10] is that our security type system covers dynamic thread creation as well as lock-based synchronization.

Figure 1 illustrates how the different modules of our analysis interact. The guarantee inference takes a program annotated with assumptions as input and adds guarantee annotations. This program is input to the assumption verifier and the security type system. A program is then accepted as secure if and only if it is accepted by the assumption verifier as well as the security type system.

Overall, our analysis is the first completely automated, static information-flow analysis that soundly enforces termination-sensitive noninterference while permitting flow-sensitive tracking of security levels for shared variables.

2   Basic Notions and Notation

2.1   Model of Computation

We consider multi-threaded programs whose threads synchronize by locks and communicate via shared memory. We focus on interleaving concurrency (i.e., one thread performs a step at a time), non-deterministic scheduling (i.e., each thread could be chosen to perform a step next), and non-re-entrant locks (i.e., a lock can only be acquired if no thread, including the acquiring thread, holds this lock). To capture the behavior of multi-threaded programs, we use two transition systems: a local labeled transition system to capture the behavior of individual threads and a global transition system to capture the behavior of multiple threads.

We assume as given a *finite set of locks Lck* and define the *set of all memory configurations* by $Mem = Var \rightarrow Val$, where *Var* is a *finite set of variables* and *Val* is a *set of values*. We leave *Var* and *Val* both under-specified.

We refer to the states and labels of local, labeled transition systems as local configurations and events, respectively. Formally, a *local transition system* is a triple $(LCnf, Eve, \rightarrow)$ where *LCnf* and *Eve* are sets and $\rightarrow \subseteq LCnf \times Eve \times LCnf$. We define the *set of local configurations* by $LCnf = CCnf \times Mem$, where *CCnf* is a *set of control configurations* that we leave under-specified for now. An *event* is a term that captures the non-local effects of a thread's computation. We define the set of all events by $Eve = \{\epsilon, \nearrow_{ccnf}, l, \neg l \mid ccnf \in CCnf, l \in Lck\}$. We use the events $\nearrow_{ccnf}$, l, and $\neg l$ to capture the creation of a new thread with initial control configuration *ccnf*, the acquisition of lock l, and the release of l, respectively. The term ϵ signals that no non-local effect occurs. We assume that termination is captured by a predicate *trm* on control configurations.

A *global transition system* is a pair $(GCnf, \twoheadrightarrow)$, where *GCnf* is a *set of global configurations* and $\twoheadrightarrow \subseteq GCnf \times GCnf$. We define *GCnf* by $GCnf = CCnf^{+} \times Mem$, i.e., a global configuration is a pair of a non-empty list of local control configurations and a memory configuration. A global configuration $\langle [ccnf_1, \ldots, ccnf_n], mem \rangle$ models a snapshot of a computation with n threads where the ith thread's state is captured by $(ccnf_i, mem)$ for $1 \leq i \leq n$. We say that a list of control configurations $[ccnf_1, \ldots, ccnf_n]$ has *terminated* (denoted $trm([ccnf_1, \ldots, ccnf_n])$) iff $trm(ccnf_i)$ holds for all $i \in \{1, \ldots, n\}$.

We assume the control configuration of a thread to capture which locks are held by this thread. To retrieve the set of acquired locks, we use a function

$locks : CCnf \rightarrow 2^{Lck}$ and inductively lift it to a function $locks : CCnf^* \rightarrow 2^{Lck}$ by $locks([]) = \emptyset$ and $locks(\overrightarrow{ccnf}++[ccnf]) = locks(\overrightarrow{ccnf}) \cup locks(ccnf)$. In a global configuration $\langle [ccnf_1, \ldots, ccnf_n], mem \rangle$, $locks(ccnf_i)$ is the set of locks acquired by the ith thread and $Lck \setminus locks([ccnf_1, \ldots, ccnf_n])$ is the set of available locks.

We say that a local transition system $(LCnf, Eve, \rightarrow)$ *handles locks properly* iff (1) $(ccnf, mem) \xrightarrow{l} (ccnf', mem')$ implies $locks(ccnf') = locks(ccnf) \cup \{l\}$,[1] (2) $(ccnf, mem) \xrightarrow{\neg l} (ccnf', mem')$ implies $locks(ccnf) = locks(ccnf') \cup \{l\}$, (3) $(ccnf, mem) \xrightarrow{\alpha} (ccnf', mem')$ and $\alpha \notin \{l, \neg l \mid l \in Lck\}$ imply $locks(ccnf') = locks(ccnf)$, and (4) $(ccnf, mem) \xrightarrow{\nearrow ccnf^*} (ccnf', mem')$ implies $locks(ccnf^*) = \emptyset$.

Let $(LCnf, Eve, \rightarrow)$ be a local transition system that handles locks properly. The global transition relation $\twoheadrightarrow \subseteq GCnf \times GCnf$ induced by this local transition system is the smallest relation that satisfies the following conditions:

1. If $(ccnf_i, mem) \xrightarrow{l} (ccnf'_i, mem')$ and $l \notin locks(\overrightarrow{ccnf}_1 ++ \overrightarrow{ccnf}_2)$
   then
   $\langle \overrightarrow{ccnf}_1 ++ [ccnf_i] ++ \overrightarrow{ccnf}_2, mem \rangle \twoheadrightarrow \langle \overrightarrow{ccnf}_1 ++ [ccnf'_i] ++ \overrightarrow{ccnf}_2, mem' \rangle$.
2. If $(ccnf_i, mem) \xrightarrow{\nearrow ccnf} (ccnf'_i, mem')$
   then
   $\langle \overrightarrow{ccnf}_1 ++ [ccnf_i] ++ \overrightarrow{ccnf}_2, mem \rangle \twoheadrightarrow \langle \overrightarrow{ccnf}_1 ++ [ccnf, ccnf'_i] ++ \overrightarrow{ccnf}_2, mem' \rangle$.
3. If $(ccnf_i, mem) \xrightarrow{\alpha} (ccnf'_i, mem')$ and $\alpha \notin \{\nearrow ccnf, l \mid ccnf \in CCnf, l \in Lck\}$
   then
   $\langle \overrightarrow{ccnf}_1 ++ [ccnf_i] ++ \overrightarrow{ccnf}_2, mem \rangle \twoheadrightarrow \langle \overrightarrow{ccnf}_1 ++ [ccnf'_i] ++ \overrightarrow{ccnf}_2, mem' \rangle$.

The first item above captures the acquisition of a lock by the thread at position $i = 1 + \sharp(\overrightarrow{ccnf}_1)$. Since the local transition system handles locks properly, a lock can only be acquired if no thread – including thread i – holds this lock. The second item captures the creation of a thread by the ith thread. Due to the proper handling of locks, newly created threads hold no locks. Finally, the third item handles all other steps of the ith thread, including the release of a lock.

We inductively define a family of relations $(\twoheadrightarrow_k)_{k \in \mathbb{N}}$ by $gcnf \twoheadrightarrow_0 gcnf$ and if $gcnf \twoheadrightarrow_k gcnf'$ and $gcnf' \twoheadrightarrow gcnf''$ then $gcnf \twoheadrightarrow_{k+1} gcnf''$. The transitive, reflexive closure of $\twoheadrightarrow$ is defined by $gcnf \twoheadrightarrow^* gcnf'$ iff $\exists k \in \mathbb{N}. gcnf \twoheadrightarrow_k gcnf'$. If $gcnf \twoheadrightarrow^* gcnf'$ then $gcnf'$ is *reachable* from $gcnf$. We define the *set of all global configurations reachable from* $gcnf$ by $gReach(gcnf) = \{gcnf' \mid gcnf \twoheadrightarrow^* gcnf'\}$.

In Sect. 2.5, we define a local transition system $(LCnf, Eve, \rightarrow)$ for a simple programming language and capture multi-threaded computations by the global transition system $(GCnf, \twoheadrightarrow)$, where $\twoheadrightarrow$ is induced by $(LCnf, Eve, \rightarrow)$.

2.2 Attacker Model and Definition of Security

We focus on confidentiality in this article. More concretely, we assume that certain variables store secrets, and we only classify a program as secure if it does

---

[1] We use $\dot{\cup}$ to denote the disjoint union of two sets, e.g., $locks(ccnf') = locks(ccnf) \dot{\cup} \{l\}$ is equivalent to $locks(ccnf') = (locks(ccnf) \cup \{l\}) \land l \notin locks(ccnf)$.

not reveal information about these secrets when it is run. We consider attackers that might be able to observe the values of all other variables both, before and after a program run. We refer to variables that initially store secrets as **high** and to variables that might be observable to the attacker as **low**.

We define a set of security levels by $Lev = \{\textbf{low}, \textbf{high}\}$ and use a function $lev : Var \rightarrow Lev$ to associate a security level with each variable. For the attacker, two memory configurations are indistinguishable if they agree on the values of all low variables. We say that $mem, mem' \in Mem$ are **low**-*equal* (denoted by $mem =_{\textbf{low}}^{lev} mem'$) iff $\forall x \in Var.\, (lev(x) = \textbf{low} \implies mem(x) = mem'(x))$ holds.

Definition 1. *A control configuration ccnf is* secure *for* $lev : Var \rightarrow Lev$ *iff*

$$\forall mem_1, mem'_1, mem_2 \in Mem.\, \forall \overrightarrow{ccnf}_1 \in CCnf^+.$$
$$\langle [ccnf], mem_1 \rangle \twoheadrightarrow^* \langle \overrightarrow{ccnf}_1, mem'_1 \rangle \wedge trm(\overrightarrow{ccnf}_1) \wedge mem_1 =_{\textbf{low}}^{lev} mem_2$$
$$\implies \exists mem'_2 \in Mem.\, \exists \overrightarrow{ccnf}_2 \in CCnf^+.$$
$$\langle [ccnf], mem_2 \rangle \twoheadrightarrow^* \langle \overrightarrow{ccnf}_2, mem'_2 \rangle \wedge trm(\overrightarrow{ccnf}_2) \wedge mem'_1 =_{\textbf{low}}^{lev} mem'_2$$

Our security definition captures possibilistic, termination-sensitive noninterference for a two-level security policy [15]. That is, if a program satisfies our security definition then the initial values of **high** variables do not influence the possibility of a **low** attacker's observations. In particular, programs that leak information via their termination behavior [4] do not satisfy Definition 1.

2.3 Dynamic Pushdown Networks

We briefly recall the result on analysis of dynamic pushdown networks (DPNs) from [9] exploited in the assumption verifier and describe the connection to our model of computation. A DPN consists of multiple instances of independent pushdown systems running in parallel. Additional instances can be created dynamically. Synchronisation is supported in the form of locks. Using finite data abstraction, DPNs can thus model concurrent programs with recursive procedures, dynamic thread creation, and synchronization with locks.

Formally, a DPN is a tuple (P, Γ, A, Δ) where P is a finite set of control states, Γ is a finite set of stack symbols, A is a finite set of actions, and $\Delta \subseteq P\Gamma \times A \times P\Gamma^*$ is a finite set of transitions. An action from $\{\nearrow_{p,\gamma} \mid p \in P, \gamma \in \Gamma\} \subseteq A$ indicates creation of a new pushdown instance with a control state p and stack symbol γ, and an action from $\{l, \neg l \mid l \in Lck\} \subseteq A$ indicates acquisition and release of a lock l. The set of acquired locks can be retrieved from a control state with the function $locks : P \rightarrow 2^{Lck}$. The set of acquired locks in a control state must be consistent with transitions, i.e. for all $(p\gamma, a, p'w') \in \Delta$ we have $locks(p') = \{l\} \, \dot\cup \, locks(p)$ if $a = l$, $locks(p) = \{l\} \, \dot\cup \, locks(p')$ if $a = \neg l$ and $locks(p) = locks(p')$ otherwise; in addition $locks(p'') = \emptyset$ if $a = \nearrow_{p'',\gamma''}$. Note that there is no re-entrant use of locks.

Configurations of a DPN are lists of pushdown instances represented as words from $DCnf = (P\Gamma^*)^+$. Let $locks(p_1 w_1 \ldots p_n w_n) = \bigcup_{i \in \{1,\ldots,n\}} locks(p_i)$.

A step of the semantics of the DPN rewrites the control state and topmost stack-symbol of one pushdown instance according to a transition rule, if allowed by the state of locks. On thread creation, a new pushdown instance is added to the left of the current instance in the configuration. Formally, the transition relation $\rightarrowtail$ is the smallest relation such that $s\,p\gamma w\,s' \rightarrowtail s\,s''\,p'w'w\,s'$ holds for all $s, s' \in DCnf, w \in \Gamma^*, (p\gamma, a, p'w') \in \Delta$ provided $l \notin locks(sp\gamma ws')$ if $a = l$ and $s'' = p''\gamma''$ if $a = \nearrow_{p'',\gamma''}$ and $s'' = \varepsilon$ otherwise.

We say that a thread uses locks in a well-nested fashion if it releases all locks in opposite order of their acquisition. Given a DPN whose threads use locks in a well-nested fashion and a regular set $B \subseteq (P \cup \Gamma)^*$, we can check effectively, whether a configuration in B is reachable from initial configuration s_0 or not, i.e., whether $\exists s \in B : s_0 \rightarrowtail^* s$ (see [9]).

In order to analyze a program from an initial configuration $\langle [ccnf], mem \rangle$, we consider a DPN $\mathcal{M}_{ccnf} = (P_{ccnf}, \Gamma_{ccnf}, A_{ccnf}, \Delta_{ccnf})$ with $P_{ccnf} \subseteq CCnf$, $ccnf \in P_{ccnf}$ and $\Gamma_{ccnf} = \{\#\}$ that satisfies the following condition: if $ccnf' \in P_{ccnf}$ and $(ccnf', mem) \xrightarrow{\alpha} (ccnf'', mem')$ then $ccnf'' \in P_{ccnf}$, $\alpha' \in A_{ccnf}$ and $(ccnf'\#, \alpha', ccnf''\#) \in \Delta_{ccnf}$, where $\alpha' = \alpha$ for $\alpha \notin \{\nearrow_{ccnf} | ccnf \in CCnf\}$, and $ccnf''' \in P_{ccnf}$ and $\alpha' = \nearrow_{ccnf''',\#}$ for $\alpha = \nearrow_{ccnf'''}$. Elements of P_{ccnf} abstract local configurations in the sense that they do not carry information about memory configurations. Correspondingly, the transitions in Δ_{ccnf} abstract steps in the local semantics. However, labelling and hence synchronisation and thread creation is preserved. We reuse the function $locks$ defined for control configurations.

The DPN $\mathcal{M}_{ccnf}$ can be used to approximate reachability of configurations starting from $\langle [ccnf], mem \rangle$ respecting synchronisation via locks and thread creation, since $\langle [ccnf], mem \rangle \rightarrowtail^* \langle [ccnf_1, \ldots, ccnf_n], mem' \rangle$ implies that $ccnf\# \rightarrowtail^* ccnf_1\# \ldots ccnf_n\#$. Hence, an unreachable configuration in the DPN translates to an unreachable configuration in the program. Since we abstract from the shared global memory, the converse direction does not hold in general.

The above approach is fitted to non-recursive programs but can easily be extended to recursive programs by using a larger stack alphabet.

2.4   Control Configurations and Modes

We specialize control configurations to triples of the form $(c, lkst, mdst)$, where c is a *command*, $lkst$ is a *lock state*, and $mdst$ is a *mode state*. In the control configuration of a thread, the command specifies how the thread's computation will continue, the lock state specifies which locks the thread currently holds, and the mode state specifies the thread's current assumptions about its environment as well as the guarantees that the thread currently provides to its environment.

We use *Com*, *LkSt*, and *MdSt* to denote the set of all commands, the set of all lock states, and the set of all mode states, respectively, i.e., $CCnf = Com \times LkSt \times MdSt$. We leave *Com* under-specified and define *LkSt* and *MdSt* below. In Sect. 2.5, we specialize *Com* for the syntax of a concrete programming language and formalize the language's semantics by a local transition system.

Formally, a lock state is a set of locks, i.e., $LkSt = 2^{Lck}$. In a control configuration $(c, lkst, mdst)$ of a thread, the lock state $lkst$ specifies which locks this thread holds. Hence, we define the function $locks$ by $locks((c, lkst, mdst)) = lkst$.

We define mode states to be functions from modes to sets of variables, i.e., $MdSt = Md \rightarrow 2^{Var}$, where $Md = \{\text{A-NR, A-NW, G-NR, G-NW}\}$ is the *set of modes*. The modes A-NR (for *no-read assumption*) and A-NW (for *no-write assumption*) represent assumptions, while the modes G-NR (for *no-read guarantee*) and G-NW (for *no-write guarantee*) represent guarantees. If $x \in mdst(\text{A-NW})$ then it is assumed that the thread's environment does not write x. Similarly, if $y \in mdst(\text{A-NR})$ then it is assumed that the thread's environment does not read the variable y. If $x \in mdst(\text{G-NW})$ and $y \in mdst(\text{G-NR})$, then the thread guarantees to not write x and to not read y, respectively. We say *a mode state $mdst$ is consistent with a mode state $mdst'$* iff $mdst(\text{A-NW}) \subseteq mdst'(\text{G-NW})$ and $mdst(\text{A-NR}) \subseteq mdst'(\text{G-NR})$, i.e., if all assumptions made by $mdst$ are matched by corresponding guarantees of $mdst'$.

We say that a local configuration $((c, lkst, mdst), mem)$ *provides its no-write guarantees* iff for all $x \in mdst(\text{G-NW})$ and $(ccnf', mem') \in LCnf$ the implication

$$((c, lkst, mdst), mem) \xrightarrow{\alpha} (ccnf', mem') \implies mem'(x) = mem(x) \qquad (1)$$

holds. Moreover, we say $((c, lkst, mdst), mem)$ *provides its no-read guarantees* iff for all $y \in mdst(\text{G-NR})$, $v \in Val$, and $(ccnf', mem') \in LCnf$ the implication

$$((c, lkst, mdst), mem) \xrightarrow{\alpha} (ccnf', mem') \qquad (2)$$
$$\implies ((c, lkst, mdst), mem[y \mapsto v]) \xrightarrow{\alpha} (ccnf', mem')$$
$$\lor ((c, lkst, mdst), mem[y \mapsto v]) \xrightarrow{\alpha} (ccnf', mem'[y \mapsto v])$$

holds. The two disjuncts on the right hand side of the implication cover the case where the variable y is written and not written, respectively, in the step. Finally, we say that a local configuration *provides its guarantees* if it provides both, its no-write guarantees and its no-read guarantees.

We say that a global configuration $\langle [ccnf_1, \ldots, ccnf_n], mem \rangle$ with $ccnf_i = (c_i, lkst_i, mdst_i)$ for each $i \in \{1, \ldots, n\}$ *justifies its assumptions* iff $mdst_j$ is consistent with $mdst_k$ for all $j, k \in \{1, \ldots, n\}$, $j \neq k$. Intuitively, this means that if one thread makes an assumption about a variable then all other threads must provide the corresponding guarantee.

Modes and mode states were introduced in [10] as a basis for rely-guarantee-style reasoning about information-flow security. The approach enables one to verify the security of multi-threaded programs in a modular fashion, based on security guarantees for each individual thread. More concretely, one statically verifies that steps of each thread only cause flows of information that comply with a given security policy. Rely-guarantee-style reasoning frees one from having to reason about arbitrary environments, one only needs to consider environments that satisfy the thread's current assumptions. Such rely-guarantee-style reasoning is sound if at each step of a computation the assumptions of all threads are justified and the guarantees of all threads are provided.

Definition 2. *A global configuration gcnf ensures a locally sound use of modes iff for each gcnf′ ∈ gReach(gcnf), where gcnf′ = ⟨[ccnf′₁, . . . , ccnf′ₙ], mem′⟩, and each i ∈ {1, . . . , n}, the local configuration (ccnf′ᵢ, mem′) provides its guarantees.*

A global configuration gcnf ensures a globally sound use of modes iff each gcnf′ ∈ gReach(gcnf) justifies its assumptions.

A global configuration gcnf ensures a sound use of modes iff gcnf ensures both, a locally sound use of modes and a globally sound use of modes.

Our semantics of modes is similar to the one in [3,10]. One original extension of rely-guarantee-style reasoning about information-flow security in this article is that we cover dynamic thread creation and synchronization with locks, which are two language features not supported by this prior work.

2.5   A Concrete Programming Language with Modes

We define an example programming language with annotations for acquiring and releasing modes. The set of *annotations* is $Ann = \{\mathsf{acq}(md, \overline{x}), \mathsf{rel}(md, \overline{x}) \mid md \in Md \wedge \overline{x} \subseteq Var\}$. An annotation $\mathsf{acq}(md, \overline{x})$ acquires the mode md for all variables in $\overline{x}$, and an annotation $\mathsf{rel}(md, \overline{x})$ releases the mode md for all variables in $\overline{x}$. To capture this formally, we define the function $updMds : MdSt \times Ann \rightarrow MdSt$ by $updMds(mdst, \mathsf{acq}(md, \overline{x})) = mdst[md \mapsto mdst(md) \cup \overline{x}]$ and $updMds(mdst, \mathsf{rel}(md, \overline{x})) = mdst[md \mapsto mdst(md) \setminus \overline{x}]$, and lift it to lists of annotations by $updMds(mdst, []) = mdst$ and $updMds(mdst, [a] ++ \overrightarrow{a}) = updMds(updMds(mdst, a), \overrightarrow{a})$.

We define the special mode state $mdst_\perp$ by $mdst_\perp(\text{A-NR}) = mdst_\perp(\text{A-NW}) = \emptyset$ and $mdst_\perp(\text{G-NR}) = mdst_\perp(\text{G-NW}) = Var$. It is minimal in the sense that it imposes no constraints on assumptions and guarantees of its environment.

We assume as given a set Exp of *expressions*, a function $eval : Exp \times Mem \rightarrow Val$ that returns the value to which an expression evaluates in a given memory, and a function $vars : Exp \rightarrow 2^{Var}$ that returns the set of all variables that appear syntactically in an expression.

The set Com_p of syntactically correct programs is defined by the grammar:

$$\odot := \epsilon \mid @\overrightarrow{a}$$

$$c_p := \mathbf{skip} \mid x := e \mid \mathbf{if}\ e\ \mathbf{then}\ c_p\ \mathbf{else}\ c_p\ \mathbf{fi} \mid \mathbf{while}\ e\ \mathbf{do}\ c_p\ \mathbf{od} \mid c_p; c_p$$
$$\mid \mathbf{spawn}(c_p) \mid \mathbf{lock}(l)\odot; c_p; \mathbf{unlock}(l)\odot \mid c_p\odot$$

where $\overrightarrow{a} \in Ann^*$, $x \in Var$, $e \in Exp$, and $l \in Lck$. The syntax ensures a well-nested use of locks. The set Com of commands is defined by the grammar:

$$c := \mathbf{stop} \mid \mathbf{lock}(l)\odot \mid \mathbf{unlock}(l)\odot \mid c; c \mid c_p$$

We define that $trm((c, lkst, mdst))$ holds iff $c = \mathbf{stop}$. That is, the symbol $\mathbf{stop}$ indicates that the computation of a thread has terminated.

The local transition system for our programming language is defined by the calculus in Fig. 2. For the rules SK, AS, SQ1, SQ2, IFT, IFF, WHT, and WHF, SP, the lock state as well as the mode state is irrelevant for the premises and both remain

$$\text{SK} \frac{}{(\textbf{skip}, \mathit{lkst}, \mathit{mdst}, \mathit{mem}) \xrightarrow{\epsilon} (\textbf{stop}, \mathit{lkst}, \mathit{mdst}, \mathit{mem})}$$

$$\text{AS} \frac{\mathit{eval}(e, \mathit{mem}) = v \qquad \mathit{mem}' = \mathit{mem}[x \mapsto v]}{(x{:=}e, \mathit{lkst}, \mathit{mdst}, \mathit{mem}) \xrightarrow{\epsilon} (\textbf{stop}, \mathit{lkst}, \mathit{mdst}, \mathit{mem}')}$$

$$\text{SQ1} \frac{(c_1, \mathit{lkst}, \mathit{mdst}, \mathit{mem}) \xrightarrow{\alpha} (c_1', \mathit{lkst}', \mathit{mdst}', \mathit{mem}') \quad c_1' \neq \textbf{stop}}{(c_1; c_2, \mathit{lkst}, \mathit{mdst}, \mathit{mem}) \xrightarrow{\alpha} (c_1'; c_2, \mathit{lkst}', \mathit{mdst}', \mathit{mem}')}$$

$$\text{SQ2} \frac{(c_1, \mathit{lkst}, \mathit{mdst}, \mathit{mem}) \xrightarrow{\alpha} (\textbf{stop}, \mathit{lkst}', \mathit{mdst}', \mathit{mem}')}{(c_1; c_2, \mathit{lkst}, \mathit{mdst}, \mathit{mem}) \xrightarrow{\alpha} (c_2, \mathit{lkst}', \mathit{mdst}', \mathit{mem}')}$$

$$\text{SP} \frac{}{(\textbf{spawn}(c), \mathit{lkst}, \mathit{mdst}, \mathit{mem}) \xrightarrow{\angle(c, \emptyset, \mathit{mdst}_\perp)} (\textbf{stop}, \mathit{lkst}, \mathit{mdst}, \mathit{mem})}$$

$$\text{IFT} \frac{\mathit{eval}(e, \mathit{mem}) = \textbf{true}}{(\textbf{if } e \textbf{ then } c \textbf{ else } c' \textbf{ fi}, \mathit{lkst}, \mathit{mdst}, \mathit{mem}) \xrightarrow{\epsilon} (c, \mathit{lkst}, \mathit{mdst}, \mathit{mem})}$$

$$\text{IFF} \frac{\mathit{eval}(e, \mathit{mem}) = \textbf{false}}{(\textbf{if } e \textbf{ then } c \textbf{ else } c' \textbf{ fi}, \mathit{lkst}, \mathit{mdst}, \mathit{mem}) \xrightarrow{\epsilon} (c', \mathit{lkst}, \mathit{mdst}, \mathit{mem})}$$

$$\text{WHT} \frac{\mathit{eval}(e, \mathit{mem}) = \textbf{true}}{(\textbf{while } e \textbf{ do } c \textbf{ od}, \mathit{lkst}, \mathit{mdst}, \mathit{mem}) \xrightarrow{\epsilon} (c; \textbf{while } e \textbf{ do } c \textbf{ od}, \mathit{lkst}, \mathit{mdst}, \mathit{mem})}$$

$$\text{WHF} \frac{\mathit{eval}(e, \mathit{mem}) = \textbf{false}}{(\textbf{while } e \textbf{ do } c \textbf{ od}, \mathit{lkst}, \mathit{mdst}, \mathit{mem}) \xrightarrow{\epsilon} (\textbf{stop}, \mathit{lkst}, \mathit{mdst}, \mathit{mem})}$$

$$\text{LK} \frac{\mathit{lkst} \,\dot\cup\, \{l\} = \mathit{lkst}'}{(\textbf{lock}(l), \mathit{lkst}, \mathit{mdst}, \mathit{mem}) \xrightarrow{l} (\textbf{stop}, \mathit{lkst}', \mathit{mdst}, \mathit{mem})}$$

$$\text{ULK} \frac{\mathit{lkst} = \mathit{lkst}' \,\dot\cup\, \{l\}}{(\textbf{unlock}(l), \mathit{lkst}, \mathit{mdst}, \mathit{mem}) \xrightarrow{\neg l} (\textbf{stop}, \mathit{lkst}', \mathit{mdst}, \mathit{mem})}$$

$$\text{AN1} \frac{(c, \mathit{lkst}, \mathit{mdst}, \mathit{mem}) \xrightarrow{\alpha} (\textbf{stop}, \mathit{lkst}', \mathit{mdst}', \mathit{mem}') \quad \mathit{mdst}'' = \mathit{updMds}(\mathit{mdst}', \vec{a})}{(c@\vec{a}, \mathit{lkst}, \mathit{mdst}, \mathit{mem}) \xrightarrow{\alpha} (\textbf{stop}, \mathit{lkst}', \mathit{mdst}'', \mathit{mem}')}$$

$$\text{AN2} \frac{(c, \mathit{lkst}, \mathit{mdst}, \mathit{mem}) \xrightarrow{\alpha} (c', \mathit{lkst}', \mathit{mdst}', \mathit{mem}') \quad c' \neq \textbf{stop}}{(c@\vec{a}, \mathit{lkst}, \mathit{mdst}, \mathit{mem}) \xrightarrow{\alpha} (c'@\vec{a}, \mathit{lkst}', \mathit{mdst}', \mathit{mem}')}$$

Fig. 2. Semantics of the programming language

unchanged. The rules LK and ULK realize acquiring and releasing a lock, respectively. The rule AN1 updates the mode state according to an annotation if the annotated command is reduced to **stop**. The rule AN2 preserves the annotation if the command is not reduced to **stop**.

Given a program c_p, we say that c_p *is secure for lev* iff $(c_p, \emptyset, \mathit{mdst}_\perp)$ is secure for *lev*, that c_p *ensures a locally sound use of modes* iff $\langle[(c_p, \emptyset, \mathit{mdst}_\perp)], \mathit{mem}\rangle$ ensures a locally sound use of modes for all $\mathit{mem} \in \mathit{Mem}$, that c_p *ensures a globally sound use of modes* iff $\langle[(c_p, \emptyset, \mathit{mdst}_\perp)], \mathit{mem}\rangle$ ensures a globally sound use of modes for all $mem \in Mem$, and that c_p *ensures a sound use of modes* iff $\langle[(c_p, \emptyset, mdst_\perp)], mem\rangle$ ensures a sound use of modes for all $mem \in Mem$.

3   A DPN-based Analysis for Sound Assumptions

We propose a two-step approach for ensuring a globally sound use of modes for a given program c_p. First, we construct a DPN that simulates c_p in the sense of Sect. 2.3. Second, we build an automaton that accepts all DPN configurations that contain a pair of inconsistent mode states. By the connection between DPN and program executions, c_p uses modes globally sound, if no such configuration is reachable in the DPN from a particular initial configuration. The techniques from [9] then enable us to determine whether this is the case.

We construct a DPN $\mathcal{M}_{ccnf}$ for the control configuration $ccnf = (c_p, \emptyset, mdst_\perp)$ as follows: Starting with $ccnf$, we collect all reachable control configurations, actions, and transitions using the rules from Fig. 2, ignoring the memory configurations. The resulting sets P_{ccnf}, A_{ccnf} and Δ_{ccnf} of control states, actions, and transitions satisfy all requirements from Sect. 2.3. Due to the syntax of programs locks are used well-nested in the DPN $\mathcal{M}_{ccnf}$ and mode states are preserved in its configurations.

For the second step, we first introduce a function that checks the mutual consistency of two mode states and returns a summary mode state.

Definition 3. *Let $MdSt_\top = MdSt \cup \{\top\}$. The function $\oplus : MdSt_\top \times MdSt_\top \to MdSt_\top$ is defined by $mdst \oplus mdst' = mdst''$ where*

- $mdst''(md) = mdst(md) \cup mdst'(md)$ *for $md \in \{\text{A-NR, A-NW}\}$ and*
  $mdst''(md) = mdst(md) \cap mdst'(md)$ *for $md \in \{\text{G-NR, G-NW}\}$*
  if $mdst \neq \top$, $mdst' \neq \top$, $mdst$ is consistent with $mdst'$, and
     $mdst'$ is consistent with $mdst$.
- $mdst'' = \top$ *otherwise.*

If the two parameter mode states are mutually consistent, the function $\oplus$ returns a regular mode state that imposes the same constraints on concurrent threads as the combination of the original mode states. That is, it makes all assumptions that at least one of the mode states makes and provides only those guarantees that both mode states provide. If one of the parameter mode states makes an assumption that the other mode state does not match with a corresponding guarantee, the function returns the special symbol $\top$.

We are now ready to define the automaton that characterizes DPN configurations containing inconsistent mode states using the function $\oplus$.

Definition 4. *For a DPN $\mathcal{M}_{ccnf} = (P_{ccnf}, \Gamma_{ccnf}, A_{ccnf}, \Delta_{ccnf})$ as described above, we define $\mathcal{A}_{ccnf} = (MdSt_\top, P_{ccnf} \cup \Gamma_{ccnf}, \delta, mdst_\perp, \{\top\})$ as the conflict automaton, where $\delta = \{(q, (c, lkst, mdst), q \oplus mdst) \mid q \in MdSt_\top, (c, lkst, mdst) \in P_{ccnf}\} \cup \{(q, \#, q) \mid q \in MdSt_\top\}$. We denote the language accepted by the automaton by $\mathcal{L}(\mathcal{A}_{ccnf})$.*

$$\text{ISK} \frac{\overrightarrow{a} = anno(\overline{x}, \emptyset, \overline{x}_r, \overline{x}_w)}{\overline{x} \vdash \emptyset, \emptyset\{\mathbf{skip}\}\overline{x}_r, \overline{x}_w : \mathbf{skip}@\overrightarrow{a}} \qquad \text{IAS} \frac{\overrightarrow{a} = anno(vars(e) \cup \overline{x}, \{x\}, \overline{x}_r, \overline{x}_w)}{\overline{x} \vdash vars(e), \{x\}\{x{:=}e\}\overline{x}_r, \overline{x}_w : x{:=}e@\overrightarrow{a}}$$

$$\text{ILO} \frac{\overrightarrow{a} = anno(\overline{x}, \emptyset, \overline{x}_r, \overline{x}_w)}{\overline{x}\vdash\emptyset, \emptyset\{\mathbf{lock}(l)\}\overline{x}_r, \overline{x}_w : \mathbf{lock}(l)@\overrightarrow{a}} \qquad \text{IUL} \frac{\overrightarrow{a} = anno(\overline{x}, \emptyset, \overline{x}_r, \overline{x}_w)}{\overline{x}\vdash\emptyset, \emptyset\{\mathbf{unlock}(l)\}\overline{x}_r, \overline{x}_w : \mathbf{unlock}(l)@\overrightarrow{a}}$$

$$\text{IIF} \frac{\overline{x} \cup vars(e) \vdash \emptyset, \emptyset\{\mathbf{skip}; c_i\}\overline{x}_r, \overline{x}_w : c_i' \text{ for all } i \in \{1,2\}}{\overline{x} \vdash vars(e), \emptyset\{\mathbf{if}\ e\ \mathbf{then}\ c_1\ \mathbf{else}\ c_2\ \mathbf{fi}\}\overline{x}_r, \overline{x}_w : \mathbf{if}\ e\ \mathbf{then}\ c_1'\ \mathbf{else}\ c_2'\ \mathbf{fi}}$$

$$\text{IWH} \frac{\overline{x} \cup vars(e) \vdash \emptyset, \emptyset\{\mathbf{skip}; c\}vars(e), \emptyset : c' \quad \overrightarrow{a} = anno(\overline{x} \cup vars(e), \emptyset, \overline{x}_r, \overline{x}_w)}{\overline{x} \vdash vars(e), \emptyset\{\mathbf{while}\ e\ \mathbf{do}\ c\ \mathbf{od}\}\overline{x}_r, \overline{x}_w : \mathbf{while}\ e\ \mathbf{do}\ c'\ \mathbf{od}@\overrightarrow{a}}$$

$$\text{ISQ} \frac{\begin{array}{c}\overline{x} \vdash \overline{x}_r', \overline{x}_w'\{c_1\}\overline{x}_r'', \overline{x}_w'' : c_1' \\ \overline{x}' \vdash \overline{x}_r'', \overline{x}_w''\{c_2\}\overline{x}_r, \overline{x}_w : c_2'\end{array}}{\overline{x} \vdash \overline{x}_r', \overline{x}_w'\{c_1; c_2\}\overline{x}_r, \overline{x}_w : c_1'; c_2'} \qquad \text{IAN} \frac{\overrightarrow{a}' = \overrightarrow{a} \restriction_A \quad \overline{x} \vdash \overline{x}_r', \overline{x}_w'\{c\}\overline{x}_r, \overline{x}_w : c'}{\overline{x} \vdash \overline{x}_r', \overline{x}_w'\{c@\overrightarrow{a}\}\overline{x}_r, \overline{x}_w : c'@\overrightarrow{a}'}$$

$$\text{ISP} \frac{\emptyset \vdash \emptyset, \emptyset\{\mathbf{skip}; c\}\emptyset, \emptyset : c' \quad \overrightarrow{a} = anno(\overline{x}, \emptyset, \overline{x}_r, \overline{x}_w)}{\overline{x} \vdash \emptyset, \emptyset\{\mathbf{spawn}(c)\}\overline{x}_r, \overline{x}_w : \mathbf{spawn}(c')@\overrightarrow{a}}$$

with $anno(\overline{x}_1, \overline{x}_2, \overline{x}_3, \overline{x}_4) = [\mathsf{acq}(\mathsf{G\text{-}NR}, \overline{x}_1), \mathsf{acq}(\mathsf{G\text{-}NW}, \overline{x}_2), \mathsf{rel}(\mathsf{G\text{-}NR}, \overline{x}_3), \mathsf{rel}(\mathsf{G\text{-}NW}, \overline{x}_4)]$

Fig. 3. Inference of guarantee annotations

The states of the automaton record the summary mode state of the partial configuration already read. Thus the initial state is the minimal mode state and transitions accepting a control state add the mode state of the process to the summary using the $\oplus$ operation. Since we are interested in the configurations with inconsistent mode states, $\top$ is the only accepting state.

DPN-reachability and globally sound use of modes are connected as follows:

Theorem 1. *Let* $ccnf = (c_p, \emptyset, mdst_\perp)$. *If* $\mathcal{L}(\mathcal{A}_{ccnf})$ *is not reachable from* $ccnf\#$ *in DPN* $\mathcal{M}_{ccnf}$, *then* c_p *ensures a globally sound use of modes.*

4 An Inference for Sound Guarantees

We propose an inference to automatically annotate a command with guarantees. Recall that the initial mode state provides all guarantees, and that mode states are updated based on annotations after the annotated command terminates. With this in mind, the intuition of our inference is that a command requests the release of guarantees that it cannot provide from the preceding command and vouches to re-acquire said guarantees. Hence, the inference propagates sets of variables which may be read or written by a command backwards.

A judgment $\overline{x} \vdash \overline{x}_r', \overline{x}_w'\{c\}\overline{x}_r, \overline{x}_w : c'$ with $\overline{x}, \overline{x}_r, \overline{x}_r', \overline{x}_w, \overline{x}_w' \subseteq Var$ and $c, c' \in Com$ of the inference is derivable with the rules in Fig. 3. The set $\overline{x}$ comprises variables for which a conditional requests that a no-read guarantee shall be re-acquired in the body of the conditional. The sets $\overline{x}_r'$ and $\overline{x}_w'$ comprise variables for which c does not provide a no-read and no-write guarantee, respectively.

The sets $\overline{x}_r$ and $\overline{x}_w$ comprise variables for which a release of the respective guarantees is requested. The resulting command c' is annotated with guarantees.

All rules, except IIF and IAN, annotate a command to re-acquire guarantees that this command cannot provide before releasing requested guarantees. The rule IIF requests that its branches re-acquire and release all guarantees. The rule IAN removes existing guarantee annotations to avoid conflicts with inferred guarantees using a projection to assumption annotations The projection $\lceil_A$ is defined by $[] \lceil_A = []$, $([a]{+}{+}\overrightarrow{a}) \lceil_A = [a]{+}{+}(\overrightarrow{a} \lceil_A)$ if $a \in \{\mathsf{acq}(md, \overline{x}), \mathsf{rel}(md, \overline{x}) \mid md \in \{\text{A-NR, A-NW}\} \wedge \overline{x} \subseteq Var\}$ and $([a]{+}{+}\overrightarrow{a}) \lceil_A = \overrightarrow{a} \lceil_A$ otherwise.

Theorem 2. *If $\emptyset \vdash \emptyset, \emptyset\{\mathbf{skip}; c'_p\}\emptyset, \emptyset : c_p$ is derivable, then c_p ensures a locally sound use of modes.*

Note that some rules add **skip** commands. These additional commands do not influence which final memories are reachable. We do this as a lightweight measure to support pre-annotations without further complicating our formalism.

5 A Type System for Information-Flow Security

We extend the security type system from [10, 18]. To this end, we define a total, reflexive order $\sqsubseteq$ on *Lev* such that **low** $\sqsubseteq$ **high**. To support flow-sensitive tracking of security levels for shared variables, we use *partial level assignments*, i.e. partial functions from $Var \rightharpoonup Lev$. For a given level assignment *lev* and a given partial level assignment Λ, a lookup $\Lambda_{lev}\langle x \rangle$ is defined by $\Lambda_{lev}\langle x \rangle = \Lambda(x)$ if $x \in pre(\Lambda)$ and $\Lambda_{lev}\langle x \rangle = lev(x)$ otherwise. Moreover, the partial type environment $\Lambda' = \Lambda \oplus_{lev} a$ is defined by $\Lambda'(x) = \Lambda_{lev}\langle x \rangle$ for all $x \in pre(\Lambda')$ and

$$
pre(\Lambda') = \begin{cases}
pre(\Lambda) \cup \{x \mid x \in \overline{x} \wedge lev(x) = \textbf{low}\} & \text{if } a = \mathsf{acq}(\text{A-NR}, \overline{x}) \\
pre(\Lambda) \cup \{x \mid x \in \overline{x} \wedge lev(x) = \textbf{high}\} & \text{if } a = \mathsf{acq}(\text{A-NW}, \overline{x}) \\
pre(\Lambda) \setminus \{x \mid x \in \overline{x} \wedge lev(x) = \textbf{low}\} & \text{if } a = \mathsf{rel}(\text{A-NR}, \overline{x}) \\
pre(\Lambda) \setminus \{x \mid x \in \overline{x} \wedge lev(x) = \textbf{high}\} & \text{if } a = \mathsf{rel}(\text{A-NW}, \overline{x}) \\
pre(\Lambda) & \text{otherwise.}
\end{cases}
$$

For **low**-variables, acquiring a no-read assumption enables floating of security levels. This allows tracking when a **low**-variable possibly stores sensitive information. For **high**-variables, acquiring a no-write assumption enables floating of security levels. This allows tracking when a **high**-variable definitely stores public information. Releasing the respective assumptions disables floating of security levels again. We lift the definition of $\oplus_{lev}$ to lists of annotations as follows $\Lambda \oplus_{lev} [] = \Lambda$ and $\Lambda \oplus_{lev} ([a]{+}{+}\overrightarrow{a}) = (\Lambda \oplus_{lev} a) \oplus_{lev} \overrightarrow{a}$.

The type system in Fig. 4 allows to derive judgements of the form $\vdash_{lev} \Lambda\{c\}\Lambda' : c'$. If such a judgment is derivable and *lev* and Λ together approximate where secrets are stored initially, then *lev* and Λ' approximate where secrets are stored after running c, provided concurrent threads behave according to the assumptions. The command c' is a **low**-slice of c, i.e. an abstraction of c in which sub-commands that do not contribute to the behaviour observable via **low**-variables

$$\text{TEX} \frac{}{\vdash_{lev,\Lambda} e : \bigsqcup_{x \in vars(e)} \Lambda_{lev}\langle x\rangle}$$

$$\text{TAH} \frac{lev(x) = \textbf{high} \quad x \notin pre(\Lambda)}{\vdash_{lev} \Lambda\{x:=e\}\Lambda : \textbf{skip}}$$

$$\text{TSK} \frac{}{\vdash_{lev} \Lambda\{\textbf{skip}\}\Lambda : \textbf{skip}}$$

$$\text{TAL} \frac{\vdash_{lev,\Lambda} e : \textbf{low} \quad lev(x) = \textbf{low} \quad x \notin pre(\Lambda)}{\vdash_{lev} \Lambda\{x:=e\}\Lambda : x:=e}$$

$$\text{TLO} \frac{}{\vdash_{lev} \Lambda\{\textbf{lock}(l)\}\Lambda : \textbf{lock}(l)}$$

$$\text{TFL} \frac{\vdash_{lev,\Lambda} e : \textbf{low} \quad x \in pre(\Lambda)}{\vdash_{lev} \Lambda\{x:=e\}\Lambda[x \mapsto \textbf{low}] : x:=e}$$

$$\text{TUL} \frac{}{\vdash_{lev} \Lambda\{\textbf{unlock}(l)\}\Lambda : \textbf{unlock}(l)}$$

$$\text{TFH} \frac{x \in pre(\Lambda)}{\vdash_{lev} \Lambda\{x:=e\}\Lambda[x \mapsto \textbf{high}] : \textbf{skip}}$$

$$\text{TWL} \frac{\Lambda \sqsubseteq \Lambda' \quad \Lambda'' \sqsubseteq \Lambda' \quad \vdash_{lev,\Lambda'} e : \textbf{low} \quad \vdash_{lev} \Lambda'\{c\}\Lambda'' : c'}{\vdash_{lev} \Lambda\{\textbf{while } e \textbf{ do } c \textbf{ od}\}\Lambda' : \textbf{while } e \textbf{ do } c' \textbf{ od}}$$

$$\text{TIL} \frac{\vdash_{lev,\Lambda} e : \textbf{low} \quad \vdash_{lev} \Lambda\{c_1\}\Lambda'' : c_1' \quad \vdash_{lev} \Lambda\{c_2\}\Lambda''' : c_2' \quad \Lambda' = \Lambda'' \sqcup \Lambda'''}{\vdash_{lev} \Lambda\{\textbf{if } e \textbf{ then } c_1 \textbf{ else } c_2 \textbf{ fi}\}\Lambda' : \textbf{if } e \textbf{ then } c_1' \textbf{ else } c_2' \textbf{ fi}}$$

$$\text{TIH} \frac{\vdash_{lev} \Lambda\{c_1\}\Lambda'' : c_1' \quad \vdash_{lev} \Lambda\{c_2\}\Lambda''' : c_2' \quad c_1' = c_2' \quad \Lambda' = \Lambda'' \sqcup \Lambda'''}{\vdash_{lev} \Lambda\{\textbf{if } e \textbf{ then } c_1 \textbf{ else } c_2 \textbf{ fi}\}\Lambda' : \textbf{skip}; c_1'}$$

$$\text{TSQ} \frac{\vdash_{lev} \Lambda\{c\}\Lambda'' : c'' \quad \vdash_{lev} \Lambda''\{c'\}\Lambda' : c'''}{\vdash_{lev} \Lambda\{c; c'\}\Lambda' : c''; c'''}$$

$$\text{TAN} \frac{\vdash_{lev} \Lambda\{c\}\Lambda' : c' \quad \forall x.\Lambda'_{lev}\langle x\rangle \sqsubseteq \Lambda''_{lev}\langle x\rangle \quad \Lambda'' = (\Lambda' \oplus_{lev} \overrightarrow{a}) \quad \overrightarrow{a}' = \overrightarrow{a}\restriction_{\text{A-NR,A-NW}}}{\vdash_{lev} \Lambda\{c@\overrightarrow{a}\}\Lambda'' : c'@\overrightarrow{a}'}$$

$$\text{TSP} \frac{\vdash_{lev} c : c'}{\vdash_{lev} \Lambda\{\textbf{spawn}(c)\}\Lambda : \textbf{spawn}(c')}$$

$$\text{TTH} \frac{\vdash_{lev} \Lambda\{c\}\Lambda : c' \quad pre(\Lambda) = \emptyset}{\vdash_{lev} c : c'}$$

with $\Lambda \sqsubseteq \Lambda'$ iff $pre(\Lambda) = pre(\Lambda')$ and $\Lambda(x) \sqsubseteq \Lambda'(x)$ for all $x \in pre(\Lambda)$

Fig. 4. Security type system

are replaced by **skip**. The rule TTH with judgment $\vdash_{lev} c : c'$ ensures that lev alone approximates where secrets are stored. If no such judgment is derivable for a command c, then a secret might influence a **low**-variable in c.

The rule TAN enables and disables flow-sensitivity for particular variables by updating the pre-image of the partial level assignment, and ensures that a secret written into a variable x with $lev(x) = \textbf{low}$ must be overwritten before disabling flow-sensitivity for x. The rules TFL and TFH track the floating security level of a variable x by updating the level of x in the partial level assignment. The rule TIH permits branching on secrets. To avoid implicit information leaks due to such branchings, TIH requires that the **low**-slices of both branches are syntactically identical. The rules TAH, TFH, and TIH perform the **low**-slicing.

Theorem 3. *If c_p ensures a sound use of modes and $\vdash_{lev} c_p : c'$ is derivable, then c_p is secure for lev.*

Theorems 1, 2, and 3 establish the soundness result for our combined analysis:

Corollary 1. *If* $\emptyset \vdash \emptyset, \emptyset\{\mathbf{skip}; c'_p\}\emptyset, \emptyset : c_p$, *and* $\vdash_{lev} c_p : c'$ *are derivable and* $\mathcal{L}(\mathcal{A}_{ccnf})$ *is not reachable from* $ccnf\#$ *in DPN* $\mathcal{M}_{ccnf}$ *for* $ccnf = (c_p, \emptyset, mdst_\perp)$, *then* c_p *is secure for lev.*

6 Applying the Analysis

We illustrate how our type system gains precision from assumptions, while the DPN-based analysis ensures soundness of the combined analysis with the example program $c_1 = \mathbf{spawn}(o2{:=}o1; o1{:=}o2); o1{:=}s1; s1{:=}s2; s2{:=}o1; o1{:=}0$ and level assignment *lev* with $lev(o1) = lev(o2) = \mathbf{low}$ and $lev(s1) = lev(s2) = \mathbf{high}$. The program c_1 may leak the value of *s1* to an observer of *o1* due to concurrent execution of both threads.

Our security type system indeed rejects c_1, because no typing rule is applicable for $o1{:=}s1$: The rule TAH cannot be applied due to $lev(o1) \neq \mathbf{high}$, the rule TAL cannot be applied due to $lev(s1) \neq \mathbf{low}$, and the rules TFL as well as TFH cannot be applied due to $o1 \notin pre(\Lambda)$ (as the pre-image of the partial level assignment is initially empty and there are no annotations in the program). Using the assumption A-NR to enable flow-sensitivity for variable *o1*, $o1{:=}s1$ can be typed using TFH. To this end the program c_1 can be annotated as follows:

$$\mathbf{spawn}(o2{:=}o1; o1{:=}o2)@[\mathsf{acq}(\mathsf{A\text{-}NR}, \{o1\})];$$
$$o1{:=}s1; s1{:=}s2; s2{:=}o1; o1{:=}0@[\mathsf{rel}(\mathsf{A\text{-}NR}, \{o1\})]$$

However the program still contains the leak and the analysis detects this. The guarantee inference transforms the command $o2{:=}o1; o1{:=}o2$ of the spawned thread with the rules ISP, ISQ, ISK, and IAS into the following command:

$$\mathbf{skip}@[\mathsf{acq}(\mathsf{G\text{-}NR}, \emptyset), \mathsf{acq}(\mathsf{G\text{-}NW}, \emptyset), \mathsf{rel}(\mathsf{G\text{-}NR}, \{o1\}), \mathsf{rel}(\mathsf{G\text{-}NW}, \{o2\})];$$
$$o2{:=}o1@[\mathsf{acq}(\mathsf{G\text{-}NR}, \{o1\}), \mathsf{acq}(\mathsf{G\text{-}NW}, \{o2\}), \mathsf{rel}(\mathsf{G\text{-}NR}, \{o2\}), \mathsf{rel}(\mathsf{G\text{-}NW}, \{o1\})];$$
$$o1{:=}o2@[\mathsf{acq}(\mathsf{G\text{-}NR}, \{o2\}), \mathsf{acq}(\mathsf{G\text{-}NW}, \{o1\}), \mathsf{rel}(\mathsf{G\text{-}NR}, \emptyset), \mathsf{rel}(\mathsf{G\text{-}NW}, \emptyset)].$$

The annotation $\mathsf{rel}(\mathsf{G\text{-}NR}, \{o1\})$ in the first line makes explicit that the thread cannot provide the guarantee to not read *o1* during its next step, i.e. during the step of $o2{:=}o1$ in the second line. By spawning the new thread and executing its annotated first **skip** step, we reach a configuration with two threads. We have $o1 \notin mdst_2(\mathsf{G\text{-}NR})$ for the mode state of the spawned thread due to the annotation $\mathsf{rel}(\mathsf{G\text{-}NR}, \{o1\})$. Furthermore, we have $o1 \in mdst_1(\mathsf{A\text{-}NR})$ for the mode state of the original thread due to the annotation $\mathsf{acq}(\mathsf{A\text{-}NR}, \{o1\})$. Hence we have a reachable configuration that does not justify its assumptions. The corresponding DPN configuration preserves the mode states and is thus accepted by our conflict automaton that accepts DPN configurations with inconsistent mode states. Since the DPN over-approximates reachablitiy of the semantics, the reachability analysis from [9] detects that this DPN configuration is reachable, i.e. it detects a possible violation of globally sound use of modes and, hence, the program is rejected.

Adding synchronization via locks to ensure mutual exclusion of the regions accessing variable *o1* finally makes the program secure and no configuration with inconsistent mode states is reachable in the semantics anymore. Since the DPN models locking precisely, the DPN analysis also no longer detects reachability of any violation of globally sound use of modes. The following version of c_1 with additional synchronization has no leak and is accepted by our analysis:

$$c_2 = \textbf{spawn}(\textbf{lock}(l); o2:=o1; o1:=o2; \textbf{unlock}(l)); \textbf{lock}(l)@[\text{acq}(\text{A-NR}, \{o1\})];$$
$$o1:=s1; s1:=s2; s2:=o1; o1:=0; \textbf{unlock}(l)@[\text{rel}(\text{A-NR}, \{o1\})]$$

Theorem 4. *Let lev be a domain assignment with lev(o1) = lev(o2) = **low** and lev(s1) = lev(s2) = **high**. Then there are c'_2, c''_2 such that $\emptyset \vdash \emptyset, \emptyset\{\textbf{skip}; c_2\}\emptyset, \emptyset : c'_2$ and $\vdash_{lev} c'_2 : c''_2$ are derivable, and $\mathcal{L}(\mathcal{A}_{ccnf})$ is not reachable from ccnf# in DPN $\mathcal{M}_{ccnf}$ for ccnf = $(c'_2, \emptyset, mdst_\perp)$. Hence, c'_2 is secure for lev.*

7   Related Work

Andrews and Reitman [1] were the first to propose a static information-flow analysis based on flow rules, yet without a soundness proof wrt. a semantic security property. In [17], Smith and Volpano proposed the first security type system with a soundness proof against termination-sensitive noninterference.

The focus for most security type systems with support for synchronization, e.g. [14,19,20], has been preventing information leaks via synchronization. To the best of our knowledge, only the analyses in [10,11,18] can exploit synchronization for their precision. In [11], barrier synchonization allows combining different proof techniques in an analysis. In [10], Mantel, Sands, and Sudbrock introduced the rely-guarantee-style reasoning and the first flow-sensitive security type system for concurrent programs. The relationship of this article to [10] has already been clarified in the introduction.

Beyond security type systems, model-checking, e.g. in [8,13], as well as program dependence graphs, e.g. in [6], have been used to verify information-flow security for concurrent programs. These techniques promise very precise results, but are not necessarily compositional. A compositional analysis reduces the conceptual complexity of the verification, opens up the possibility to re-use analysis results of components, and, thus, can contribute to the scalability of an analysis. Our type system and our guarantee inference are compositional, meaning they can be applied to individual threads. Only our DPN-based analysis, which verifies the assumptions exploited by the type system for the actual program composed of multiple threads, is a whole-program analysis.

8   Conclusion

We automated a modular information-flow analysis for multi-threaded programs with a novel combination of a security type system and a reachability analysis based on DPNs. The combined analysis is sound wrt. termination-sensitive noninterference. The security type system supports flow-sensitive tracking of security levels for shared variables in the analysis of a given thread by exploiting assumptions about accesses to said variables by other threads. Using a conceptual example, we illustrated how the modules of our analysis interact and how synchronization with locks can contribute to the precision of our analysis.

Lifting the analysis to a realistic language with recursive procedure calls and dynamically allocated data structures is an open task for future work. Finally, we would like to implement our analysis and evaluate it in practice.

Acknowledgments. This work was funded by the DFG under the projects RSCP (MA 3326/4-1/2/3) and IFC4MC (MU 1508/2-1/2/3) in the priority program RS^3 (SPP 1496) and under project OpIAT (MU 1508/1-1/2).

References

1. Andrews, G., Reitman, R.: An axiomatic approach to information flow in programs. ACM Trans. Program. Lang. Syst. **2**(1), 56–76 (1980)
2. Arden, O., Chong, S., Liu, J., Myers, A.C., Nystrom, N., Vikram, K., Zdancewic, S., Zhang, D., Zheng, L.: Jif. Software release: http://www.cs.cornell.edu/jif/ (2014)
3. Askarov, A., Chong, S., Mantel, H.: Hybrid monitors for concurrent noninterference. In: 28th IEEE Computer Security Foundations Symposium, pp. 137–151 (2015)
4. Askarov, A., Hunt, S., Sabelfeld, A., Sands, D.: Termination-insensitive noninterference leaks more than just a bit. In: 13th European Symposium on Research in Computer Security, pp. 333–348 (2008)
5. Broberg, N., van Delft, B., Sands, D.: Paragon for practical programming with information-flow control. In: Shan, C. (ed.) APLAS 2013. LNCS, vol. 8301, pp. 217–232. Springer, Heidelberg (2013)
6. Giffhorn, D., Snelting, G.: A new algorithm for low-deterministic security. International Journal of Information Security pp. 1–25 (2014)
7. Hammer, C., Snelting, G.: Flow-sensitive, context-sensitive, and object-sensitive information flow control based on program dependence graphs. Int. J. Inf. Secur. **8**(6), 399–422 (2009)
8. Huisman, M., Blondeel, H.-C.: Model-checking secure information flow for multi-threaded programs. In: Mödersheim, S., Palamidessi, C. (eds.) TOSCA 2011. LNCS, vol. 6993, pp. 148–165. Springer, Heidelberg (2012)
9. Lammich, P., Müller-Olm, M., Wenner, A.: Predecessor sets of dynamic pushdown networks with tree-regular constraints. In: Bouajjani, A., Maler, O. (eds.) CAV 2009. LNCS, vol. 5643, pp. 525–539. Springer, Heidelberg (2009)
10. Mantel, H., Sands, D., Sudbrock, H.: Assumptions and guarantees for compositional noninterference. In: 24th IEEE Computer Security Foundations Symposium, pp. 218–232 (2011)
11. Mantel, H., Sudbrock, H., Krauߟer, T.: Combining different proof techniques for verifying information flow security. In: Puebla, G. (ed.) LOPSTR 2006. LNCS, vol. 4407, pp. 94–110. Springer, Heidelberg (2007)
12. Myers, A.C.: JFlow: practical mostly-static information flow control. In: 26th ACM SIGPLAN-SIGACT Symposium on Principles of Programming Languages, pp. 228–241 (1999)
13. Ngo, T.M., Stoelinga, M., Huisman, M.: Confidentiality for probabilistic multi-threaded programs and its verification. In: Jürjens, J., Livshits, B., Scandariato, R. (eds.) ESSoS 2013. LNCS, vol. 7781, pp. 107–122. Springer, Heidelberg (2013)

14. Sabelfeld, A.: The impact of synchronisation on secure information flow in concurrent programs. In: Bjørner, D., Broy, M., Zamulin, A.V. (eds.) PSI 2001. LNCS, vol. 2244, pp. 225–239. Springer, Heidelberg (2001)
15. Sabelfeld, A., Myers, A.C.: Language-based information-flow security. IEEE J. Sel. Areas Commun. **21**(1), 5–19 (2003)
16. Sabelfeld, A., Sands, D.: Probabilistic noninterference for multi-threaded programs. In: 13th IEEE Computer Security Foundations Workshop, pp. 200–214 (2000)
17. Smith, G., Volpano, D.: Secure information flow in a multi-threaded imperative language. In: 25th ACM SIGPLAN-SIGACT Symposium on Principles of Programming Languages, pp. 355–364 (1998)
18. Sudbrock, H.: Compositional and Scheduler-Independent Information Flow Security. Ph.D. thesis, Technische Universität Darmstadt, Germany (2013)
19. Terauchi, T.: A type system for observational determinism. In: 21st IEEE Computer Security Foundations Symposium, pp. 287–300 (2008)
20. Vaughan, J., Millstein, T.: Secure information flow for concurrent programs under total store order. In: 25th IEEE Computer Security Foundations Symposium, pp. 19–29 (2012)

Automated Verification

Checking Java Assertions Using Automated Test-Case Generation

Rafael Caballero[1]([✉]), Manuel Montenegro[1], Herbert Kuchen[2],
and Vincent von Hof[2]

[1] University Complutense de Madrid, Madrid, Spain
`rafacr@ucm.es`
[2] Institute of Information Systems, University of Münster, Münster, Germany

Abstract. We present a technique for checking the validity of Java assertions using an arbitrary automated test-case generator. Our framework transforms the program by introducing code that detects whether the assertion conditions are met by every direct and indirect method call within a certain depth level. Then, any automated test-case generator can be used to look for input examples that falsify the conditions. We show by means of experimental results the effectiveness of our proposal.

Keywords: Assertions · Conditions · Test-cases · Java · Test-case generation

1 Introduction

Using assertions is a common programming practice, and especially in the case of what is known as 'programming by contract' [5], where they can be used e.g. to formulate pre- and postconditions of methods as well as invariants of loops. Assertions in Java [6] are used for finding errors in an implementation at runtime during the test phase of the development cycle. If the condition in an `assert` statement is evaluated to false during program execution, an *AssertionException* is thrown.

The goal of our work is to use automated test-case generators for detecting assertion violations. Observe that, in contrast to model checking, automated test-case generators are not complete and thus our proposal may miss possible assertion violations, but as our experiments show it works quite well in practice and is helpful as a first approach during program development before using model checking. The overhead of an automated test-case generator is smaller than for full model checking, since data and/or control coverage criteria known from testing are used as a heuristic to reduce the search space. However, finding an input for a method $m()$ that falsifies some assertion in the body of $m()$ is not enough. For instance, in the case of preconditions it is important to observe whether the methods calling $m()$ ensure that the call arguments satisfy the precondition.Thus, we extend the proposal to indirect calls of these methods (up to a fixed level of indirection), allowing checking the assertions in the context of the whole program.

© Springer International Publishing Switzerland 2015
M. Falaschi (Ed.): LOPSTR 2015, LNCS 9527, pp. 221–226, 2015.
DOI: 10.1007/978-3-319-27436-2_13

```
1   public class Sqrt {
2     static double eps = 0.00001;
3
4     public double sqrt(double r){
5       double a, a1 = r + eps;
6       do { a = a1;
7         a1 = a+r/a/2.0; //erroneous!
8         assert a==1.0 ||
9             (a1>1.0 ? a1 < a
10                    : a1 > a);}
11      while (Math.abs(a - a1) >= eps);
12      return a1;
13    }}
14  public class Circle {
15    double area;
16
17    Circle(double area) {this.area = area; }
18
19    public double getRadius() {
20      assert area>=0;
21      return Sqrt.sqrt(area/Math.PI); }}
```

Fig. 1. Java method sqrt, corresponding control-flow graph, and class Circle.

In order to fulfill these goals we propose a technique based on a source-to-source transformation that converts the assertions into if statements and changes the return type of methods to represent the path of calls leading to an assertion violation as well as the normal results of the original program. Converting the assertions into a program control-flow statement is very useful for white-box, path-oriented test-case generators, which determine the program paths leading to some selected statement and then generate input data to traverse such a path (see [2] for a recent survey on the different types of test-case generators). Thus, our transformation allows this kind of generators to include the assertion conditions into the sets of paths to be covered.

2   Assertions and Automated Test-Case Generation

Java assertions ensure at runtime (if executed with the right option) that the program state fulfills certain restrictions. Figure 1 shows our running example. The radius of a circle is computed based on an erroneous implementation of the sqrt method (a1 = a+r/a/2.0; should be a1 = (a+r/a)/2.0;).

Our idea is to use a test-case generator to detect possible violations of the occurring assertions. A test-case generator is typically based on some heuristic which reduces its search space dramatically. Often it tries to achieve a high coverage of the control and/or data flow.

EvoSuite [4] generates test cases also for code with assert conditions. However, its search-based approach does not always generate test cases exposing assertion violations. In particular, it has difficulties with indirect calls such as the assertion in Sqrt.sqrt after a call from Circle.getRadius. A reason is that EvoSuite does not model the call stack. Thus, the test cases generated by EvoSuite for Circle.getRadius only expose one of the two possible violations, namely the one related to a negative area.

```
public abstract class MayBe<T> {

    public static class Value<T> extends MayBe<T> { // ...
    public static class CondError<T> extends MayBe<T> { // ...

    // did the method return a normal value (no violation)?
    abstract public boolean isValue();

    // value returned by the method.
    abstract public T getValue();

    // No condition violation detected. Return the same value
    // as before the instrumentation.
    public static <K> MayBe<K> createValue(K value) {
        return new Value<K>(value);  }

    // an assert condition is not verified
    public static <T> MayBe<T> generateError(String method,
                                    int position) {
        return new CondError<T>(new Call(method, position));}

    // method calls another method whose precondition or
    // postcondition is not satisfied.
    public static <T,S> MayBe<T> propagateError(String method,
                           int position , MayBe<S> error){
        return new CondError<T>(new Call(method, position),
                               (CondError<S>) error);}
}
```

Fig. 2. Class MayBe<T>: new result type for instrumented methods.

There are other test-data generators such as JPet [1] that do not consider assert statements and thus cannot generate test cases for them. In the sequel, we present a program transformation that allows both EvoSuite and JPet to detect both possible assertion violations.

3   Program Transformation

The idea of the program transformation is to instrument the code in order to obtain special output values that represent possible violations of assertion conditions. Then, an automatic test-case generator is employed to obtain the inputs that produce these special values. In our case the instrumented methods employ the class MayBe<T> of Fig. 2. The overall idea is that a method returning a value of type T in the original code returns a value of type MayBe<T> in the instrumented code. MayBe<T> is in fact an abstract class with two subclasses, Value<T> and CondError<T>. Value<T> represents a value with the same type as in the original code, and it is used via the method MayBe.createValue whenever no assertion violation has been found. If an assertion condition is not satisfied, a CondError value is returned. There are two possibilities:

- The assertion is in the same method. Suppose it is the i-th assertion in the body of the method following the textual order. In this case, the method returns MayBe.generateError(name,i); with name the method name. The purpose of the method generateError is to create a new CondError object.

```
public class Sqrt {

  static double eps = 0.000001;

  public static MayBe<Double> sqrtCopy(double r){
    double a, a1 = 1.0;
    a = a1;
    a1 = a+r/a/2.0;
    double aux = Math.abs(a-a1);
    while (aux >= eps);{
      a = a1;
      a1 = a+r/a/2.0;
      if (!(a==1.0 || (a1>1.0 ? a1<a : a1>a)))
        return MayBe.generateError("sqrt", 2);
      aux = Math.abs(a-a1); }
    return MayBe.createValue(a1);
} }

public class Circle {
  double area;
  Circle(double area) {this.area = area;}

  public MayBe<Double> getRadius() {
    if (!(area>=0))
      return MayBe.generateError("getRadius", 1);
    MayBe<Double> r = Sqrt.sqrtCopy(area/Math.PI);
    if (!r.isValue())
      return MayBe.propagateError("getRadius", 2, r);
    return r;
} }
```

Fig. 3. Transformed running example.

Observe that the constructor of `CondError` receives as parameter a `Call` object. This object represents the point where a condition is not verified, and it is defined by the parameters already mentioned: the name of the method, and the position i.

- The method detects that an assertion violation has occurred indirectly through the i-th method call in its body. Then, the method needs to extend the path and propagate the error. This is done using a call `propagateError(name,i,error)`, where `error` is the value to propagate. The corresponding constructor of class `CondError` adds the new call to the path.

Figure 3 shows the transformed running example. The methods not related (in)directly to assertions, e.g. the constructor of `Circle`, remain unchanged. Due to the lack of space, we omit the treatment of inheritance here. It can be found in [3].

4   Experiments

We have evaluated a few examples with different test-case generators with and without our program transformation. We have also developed a prototype that performs this transformation automatically. It can be found at https://github.com/wwu-ucm/assert-transformer, whereas the aforementioned examples can be found at https://github.com/wwu-ucm/examples.

Table 1. Detecting assertion violations.

		EvoSuite		JPet	
Method	Total	P	P^T	P	P^T
Circle.getRadius	2	1	2	0	2
BloodDonor.canGiveBlood	2	0	2	0	2
TestTree.insertAndFind	2	0	2	0	2
Kruskal	1	1	1	0	1
Numeric.foo	2	1	2	0	2
TestLibrary.test*	5	0	5	0	5
MergeSort.TestMergeSort	2	0	1	0	1
java.util.logging.*	5	0	2	-	-

Table 2. Control and data-flow coverage in percent.

	Binary tree		Blood donor		Kruskal		Library		MergeSort		Numeric		StdDev		Circle	
	P	P^T	P	P^T	P	P^T	P	P^T	P	P^T	P	P^T	P	P^T	P	P^T
EvoSuite	90	95	83	91	95	100	63	92	82	82	76	82	71	71	80	100
JPet	–	89	–	99	–	49	–	20	–	87	–	82	–	74	–	100

We have used two test-case generators, JPet and EvoSuite, for exposing possible assertion violations. As can be seen in Table 1, almost all possible assertion violations could be detected. Moreover, our program transformation typically improves the detection rate, since it makes the control flow more explicit than the usual assertion-violation exceptions. Column *Total* displays the number of possible assertion violations. Column P shows the number of detected assertion violations using the test-case generator and the original program, while column P^T displays the number of detected assertion violations after applying the transformation. Notice that JPet cannot find any assertion violation without our transformation, since it does not support assertions. For large examples such as the JDK logging package (6500 LOC), the configuration of JPet is tedious. As a consequence, this example has not been processed by this tool.

Our program transformation typically requires only a few seconds and even for larger programs such as the JDK 6 logging package the transformation finishes in 18.2 s. The runtime of our analysis depends on the employed test-case generator and the considered example. It can range from a few seconds to several minutes (Table 2).

5   Conclusions

We have presented an approach to use test-case generators for exposing possible assertion violations in Java programs. Our approach is a compromise between the usual detection of assertion violations at runtime and the use of a full model checker. Since test-case generators are guided by heuristics such as control- and data-flow coverage, they have to consider a much smaller search space than a model checker and can hence deliver results much more quickly.

Additionally, we have developed a program transformation which replaces assertions by computations which explicitly propagate violation information through an ordinary computation involving nested method calls. In case of a violation, our transformation makes the control flow more explicit than the usual assertion-violation exceptions. This helps the test-case generators to reach a higher coverage of the code and enables more assertion violations to be exposed and detected. Additionally, the transformation allows to use test-case generators such as JPet which do not support assertions.

We have presented some experimental results demonstrating that our approach helps indeed to expose assertion violations and that our program transformation improves the detection rate.

Acknowledgements. This work has been supported by the German Academic Exchange Service (DAAD, 2014 Competitive call Ref. 57049954), the Spanish MINECO project CAVI-ART (TIN2013-44742-C4-3-R), Madrid regional project N-GREENS Software-CM (S2013/ICE-2731) and UCM grant GR3/14-910502.

References

1. Albert, E., Cabanas, I., Flores-Montoya, A., Gómez-Zamalloa, M., Gutierrez, S.: jPET: An automatic test-case generator for Java. In: 18th Working Conference on Reverse Engineering, WCRE 2011, Limerick, Ireland, October 17–20, 2011, pp. 441–442. (2011)
2. Anand, S., Burke, E., Chen, T.Y., Clark, J., Cohen, M.B., Grieskamp, W., Harman, M., Harrold, M.J., McMinn, P.: An orchestrated survey on automated software test case generation. J. Syst. Softw. **86**(8), 1978–2001 (2013)
3. Caballero, R., Montenegro, M., Kuchen, H., von Hof, V.: A program transformation for converting java assertions into control-flow statements. Technical report 24, ERCIS (2015)
4. Galeotti, J.P., Fraser, G., Arcuri, A.: Improving search-based test suite generation with dynamic symbolic execution. In: IEEE International Symposium on Software Reliability Engineering (ISSRE), pp. 360–369. IEEE (2013)
5. Meyer, B.: Object-oriented Software Construction, 2nd edn. Prentice-Hall Inc., Upper Saddle River (1997)
6. Oracle. Programming with Assertions (2014). https://docs.oracle.com/javase/jp/8/technotes/guides/language/assert.html

A Generic Intermediate Representation for Verification Condition Generation

Manuel Montenegro[✉], Ricardo Peña, and Jaime Sánchez-Hernández

Universidad Complutense de Madrid, Madrid, Spain
montenegro@fdi.ucm.es, {ricardo,jaime}@sip.ucm.es

Abstract. As part of a platform for computer-assisted verification, we present an intermediate representation of programs that is both language independent and appropriate for the generation of verification conditions. We show how many imperative and functional languages can be translated to this generic intermediate representation, and how the generated conditions reflect the axiomatic semantics of the original program. At this representation level, loop invariants and preconditions of recursive functions belonging to the original program are represented by assertions placed at certain edges of a directed graph.

The paper defines the generic representation, sketches the transformation algorithms, and describes how the places where the invariants should be placed are computed. Assuming that, either manually or assisted by the platform, the invariants have been settled, it is shown how the verification conditions are generated. A running example illustrates the process.

Keywords: Verification platforms · Intermediate representation · Verification conditions · Program transformation

1 Introduction

In the last few years, verification platforms are becoming more and more popular [1,10,15]. Their success is in part due to the increasing power of the underlying proving machinery, the SMT solvers [7,8]. In these platforms, the user is responsible for giving the source program, its specification in the form of a precondition and a postcondition, and the invariant assertion of each loop. The platform gives support for analysing and proving termination, for generating the verification conditions (VC), and for automatically proving them, whenever this is possible.

A possible drawback is that the source language is usually fixed by the platform and it consists of a restricted subset of a real-life one. For instance, Dafny supports object-oriented programming but not inheritance. WhyML does not support object orientation, nor even has a heap.

Work partially supported by the Spanish MINECO project CAVI-ART (TIN2013-44742-C4-3-R), Madrid regional project N-GREENS Software-CM (S2013/ICE-2731) and UCM grant GR3/14-910502.

© Springer International Publishing Switzerland 2015
M. Falaschi (Ed.): LOPSTR 2015, LNCS 9527, pp. 227–243, 2015.
DOI: 10.1007/978-3-319-27436-2_14

The purpose of our platform CAVI-ART[1] is a bit more ambitious. On the one hand, it addresses real-life languages and will support most of, or ideally all, their complexities and subtleties. Additionally, it will cover both imperative, possibly object-oriented ones, such as C, C++ and Java, and functional ones such as Erlang, SML and Haskell. On the other hand, the platform will assist the user in discovering the loop invariant assertions, or equivalently, the preconditions of the recursive functions. The remaining platform assistance will be similar to that of the other platforms. In fact, we plan to reuse the infrastructure of Why3 to interface different SMTs and proof assistants, by expressing our VCs in the Why3 assertion language.

In Fig. 1 we show a picture of the whole project. A key aspect of it is designing an intermediate representation (IR) of programs to which source programs, written in a variety of languages, can be transformed. Once programs have undergone this transformation, the remaining activities —invariant synthesis, termination analysis, VC generation, VC proving— can be performed in a language-independent way. This transformation yields an abstraction of the control and data flow of the program that relies on a set of language-dependent primitive functions, which are defined via axioms and can be reused among different languages. Moreover, some of them are already present in Why3's standard library of theories, which includes definitions of integers, lists, arrays, real numbers, etc. and their associated functions.

The platform is under construction. We have completed the design, the IR, and a front-end for Java. Our current work mainly focuses on invariant synthesis. In this paper we describe such a generic IR, and show how VCs can be generated from it, guaranteeing that should all the VCs be discharged by the provers, then the original program satisfies all assertions. A key step in mapping imperative programs to the IR is transforming iteration to recursion, so that both are dealt with uniformly. A second step is to detect where invariant assertions would be needed in the resulting IR. Once these assertions have been provided, either by the user or by the platform itself, the VC generation is done automatically.

The plan of the paper is as follows: in Sect. 2 we describe the transformation of several imperative features such as primitive and structured types, classes, and the heap to a common framework. Then we explain how to abstract the control by generating a Control Flow Graph (CFG). In Sect. 3, we briefly remind how functional languages are compiled to a small core representation, which usually is a slight extension of the λ-calculus. In Sect. 4, we present and justify our IR, and give an axiomatic semantics to it by means of weakest preconditions. Section 5 describes the algorithm transforming the CFG to the IR, and detects the locations of the invariants. Section 6 explains the VC extraction algorithm. Finally, Sect. 7 draws some conclusions and reviews the related work.

---

[1] CAVI-ART stands for Computer Assisted VaIIdation by Analysis, tRansformation and Testing.

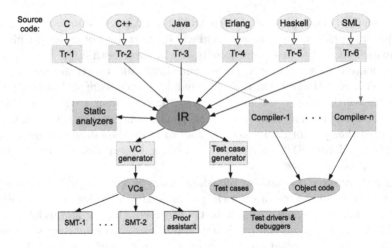

Fig. 1. CAVI-ART project overview

2   Imperative Languages

The computation model of imperative languages is given by the execution of a sequence of statements that change the state of the program. Among the diversity of the features provided by modern imperative languages (such as Java, Javascript, Python, etc.) there are two which are shared by most of them: destructive assignment and explicit management of control flow. However, languages differ in the kind of basic values that can be assigned to a variable, and the choice of control flow constructions (loops, exceptions, method calls, delegates, etc.) In the following, we shall abstract their common parts in order to determine the constructions needed by the IR. We also identify the language-specific components, so that the IR will be parametric on them.

Example 1. As a running example, let us consider in this paper the following Java implementation of the insertion sort algorithm:

```java
public void insertionSort(int[] v) {
  for (int i = 0; i < v.length; i++) {
    int e = v[i];
    int j = i - 1;
    while ((j >= 0) && (v[j] > e)) {
      v[j+1] = v[j];
      j = j - 1;
    }
    v[j+1] = e;
  }
}
```

Basic Values. For each language we identify its set of basic values. We classify them into different categories, which will subsequently be mapped to theory types of the underlying proof system.

For a given language, we consider a set of *value categories* $\{\beta_1, \ldots, \beta_n\}$, each one is a pair $\langle B_\beta, \equiv_\beta \rangle$, where B_β is the set of values contained within the category β, and $\equiv_\beta$ is an equivalence relation on these values. This relation is necessary for performing case distinction on the values at the IR level. For every language, we assume the existence of a category $\beta_{\mathbf{Bool}}$ with the set $B_{\mathbf{Bool}} = \{true, false\}$ and the usual equivalence relation.

For instance, we use in Java the set of types given by the semantics of *Jinja* [14]: booleans, integers, pointers, null reference and unit type. We also include the category of floating point numbers and arrays, since some solvers (e.g. Z3 [7]) provide direct support for them.

Built-in Operators and Functions. This is another language-dependent component. We encode them in the IR as functions whose behaviour is defined by a set of axioms. Therefore, for each programming language we define the set of primitive functions and axioms. Both can be specified in terms of already existing theories.

In the case of the translation from Java into the IR, several primitive functions are based on their counterparts defined in the Why3 Standard Library. For instance, we associate the category β_{int} of integer values with the int type defined within Why3's Int theory. The integer-based operators (such as <=, ==, +, etc.) are mapped into its corresponding counterparts in this theory. An analogous association is made with booleans and real numbers. Arrays are also translated into the type array defined within the Array theory of Why3, defined as follows:

```
1 type array 'a model { length : int; mutable elts : map int 'a }
2     invariant { 0 <= self.length }
```

The definition of the built-in operations on arrays is more involved, since a simple access to an array may result in a NullPointerException or an ArrayIndexOutOfBoundsException. We consider two different policies:

- **Safe Array Access Assumption.** The built-in function sel-array has a precondition asserting that the array is not null and that the index lies within the bounds. If this holds, then the selection yields a valid result.

$$\{\mathcal{H}(p) = Array\ a \wedge 0 \leq i < a.length\}\ \texttt{sel-array}(\mathcal{H},\ p,\ i)\ \{res = \texttt{get}\ a\ i\}$$

In this specification the $\mathcal{H}$ denotes a heap, p a heap location, and get denotes the actual array access function defined in the Why3 library. In a similar way we define mod-array, which yields the heap resulting from modifying an array in a given position.

- **Array Access with Exceptions.** We extend the specification of sel-array by considering the possibility that the array access may throw an exception. However, since exception handling is considered as a language-dependent feature, exceptions should not be part of the IR. Exception management is handled with a special type which is similar to the Either type of Haskell:

```
1 data opt_result = Ok value | Exception loc
```

$$
\begin{array}{llll}
blk & ::= & stm_1; \ldots; stm_n; jump & \{\text{ instruction BB }\} \\
& | & \textbf{return } x & \{\text{ exit BB, } x \text{ is a variable }\} \\
stm & ::= & x = e & \{\text{ assignment (single variable) }\} \\
& | & (x_1, \ldots, x_n) = e & \{\text{ assignment (several variables) }\} \\
jump & ::= & \textbf{case } x \textbf{ of } c_1 \rightarrow n_1 \ldots c_m \rightarrow n_m & \{\text{ conditional jump } (n_1, \ldots, n_m \in \mathbb{N}) \} \\
& | & \textbf{goto } n & \{\text{ unconditional jump } (n \in \mathbb{N}) \} \\
e & ::= & a & \{\text{ atom }\} \\
& | & f(a_1, \ldots, a_n) & \{\text{ function application }\} \\
a & ::= & c & \{\text{ literal }\} \\
& | & x & \{\text{ variable }\}
\end{array}
$$

Fig. 2. Structure of CFG blocks.

In this definition `value` denotes a union type for basic values, and `loc` is the type of heap locations. Both definitions are language-dependent.

The first policy is simpler, and it works if the prover can establish the validity of all array accesses contained within the method. If it cannot, the correctness of the method is not proved. With the second policy the postconditions of the method can be more precise and assert facts regarding exceptions (for instance, the reasons of an exception being thrown), but makes the resulting IR code more complicated. For the sake of simplicity, we consider the first policy in our running example.

Heap Management. The presence of a mutable memory heap plays an essential role in imperative programs. As a consequence of its physical representation in the memory, virtually all languages consider a heap $\mathcal{H}$ as a mapping from locations to values. The language-dependent element here is the kinds of values represented in the heap. In Java we follow the approach of [14] (extended with array values) and define the set of heap values as follows:

```
1  data heap_value = Array (array value)
2                  | Object string (map (string, string) value)
```

where an object instance contains a class name and a map from pairs (p, c) to pointers. In these pairs p denotes the name of an attribute and c the name of the class to which the attribute belongs.

In order to specify heap modifications we follow the same approach as in the previous section; they are managed as language-dependent built-in functions, each one with a formal specification via pre- and post-conditions. Therefore every heap-related operation subject to axiomatization, such as method calls, dynamic dispatch, etc. can be used in the IR. Since we avoid the existence of a mutable state, the operations modifying the heap are pure, in the sense that they yield another heap with the corresponding changes.

Control Flow. In order to handle this feature in a language independent way, the source program is transformed into a *control-flow graph* representation (CFG) [2].

In this graph each node is a *basic block* (BB) containing a sequence of program instructions without jumps between them (except calls to other functions or methods).

The information inside a BB is defined by the grammar given in Fig. 2. A BB can be an exit block (**return**) or contain a sequence of basic statements followed by a jump instruction. Statements are assignments whose right-hand side can be a literal, variable or a function application. In the latter case, only atomic arguments are allowed. This requires a flattening transformation on the original program and the addition of new assignments. Jump instructions refer to other BBs in the CFG, each of which is identified by a natural number. Thus, a CFG is a set of numbered BBs $\{(n_1, blk_1), \ldots, (n_r, blk_r)\}$. We can attach to each CFG an assertion which must be satisfied by every execution of the function being analysed. This is useful for specifying loop invariants.

Example 2. The transformation of our insertion sort example into a CFG yields the result shown in Fig. 3, where array accesses and basic operations have been replaced by flattened function calls. A new variable $\mathcal{H}$ is introduced to denote explicitly the heap.

$[1]:\ i = 0$
$\quad$ **goto** $[2]$
$[2]:\ x_1 = \mathtt{len}(v)$
$\quad b = \mathtt{<}(i, x_1)$
$\quad$ **case** b **of**
$\qquad true \rightarrow [3]$
$\qquad false \rightarrow [7]$
$[3]:\ e = \mathtt{sel\text{-}array}(\mathcal{H}, v, i)$
$\quad j = \mathtt{-}(i, 1)$
$\quad$ **goto** $[4]$

$[4]:\ b_1 = \mathtt{>=}(j, 0)$
$\quad x_2 = \mathtt{sel\text{-}array}(\mathcal{H}, v, j)$
$\quad b_2 = \mathtt{>}(x_2, e)$
$\quad b_3 = \mathtt{\&\&}(b_1, b_2)$
$\quad$ **case** b_3 **of**
$\qquad true \rightarrow [5]$
$\qquad false \rightarrow [6]$

$[5]:\ x_3 = \mathtt{sel\text{-}array}(\mathcal{H}, v, j)$
$\quad x_4 = \mathtt{+}(j, 1)$
$\quad \mathcal{H} = \mathtt{mod\text{-}array}(\mathcal{H}, v, x_4, x_3)$
$\quad j = \mathtt{-}(j, 1)$
$\quad$ **goto** $[4]$
$[6]:\ x_5 = \mathtt{+}(j, 1)$
$\quad \mathcal{H} = \mathtt{mod\text{-}array}(\mathcal{H}, v, x_5, e)$
$\quad i = \mathtt{+}(i, 1)$
$\quad$ **goto** $[2]$
$[7]:$ **return** $\mathcal{H}$

Fig. 3. CFG blocks of the *insertionSort* algorithm.

Our next step is translating the CFG of the input program into a set of mutually recursive functions, from which the verification conditions will be extracted. In order to obtain a set of functions, we dispose of destructive assignment by transforming our program into *Static Single Assignment* form (SSA) [3]. After this, each program variable is assigned exactly once, and subsequent assignments are done to different *versions* of the variable, each one having a different name. In our case, the SSA transformation is applied locally to each BB. Instead of having ϕ functions in confluence nodes (as usual in SSA), the transformation performs a liveness analysis at the beginning of each node. Let LV_i be the set of live variables at node i (before applying SSA transformation). After applying the local transformation to each node, we have to compute, for every node j pointing to i, a substitution $\theta_{j,i}$ that maps each variable $x \in LV_i$ to the last version of that variable occurring in j.

Example 3. The liveness analysis on the CFG of Example 2 produces:

$$LV_1 = \{v, \mathcal{H}\} \quad LV_3 = \{v, i, \mathcal{H}\} \quad LV_5 = \{v, i, j, e, \mathcal{H}\} \quad LV_7 = \{\mathcal{H}\}$$
$$LV_2 = \{v, i, \mathcal{H}\} \quad LV_4 = \{v, i, j, e, \mathcal{H}\} \quad LV_6 = \{v, i, j, e, \mathcal{H}\}$$

The translation of each BB into SSA leads to the CFG represented in Fig. 4, in which the left-hand sides $\mathcal{H}$ and j of the BB [5] have been replaced by $\mathcal{H}_1$ and j_1, respectively. The corresponding mapping from [5] to [4] would be $\theta_{5,4} = [v \mapsto v, i \mapsto i, j \mapsto j_1, e \mapsto e, \mathcal{H} \mapsto \mathcal{H}_1]$.

After this transformation, we translate each BB i of the CFG into a recursive function receiving as arguments the variables in LV_i. Its definition is the sequence of BB statements, and the jump branches are calls to the adjacent BBs by using the respective substitutions. This will be shown in Sect. 5.

3   Functional Languages

Functional languages are radically different to imperative ones, as they provide neither destructive variable assignment, nor control flow management. There is no notion of state, as it is the case in the imperative paradigm. Their main features are pattern matching for function definition arguments, higher-order functions, recursive definitions for data types and functions, and lambda abstractions.

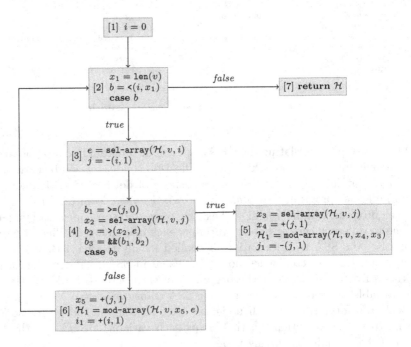

Fig. 4. Representation of the CFG/SSA of the insertion sort algorithm

In addition, some of them are polymorphic and strongly typed, and provide a type inference algorithm (Haskell, ML). A few of them are lazy (Haskell).

From a theoretical point of view, functional languages emerge from the λ-calculus, which in fact can be seen as a minimalistic core language for all of them. But from a practical point of view, the core for real functional languages usually includes constructor application, **let**-expressions for local definitions, recursive **letrec**-expressions, and **case**-expressions, a normalized form of pattern matching. This is the case of the enriched lambda-expressions used in [16], and also in the core languages of the Glasgow Haskell Compiler [18], ML [17] and Erlang [4].

Sometimes it may be useful to enrich the core syntax in order to facilitate the compiler code generation, or to reduce it in order to simplify formal reasoning. For example, λ-abstractions can be removed from the core with the well-known *lambda lifting* transformation [16], which transforms λ-abstractions into ordinary (named) functions. Also, the applicative notation can be flattened in order to avoid complex nesting of expressions. Moreover, nested pattern matching can be compiled in such a way [16], that it is converted in a sequence of nested **case**-expressions, each one with flat and mutually exclusive patterns, and covering all datatype constructors.

4   The Intermediate Representacion

From the precedent sections, it is clear that a minimal (core) functional language can serve both to represent imperative programs which have undergone an SSA transformation and functional programs which have been previously desugared. The minimal common elements of this core language are the followings:

- Sequential **let** expressions, which also represent imperative SSA assignments.
- Recursive **letrec** expressions, needed to define mutually recursive functions.
- λ-abstractions and applications, needed to define and apply functions.
- **case** expressions, which can serve both to mimic imperative **switch** statements and to express functional pattern matching.

In addition to this, imperative languages need support for structured data types such as arrays and records, and functional languages need support for algebraic data types, polymorphism, and higher-order. Taking all this into account, we have defined an intermediate representation (IR) that gives support to most of the features one can find in imperative and functional languages. In Fig. 5 we show the abstract syntax of our IR.

We justify some of the decisions leading to this IR. Firstly, the arguments of applications and **case** discriminants are restricted to be atoms. This facilitates the renaming of predicate arguments when propagating assertions, and also makes the definition of weakest preconditions for the **case** construction simpler. We make note that an **if** construction is not needed as it is a particular instance of **case**. Secondly, function definitions are confined to be in a **letfun** expression, and they are by default mutually recursive. A **letfun** can be considered as syntactic sugar for a functional **letrec** expression in which each variable is bound

```
a ->                                -- atom
    c                               -- constant
    | x                             -- variable
ae ->                               -- atomic expression
    a                               -- atom
    | f a1 ... an                   -- primitive operator/function application
    | C a1 ... an                   -- constructor application
e ->                                -- structured expression
    ae                              -- atomic expression
    | let p = ae in e               -- sequential let
    | letfun fd1 ... fdn in e       -- function definition block
    | case a of alt1 ... altn       -- algebraic type or primitive type case
              [ _ -> e]             -- optional default clause
alt ->                              -- case alternative
    C x1 ... xr -> e                -- algebraic type alternative
    c           -> e                -- primitive type alternative
p ->                                -- pattern
    x                               -- variable pattern
    | (x1,...,xn)                   -- tuple pattern
fd -> f x1 ... xn = e               -- function definition. The name f is global
```

Fig. 5. Abstract syntax of the CAVI-ART Intermediate Representation

to a lambda abstraction. Thirdly, expressions are in the so-called A-normal form [11]. In particular, this implies that in **let** bindings, applications occur as stand alone expressions. Finally, **case** patterns are flat and they exclude each other, so that only one alternative is possible. If a **case** does not include an alternative for each constructor, it necessarily has a default clause. The purpose of all these restrictions is again to facilitate the definition of weakest preconditions and the generation of verification conditions.

The IR is strongly typed and the type system is polymorphic in a Hindley-Milner style, similar to that of the logical language Why3 [9] in which the assertions are expressed. This type system supports both polymorphic functional languages such as SML and Haskell, untyped functional languages such as Erlang, monomorphic imperative languages such as C, and polymorphic imperative languages such as C++ or Java.

Arrays and records are not built-in data types of the IR, but they can be defined in a language-specific way as explained in Sect. 2 for arrays, and similarly for records. Algebraic data types (ADT) can be defined in the IR, and pattern matching is supported by **case** expressions. All these types (i.e. arrays, records and ADTs), and their primitive operators, are directly supported by the SMTs underlying the CAVI-ART platform. They contain a rich set of axioms allowing to reason about the formulas using them.

Other features which are present in a particular language but not in others, can be mapped to the IR by the front-end of each particular language, either by introducing new primitive types and operators, supported by their corresponding theories, or by representing them in the IR built-in types. An example of this is the mapping of the OO-language heap into an array variable that is passed around as an additional argument of methods, as it has been illustrated in our running example.

The definition of the axiomatic semantics of the IR, given as weakest preconditions, is as follows:

$$wp(\textbf{let } x = e_1 \textbf{ in } e_2, R) \stackrel{\text{def}}{=} Dom(e_1) \wedge (x = e_1) \rightarrow wp(e_2, R)$$

$$wp(\textbf{case } a \textbf{ of } \ldots C \; x_1 \cdots x_n \rightarrow e \ldots, R) \stackrel{\text{def}}{=} (a = C \; x_1 \cdots x_n) \rightarrow wp(e, R)$$
$$\wedge \text{ the remaining alternatives}$$

$$wp(\textbf{case } a \textbf{ of } \textit{true} \rightarrow e_1; \textit{false} \rightarrow e_2, R) \stackrel{\text{def}}{=} (a \rightarrow wp(e_1, R)) \wedge (\neg a \rightarrow wp(e_2, R))$$

For many primitive applications (e.g. $e \equiv x + y$), $Dom(e)$ is assumed to be *true*. But some others are partial functions and require a precondition. Function definitions are assumed to be annotated with their respective precondition and postcondition. Let $f \; x_1 \cdots x_n = e$ be the definition of a function f, and let respectively $Q(x_1, \ldots, x_n)$ and $R(x_1, \ldots, x_n, \textit{res})$ be its precondition and postcondition, where *res* stands for f's result. Then, in an application such as $f(a_1, \ldots, a_n)$, it must be proved that $Q(a_1, \ldots, a_n)$ holds before reaching this call, and it can be assumed that $R(a_1, \ldots, a_n, \textit{res})$ holds when f returns.

In principle, we do not need to define an operational semantics for the IR, since its aim is not to be executed, but rather to be used for verification. When we define $wp(e, R) \stackrel{\text{def}}{=} Q$ for an expression e with free variables $\overline{x}$, and predicates $Q(\overline{x})$ and $R(\overline{x}, v)$, we mean as usual that the set of all the initial states for the variables $\overline{x}$ guaranteeing that the value v to which e is evaluated satisfies $R(\overline{x}, v)$ is exactly that specified by $Q(\overline{x})$. This logical definition is supposed to capture the semantics of e independently of the details of its evaluation. In this way, it is not important whether the evaluation mechanism is imperative or functional, whether there is, or is not, internal sharing during e's evaluation, or even whether the evaluation order is lazy or eager.

5   Determining the Invariant Locations

In order to set the invariant conditions in the appropriate places, the CFG must be structured according to its strongly connected components (SCC) and subcomponents. Formally, the *Connected Components Structure* (CCS) of a graph G is a list of components, where each one is either a single node, or a pair with an *entry point* and a list of components:

$$CCS \quad ::= [COMP]$$
$$COMP ::= node \mid (\textit{entry\_node}, CCS)$$

This structure is built up by computing the maximal SCCs of a graph, then the SCCs inside these components, and so on. The resulting structure contains all the nodes of the graph grouped according to the loops of the original program. For any pair (*entry_node*, *ccs*) in the structure, the subgraph of G corresponding to the nodes of *ccs* is a connected one, and *entry_node* is the only entry point to this subgraph. Moreover, for any component c of a CCS, except for the outermost one, there is a component c' in the immediate prior level which contains some node connected to the entry node of c.

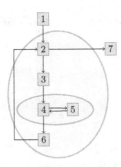

Fig. 6. CCS for the CFG of Fig. 4

Example 4. Considering the CFG of Fig. 4, and disregarding for simplicity the node contents, the CCS is $[1, (2, [3, (4, [5]), 6]), 7]$, which is represented in Fig. 6. We get the external components: 1, 7, and the node set $[2, 3, 4, 5, 6]$, which has 2 as entry point, and then another internal component with the nodes $[4, 5]$, which has 4 as entry point. The invariants should be placed before the entry points 2 and 4, which correspond to the entry points of the loops of Example 1.

```
1   cfg_to_ccs(G):
2   if G is a single node then
3       return [G]
4   else
5       In ← entry_point(G)
6       G' ← G − {In}
7       Comps ← strongly_connected_components(G')
8       [C₁,...,Cₙ] ← sort(Comps)
9       for all Cᵢ in [C₁,...,Cₙ] do
10          Gᵢ ← subraph of G with the nodes of Cᵢ
11          CCSᵢ ← cfg_to_ccs(Gᵢ)
12      end for
13      return (In,[CCS₁,...,CCSₙ])
14  end if
```

Fig. 7. Algorithm computing the CCS of a graph

The function *cfg_to_ccs(G)* of Fig. 7 computes the CCS of a given control flow graph G. It decomposes the graph into connected subgraphs in successive recursive calls, until it reaches the base case of a single node (line 2). Otherwise, it searches for the entry point of the graph (line 5) which is guaranteed to exists since the graph is a CFG. Then, it considers the subgraph G' (line 6) obtained by removing the entry point and its edges, and computes their strongly connected components (see [5]) of this subgraph (line 7). These components are then sorted by the function *sort* (line 8) as follows:

- Collapse each strongly connected component into a single node;
- Compute a topological sort of the resulting graph in a list (this sorting is always possible as cycles have been collapsed in the previous step);
- Uncollapse the strongly connected components;

Then for each list component (line 9), the algorithm obtains the corresponding subgraph (line 10) and the corresponding CCS for them (line 11). Finally, it returns the list of components, together with their entry point (line 13).

6 Generating the Verification Conditions

The verification of a complex program is usually done in a modular way, procedure by procedure. Indeed, this is the whole purpose of defining pre-post assertions for every user procedure: to make it possible the verification of each one independently of the others. We concentrate then in the activities associated to generating the VCs for a single user procedure. By this we mean a user unit, together with its pre-post assertions, disregarding whether it comes from an imperative or a functional input language.

After the transformation of Sect. 5, invariant assertions are placed as preconditions of some IR nodes. The CAVI-ART platform will help the user in this task, either by synthesizing parts of the invariants, or by completing the incomplete ones given by the user. The description of this part of the project is beyond the purpose of this paper. In what follows, we assume that the invariants have been placed by someone in the locations computed by the algorithm of Sect. 5.

Summarizing the result of the transformations described in Sects. 2 and 3, given a procedure we get an IR consisting of:

1. A function definition for every basic block (BB).
2. Each BB consists of a sequence of **let** expressions, ended in a jump. Each **let** binding is either an atom, or an application. A jump is simple, or it is a **case** with a simple jump at each of its branches. A simple jump to the *exit* node consists of a tuple expression returning the relevant variables. Otherwise, it is a call to another BB, passing the relevant variables as arguments.
3. The postcondition assertion, annotated in every arc to the *exit* node.
4. The precondition assertion, annotated in the only arc leaving the *entry* node.
5. An invariant assertion as precondition of the entry node of every CCS.

The IR may have a hierarchical structure reflecting the decomposition of an imperative CFG into its constituent CCSs. In this section, we look at it as a flat set of BBs recursively calling to each other, or as a control flow graph consisting of a set of nodes and a set of directed arcs between them.

Example 5. In Fig. 8 we show the flattened version of the IR corresponding to the CFG/SSA of the example of Fig. 4. In that IR, the locations of assertions I_1 and I_2 —i.e. the preconditions of f_2 and f_4— are indicated, and they correspond to the invariants. The precondition Q and the postconditon R are also indicated. For this example, a typical postcondition R will assert that the output vector is sorted and that it is a permutation of the original one. A typical invariant I_1 of the **for** loop will assert also the second property, and that sortedness holds up to the element in position $i - 1$. Invariant I_2 is a bit more involved.

```
{Q(v, H)}
insertionSort v H =
letfun
    f₁ v H =        let i = 0 in f₂ v i H
    {I₁(v, i, H)}
    f₂ v i H =                                  f₅ v i j e H = let x₃ = sel-array(H, v, j) in
                    let x₁ = len(v) in                         let x₄ = +(j, 1) in
                    let b = <(i, x₁) in                        let H₁ = mod-array(H, v, x₄, x₃) in
                    case b of true → f₃ v i H                  let j₁ = -(j, 1) in f₄ v i j₁ e H₁
                             false → H         f₆ v i j e H = let x₅ = +(j, 1) in
    f₃ v i H =      let e = sel-array(H, v, i) in              let H₁ = mod-array(H, v, x₅, e) in
                    let j = -(i, 1) in f₄ v i j e H            let i₁ = +(i, 1) in f₂ v i₁ H₁
    {I₂(v, i, j, e, H)}
    f₄ v i j e H =  let b₁ = >=(j, 0) in        in f₁ v H
                    let x₂ = sel-array(H, v, j) in   {R(v, H, res)}
                    let b₂ = >(x₂, e) in
                    let b₃ = &&(b₁, b₂) in
                    case b₃ of true → f₅ v i j e H
                             false → f₆ v i j e H
```

Fig. 8. IR of the insertion sort algorithm

The VC generation has two phases: (1) Assertion propagation, and (2) VC extraction.

Assertion Propagation. Let us start with a simple case, a BB having a simple jump at its end, and the rest of the BB consisting of a **let** sequence in which each bound expression is a primitive operator application, i.e. it has the form:

$$\text{let } x_1 = e_1 \text{ in } \ldots \text{ let } x_n = e_n \text{ in } \{Q\} f \, \bar{a}$$

where each e_i represents a primitive application. Moreover, we know the assertion Q that must hold in the output arc, i.e. the precondition $Q(\bar{y})$ of function f. Then, the precondition Q_1 propagated to the beginning of this BB, assuming that $Dom(e_i) = true$ for all i, is simply:

$$Q_1 \equiv (x_1 = e_1) \to \cdots \to (x_n = e_n) \to Q(\bar{a})$$

Let us assume now that the i-th bound expression of the BB is a call $g \, \overline{a'}$ to an external function g for which we know its precondition $Q_g(\bar{y})$, and its postcondition $R_g(\bar{y}, res)$. Then, the propagation is split into two parts:

$$Q_1 \equiv (x_1 = e_1) \to \cdots \to (x_{i-1} = e_{i-1}) \to Q_g(\overline{a'})$$
$$R_1 \equiv (x_i = res) \to (x_{i+1} = e_{i+1}) \to \cdots \to (x_n = e_n) \to Q(\bar{a})$$

The following VC is also generated: $(x_1 = e_1) \to \cdots \to (x_{i-1} = e_{i-1}) \to R_g(\overline{a'}, res) \to R_1$. We proceed in a similar when more than one external call is present in the BB.

If the BB ends in a jump such as **case** a **of** $true \to \{Q_1\} \ldots false \to \{Q_2\} \ldots$, where we know the assertions Q_1 and Q_2 holding at each output jump, the assertion propagated just before the **case** is: $(a \to Q_1) \wedge (\neg a \to Q_2)$. The rest of the backwards propagation through the BB is as above.

Finally, if the BB ends in a jump such as **case** a **of** $\ldots C_i\, x_1 \cdots x_n \to \{Q_i\} \ldots$, where we know the assertion Q_i holding at each output jump, the asssertion propagated just before the **case** is:

$(a = C_i\, x_1 \cdots x_n \to Q_i) \wedge \ldots$ a similar conjunction for each remaining branch

If a default clause is present, the branch assertion Q at this jump is known, and there are k prior branches, the conjunction for this clause is a bit more complex:

$$(a \neq C_1\, x_{11} \cdots x_{1n_1}) \wedge \cdots \wedge (a \neq C_k\, x_{k1} \cdots x_{1n_k}) \to Q$$

VC Extraction. After the propagation phase, we get an assertion propagated just before every BB body, and also some VCs coming from the calls to external procedures. The remaining VCs belong to one of the two following cases:

1. The user procedure precondition Q_P must be stronger than or equal to the assertion Q propagated to the single arc leaving the *entry* node, i.e. the verification condition $Q_P \to Q$ is generated.
2. If the BB precondition is an invariant I, then this invariant must be stronger than or equal to the assertion Q propagated just before the BB body, i.e. the verification condition $I \to Q$ is generated for each invariant I.

Example 6. For the example of Fig. 8, the following VCs are generated:

1. $Q(v, \mathcal{H}) \to (i = 0) \to I_1(v, i, \mathcal{H})$
2. $I_1(v, i, \mathcal{H}) \to (x_1 = \texttt{len}(v)) \to (b = i < x_1) \to \neg b \to R(v, \mathcal{H}, \mathcal{H})$
3. $I_1(v, i, \mathcal{H}) \to (x_1 = \texttt{len}(v)) \to (b = i < x_1) \to b \to$
   $(\mathcal{H}(v) = Array\ a \wedge 0 \leq i < a.length)$
4. $I_1(v, i, \mathcal{H}) \to (x_1 = \texttt{len}(v)) \to (b = i < x_1) \to b \to (e = \texttt{get}\ a\ i) \to$
   $(j = i - 1) \to I_2(v, i, j, e, \mathcal{H})$
5. $I_2(v, i, j, e, \mathcal{H}) \to (b_1 = j \geq 0) \to (\mathcal{H}(v) = Array\ a \wedge 0 \leq j + 1 < a.length)$
6. $I_2(v, i, j, e, \mathcal{H}) \to (b_1 = j \geq 0) \to (x_2 = \texttt{get}\ a\ j) \to (b_2 = x_2 > e) \to$
   $(b_3 = b_1\ \&\&\ b_2) \to b_3 \to (x_3 = \texttt{get}\ a\ j) \to (x_4 = j + 1) \to$
   $(a_1 = \texttt{set}\ a\ x_4\ x_3) \to (\mathcal{H}_1 = \texttt{set}\ \mathcal{H}\ v\ a_1)) \to (j_1 = j - 1) \to I_2(v, i, j_1, e, \mathcal{H}_1)$
7. $I_2(v, i, j, e, \mathcal{H}) \to (b_1 = j \geq 0) \to (\mathcal{H}(v) = Array\ a \wedge 0 \leq j + 1 < a.length)$
8. $I_2(v, i, j, e, \mathcal{H}) \to (b_1 = j \geq 0) \to (x_2 = \texttt{get}\ a\ j) \to (b_2 = x_2 > e) \to$
   $(b_3 = b_1\ \&\&\ b_2) \to \neg b_3 \to (x_5 = j + 1) \to (a_1 = \texttt{set}\ a\ x_5\ e) \to$
   $(\mathcal{H}_1 = \texttt{set}\ \mathcal{H}\ v\ a_1)) \to (i_1 = i + 1) \to I_1(v, i_1, \mathcal{H}_1)$

When Q, I_1, I_2 and R are replaced by actual predicates, the resulting VCs could be automatically discharged by a platform such as Why3. Its gallery of verified programs (see http://why3.lri.fr/.), includes an insertion sort algorithm with VCs very similar to ours which are easily discharged.

7   Conclusions and Related Work

Many other intermediate representations of programs have been defined with different purposes. Restricting us to IRs for verification platforms, it has become popular the so-called IVLs (*Intermediate Verification Languages*). An example of these is Boogie2, used in Dafny [15]. Its semantics is given in terms of sets of traces, and it is very much tied to imperative languages. Its type system is more powerful than Hindley-Milner polymorphism, and this feature has shown to be very convenient for modeling the OO-languages heap. But it is not clear how functional languages could be mapped to it. The Why3 platform [10] offers WhyML as IR. In fact, this is a high level language, a kind of Standard ML with state and loops, but it has been used as IR for verifying C and Java programs. Due to its lack of support, some features of these languages, notably the heap, has been modeled in an awkward way.

A third related IR is LLVM[2]. Its purpose is to serve as IR for both imperative and functional languages in order to promote portability of these languages to different machines, interoperability between different paradigms, and to take profit of common static analyses targeted towards runtime performance. At first, we considered LLVM as IR for out platform, but we did not like the way in which functional languages can be translated into it. For instance, the case distinction provided by our `case` is closer to the pattern matching translation of Haskell (based on data types) and can be subsequently translated into a Why3 theory in a straightforward way. In the LLVM, however, we would need to perform a case distinction (`switch`) on the tag of the constructor and then assign the variables bound by each pattern in each branch. In addition, the LLVM provides a considerable amount of low level operations, whereas our purpose was quite the opposite: provide a reduced set of language dependent primitive functions whose behaviour will be specified in a language dependent theory. This allows us to express their properties in a way that is closer to the source language. The same applies to array indexing, which is built in the LLVM IR via the `getelementptr`/`extractvalue` instructions, whereas in our approach is another language dependent primitive whose behaviour may vary between different languages, especially when indexing beyond the array bounds.

A last related formalism is that of Constrained Horn Clauses (CHC) [12,13]. They have been successfully used to express properties that a program must satisfy, such as termination or functional correctness. There are sophisticated algorithms which may decide, whenever this is possible, whether a CHC set is satisfiable or not, and hence whether the desired property holds. In this sense, CHC can be seen as a machinery complementary to that of SMT solvers, in order to automatically verify properties. It is more questionable whether CHC can play the role of our IR. In [6] it is shown how to encode the semantics of a subset of C into CHC programs, and how to transform a C program into a semantically equivalent CHC one by specializing the semantics with respect to

---

[2] LLVM stands for Low Level Virtual Machine. See http://llvm.org/.

the C source. For our purposes, however, this kind of representation is too low level and introduces many details which obscure the generation of verification conditions based on assertions and weakest preconditions.

Our IR supports most features of imperative and functional languages, including all varieties of control statements, exceptions, recursion, object orientation, heap modeling, arrays, algebraic data types, pattern matching, polymorphism, and higher-order. Moreover, the VCs we generate are very much adapted to what current SMTs expect. For the moment, we do not support concurrency and reflection in the sense of languages such as Java.

References

1. Ahrendt, W., Beckert, B., Bruns, D., Bubel, R., Gladisch, C., Grebing, S., Hähnle, R., Hentschel, M., Herda, M., Klebanov, V., Mostowski, W., Scheben, C., Schmitt, P.H., Ulbrich, M.: The key platform for verification and analysis of java programs. In: Giannakopoulou, D., Kroening, D. (eds.) VSTTE 2014. LNCS, vol. 8471, pp. 55–71. Springer, Heidelberg (2014)
2. Allen, F.E.: Control flow analysis. In: Proceedings of a Symposium on Compiler Optimization, pp. 1–19. ACM, New York (1970)
3. Appel, A.W., Palsberg, J.: Modern Compiler Implementation in Java, 2nd edn. Cambridge University Press, New York (2003)
4. Carlsson, R.: An introduction to core erlang. In: Proceedings of the PLI01 Erlang Workshop (2001)
5. Cormen, T.H., Leiserson, C.E., Rivest, R.L., Stein, C.: Introduction to Algorithms, 3rd edn. The MIT Press, Cambridge (2009)
6. De Angelis, E., Fioravanti, F., Pettorossi, A., Proietti, M.: Semantics-based generation of verification conditions by program specialization. In: PPDP 2015, pp. 91–102. ACM (2015)
7. de Moura, L., Bjørner, N.S.: Z3: An efficient SMT solver. In: Ramakrishnan, C.R., Rehof, J. (eds.) TACAS 2008. LNCS, vol. 4963, pp. 337–340. Springer, Heidelberg (2008)
8. Deters, M., Reynolds, A., King, T., Barrett, C.W., Tinelli, C.: A tour of CVC4: how it works, and how to use it. In: FMCAD 2014, p. 7. IEEE (2014)
9. Filliâtre, J.-C.: One logic to use them all. In: Bonacina, M.P. (ed.) CADE 2013. LNCS, vol. 7898, pp. 1–20. Springer, Heidelberg (2013)
10. Filliâtre, J.-C., Paskevich, A.: Why3 — Where programs meet provers. In: Felleisen, M., Gardner, P. (eds.) ESOP 2013. LNCS, vol. 7792, pp. 125–128. Springer, Heidelberg (2013)
11. Flanagan, C., Sabry, A., Duba, B.F., Felleisen, M.: The essence of compiling with continuations. In: PLDI 1993, pp. 237–247. ACM (1993)
12. Gallagher, J.P., Kafle, B.: Analysis and transformation tools for constrained horn clause verification. CoRR, abs/1405.3883 (2014)
13. Grebenshchikov, S., Lopes, N.P., Popeea, C., Rybalchenko, A.: Synthesizing software verifiers from proof rules. In: PLDI 2012, pp. 405–416. ACM (2012)
14. Klein, G., Nipkow, T.: A machine-checked model for a java-like language, virtual machine, and compiler. ACM Trans. Program. Lang. Syst. 28(4), 619–695 (2006)
15. Leino, K.R.M.: Developing verified programs with dafny. In: Brosgol, B., Boleng, J., Taft, S.T. (eds.) HILT, pp. 9–10. ACM (2012)

16. Jones, S.L.P., Lester, D.R.: Implementing functional languages (Prentice Hall international series in computer science). Prentice Hall, New York (1992)

17. Rémy, D.: Using, understanding, and unraveling the OCaml language from practice to theory and vice versa. In: Barthe, G., Dybjer, P., Pinto, L., Saraiva, J. (eds.) APPSEM 2000. LNCS, vol. 2395, p. 413. Springer, Heidelberg (2002)

18. Team, G.: Glasgow Haskell Compiler core Language. https://ghc.haskell.org/trac/ghc/wiki/Commentary/Compiler/CoreSynType. Accessed 30 April 2015

Combining Top-Down and Bottom-Up Techniques in Program Derivation

Dipak L. Chaudhari[✉] and Om Damani

Indian Institute of Technology Bombay, Mumbai, India
{dipakc,damani}@cse.iitb.ac.in

Abstract. The traditional stepwise refinement based program derivation methodologies are primarily top-down. Strictly following the top-
down program derivation approach may require backtracking resulting
in rework. Moreover, the top down approach does not directly help in sug-
gesting the next course of action in case of a failed derivation attempt. In
this work we seamlessly incorporate a bottom up assumption propagation
technique into a primarily top down derivation methodology. We present
new tactics for back-propagating the assumptions made during the top-
down phase. These tactics help in reducing the guesswork during the
derivations. We have implemented these tactics in a program derivation
system. With the help of simple examples, we show how this approach
is useful for avoiding backtracking thereby simplifying the derivations.

Keywords: Calculational style · Program derivation

1 Introduction

In the calculational style of programming [10,13,14], programs are systematically derived from their formal specifications in a top-down manner. At each
step, a derivation rule is applied to a partially derived program at hand, finally
resulting in the fully derived program. Although systematic, this approach can
still be considered as informal. The refinement calculus [1,15] formalizes this
top-down derivation approach. It provides a set of formally verified refinement
rules (transformations).

At an intermediate stage in a top down derivation, users have to select an
appropriate refinement rule by analyzing the structure of the specification under
consideration. However it is not always possible to come up with the right choice
on the first attempt. Users often need to backtrack and try out different rules.
The failed attempts, however, often provide added insight which help, to some
extent, in deciding the future course of action. In the words of Morgan [15]:
"excursions like the above ... are not fruitless...we have discovered that we need
the extra conjunct in the precondition, and so we simply place it in the invariant and try again." Although the failed attempts are *not fruitless* and provide
the required insight, the *trying again* results in rework. The derived program
fragments (and the discharged proof obligations) need to be recalculated (redis-
charged) during the next attempt. The failed attempts also break the flow of the

© Springer International Publishing Switzerland 2015
M. Falaschi (Ed.): LOPSTR 2015, LNCS 9527, pp. 244–258, 2015.
DOI: 10.1007/978-3-319-27436-2_15

derivations and make them difficult to organize. Moreover, the learnings from the failed attempt are not directly applicable; some guesswork is needed in deciding the future course of action.

The non-linear and lengthy derivations are the major hindrance in widespread adoption of the calculational derivation methodology. In our earlier work [6], we developed a system called CAPS[1] (Calculational Assistant for Programming from the Specifications). The CAPS system provides, among other features, support for backtracking and branching by maintaining the complete derivation tree. Although these features helps in managing the non-linear derivations (along with the failed attempts), the problem of rework still remains. Users have to repeat most of the steps carried out during the failed attempt with slight modifications. In the manual derivations (e.g. as in [14]), users do not actually backtrack and redo the complete derivations; they just figure out the impact of the modifications and add relevant program fragments to maintain the correctness. However, without proper formalization and tool support, this approach remains error prone.

Tools supporting the refinement based formal program derivation (Cocktail [12], Refine [16], Refinement Calculator [4] and PRT [5]) mostly follow the top-down methodology. Not much emphasis has been given on avoiding the unnecessary backtrackings. The refinement strategies cataloged by these tools help to some extent in avoiding the common pitfalls. However, a general framework for allowing the users to assume predicates and then propagating these predicates to appropriate location is missing.

In this work, we have seamlessly incorporated the bottom-up techniques into a top-down derivation methodology in order to avoid the unnecessary backtrackings and the associated rework. We present derivation tactics for capturing the assumptions made during the top-down phase and subsequently back-propagating these assumptions to appropriate program locations. We have implemented this approach in the CAPS system. With the help of small examples, we explain how this approach avoids unnecessary backtracking, reduces guesswork, and results in simpler derivations in the CAPS system.

2   Motivating Example

In this section, we present a sketch of the calculational derivation for a simple program performed in a top-down manner. We discuss how the top-down approach is insufficient to capture the natural flow of the derivation and results in additional guesswork and rework. Consider the following programming task (adapted from exercise 4.3.4 in [14]. The informal derivation of this problem also appears in [6]).

Let $f[0..N)$ be an array of booleans where N is a natural number. Derive a program for the computation of a boolean variable r such that r is true iff all the true values in the array come before all the alse values.

---

[1] CAPS is available at http://www.cse.iitb.ac.in/~damani/CAPS.

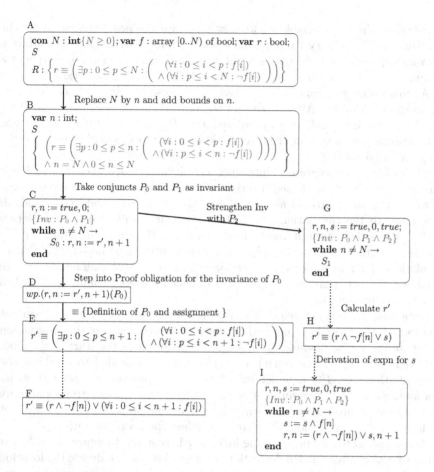

Fig. 1. Sketch of the top-down derivation of the motivating example. P_0 : $(r \equiv (\exists p : 0 \leq p \leq n : ((\forall i : 0 \leq i < p : f[i]) \wedge (\forall i : p \leq i < n : \neg f[i]))))$ $P_1 : 0 \leq n \leq N$; $P_2 : s \equiv (\forall i : 0 \leq i < n : f[i])$

Figure 1 depicts the derivation process for this program. We start the derivation by providing the formal specification (node A) of the program. We then apply the *Replace Constant by a Variable* [14] heuristic to replace the constant N with a fresh variable n as shown in node B. We follow the general guideline of adding bounds on the introduced variable n by adding a conjunct $P_1 : 0 \leq n \leq N$ to the postcondition. Although this conjunct looks redundant due to existence of the stronger predicate $n = N$, it is used later and becomes part of the loop invariant. We then apply the *Take Conjuncts as Invariants* heuristics to select conjuncts P_0 and P_1 as invariants and negation of the remaining conjunct $n = N$ as the guard of the while loop. To ensure termination, we choose to increment variable n by 1 and envision an assignment $r, n := r', n + 1$, where r' is a metavariable. The partially derived program at this stage is shown in node C.

To calculate the metavariable r', we now step into the proof obligation for the invariance of P_0 and try to manipulate the formula with the aim of finding a program expression for r'. After several formula transformations, we arrive at a formula $(r' \equiv (r \land \neg f[n]) \lor (\forall i : 0 \leq i < n + 1 : f[i]))$ shown in node F.

At this point, we realize that we can not represent r' in terms of the existing program variables. After analyzing the derivation, we speculate that if we introduce a fresh variable (say s) and maintain $P_2 : s \equiv (\forall i : 0 \leq i < n : f[i])$ as an additional loop invariant then we would be able to express r' in terms of the program variables.

We backtrack to program C, introduce a fresh variable s, and envision a *While* program with the strengthened invariant. For the derivation of the program S_1, we follow the same process as that of S_0. The steps from node G to node H correspond to the calculation of r'. These steps are similar to the calculation of r' in the failed attempt (node E to node F). However, this time, we are able to instantiate r' with the help of the newly introduced invariant P_2. After calculation of r', we proceed further for the derivation of assignment for the variable s. The program can be improved further by strengthening the guard to ensure early termination.

Note that we did not select $s \equiv (\forall i : 0 \leq i < n + 1 : f[i])$ as an invariant even though the formula is required at node F. This comes from the observation that it would not be possible to establish the invariant at the start of the loop. Since n is initially 0, assignment $s := f[0]$ would be needed to establish the invariant. However, $f[0]$ is undefined when $N = 0$. Instead we added P_2 as an invariant. Selection of this formula needs foresight that the occurrences of n are textually substituted by $n + 1$ during the derivation (step D-E), so we will get the formula we want at node F, if we strengthen the invariant with P_2.

As we saw in this example, some ingenuity is required to figure out the next course of action after a failed derivation attempt. We need to decide the location from where to branch and what new things to try. The backtracking results in rework and breaks the linear flow of the derivation making the derivation complex.

3   Mixing Top-Down and Bottom-Up Approaches

In this section, we first describe the derivation methodology adopted in CAPS and then present our approach for incorporating the bottom-up reasoning in a primarily top-down approach.

3.1   Derivation Methodology

For representing a program fragment and its specification, we use an extension of the Guarded Command Language (GCL) [9] called *AnnotatedProgram*. It is obtained by augmenting each program construct in the GCL with its precondition and postcondition. It is different from the Hoare triple in a sense that, in addition to the program, every subprogram is also annotated with the pre- and post-conditions. We also introduce a new program construct *UnkProg* to represent an unsynthesized program fragment. Annotated program with a precondition α, a postcondition β, and body S is represented as $\{\alpha\}\,S\,\{\beta\}$. We use the formulas in sorted first-order predicate logic for expressing the precondition and the postcondition of the programs. We adopt the Eindhoven notation [2] for representing the quantified formulas.

Users start a derivation by providing a formal specification of a program and then incrementally transform it into a fully synthesized annotated program by applying predefined transformation rules called *Derivation Tactics*. The complete derivation history is recorded in the form of a *Derivation Tree*. The system provides various features like structured derivations, stepping into subcomponents, and backtracking. The system automates most of the mundane tasks and employs the automated theorem provers Alt-Ergo [7], CVC3 [3], SPASS [17] and Z3 [8] for discharging proof obligations. The Why3 tool [11] is used to interface with these theorem provers.

Nature of the Transformation Rules. In the stepwise refinement based approaches [1,15], a formal specification is incrementally transformed into a concrete program. A specification (pre- and post-conditions) is treated as an abstract program (called a specification statement). At any intermediate stage during the derivation, a program might contain specification statements as well as executable constructs. The traditional refinement rules are transformations that convert a specification statement into another program which may in turn contain specifications statements and the concrete constructs. In the conventional approach, once a specification statement is transformed into a concrete construct, its pre- and post-conditions are not carried forward.

In contrast to the conventional approach, we maintain the specifications of all the subprogram (concrete as well as unsynthesized). This allows us to provide rules which transform any correct program (not just a specification statement) into another correct program. These rules try to reuse the already derived program fragments and utilize the already discharged proof obligations to ensure correctness.

Program and Formula Modes. The CAPS system provides tactics for transforming partially derived programs as well as the proof obligation formulas. These two modes are referred as the *Program Mode* and the *Formula Mode* respectively. Users can envision missing program fragments in terms of metavariables which are then derived by manipulating the proof obligation formulas. The *StepIntoPO* (Step Into Proof Obligation) tactic is used to transition from programs to corresponding proof obligation formulas. On applying the tactic to an annotated program containing metavariables, a new formula node representing the proof obligations (verification conditions) is created in the derivation tree. This formula is then incrementally transformed to a form, from which it is easier to instantiate the metavariables. After successfully discharging the proof obligation and instantiating all the metavariables, a tactic called *StepOut* is applied to get an annotated program with all the metavariables replaced by the corresponding instantiations.

3.2   Incorporating the Bottom-Up Approach

In order to incorporate the bottom-up approach in the primarily top-down methodology, we need a way to accumulate assumptions made during the derivation and then to propagate these assumptions upstream. After propagating the assumptions to appropriate location in the derived program, user can introduce appropriate program constructs to establish the assumptions.

The bottom-up phase has three main steps.

- **Assume:** To derive an annotated program $\{\alpha\}\, UnkProg(1)\, \{\beta\}$, we envision an assignment containing metavariables and step into the proof obligation for the program. We then try to simplify the formula with the objective of guessing the expressions for the metavariables. However, to do so, imagine that we need to assume θ. Instead of backtracking, we just accumulate the assumption and proceed further to derive a program S. In the derived annotated program (Fig. 2), $assume(\theta)$ establishes the assumed predicate θ while preserving α. For brevity, we abbreviate the statement $assume(\theta)$ as $A(\theta)$.
- **Propagate:** We may not want to materialize the program to establish θ at the current program location. We then propagate the assumption upstream to an appropriate program location. Depending on the program constructs through which the assumption is propagated, the assumed predicate at the new upstream location might be different from the one being propagated.
- **Realize:** Materialize the *assume* statement by converting it to an unknown program fragment which can be derived subsequently from its specification.

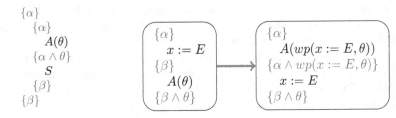

Fig. 2. Result of assuming precondition θ in the derivation of $\{\alpha\}\, UnkProg(1)\, \{\beta\}$

Fig. 3. *AssignmentUp* tactic.

4   Propagating and Establishing Assumptions

The propagation step mentioned in the previous section is an important step in the bottom up phase. We have developed transformation rules for propagating the assumptions upstream through various program constructs. Some of these rules also establish the assumptions after propagating them. The transformation rules transform an annotated program (source program) into another annotated program (target program) with the same specification (i.e. with the same precondition and postcondition). The transformation rules are verified for correctness:

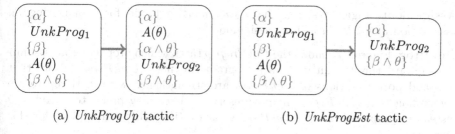

(a) *UnkProgUp* tactic   (b) *UnkProgEst* tactic

Fig. 4. *UnkProg*tactics

if the source program is correct, then the transformed program is also correct. To prove correctness of a rule, we prove the validity of the formula $PO(S) \Rightarrow PO(T)$ where $PO(S)$ and $PO(T)$ are the proof obligations of the source program S and target program T respectively. The transformation rules are implemented in the CAPS system as tactics. Some of the tactics have associated applicability conditions (also called as *proviso*). A tactic can be applied only when the associated proviso is discharged successfully.

4.1   Atomic Constructs

Atomic constructs are the program constructs that do not have subprograms. In this section, we present some rules for the *Assignment* and *UnkProg* constructs.

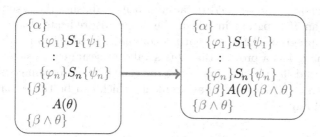

Fig. 5. *CompositionIn* tactic

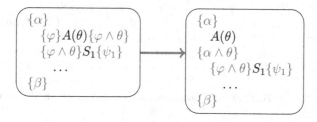

Fig. 6. *CompositionOut* tactic

Assignment. Figure 3 shows the *AssignmentUp* tactic for propagating an assumption upwards through an assignment.

UnkProg. Figure 4(a) shows the *UnkProgUp* tactic which propagates an assumption upward through an unknown program fragment ($UnkProg_1$). Note that pre- and post-conditions of $UnkProg_2$ are strengthened with θ. Here, we are demanding that $UnkProg_2$ should preserve θ. User may prefer to establish θ instead of propagating. The *UnkProgEst* tactic (Fig. 4(b)) can be used for this purpose.

We have not presented the rules for the simple constructs like *skip* and *assume*, since the propagation rules for these constructs are simple.

4.2 Composition

Figure 5 shows a *Composition* program which is composed of another *Composition* and an *assume(θ)* statement. The *CompositionIn* tactic can be used to propagate the assumption θ inside the *Composition* construct. The assumption can then be propagated upwards through the subprograms of the composition (S_n to S_1) using appropriate rules. The *CompositionOut* tactic (Fig. 6) propagates the *assume* statement before the composition statement.

The CAPS system supports nested composition constructs (Composition constructed out of other compositions). Although a nested composition can be collapsed to form a single composition, this construct is useful when we want to apply a tactic to a subcomposition.

Figure 7 shows the *CompoToIf* tactic which establishes the assumption θ by introducing an *if* program in which the assumed predicate θ appears as the guard of the program. Another guarded command is added to handle the other case. This tactic has a proviso that θ is a valid program expression. This tactic allows users to delay the decision about the type of the program constructs. For example, users may envision an assignment, which can be turned later into an *if* program if required.

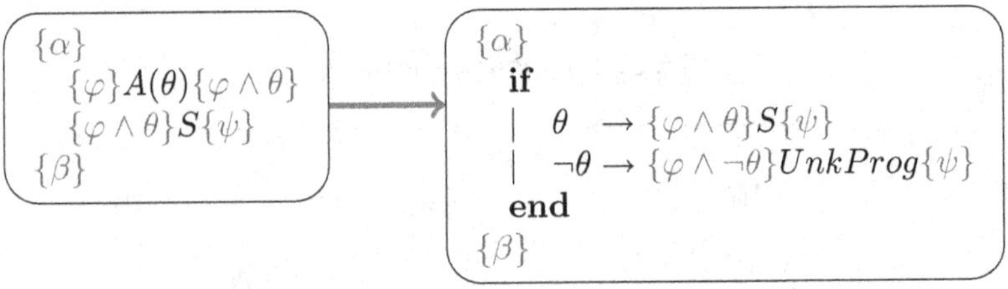

Fig. 7. *CompoToIf* tactic: Transforms a *composition* to an *if* program.

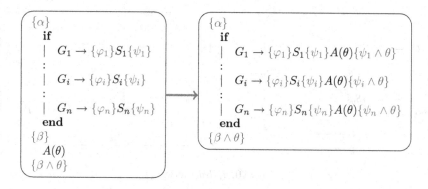

Fig. 8. *IfIn* tactic.

4.3   If

Figure 8 shows the *IfIn* tactic. An *assume* statement that appears after the *if* construct in the source program is pushed inside the *if* construct in the target program. In the target program, θ is assumed at the end of every guarded command.

Figure 9 shows the *IfOut* tactic. In the source program, θ is assumed before the subprogram S_i, whereas in the target program, θ^* is assumed before the *if* program. Note that θ^* (which is defined as $(G_i \Rightarrow \theta)$) is weaker than θ. As a result of assuming θ^* before the *if* construct, we also strengthen the precondition of the other guarded commands. This strengthening of the precondition is beneficial for the unsynthesized program fragments as it may make the task of derivation simpler.

Instead of propagating the assumption, it can be established by strengthening the guard. This can be achieved by applying the *IfGrd* tactic (Fig. 10). An additional guarded command needs to be added to the *if* program to preserve correctness. This tactic does not propagate the assumption; instead it establishes it.

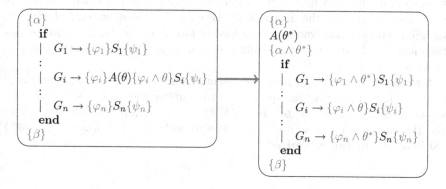

Fig. 9. *IfOut* tactic. ($\theta^* \triangleq G_i \Rightarrow \theta$)

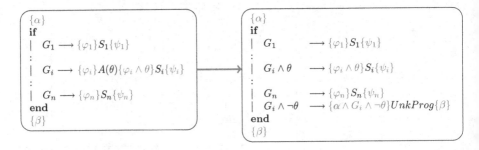

Fig. 10. *IfGrd* tactic

4.4   While

The assumption propagation tactics involving the *While* construct are more complex than those for the other constructs since strengthening an invariant strengthens the precondition as well as the postcondition of the loop body.

***WhileIn* Tactic.** Figure 11 shows the *WhileIn* tactic. The source program has an assumption after the while loop. In order to propagate the assumption θ upward, we strengthen the invariant of the while loop with $\neg G \Rightarrow \theta$. This is the weakest formula that will assert θ after the while loop. We add an *assume* statement after the loop body to maintain the invariant and another *assume* statement before the loop to establish the invariant at the entry of the loop.

***WhileStrInv* Tactic.** Figure 12 shows the *WhileStrInv* tactic. In the source program, the predicate θ is assumed at the start of the loop body. To make θ valid at the start of the loop body S, we strengthen the invariant with $(G \Rightarrow \theta)$. An *assume* statement $A(G \Rightarrow \theta)$ is added after the loop body to ensure that invariant is preserved. Another *assume* statement is added before the while loop to establish the invariant at the entry of the loop.

***WhilePostStrInv* Tactic.** Figure 13 shows the *WhilePostStrInv* tactic. There are two steps in this tactic. In the first step, postcondition of the program S is strengthened with θ^* which is the strongest postcondition of θ with respect to S. In the second step, the invariant of the while loop is strengthened with θ^*. An unknown program fragment is added before S to establish θ. An *assume* statement is added before the while program to establish θ^* at the entry of the loop.

Strongest postconditions involve existential quantifiers. We have implemented heuristics for eliminating the quantifiers to simplify the formulas. In this tactic, we have defined θ^* to be the $sp(S, \theta)$. However, any formula θ^w weaker than the strongest postcondition will also work as long as the program $\{\varphi \wedge \theta^w\}$ $UnkProg\{\varphi \wedge \theta\}$ can be derived.

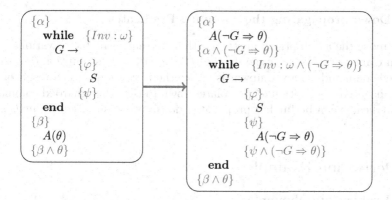

Fig. 11. *WhileIn* tactic: strengthens the invariant with $\neg G \Rightarrow \theta$

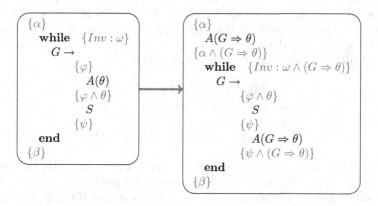

Fig. 12. *WhileStrInv* tactic: strengthens the invariant with $G \Rightarrow \theta$

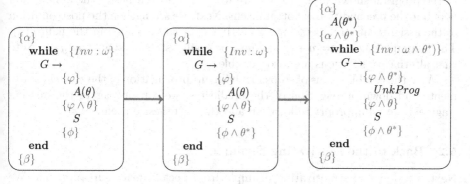

Fig. 13. *WhilePostStrInv* tactic: strengthens the loop invariant with θ^* where $\theta^* \triangleq sp(S, \theta)$

4.5   Down-propagating the Assumed Predicates

As we move the assumptions upstream, they become available to various down-stream constructs. For example, in the *IfOut* tactic the assumption θ in the ith guarded command is moved upwards before the *if* construct. As a result of this, the propagated assumption θ^* percolates down to the other guarded commands. The predicates can be further propagated downwards using the *StrengthenPost* tactic.

5   Derivation Examples

5.1   Evaluating Polynomials

A typical derivation involves interleaved instances of up-propagation of the *assume* statements and down-propagation of the assumed predicates. To demonstrate this, we present some of the steps from the derivation of a program for evaluating a polynomial whose coefficients are stored in an array (also called *Horner's rule*). The program is specified as follows.

$$\textbf{con } A[0..N) \text{ array of int } \{N \geq 0\};$$
$$\textbf{con } x : \text{int}; \textbf{var } r : \text{int};$$
$$S$$
$$\left\{R : r = \left(\sum i : 0 \leq i < N : c[i] * x^i\right)\right\}$$

We skip the initial tactic applications and directly jump to the program shown in Fig. 14(a). The user has already assumed predicate $P_2 : y = x^n$ during the calculation of r' (not shown). We next apply the *WhileStrInv* tactic to stengthen the invariant with P_2 to arrive at program shown in the figure (b). We then propagate the *assume* statement upwards through $n := n + 1$ to arrive at the program shown in figure (c). We would like to synthesize the assumption here but the precondition is not sufficient. Next, we strengthen the postcondition of the assignment statement for r to arrive at program shown in the figure (d). The assumption $P_2(n := n + 1)$ can now be easily established as $y := y * x$. Note that alternative solutions are also possible.

With the combinations of steps involving up-propagation of the *assume* statements and down-propagation of the predicates, we can propagate the missing fragments to an appropriate location and then synthesize them.

5.2   Back to the Motivating Example

Next, we derive the motivating example from Sect. 2 using our approach. We start from formula F in Fig. 1. At this point, we are not able to express the formula $tt(n + 1)$ (where $tt(n) \triangleq (\forall i : 0 \leq i < n : f[i])$) as a program expression. Instead of backtracking, we introduce a fresh variable s and assume the formula

$s \equiv tt(n+1)$ and proceed further with the calculation.

$$\dots$$

$$r' \equiv (r \wedge \neg f[n]) \vee (\forall i : 0 \leq i < n+1 : f[i])$$
$$\equiv \{ \text{ Introduce variable } s \text{ and assume } s \equiv (\forall i : 0 \leq i < n+1 : f[i]) \}$$
$$r' \equiv (r \wedge \neg f[n]) \vee s$$
$$\equiv \{ \text{Step out from formula mode} \}$$

After stepping out from the formula mode, we arrive at the while loop where the body of the loop contains the *assume* statement.

$$\{P_0 \wedge P_1\}$$
$$\textbf{while} \quad \{Inv : P_0 \wedge P_1\}$$
$$n \neq N \rightarrow$$
$$\{P_0 \wedge P_1 \wedge n \neq N\}$$
$$\textbf{assume}(s \equiv tt(n+1))$$
$$\{P_0 \wedge P_1 \wedge n \neq N \wedge s \equiv tt(n+1)\}$$
$$r, n := (r \wedge \neg f[n]) \vee s, n+1$$
$$\{P_0 \wedge P_1\}$$
$$\textbf{end}$$
$$\{R\}$$

We can establish the assumption at the current location however that would be expensive since we would need to traverse the array inside the loop. We can apply the *WhileStrInv* tactic or the *WhilePostStrInv* tactic. Applying the *WhileStrInv* would add $n \neq N \Rightarrow s \equiv tt(n+1)$ as an invariant. With this invariant the initialization problem discussed in Sect. 2 does not arise and this choice results in a different solution. Here, we apply the *WhilePostStrInv* tactic which adds $s \equiv tt(n)$ as an invariant. By applying this tactic, we arrive at the following program.

$$\{P_0 \wedge P_1\}$$
$$\textbf{assume}(s \equiv tt(n))$$
$$\{P_0 \wedge P_1 \wedge s \equiv tt(n)\}$$
$$\textbf{while} \quad \{Inv : P_0 \wedge P_1 \wedge s \equiv tt(n)\}$$
$$n \neq N \rightarrow$$
$$\{P_0 \wedge P_1 \wedge n \neq N \wedge s \equiv tt(n)\}$$
$$UnkProg$$
$$\{P_0 \wedge P_1 \wedge n \neq N \wedge s \equiv tt(n+1)\}$$
$$r, n := (r \wedge \neg f[n]) \vee s, n+1$$
$$\{P_0 \wedge P_1 \wedge s \equiv tt(n)\}$$
$$\textbf{end}$$
$$\{R\}$$

We can now proceed further with the derivation of the *UnkProg* fragment and the initialization *assume* statement as usual.

Using the bottom-up assumption propagation technique, we could maintain the natural flow of the derivation. This derivation reduces the guesswork and avoids unnecessary branching.

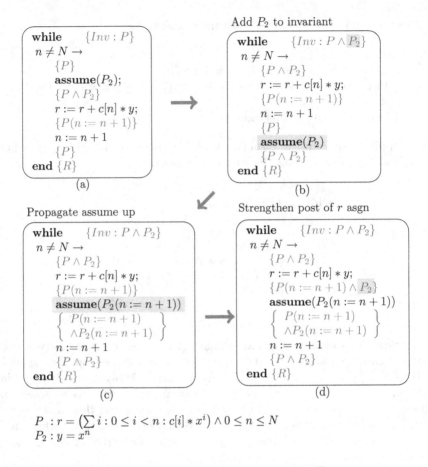

Fig. 14. Some steps in the derivation of a program for the *Horner's rule*. Invariant initializations at the entry of the loop are not shown.

6   Conclusion

We have developed tactics (rules) for up-propagating the information assumed during the top down phase. These tactics have been implemented in the CAPS system. With the help of simple examples we have demonstrated that the seamless integration of the bottom-up and top-down techniques help in reducing the unnecessary backtrackings and associated rework. The methodology also helps in reducing the guesswork involved in the derivations by allowing the user to delay decisions.

Acknowledgements. The work of the first author was supported by the Tata Consultancy Services (TCS) Research Fellowship and a grant from the Ministry of Human Resource Development, Government of India.

References

1. Back, R.J., von Wright, J.: Refinement Calculus: A Systematic Introduction. Graduate Texts in Computer Science. Springer, Berlin (1998)
2. Backhouse, R., Michaelis, D.: Exercises in quantifier manipulation. In: Uustalu, T. (ed.) MPC 2006. LNCS, vol. 4014, pp. 69–81. Springer, Heidelberg (2006)
3. Barrett, C.W., Tinelli, C.: CVC3. In: Damm, W., Hermanns, H. (eds.) CAV 2007. LNCS, vol. 4590, pp. 298–302. Springer, Heidelberg (2007)
4. Butler, M., Långbacka, T.: Program derivation using the refinement calculator. In: von Wright, J., Harrison, J., Grundy, J. (eds.) TPHOLs 1996. LNCS, vol. 1125. Springer, Heidelberg (1996)
5. Carrington, D., Hayes, I., Nickson, R., Watson, G.N., Welsh, J.: A tool for developing correct programs by refinement. Technical report (1996). http://espace.library.uq.edu.au/view/UQ:10768
6. Chaudhari, D.L., Damani, O.: Automated theorem prover assisted program calculations. In: Albert, E., Sekerinski, E. (eds.) IFM 2014. LNCS, vol. 8739, pp. 205–220. Springer, Heidelberg (2014)
7. Conchon, S., Contejean, E.: The alt-ergo automatic theorem prover (2008). http://alt-ergo.lri.fr
8. de Moura, L., Bjørner, N.S.: Z3: An efficient SMT solver. In: Ramakrishnan, C.R., Rehof, J. (eds.) TACAS 2008. LNCS, vol. 4963, pp. 337–340. Springer, Heidelberg (2008)
9. Dijkstra, E.W.: Guarded commands, nondeterminacy and formal derivation of programs. Commun. ACM **18**(8), 453–457 (1975)
10. Dijkstra, E.W.: A Discipline of Programming. Prentice Hall, Englewood Cliffs (1997)
11. Filliâtre, J.-C., Paskevich, A.: Why3 — Where programs meet provers. In: Felleisen, M., Gardner, P. (eds.) ESOP 2013. LNCS, vol. 7792, pp. 125–128. Springer, Heidelberg (2013)
12. Franssen, M.: Cocktail: A tool for deriving correct programs. In: Workshop on Automated Reasoning (1999)
13. Gries, D.: The Science of Programming, 1st edn. Springer-Verlag New York Inc., Secaucus (1987)
14. Kaldewaij, A.: Programming: The Derivation of Algorithms. Prentice-Hall Inc., Upper Saddle River (1990)
15. Morgan, C.: Programming from Specifications. Prentice-Hall Inc., Englewood Cliffs (1990)
16. Oliveira, M., Xavier, M., Cavalcanti, A.: Refine and gabriel: support for refinement and tactics. In: Proceedings of the Second International Conference on Software Engineering and Formal Methods, 2004, SEFM 2004, pp. 310–319. IEEE (2004)
17. Weidenbach, C., Brahm, U., Hillenbrand, T., Keen, E., Theobald, C., Topic, D.: SPASS Version 2.0. In: Voronkov, A. (ed.) CADE 2002. LNCS (LNAI), vol. 2392, pp. 275–279. Springer, Heidelberg (2002)

Program Debugging and Slicing

A Generalized Model for Algorithmic Debugging

David Insa and Josep Silva[(✉)]

Universitat Politècnica de València, Valencia, Spain
{dinsa,jsilva}@dsic.upv.es

Abstract. Algorithmic debugging is a semi-automatic debugging technique that is present in practically all mature programming languages. In this paper we claim that the state of the practice in algorithmic debugging is a step forward compared to the state of the theory. In particular, we argue that novel techniques for algorithmic debugging cannot be supported by the standard internal data structures used in this technique, and a generalization of the standard definitions and algorithms is needed. We identify two specific problems of the standard formulation and implementations of algorithmic debugging, and we propose a reformulation to solve both problems. The reformulation has been done in a paradigm-independent manner to make it useful and reusable in different programming languages.

1 Introduction

One of the most important debugging techniques is *Algorithmic Debugging* (AD) [27]. This technique has experienced a significant advance in the last decade. Concretely, new techniques have been proposed to improve performance [9,15], to improve scalability [11], to improve interaction with the user [6], and to improve GUIs [12,13]. The maturity of these techniques has eventually led to the integration of algorithmic debuggers into sophisticated programming environments. Two interesting cases are [11,12], which combine AD with the standard debugging perspective of Eclipse [1]. The main advantage of AD is its high level of abstraction. It is even possible to debug a program without looking at the source code.

Example 1. Let us assume the existence of a buggy Java code composed of three methods: int add(int x, int y) sums its two arguments, boolean isEven (int x) returns true if its only argument is even, or false otherwise; and, int sumNumbers(int[] array, String eo) takes an array of integers and sums the elements that are even or odd depending on the value of the second argument. Therefore, with the following method invocation:

```
int[] array = { 1, 2, 3 };
int sum = sumNumbers(array, "odd");
```

the result should be 4. Nevertheless, due to a bug in the code, the result is 3.

Thanks to AD, with only this information (without knowing anything about the source code) we can identify the buggy method. For instance, if we debug

© Springer International Publishing Switzerland 2015
M. Falaschi (Ed.): LOPSTR 2015, LNCS 9527, pp. 261–276, 2015.
DOI: 10.1007/978-3-319-27436-2_16

this program with the Hybrid Debugger for Java (HDJ),[1] we obtain the following debugging session (questions are generated by HDJ, and answers are provided by the user):

```
Starting Debugging Session:
    (1) sumNumbers({1,2,3},"odd")=3? No
    (2) isEven(2)=true? Yes
    (3) isEven(3)=false? Yes
    (4) add(0,3)=3? Yes
Bug found in method "sumNumbers" with the call "sumNumbers({1,2,3},"odd")".
```

Hence, an AD session is just a dialogue where the debugger asks questions and the user answers them. The Java code associated with this debugging session is depicted in Fig. 1.

```
                                  (7)  int sumNumbers(int[] array, String eo) {
                                  (8)      int sum = 0;
(1) int add(int x, int y) {       (9)      for (int i = 1; i < array.length; i++) {
(2)     return x + y;             (10)         if (eo.equals("even") && isEven(array[i]))
(3) }                             (11)             sum = add(sum, array[i]);
                                  (12)         if (eo.equals("odd") && !isEven(array[i]))
(4) boolean isEven(int x) {       (13)             sum = add(sum, array[i]);
(5)     return (x % 2) == 0;      (14)      }
(6) }                             (15)      return sum;
                                  (16) }
```

Fig. 1. Java program to sum the even or odd numbers of an array

What the debugger internally does is to generate a data structure that represents the execution of the program. This data structure, often called Execution Tree (ET), is depicted in Fig. 2. The ET has a node for each method invocation.[2] Each node normally contains a reference to the method that is being executed, the value of its arguments, the old and new values of the variables that may be changed within the execution, and its returned value. The debugger just traverses the ET asking the user about the validity of the nodes (i.e., nodes are marked as correct or wrong) until a buggy node is found. A node is buggy when it is wrong, and all of its children (if any) are correct.

The main properties of AD are the following:

Theorem 1 (Correctness of AD [23]**).** *Given an ET with a buggy node n, the method associated with n contains a bug.*

Theorem 2 (Completeness of AD [27]**).** *Given an ET with a bug symptom (i.e., the root is wrong), provided that all the questions generated by the debugger are correctly answered, then, a bug will eventually be found.*

---

[1] http://www.dsic.upv.es/~jsilva/HDJ/.

[2] In the ET, nodes represent computations. Hence, depending on the underlying paradigm, they can represent methods, functions, procedures, clauses, etc. Our discussions in this paper can be applied to both the imperative and the declarative paradigms, but, for the sake of concreteness, we will focuss the discussion on the imperative paradigm and our examples on Java.

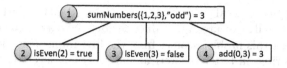

Fig. 2. ET generated for the program in Fig. 1

1.1 Contributions of This Work

In this work, we propose a new redefinition of AD in such a way that: (i) It is paradigm- and language-independent, and thus it is reusable by other researchers. (ii) It is a conservative generalization of the traditional formulation of AD, in such a way that many previous AD techniques are a particular case of this new formulation. (iii) It is formulated in a way that definitions of the data structures, properties, strategies, and algorithms are specified separately, so that they can be reused and/or concretized in a particular case. (iv) It states that the output of an algorithmic debugger should contain dynamic information (i.e., it should not include non-executed code). And, (v) it allows the debugger to ask questions about a code inside a method (and not only about the whole method).

2 Some Problems Identified in Current Algorithmic Debuggers

We have been actively working in the area of AD for the last 10 years. This paper somehow summarizes and criticizes our own work to make a step forward. We claim that almost all current algorithmic debuggers—at least all that we know, including the most extended, which we compared in [7], and including our own implementations—have fundamental problems that were somehow inherited from the original formulation of AD [27].

In particular, we claim that the original formulation of AD, and most of the later definitions and implementations are obsolete with respect to the last advances on the practical side of AD. For instance, two important problems of the standard definitions of AD are the granularity and the static nature of the found errors (AD reports a whole routine as buggy). We can illustrate these problems observing again the debugging session of Example 1: The whole method sumNumbers is pointed out as buggy. This is very imprecise specially if sumNumbers were a method with a lot of code. However, AD researchers and developers are used to this behavior, and they would argue that this is the normal output of any algorithmic debugger. However, from an engineering perspective, this is quite surprising because the analysis performed by the debugger is by definition dynamic (in fact, the whole program is actually executed). Hence, the debugger should know that line 11 of Fig. 1 is never executed, and thus it should not be reported as buggy. This leads us to our first proposition: The information reported by an algorithmic debugger should be dynamic instead of static.

That is, the output of the algorithmic debugger should be the part of the method that has been actually executed to produce the bug, instead of the whole method.

We think that this problem comes from the first implementations of AD and it has been inherited in latter theoretical and practical developments. In fact, if we execute this program with the debuggers: Buddha [26], DDT [4], Freja [21], Hat-Delta (and its predecessor Hat-Detect) [8], B.i.O. [3], Mercury's Algorithmic Debugger [19], Münster Curry Debugger [18], Nude [20], DDJ [13], and HDJ [11], they all would output the whole sumNumbers as buggy together with a counterexample that produces the bug (the found buggy node). Unfortunately, none of these debuggers make further use of the counterexample. An option would be that the debuggers use dynamic program slicing (to be precise, dynamic chopping) [30] to minimize the code shown as buggy.

Traditionally, AD reports a whole method as buggy. To reduce the granularity of the reported errors, new techniques have appeared (see, e.g., [5,16]) that allow for debugging inside a method. Unfortunately, the standard definition of ET is not prepared for that. In fact, some of the recent transformations defined for AD do not fit in the traditional definition of the data structures used in this discipline. For instance, the *Tree Balancing* technique presented in [15], or the *zooming* technique presented in [5] cannot be represented with standard AD data structures such as the *Evaluation Dependence Tree* [24].

This lack of a common theoretical framework with standard data structures that are powerful enough as to represent recent developments makes researchers to reinvent the wheel once and again. In particular, we have observed that researchers (including ourselves) have produced local and partial formalizations to define their debuggers for a particular language and/or implementation (see, e.g., [6,15,16]). These theoretical developments are hardly reusable in other languages, and thus, they only serve as a formal description of their system, or as a means to prove results.

3   Related Work

Algorithmic debugging has been applied to all mature languages. All current implementations use a sort of ET to represent computations. Even in those lazy implementations of AD where the execution of the front-end and the back-end is interleaved (see, e.g., [22]), the construction of the ET is needed before the program can be debugged. Along the years each paradigm has adopted a well-defined and studied data structure to represent the ET.

3.1   A Little Bit of History

Algorithmic debugging started in the seminal work by Shapiro with the notion of contradiction backtracking using "crucial experiments" within Popper's philosophical dictum of conjectures and refutations [28]. Hence, the first notion of

ET appeared in the context of the logic paradigm. Shapiro used refutation trees as ETs. Later implementations of AD in the logic paradigm such as NU-Prolog [2,31] also used refutation trees.

In the context of the functional paradigm, the data structure used was proposed by Henrik Nilsson and Jan Sparud: The Evaluation Dependence Tree (EDT). They first proposed this data structure as a record of the execution [24], and then, as an appropriate ET for AD [25]. The EDT is particularly useful to represent lazy computations by hiding non-computed terms. In fact, the own EDT can be computed lazily as in [22]. The most successful implementations of AD for the functional paradigm are based on the EDT. Notable cases are the Freja [21], Hat-Delta [8], and Buddha [26] debuggers.

In multi-paradigm languages such as Mercury, TOY, or Curry, the ET is also represented with either a proof tree or an EDT. Examples of debuggers for these languages are the Mercury Debugger [19], the Münster Curry Debugger [18], DDT [4], and B.i.O [3].

In the imperative paradigm, a redefinition of the EDT was used. It has been often called *Execution Tree* [10,13], but, conceptually, it is equivalent to the EDT, and it can be seen as a dynamic version of the *Call Graph* where every single call generates a different node in the graph, and thus no cycles are possible (i.e., it is a tree).

3.2   Modern Implementations

All the debuggers mentioned in the previous sections are somehow "standard" in the sense that they are based on the standard definition of the ET (either the refutation trees or the EDT). However, in the last 5 years, there has been a new trend in AD tools: Researchers have implemented new techniques that go beyond the standard definition of the ET. Contrarily to the previously described tools, modern algorithmic debuggers are not standalone tools. They are plugins that can be integrated as part of an IDE. Examples of these debuggers are JHyde [12] and HDJ [11], being both of them part of Eclipse. Precisely because they are integrated into a development environment, they have direct access to dynamic information—they can even manipulate the JVM at runtime—that can be used to enhance the debugging sessions. In particular, the following techniques go beyond the standard ET: (i) *Tree compression* hides nodes of the ET (it breaks the standard parent-child relation in the ET). (ii) *Tree balancing* introduces new artificial nodes in the ET (it breaks the standard definition of ET node). (iii) *Loop expansion* and (iv) *ET zooming* decompose ET nodes (they break the standard definition of ET node). We are not aware of any definition of ET able to represent the previous four techniques.

4   Paradigm-Independent Redefinition of Algorithmic Debugging

Some of our last developments for AD cannot be formalized with the standard AD formulation. In a few cases, we just skipped the formalization of our technique and provided an implementation. In other cases we wanted to prove some properties, and thus we formalized (for one specific language, e.g., Java) the part of the system affected by those properties. Other developments were done for other paradigms, e.g., the functional paradigm, and we also formalized a different part of the system with different data structures. We have observed the same behavior in other researchers, and clearly, this is due to the lack of a standard solution.

We want to provide a definition of AD that is paradigm-independent (i.e., it can be used by either imperative- or declarative- languages). From the best of our knowledge, there does not exist such a formal definition of AD. Hence, in this section we formulate AD in an abstract way. The main generalization of our new formulation is to consider that ET nodes are not necessarily routines as in previous definitions (see, e.g., [24]). Contrarily, we allow ET nodes to contain any piece of code. This permits AD to report any code as buggy, and not only routines, thus potentially reducing the granularity of the reported errors to single expressions.

In the following, we will only call *Execution Tree* to our new definition, and we will call *Routine Tree* (RT) to the traditional definition (that we also formalize in the next sections). Because our new definition is a conservative generalization, the RT is a particular case of the ET as it can be observed in the UML model of Fig. 3.

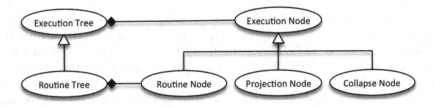

Fig. 3. UML model representing the structure of the execution tree

Observe that an execution node can be specialized depending on the piece of code it represents. In particular, we specialize three kinds of execution nodes named *Routine Node*, *Projection Node*, and *Collapse Node*. They correspond to definitions that already exist in the literature (see [10,15]), but other kinds of nodes could appear in the future.

4.1   The Execution Tree

In this section we introduce some notation and formalize the notion of Execution Tree used in the rest of the paper. We want to keep the discussion and definitions in this section paradigm-independent. Hence, we consider programs as state transition systems.

Definition 1 (Program). *A program* $P = \{W, I, R, C\}$ *consists of:*

- *W: A set of* states.
- *I: A set of* starting states, *such that* $I \subseteq W$.
- *R: A* transition relation, *such that* $R \subseteq W \times W$.
- *C: A* source code, *composed of a set of statements.*

Definition 2 (Computation). *A computation is a maximal sequence of states* $s_1, s_2, \ldots$ *such that:*

- s_1 *is a starting state, i.e.,* $s_1 \in I$.
- $(s_i, s_{i+1}) \in R$ *for all* $i \geq 1$ *(and* $i \leq n-1$, *if the sequence is of the finite length* n).

A finite segment $s_i, s_{i+1}, \ldots, s_j$ *where* $1 \leq i < j \leq n$ *is called a* subcomputation.

In the source code of a program, we consider statements[3] as the basic execution unit. Therefore, in the following, the source code of a program P is a set of statements $st_1, st_2, \ldots, st_n$ that produces the computation $s_0, s_1, \ldots, s_m$ for a given starting state s_0. We cannot provide a specific model of computation if we want to be paradigm-independent, thus we do not define the relation between statements and the transition relation R. This is possible (and convenient) thanks to the abstract nature of algorithmic debugging. In particular, algorithmic debugging only needs an initial state, a code, and a final state to identify bugs. No matter how the code makes the transition from the initial state to the final state. The user will decide whether this transition is correct or not.

Because the considered execution unit is the statement, it is possible to identify a bug in a single statement. This contrasts with traditional algorithmic debugging where routines are the execution units, and thus a whole routine is always reported as buggy.

We also use the notion of *code fragment* of a program P, which refers to any subset of statements in the source code C of P that produces a subcomputation $s_i, \ldots, s_j$ with $0 \leq i < j \leq m$. Code fragments often represent functions or loops in a program, but they can also represent blocks, single statements, or even function calls together with the whole called function.

Intuitively, not all the statements in a given code c that produces a computation $\mathcal{C}$ are actually executed. Some parts of the code are not needed to produce the computation (e.g., because they are dead code, because some condition does not hold, etc.). The projection of c modulo $\mathcal{C}$ is a subset of c where the unneeded code in c to produce $\mathcal{C}$ has been removed. Projections are often computed with dynamic slicing [30].

Definition 3 (Code Projection). *Given a code fragment* c *and a computation* $\mathcal{C}_c = s_0, \ldots, s_n$ *produced by* c *from a given initial state* s_0, *a projection of* c *modulo* $\mathcal{C}_c$ *is a code fragment that contains the minimum subset of* c *needed to produce the computation* $\mathcal{C}_c$.

---

[3] Note the careful use of the word "statement" to refer to either imperative instructions, declarative expressions, etc.

We assume that each state in W is composed of pairs variable-value. The initial and final states, s_i and s_j, describe the effects of a given code fragment c. All three together form a *code behavior*.

Definition 4 (Code Behavior). *Given a code fragment c and a computation $C_c = s_0, \ldots, s_n$ produced by c from a given initial state s_0, the code behavior of C_c is a triple $(s_0, \mathcal{P}_{C_c}(c), s_n)$, where $\mathcal{P}_{C_c}(c)$ is the projection of c modulo C_c.*

Code behavior corresponds to the questions asked by the debugger. These questions are along the lines of *Should the code c with the initial state s_0 produce the final state s_n?*, or *Code c produced s_n from s_0, is that correct?* Many previous definitions of AD (see, e.g., [5,22]) define the code behavior as the triple (s_0, c, s_n), which corresponds to the execution of a routine c, and usually the debugger only needs to show the call to c instead of showing both the call to c and the own routine c. Definition 4, however, introduces two important novelties:

- It allows c to be any code fragment, and not only a routine.
- It substitutes c by a projection of c modulo C_c, thus the code associated with a code behavior only contains the code actually needed to produce that behavior.

This dynamic notion is much more precise than the usual static notion that considers (the complete code of) a routine.

Definition 5 (Intended Model). *Given a program $P = \{W, I, R, C\}$, an intended model $\mathcal{M}$ for P is a set of tuples $(s_i, \mathcal{P}(c), s_j)$ where $s_i, s_j \in W$ and $\mathcal{P}(c)$ is a projection of a code fragment $c \subseteq C$.*

Each tuple of the form $(s_i, \mathcal{P}(c), s_j)$ specifies that the execution of code $\mathcal{P}(c)$ from state s_i leads to state s_j. Intuitively, an intended model of a program contains the set of code behaviors that are correct with respect to what the programmer had in mind when he programmed these codes. It is used as a reference point against which one can compare computations to determine whether they are correct or wrong.

We are now in a position to define the nodes of an execution tree.

Definition 6 (Execution Node). *Let $P = \{W, I, R, C\}$ be a program. Let C_c be a computation produced by a code fragment $c \subseteq C$. Let $\mathcal{M}$ be an intended model for P. The execution node induced by C_c is a pair $(\mathcal{B}, \mathcal{S})$ where:*

1. *$\mathcal{B}$ is the code behavior of C_c, and*
2. *$\mathcal{S}$ is the state of the node, which can be either:*
   - *undefined, or*
   - *the correctness of $\mathcal{B}$ with respect to $\mathcal{M}$:* $\begin{cases} \text{correct} & \text{if } \mathcal{B} \in \mathcal{M} \\ \text{wrong} & \text{if } \mathcal{B} \notin \mathcal{M} \end{cases}$

Observe that an execution node contains (inside $\mathcal{B}$) the source code $\mathcal{P}_{C_c}(c)$ responsible of the computation it represents. Hence, if this node is eventually declared as buggy, its associated code is uniquely identified. This definition of execution node is general enough as to represent previous nodes that are used in different techniques. For instance, if the code of the node is a function, it can be represented as a *routine node*. Similarly, *projection nodes* and *collapse nodes*, introduced in [15], are special nodes that agglutinate the code of several other nodes. Clearly, they are also particular cases of our general definition.

In order to properly define execution trees, we need to define first a relation between execution nodes that specifies the parent-child relation.

Definition 7 (Execution Nodes Dependency). *Let N be a set of execution nodes. Given an execution node $n_c \in N$ induced by a computation C_c, and an execution node $n_{c'} \in N$ induced by a subcomputation $C_{c'}$ of C_c, we say that n_c directly depends on $n_{c'}$ (expressed as $n_c \xrightarrow{N} n_{c'}$) if and only if there does not exist an execution node $n_{c''} \in N$ induced by subcomputation $C_{c''}$ of C_c, such that $C_{c'}$ is a subcomputation of $C_{c''}$.*

Observe that this dependency relation is intransitive, which is needed to define the parent-child relation in a tree. Hence, provided that we have three execution nodes, n_1, n_2, n_3, if $n_1 \xrightarrow{N} n_2 \xrightarrow{N} n_3$ then $n_1 \not\xrightarrow{N} n_3$.

Example 2 Given the following program:

```
CODE:
x++; y++; x=x+y;
```

and the initial state (x=1,y=2) we can generate the following execution nodes (among others):

ET NODES:

	(initial state)	code	(end state)
node 1:	(x=1,y=2)	x++; y++; x=x+y;	(x=5,y=3)
node 2:	(x=1,y=2)	x++; y++;	(x=2,y=3)
node 3:	(x=2,y=3)	x=x+y;	(x=5,y=3)
node 4:	(x=2,y=2)	y++;	(x=2,y=3)

with N={node 1, node 2, node 3, node 4}
we have $node 1 \xrightarrow{N} node\ 2 \xrightarrow{N} node\ 4$ and $node\ 1 \xrightarrow{N} node\ 3$

Finally, we define an *execution tree*. It essentially represents the execution of a code in a structured way where each node represents a sub-execution of its parent. Formally,

Definition 8 (Execution Tree). *Let C_c be a computation produced by a code fragment c. An Execution Tree (ET) of C_c is a tree $T = (N, E)$ where:*

- *$\forall n \in N$, n is the execution node induced by a subcomputation of C_c,*
- *The root of the ET is the execution node induced by C_c,*
- *$\forall (n_1, n_2) \in E \ . \ n_1 \xrightarrow{N} n_2$.*

This definition is a generalization of the usual call tree (CT), which in turn comes from the refutation trees initially defined for AD in [27,28]. One important difference between them is that, given a computation C_c produced by a code fragment c, the CT associated with C_c is unique because it is only formed of routine nodes. In contrast, there exist different valid ETs associated with C_c due to the flexibility introduced by the execution nodes (i.e., with routine nodes only one set N is possible, while with execution nodes different sets N are possible). This flexibility of having several possible valid ETs to represent one computation is interesting because it leaves room for transforming the ET and still being an ET. Contrarily, the CT cannot be transformed because it would not be a CT anymore.

Once the ET is built, the debugger traverses the ET asking the oracle about the correctness of the information stored in each node. Using the answers, the debugger identifies a *buggy node* that is associated with a *buggy code* of the program. We can now formally define the notion of buggy node.

Definition 9 (Buggy Node). *Let $T = (N, E)$ be an execution tree. A buggy node of T is an execution node $n = (\mathcal{B}, \mathcal{S}) \in N$ where:*

(i) $\mathcal{S} = $ wrong, and
(ii) $\forall n' = (\mathcal{B}', \mathcal{S}') \in N$, $(n, n') \in E$. $\mathcal{S}' = $ correct.
Moreover, we say that a buggy node n is traceable *if and only if:*
(iii) $\forall n' = (\mathcal{B}', \mathcal{S}') \in N$, $(n', n) \in E^*$. $\mathcal{S}' = $ wrong.

We use E^* to refer to the symmetric and transitive closure of E. This is the usual definition of buggy node (see, e.g., [23]): a wrong node with all its children correct. We also introduce the notion of *traceable*. Roughly, traceable buggy nodes are those buggy nodes that may be directly responsible of the wrong behavior of the program (their effects are visible in the root of the tree). This property makes them debuggable by all AD strategies that are variants of Top-Down (see [29]).

Lemma 1 (Buggy Code). *Let T be an ET with a buggy node $((s, d, s'), \mathcal{S})$ whose children are $((s_1, d_1, s_1'), \mathcal{S}_1)$, $((s_2, d_2, s_2'), \mathcal{S}_2) \dots ((s_n, d_n, s_n'), \mathcal{S}_n)$. Then,*
$$d \setminus \bigcup_{1 \leq i \leq n} d_i \text{ contains a bug.}$$

Note that we use (s, d, s') meaning $(s, \mathcal{P}_{C_c}(c), s')$ for some c, and $\setminus$ is the set difference operator.

Proof (Buggy Code). Trivial adaptation from the proof by Lloyd [17] for Prolog.

Lemma 1 illustrates what (buggy) code should be shown to the user. When a buggy node is detected, the (buggy) code shown to the user is the code of the buggy node minus the code of its children.

4.2   Routine Tree

In this section we formalize the notion of RT used in most AD literature as a particular case of the ET. We call *routine tree* to this specialization of the ET to make explicit its multi-paradigm nature, because routines can refer to functions, procedures, methods, predicates, etc. We first define a *routine node*, which is a specialization of an execution node.

Definition 10 (Routine Node). *A routine node is an execution node* $((s_0, \mathcal{P}_{\mathcal{C}_c}(c), s_n), \mathcal{S})$ *where code fragment c only contains:*

– *a routine call r, together with*
– *all the code of the routines directly or indirectly called from r.*

Therefore, in a routine node, s_0 and s_n are, respectively, the states just before and after the execution of the called routine. Almost all implementations reduce c to the routine call, and they skip the code of the own routine.

Definition 11 (Routine Tree). *A routine tree is an execution tree where all nodes are routine nodes.*

4.3   Search Strategies for AD

Once the ET is built, AD uses a search strategy to select one node. During many years, the main goal of most AD researchers has been the definition of better strategies to reduce the search space after every answer, and to reduce the complexity of the questions. A survey of search strategies for AD can be found in [29]. In our formalization, a search strategy is just a function that analyzes the ET and returns an execution node (either the next node to ask, or a buggy node).

Definition 12 (Strategy). *A search strategy is a function whose input is an execution tree* $T = (N, E)$ *and whose output is an execution node* $n = (\mathcal{B}, \mathcal{S}) \in N$ *such that:*

1. $\mathcal{S} =$ undefined, *or*
2. *n is a buggy node.*

4.4   AD Transformations

Some of the last research developments in AD have focussed on the definition of transformations of the ET. The goal of these transformations is to improve the structure of the ET before the debugging session starts, so that search strategies become more efficient. Some of these transformations cannot be applied to a routine tree. For this reason, we include this section to classify the kinds of transformations that have been defined so far, and establish a hierarchy so that future transformations can be also classified in.

There exist three essential elements in the front-end of an algorithmic debugger. The modification of any of them can lead to a different final output of the front-end (i.e., a different ET). Therefore, we classify the transformations in three different levels:

– Transformations of the *source code*: Transformations of the source code such as *inlining* are used to reduce the size of the ET by hiding routines. Contrarily, transformations such as *loops to recursion* [14] are used to augment the size of the ET to reduce the granularity of the reported buggy code (a loop instead of a routine). In general, users should not be aware of the internal transformations applied by the debugger, thus the code fragment shown to the user should be the original code.

– Transformations of the *execution*: Transforming the way in which the source code is executed can change the generated ET. One example is changing eager evaluation by lazy evaluation. Another example is passing arguments by value instead of passing them by reference. We are not aware of any implementation that includes this kind of transformations.

– Transformations of the *ET*: Transforming the ET can significantly reduce the number of questions generated. In general, the ET is transformed with the aim of making search strategies to behave as a dichotomic search. Hence, they try to produce balanced ETs [15], or also deep trees that can be cut in the middle. Other transformations such as *Tree compression* [9] try to avoid the repetition of questions about the same routine, or try to improve the understandability of questions. This is the case of the *Node simplification* transformation, which reduces all terms to normal form [6].

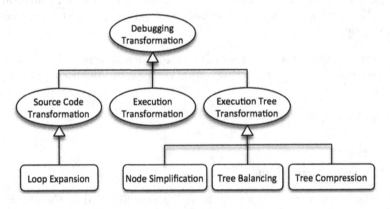

Fig. 4. AD transformations hierarchy

In Fig. 4 we classify four AD transformations already available in the state of the art. Two of them, *tree balancing* and *loop expansion* produce ETs that are not routine trees.

4.5   An AD Scheme

Finally, we describe Algorithm 1, a general schema of an algorithmic debugger that includes all phases, from the generation of the ET to the reported bug. This algorithm gives an idea of how and when, the ET, the transformations, the oracle, and the search strategies participate in the whole debugging process.

Algorithm 1. Main algorithm of an Algorithmic Debugger

Input: A program P and its input i.
Output: A buggy code c in P, or $\perp$ if no bug is detected in P.
Initializations: $\mathcal{A} = \emptyset$ // Set of answers provided by the oracle

begin
  1) $T = getExecutionTree(P, i)$
  2) $n = debugTree(T)$
  3) **return** $getCode(n, T)$
end

function $debugTree(T = (N, E))$
begin
  1) **while** $(\exists(\mathcal{B}', \mathcal{S}') \in N, \mathcal{S}' = undef \vee wrong)$
  2) $(\mathcal{B}, \mathcal{S}) = selectNode(T)$ // Strategy
  3)   **if** $(\mathcal{S} = wrong)$ **then**
  4)   **return** $(\mathcal{B}, \mathcal{S})$
  5) $answer = askOracle(\mathcal{B})$
  6) $\mathcal{A} = \mathcal{A} \cup (\mathcal{B}, answer)$
  7) $updateStates(\mathcal{A}, N)$
  8) $T = executionTreeTransformations(T)$
  9) **return** $\perp$
end

function $getExecutionTree(P, i)$
begin
  1) $P' = sourceCodeTransformations(P)$
  2) $\mathcal{E}_{P'} = executeProgram(P', i)$
  3) $\mathcal{E}'_{P'} = executionTransformations(\mathcal{E}_{P'})$
  4) $T = generateExecutionTree(\mathcal{E}'_{P'})$
  5) $T' = executionTreeTransformations(T)$
  6) **return** T'
end

function $getCode(n, T = (N, E))$
begin
  1) **if** $(n = ((s_0, d, s_n), \mathcal{S}))$ **then**
  2)   **return** $d \setminus \bigcup_{(n, ((s'_0, d_i, s'_n), \mathcal{S}')) \in E} d_i$
  3) **return** $\perp$
end

The main function performs the two phases of AD (Lines 1-2) and then returns a buggy code of the program (Line 3). In the first phase (*getExecution-Tree* function) the ET is created performing all possible transformations in the source code (Line 1), in the execution (Line 3) and in the ET (Line 5). Once the ET is created, the second phase (*debugTree* function) starts. During this phase, the debugger traverses the ET selecting nodes with a search strategy (Line 2). The *selectNode* function is an implementation of one of the search strategies in the literature. There has been a lot of research for more than a decade concerning which should be the node to ask. A survey can be found in [29]. No matter what strategy is used, *selectNode* returns a node to ask (the state of the node is *undefined*), or a buggy node (the state of the node is *wrong* (Line 3)). Once a node has been selected, the debugger asks the oracle about its correctness (Line 5). The oracle provides the intended interpretation to the algorithm. With the answer of the oracle, the debugger updates the state of the nodes of the ET (Lines 5-7). Note that the answer of the oracle can affect the state of several nodes. This effectively changes the information of the ET, and thus, at this moment, a new ET transformation could be used to optimize the ET (Line 8). Then, the process is repeated selecting more nodes. When the strategy finds a buggy node (Lines 3-4) or it cannot select more nodes (Line 1) the second phase finishes and the debugger returns (see *getCode* function) the buggy code associated with the found buggy node (see Lemma 1), or it returns a message indicating that there does not exist a bug (it is indicated with $\perp$ in Line 3), respectively. The last case happens, e.g., when all nodes are reported as *correct*.

5   Conclusion

In this paper we report about some of the problems identified in the current state of the art of AD. One of the problems identified is that much of the recent work in the area does not fit into the standard notions and definitions of AD. In particular, we claim that practically all current definitions of the ET are obsolete with respect to the new proposed techniques.

To solve this situation we propose a generalization of AD able to represent all existent AD transformations. We make this abstraction considering theoretical developments done for a particular language or technique that are generalized, but also considering novel implementations of AD that include techniques that have not been formalized.

The main objectives of this work are two: First, putting together different ideas that have appeared in many works of AD. Putting these ideas together provides a wide perspective that allows us to make a step forward in the abstraction and generalization of the theoretical side of AD. In addition, it allows for classifications and taxonomies to help understanding the state of the art. Second, our new formulation of AD tries to save time. Many researchers have defined once and again similar concepts used in different languages and tools. We provide a paradigm-independent definition that is general enough to represent all current techniques, and it can be easily instantiated to any particular language.

Our plan for the immediate future work is to extend the model to also consider concurrency. Our ETs can represent concurrency, but Algorithm 1 completely ignores it. We want to study how concurrency should be represented, asked about, answered, and presented to the user when a bug is found. We will initially implement a debugger for concurrent programs. Then, we will try to generalize the model.

Acknowledgements. This work has been partially supported by the EU (FEDER) and the Spanish *Ministerio de Economía y Competitividad (Secretaría de Estado de Investigación, Desarrollo e Innovación)* under Grant TIN2013-44742-C4-1-R and by the *Generalitat Valenciana* under Grant PROMETEOII/2015/013. David Insa was partially supported by the Spanish Ministerio de Educación under FPU Grant AP2010-4415. The authors thank the anonymous referees of LOPSTR for their constructive feedback that has contributed to improve this work. They also thank Ehud Shapiro for providing historical information about algorithmic debugging.

References

1. Eclipse (2003). http://www.eclipse.org/
2. Barbour, T., Naish, L.: Declarative debugging of a logical-functional language. Technical report, University of Melbourne (1994)
3. Braßel, B., Siegel, H.: Debugging lazy functional programs by asking the oracle. In: Chitil, O., Horváth, Z., Zsók, V. (eds.) IFL 2007. LNCS, vol. 5083, pp. 183–200. Springer, Heidelberg (2008)
4. Caballero, R.: A declarative debugger of incorrect answers for constraint functional-logic programs. In: Proceedings of the 2005 ACM-SIGPLAN Workshop on Curry and Functional Logic Programming (WCFLP 2005), pp. 8–13. ACM Press, New York, USA (2005)

5. Caballero, R., Martin-Martin, E., Riesco, A., Tamarit, S.: EDD: A declarative debugger for sequential erlang programs. In: Ábrahám, E., Havelund, K. (eds.) TACAS 2014 (ETAPS). LNCS, vol. 8413, pp. 581–586. Springer, Heidelberg (2014)
6. Caballero, R., Riesco, A., Verdejo, A., Martí-Oliet, N.: Simplifying questions in maude declarative debugger by transforming proof trees. In: Vidal, G. (ed.) LOP-STR 2011. LNCS, vol. 7225, pp. 73–89. Springer, Heidelberg (2012)
7. Cheda, D., Silva, J.: State of the practice in algorithmic debugging. Electron. Notes Theor. Comput. Sci. **246**, 55–70 (2009)
8. Davie, T., Chitil, O.: Hat-delta: one right does make a wrong. In: Butterfield, A., (ed.) Proceedings of the 17th International Workshop on Implementation and Application of Functional Languages (IFL 2005), p. 11, September 2005
9. Davie, T., Chitil, O.: Hat-delta: One right does make a wrong. In: Proceedings of the 7th Symposium on Trends in Functional Programming (TFP 2006), April 2006
10. Fritzson, P., Shahmehri, N., Kamkar, M., Gyimóthy, T.: Generalized algorithmic debugging and testing. ACM Lett. Program. Lang. Syst. (LOPLAS) **1**(4), 303–322 (1992)
11. González, J., Insa, D., Silva, J.: A new hybrid debugging architecture for eclipse. In: Gupta, G., Peña, R. (eds.) LOPSTR 2013, LNCS 8901. LNCS, vol. 8901, pp. 183–201. Springer, Heidelberg (2014)
12. Hermanns, C., Kuchen, H.: Hybrid debugging of java programs. In: Escalona, M.J., Cordeiro, J., Shishkov, B. (eds.) ICSOFT 2011. CCIS, vol. 303, pp. 91–107. Springer, Heidelberg (2013)
13. Insa, D., Silva, J.: An algorithmic debugger for java. In: Proceedings of the 26th IEEE International Conference on Software Maintenance (ICSM 2010), pp. 1–6 (2010)
14. Insa, D., Silva, J.: Automatic transformation of iterative loops into recursive methods. Inf. Soft. Technol. **58**, 95–109 (2015)
15. Insa, D., Silva, J., Riesco, A.: Speeding up algorithmic debugging using balanced execution trees. In: Veanes, M., Viganò, L. (eds.) TAP 2013. LNCS, vol. 7942, pp. 133–151. Springer, Heidelberg (2013)
16. Insa, D., Silva, J., Tomás, C.: Enhancing declarative debugging with loop expansion and tree compression. In: Albert, E. (ed.) LOPSTR 2012. LNCS, vol. 7844, pp. 71–88. Springer, Heidelberg (2013)
17. Lloyd, J.: Declarative error diagnosis. New Gener. Comput. **5**(2), 133–154 (1987)
18. Lux, M.: Münster Curry User's Guide, May 2006. http://danae.uni-muenster.de/lux/curry/user.pdf,
19. MacLarty, I.D.: Practical Declarative Debugging of Mercury Programs. Ph.D. thesis, University of Melbourne (2005)
20. Naish, L., Dart, P.W., Zobel, J.: The NU-Prolog debugging environment. In: Porto, A. (ed.) Proceedings of the 6th International Conference on Logic Programming (ICLP 1989), pp. 521–536. Lisboa, Portugal (1989)
21. Nilsson, H.: Declarative Debugging for Lazy Functional Languages. Ph.D. thesis, Linköping, Sweden, May 1998
22. Nilsson, H.: How to look busy while being as lazy as ever: the implementation of a lazy functional debugger. J. Funct. Program. **11**(6), 629–671 (2001)
23. Nilsson, H., Fritzson, P.: Algorithmic debugging for lazy functional languages. J. Funct. Program. **4**(3), 337–370 (1994)
24. Nilsson, H., Sparud, J.: The evaluation dependence tree: an execution record for lazy functional debugging. Technical report, Department of Computer and Information Science, Linköping (1996)

25. Nilsson, H., Sparud, J.: The evaluation dependence tree as a basis for lazy functional debugging. Autom. Softw. Eng. 4(2), 121–150 (1997)
26. Pope, B.: A Declarative Debugger for Haskell. Ph.D. thesis, The University of Melbourne, Australia (2006)
27. Shapiro, E.: Algorithmic Program Debugging. MIT Press, Cambridge (1982)
28. Shapiro, E.Y.: Inductive inference of theories from facts. Technical report RR 192, Yale University (New Haven, CT US) (1981)
29. Silva, J.: A survey on algorithmic debugging strategies. Adv. Eng. Softw. 42(11), 976–991 (2011)
30. Silva, J.: A vocabulary of program slicing-based techniques. ACM Comput. Surv. 44(3), 1–12 (2012)
31. Thompson, B., Naish, L.: A guide to the nu-prolog debugging environment. Technical report, University of Melbourne (1997)

Concolic Execution in Functional Programming by Program Instrumentation

Adrián Palacios and Germán Vidal[✉]

MiST, DSIC, Universitat Politècnica de València,
Camino de Vera, s/n, 46022 Valencia, Spain
{apalacios,gvidal}@dsic.upv.es

Abstract. Concolic execution, a combination of concrete and symbolic execution, has become increasingly popular in recent approaches to model checking and test case generation. In general, an interpreter of the language is augmented in order to also deal with symbolic values. In this paper, in contrast, we present an alternative approach that is based on a program instrumentation. Basically, the execution of the instrumented program in a standard environment produces a sequence of events that can be used to reconstruct the associated symbolic execution.

1 Introduction

Software testing is one of the most widely used approaches for program validation. In this context, symbolic execution [8] was introduced as an alternative to random testing—which usually achieves a poor code coverage—or the complex and time-consuming design of test-cases by the programmer or software tester. In symbolic execution, one replaces the input data by symbolic values. Then, at each branching point of the execution, all feasible paths are explored and the associated contraints on symbolic values are stored. Symbolic states thus include a so called *path condition* with the constraints stored so far. Test cases are finally produced by solving the constraints in the leaves of the symbolic execution tree, which is typically incomplete since the number of states is often infinite.

Unfortunately, both the huge search space and the complexity of the constraints make test case generation based on symbolic execution difficult to scale. For instance, as soon as the path condition cannot be proved satisfiable, the execution of this branch is terminated in order to ensure soundness, giving rise to a poor coverage in many cases.

Concolic execution [4,11] is a recent proposal that combines *conc*rete and *symb*o*lic* execution, and overcomes some of the drawbacks of previous approaches.

This work has been partially supported by the EU (FEDER) and the Spanish *Ministerio de Economía y Competitividad* under grant TIN2013-44742-C4-1-R and by the *Generalitat Valenciana* under grant PROMETEOII2015/013.

Adrián Palacios—Partially supported by the EU (FEDER) and the Spanish *Ayudas para contratos predoctorales para la formación de doctores de la Secretaría de Estado de Investigación, Desarrollo e Innovación del Ministerio de Economía y Competitividad* under FPI grant BES-2014-069749.

M. Falaschi (Ed.): LOPSTR 2015, LNCS 9527, pp. 277–292, 2015.
DOI: 10.1007/978-3-319-27436-2_17

Essentially, concolic execution takes a program and some (initially random) concrete input data, and performs both a concrete and a symbolic execution that mimics the steps of the concrete execution. In this context, symbolic execution is simpler since we know the execution path that must be followed (the same of the concrete execution). Moreover, if the path condition becomes too complex and the constraint solver cannot prove its satisfiability, we can still push some concrete data from the concrete execution, thus simplifying it and often allowing the symbolic execution to continue. This technique forms the basis of some model checking and test-case generation tools (see, e.g., SAGE [5] and Java Pathfinder [10]). Test cases produced with this technique usually achieve a better code coverage than previous approaches based solely on symbolic execution. Moreover, it scales up better to complex or large programs.

Despite its popularity in the imperative and object-oriented programming paradigms, we can only find a few preliminary approaches to concolic execution in the context of functional and logic programming. To the best of our knowledge, the first approach for a high-level declarative programming language is [13], which presented a concolic execution scheme for logic programs, which was only aimed at a simple form of *statement* coverage. This approach was later extended and improved in [9]. In the context of functional programming, [12] introduced a formalization of both concrete and symbolic execution for a simple subset of the functional and concurrent language Erlang [1], but the concolic execution procedure was barely sketched. More recently, [3] presented the design and implementation of a concolic testing tool for a complete functional subset of Erlang (i.e., the concurrency features are not considered in the paper). The tool, called CutEr, is publicly available from https://github.com/aggelgian/cuter.

However, the essential component of all these approaches is an interpreter augmented to also deal with symbolic values. In contrast, in this paper, we consider whether concolic execution can be performed by *program instrumentation*. We answer positively this question by introducing an stepwise approach based on *flattening* the initial program so that the return value of every expression is a pattern, and then instrumenting the resulting program so that its execution outputs a stream of events which suffice to reconstruct the associated symbolic execution. The main advantage w.r.t. the traditional approach to concolic execution is that the instrumented program can be run in any environment, even non-standard ones. For instance, one could run the instrumented program in a model checking environment like Concuerror [6] so that its execution would produce the sequences of events for all relevant interleavings, which might be useful for combining concolic testing and model checking.

The paper is organized as follows. Section 2 presents the considered language. Then, in Sect. 3, we present the instrumented semantics that outputs a sequence of events for each concrete execution. Section 4 introduces a program instrumentation that produces the same sequence of events but using the standard semantics. Section 5 presents a Prolog procedure for reconstructing the associated symbolic execution from the sequence of events. Finally, Sect. 6 concludes and points out some directions for further research.

$$pgm ::= \text{a}/n = \textsf{fun } (X_1, \ldots, X_n) \to e. \mid pgm \; pgm$$

$$\textsf{Exp} \ni e ::= \text{a} \mid X \mid [\,] \mid [e_1|e_2] \mid \{e_1, \ldots, e_n\} \mid \textsf{apply } e_0 \; (e_1, \ldots, e_n)$$
$$\mid \textsf{case } e \textsf{ of } clauses \textsf{ end} \mid \textsf{let } p = e_1 \textsf{ in } e_2 \mid \textsf{do } e_1 \; e_2$$

$$clauses ::= p_1 \to e_1; \ldots; p_n \to e_n$$

$$\textsf{Pat} \ni p ::= [p_1|p_2] \mid [\,] \mid \{p_1, \ldots, p_n\} \mid \text{a} \mid X$$

$$\textsf{Value} \ni v ::= [v_1|v_2] \mid [\,] \mid \{v_1, \ldots, v_n\} \mid \text{a}$$

Fig. 1. Core erlang syntax

2   The Language

In this section, we introduce the language considered in this paper. Our language is inspired in the concurrent functional language Erlang [1], which has a number of distinguishing features, like dynamic typing, concurrency via asynchronous message passing or hot code loading, that make it especially appropriate for distributed, fault-tolerant, soft real-time applications. Erlang's popularity is growing today due to the demand for concurrent services. But this popularity will also demand the development of powerful testing and verification techniques, thus the opportunity of our research.

Despite the fact that we plan to deal with full Erlang in the future, in this paper we only consider a functional subset of *Core Erlang* [2], an intermediate language used internally by the compiler.

The basic objects of the language are variables (denoted by $X, Y, \ldots \in \textsf{Var}$), atoms (denoted by a, b, ...) and constructors (which are fixed in Erlang to lists, tuples and atoms); defined functions are named using atoms too (we will use, e.g., f/n, $\text{g}/m, \ldots$). The syntax for Core Erlang programs and expressions obeys the rules shown in Fig. 1. Programs are sequences of function definitions. Each function f/n is defined by a rule $\textsf{fun } (X_1, \ldots, X_n) \to e.$ where $X_1, \ldots, X_n$ are distinct variables and the body of the function, e, can be an atom, a process identifier, a variable, a list, a tuple, a function application, a case distinction, a let expression or a do construct (i.e., do $e_1 \; e_2$ evaluates sequentially e_1 and, then, e_2, so the value of e_1 is lost). Patterns are made of lists, tuples, atoms, and variables. Values are similar to patterns but cannot contain variables.

Example 1. Consider the Erlang function (left) and its translation to Core Erlang (right) shown in Fig. 2, where some minor simplifications have been applied. Observe that Erlang's sequence operator "," is translated to a do operator when no value should be passed (using pattern matching) to the next elements in the sequence, and to a let expression otherwise. Note also that, despite the fact that this is not required by the syntax, some function applications are *flattened* in order to avoid nested applications. For this purpose, some additional let expressions are introduced. Moreover, additional default alternatives are added to each case expression in order to catch pattern matching errors, so it is common to have overlapping patterns in the clauses of a case construct.

$$f(X,Y) \rightarrow g(X),$$
$$\quad \text{case } h(X) \text{ of}$$
$$\quad\quad a \rightarrow A = h(Y),$$
$$\quad\quad\quad g(A);$$
$$\quad\quad b \rightarrow g(h([\,]))$$
$$\text{end.}$$

$$f/2 = \text{fun } (X,Y) \rightarrow \text{do apply } g/1 \ (X),$$
$$\quad \text{case apply } h/1 \ (X) \text{ of}$$
$$\quad\quad a \rightarrow \text{let } Z = \text{apply } h/1 \ (Y)$$
$$\quad\quad\quad \text{in apply } g/1 \ (Z);$$
$$\quad\quad b \rightarrow \text{let } V = \text{apply } h/1 \ ([\,])$$
$$\quad\quad\quad \text{in apply } g/1 \ (V);$$
$$\quad\quad W \rightarrow \text{fail}$$
$$\text{end.}$$

Fig. 2. Erlang function and its translation to Core Erlang

$$pgm ::= a/n = \text{fun } (X_1, \ldots, X_n) \rightarrow \text{let } X = e \text{ in } X. \ | \ pgm \ pgm$$

$$\text{Exp} \ni e ::= a \ | \ X \ | \ [\,] \ | \ [p_1|p_2] \ | \ \{p_1, \ldots, p_n\} \ | \ \text{let } p = e_1 \text{ in } e_2 \ | \ \text{do } e_1 \ e_2$$
$$\quad | \ \text{let } p = \text{apply } p_0 \ (p_1, \ldots, p_n) \text{ in } e \ | \ \text{let } p_1 = \text{case } p_2 \text{ of } clauses \text{ end in } e$$

$$clauses ::= p_1 \rightarrow e_1; \ldots; p_n \rightarrow e_n$$

$$\text{Pat} \ni p ::= [p_1|p_2] \ | \ [\,] \ | \ \{p_1, \ldots, p_n\} \ | \ a \ | \ X$$

$$\text{Value} \ni v ::= [v_1|v_2] \ | \ [\,] \ | \ \{v_1, \ldots, v_n\} \ | \ a$$

Fig. 3. Flat language syntax

As we will see later, for our instrumentation to be correct, we require some additional constraints on the syntax of programs. Basically, we require the following:

- both the name and the arguments of a function application must be patterns,
- the return value of a function must be a pattern,
- the argument of a case expression must be a pattern, and
- both function applications and case expressions can only occur in the right-hand side of a let expression.

The new constraints are needed in order to keep track of the intermediate values returned by expressions. These values are stored in a pattern, which can then be used by other expressions or returned as the result of a function application.

The restricted syntax is shown in Fig. 3. In the following, we call the programs fulfilling this syntax *flat programs*. In practice, one can transform (purely functional) Core Erlang programs to our flat syntax using a simple pre-processing transformation. Furthermore, in the flat language we also require the *bound* variables in the body of the functions to have unique, fresh names. This is not strictly necessary, but it simplifies the presentation by avoiding the use of context scopes associated to every let expression, etc. (as in [7], where the *last* binding of a variable in the environment should be considered to ensure that the right scope is used). We denote with $\overline{o_n}$ a sequence of objects $o_1, \ldots, o_n$. $Var(e)$ denotes the set of variables appearing in an expression e, and we say that e is *ground* if $Var(e) = \emptyset$.

In the following, we use the function bv to gather the bound variables of an expression:

Definition 1 (Bound Variables, bv). *Let e be an expression. The function* bv*(e) returns the set of bound variables of e as follows:*

$$
\mathsf{bv}(e) = \begin{cases}
\{\,\} & \text{if } e \in \mathsf{Pat} \\
\mathcal{V}ar(p) \cup \mathsf{bv}(e') & \text{if } e \equiv \mathsf{let}\ p = \mathsf{apply}\ p_0\ (p_1, \ldots, p_n)\ \mathsf{in}\ e' \\
\begin{aligned} &\mathcal{V}ar(p_0) \cup \ldots \cup \mathcal{V}ar(p_n) \\ &\cup\ \mathsf{bv}(e_1) \cup \ldots \cup \mathsf{bv}(e') \end{aligned} & \text{if } e \equiv \mathsf{let}\ p_0 = \mathsf{case}\ p\ \mathsf{of}\ \overline{p_n \to e_n}\ \mathsf{end}\ \mathsf{in}\ e' \\
\mathcal{V}ar(p) \cup \mathsf{bv}(e_1) \cup \mathsf{bv}(e_2) & \text{if } e \equiv \mathsf{let}\ p = e_1\ \mathsf{in}\ e_2 \\
\mathsf{bv}(e_1) \cup \mathsf{bv}(e_2) & \text{if } e \equiv \mathsf{do}\ e_1\ e_2
\end{cases}
$$

where, in the fourth case, we assume that e_1 is neither an application nor a case expression (i.e., it is a pattern or another let expression).

3   Instrumented Semantics

In this section, we present an instrumented semantics for flat programs that produces a sequence of events that will suffice to reconstruct the associated symbolic execution. Essentially, we need to keep track of function calls, returns, let bindings and case selections.

First, let us note that the produced events will not show the actual run time values of the program variables, since they will not help us to reconstruct the associated symbolic execution. Rather, the events always include the static variable names. Therefore, in order to avoid variable name clashes, we will consider that variable names are *local* to every event. As a consequence, the two first elements of all events are *params* and *vars* denoting the list of parameters and the list of bound variables in the current function, respectively. These elements will be matched with the current values in the symbolic execution built so far in order to set the right environment for the operation represented by the event. See Sect. 5 for more details.

We consider the following events, which will suffice to reconstruct the symbolic execution:

– The first event, call($params,vars,p, [p_1, \ldots, p_n]$), is associated to a function application let $p = $ apply $p_0\ (p_1, \ldots, p_n)$ in e. Here, $[p_1, \ldots, p_n]$ are the arguments of the function call, and p will be used to store the *return value* of the function call.

– The second event is exit($params,vars,p$), where p is the pattern used to store the return value of the function body. We will produce an *exit* event at the end of every function.

– The next event is bind($params, vars, p, p'$), which binds the pattern p from a generic let expression (i.e., a let expression whose argument is neither an application nor a case expression) to the return value p' of that expression (see function ret below).

– Finally, for each expression of the form

$$\mathsf{let}\ p = \mathsf{case}\ p_0\ \mathsf{of}\ p_1 \to e_1; \ldots; p_n \to e_n\ \mathsf{end}\ \mathsf{in}\ e$$

we have two associated events. The first one is

$$\mathsf{case}(params, vars, i, p_0, p_i, [(1, p_0, p_1), \ldots, (n, p_0, p_n)])$$

Here, we store the position of the selected branch, i, the case argument p_0, the selected pattern p_i, as well a list with all case branches, which will become useful for producing alternative input data in the context of concolic testing. The second event is $\mathsf{exitcase}(params, vars, p, p')$, where p' is the return value of the selected branch (see below).

Before presenting the instrumented semantics, we need the following auxiliary function that identifies the *return value* of an expression:

Definition 2 (Return Value, ret). *Let e be an expression. We let $\mathsf{ret}(e)$ denote the return value of e as follows:*

$$\mathsf{ret}(e) = \begin{cases} e & \text{if } e \in \mathsf{Pat} \\ \mathsf{ret}(e') & \text{if } e \equiv \mathsf{let}\ p\ = \mathsf{apply}\ p_0\ (p_1, \ldots, p_n)\ \mathsf{in}\ e' \\ \mathsf{ret}(e') & \text{if } e \equiv \mathsf{let}\ p_0 = \mathsf{case}\ p\ \mathsf{of}\ \overline{p_n \to e_n}\ \mathsf{end}\ \mathsf{in}\ e' \\ \mathsf{ret}(e_2) & \text{if } e \equiv \mathsf{let}\ p = e_1\ \mathsf{in}\ e_2 \\ \mathsf{ret}(e_2) & \text{if } e \equiv \mathsf{do}\ e_1\ e_2 \end{cases}$$

where, in the fourth case, we assume that e_1 is neither an application nor a case expression (i.e., it is a pattern or another let expression).

Note that function ret is not well defined for arbitrary programs, e.g., $\mathsf{ret}(\mathsf{let}\ p = e\ \mathsf{in}\ \mathsf{apply}\ e_0\ (e_1, \ldots, e_n))$ is undefined. Extending the definition to cover this case would not help too since returning an expression which is not a pattern—like $\mathsf{apply}\ e_0\ (e_1, \ldots, e_n)$—would not be useful to reconstruct the symbolic execution (where the program is not available, only the sequence of events). This is why we transform the original programs to the flat form. In this case, it is immediate to see from the syntax in Fig. 3 that ret would always return a pattern for all program expressions.

The instrumented semantics for flat programs is formalized in Fig. 4 following the style of a natural (big-step) semantics [7]. Observe that we do not need *closures* (as it is common in the natural semantics) since we do not allow fun expressions in the body of a function in this paper. Here, we use an *environment* θ—i.e., a mapping from variables to patterns—because we need to know the static values of the variables for the instrumentation (e.g., we use the case argument that appears statically in the program, rather than the instantiated run time value). The main novelty is that, for the instrumentation, we also need to keep track of the function where an expression occurs. For this purpose, we also introduce a simple context π that stores this information, i.e., for a given function fun $(X_1, \ldots, X_n) \to e$ we store a tuple $\langle [X_1, \ldots, X_n], [\mathsf{bv}(e)] \rangle$. The environment is only updated in function applications, where $[\mathsf{bv}(e)]$ denotes a list with the variables returned by $\mathsf{bv}(e)$.

Let us briefly explain the rules of the semantics. Statements have the form $\pi, \theta \vdash e \Downarrow_\tau p$, where π is the aforementioned context, θ is a substitution (the

$$\pi, \theta \vdash p \Downarrow_\epsilon p\theta$$

$$\frac{\langle vs, ps \rangle, \theta \vdash p_0 \Downarrow_\epsilon f/m \quad \dots \quad \langle vs, ps \rangle, \theta \vdash p_m \Downarrow_\epsilon p'_m}{\langle [\overline{Y_m}], [\mathsf{bv}(e_2)] \rangle, \theta \cup \sigma \vdash e_2 \Downarrow_{\tau_1} p' \quad \langle vs, ps \rangle, \theta \cup \sigma' \vdash e \Downarrow_{\tau_2} p''}{\langle vs, ps \rangle, \theta \vdash \mathsf{let}\ p = \mathsf{apply}\ p_0\ (\overline{p_m})\ \mathsf{in}\ e \Downarrow_{\mathsf{call}(vs, ps, p, [\overline{p_m}]) + \tau_1 + \mathsf{exit}([\overline{Y_m}], [\mathsf{bv}(e_2)], p''_2) + \tau_2}\ p''}$$
$$\text{if } f/m = \mathsf{fun}\ (\overline{Y_m}) \to e_2 \in \mathsf{pgm},\ \mathsf{ret}(e_2) = p''_2,$$
$$\mathsf{match}(\overline{Y_m}, \overline{p'_m}) = \sigma,\ \mathsf{match}(p, p') = \sigma'$$

$$\frac{\langle vs, ps \rangle, \theta \vdash p_0 \Downarrow_\epsilon p'_0 \quad \langle vs, ps \rangle, \theta \cup \sigma \vdash e_i \Downarrow_{\tau_1} p'_i \quad \langle vs, ps \rangle, \theta \cup \sigma' \vdash e \Downarrow_{\tau_2} p'}{\langle vs, ps \rangle, \theta \vdash \mathsf{let}\ p = \mathsf{case}\ p_0\ \mathsf{of}\ clauses\ \mathsf{end}\ \mathsf{in}\ e \Downarrow_{\mathsf{case}(vs, ps, i, p_0, p_i, alts) + \tau_1 + \mathsf{exitcase}(vs, ps, p, p'_i) + \tau_2}\ p'}$$
$$\text{if } clauses = p_1 \to e_1; \dots; p_m \to e_m,\ \mathsf{cmatch}(p'_0, clauses) = (i, p_i, \sigma),$$
$$alts = [(1, p_0, p_1), \dots, (m, p_0, p_m)],\ \mathsf{ret}(e_i) = p'_i,\ \mathsf{match}(p, p'_i) = \sigma'$$

$$\frac{\pi, \theta \vdash e_1 \Downarrow_{\tau_1} p'_1 \quad \pi, \theta \cup \sigma \vdash e_2 \Downarrow_{\tau_2} p}{\pi, \theta \vdash \mathsf{let}\ p_1 = e_1\ \mathsf{in}\ e_2 \Downarrow_{\tau_1 + \mathsf{bind}(vs, ps, p_1, \mathsf{ret}(e_1)) + \tau_2}\ p} \quad \text{if } \mathsf{match}(p_1, p'_1) = \sigma$$

$$\frac{\pi, \theta \vdash e_1 \Downarrow_{\tau_1} p_1 \quad \pi, \theta \vdash e_2 \Downarrow_{\tau_2} p_2}{\pi, \theta \vdash \mathsf{do}\ e_1\ e_2 \Downarrow_{\tau_1 + \tau_2}\ p_2}$$

Fig. 4. Flat language instrumented semantics

environment), e is an expression, τ is a sequence of events, and p is a pattern—the value of e.

The first rule deals with patterns (including variables, atoms, tuples and lists). Here, the evaluation just proceeds by applying the current environment θ to the pattern p to bind its variables (if any), which is denoted by $p\theta$. The associated sequence of events is ϵ denoting an empty sequence.

The next rule deals with function applications. In this case, the context is necessary for setting the first and second parameters of call and exit events. Note that since we only consider flat programs, both the function name and the arguments are patterns; thus, their evaluation amounts to binding their variables using the current environment, which explains why the associated sequences of events are ϵ. Note also that, when recursively evaluating the body of the function, we update the context with the information of the function called. The bound variables are collected using the function bv; and, as mentioned before, in the flat language we assume that they all have different, fresh names. Observe that the subcomputation for evaluating the body of the function called is preceded by the call event and followed by an exit event. Here, we use the auxiliary function match to compute the matching substitution (if any) between two patterns, i.e., $\mathsf{match}(p_1, p_2) = \sigma$ if $\mathcal{D}om(\sigma) \subseteq \mathcal{V}ar(p_1)$ and $p_1\sigma = p_2$, and fail otherwise. In this rule, $\mathsf{match}(\overline{Y_m}, \overline{p'_m})$ just returns the substitution $\{Y_1 \mapsto p'_1, \dots, Y_m \mapsto p'_m\}$. The *update* of an environment θ using σ is denoted by $\theta \cup \sigma$. Formally, $\theta \cup \sigma = \delta$ such that $X\delta = \sigma(X)$ if $X \in \mathcal{D}om(\sigma)$ and $X\delta = X\theta$ otherwise (i.e., σ has higher priority than θ). Observe that we use the evaluated patterns $p'_1, \dots, p'_m$ to update the environment, but the original, static patterns $p_1, \dots, p_m$ in the call event.

The next rule is used to evaluate case expressions. Here, we produce case and exitcase events that also include the parameter variables of the function and the bound variables. For selecting the matching branch of the case expression, we use the auxiliary function cmatch that is defined as follows: $\text{cmatch}(p, p_1 \rightarrow e_1; \ldots; p_n \rightarrow e_n) = (i, p_i, \sigma)$ if $\text{match}(p, p_i) = \sigma$ for some $i \in \{1, \ldots, n\}$ and $\text{match}(p, p_j) = \text{fail}$ for all $j < i$. Informally speaking, cmatch selects the first matching branch of the case expression, which follows the usual semantics of Erlang. As in the previous rule, note that we use p'_0 in cmatch but the original, static pattern p_0 in the case event.

The following rule is used to evaluate let expressions. It produces a single bind event which includes, as usual, the parameter variables of the function and the bound variables. Finally, the last rule deals with do expressions. Here, we proceed as expected and return the concatenation of the sequences of events produced when evaluating the subexpressions.

In the following, without loss of generality, we assume that the entry point to the program is always the distinguished function main/n.

Definition 3 (Instrumented Execution). *Given a flat program pgm and an initial expression,* apply main/n $(p_1, \ldots, p_n)$, *with $main/n = \text{fun} (X_1, \ldots, X_n) \rightarrow e \in \text{pgm}$, its evaluation is denoted by $\langle [\overline{X_n}], [\text{bv}(e)] \rangle, \theta \vdash e \Downarrow_\tau v$, where $\theta = \{X_1 \mapsto p_1, \ldots, X_n \mapsto p_n\}$ is a substitution, v is the computed value and $\tau + \text{exit}([\overline{X_n}], [\text{bv}(e)], \text{ret}(e))$ is the associated sequence of events.*

Example 2. Let us consider the flat program shown in Fig. 5. An example computation for apply $\text{main}/1$ ([a]) with the instrumented semantics is shown in Fig. 6. Therefore, the associated sequence of events[1] is the following:

$$\text{call}([X], [W], W, [X, X])$$
$$\text{case}([X, Y], [W_1, H, T, W_2], 2, X, [H|T], [(1, X, [\,]), (2, X, [H|T])])$$
$$\text{call}([X, Y], [W_1, H, T, W_2], W_2, [T, Y])$$
$$\text{case}([X, Y], [W_1, H, T, W_2], 1, X, [\,], [(1, X, [\,]), (2, X, [H|T])])$$
$$\text{exitcase}([X, Y], [W_1, H, T, W_2], W_1, Y)$$
$$\text{exit}([X, Y], [W_1, H, T, W_2], W_1)$$
$$\text{exitcase}([X, Y], [W_1, H, T, W_2], W_1, [H|W_2])$$
$$\text{exit}([X, Y], [W_1, H, T, W_2], W_1)$$
$$\text{exit}([X], [W], W)$$

Let us remind that variable names are *local* to each event. Also, observe that the events do not need to store the names of the invoked functions since we are only interested in the sequence of pattern matching operations, as we will see in Sect. 5.

Note that the semantics is a conservative extension of the standard semantics in the sense that the generation of events does not affect the evaluation, i.e., if we

---

[1] Note that the flat program is not syntactically correct according to Fig. 3 since the right-hand side of the functions do not have the form let $X = e$ in X with e a pattern, a let binding or a do expression. Here, we keep this simpler formulation for clarity, and it also simplifies the sequence of events by avoiding some redundant bind events.

$$\mathsf{main}/1 \;=\; \mathsf{fun}\; (X) \to \mathsf{let}\; W = \mathsf{apply}\; \mathsf{app}/2\; (X, X) \;\mathsf{in}\; W$$

$$\mathsf{app}/2 \;=\; \mathsf{fun}\; (X, Y) \to \mathsf{let}\; W_1 = \mathsf{case}\; X \;\mathsf{of}$$
$$[\,] \to Y$$
$$[H|T] \to \mathsf{let}\; W_2 = \mathsf{apply}\; \mathsf{app}/2\; (T, Y) \;\mathsf{in}\; [H|W_2]$$
$$\mathsf{end}$$
$$\mathsf{in}\; W_1$$

Fig. 5. Example flat program

remove the context information and the events labeling the arrows, we are back to the standard semantics of an eager functional language essentially equivalent to that in [7].

We will show a method for constructing the associated symbolic execution (as well as its potential alternatives) in Sect. 5.

4   Program Instrumentation

In this section, we present a program transformation that instruments a program so that its standard execution will return the same sequence of events produced with the original program and the instrumented semantics of Fig. 4.

For this purpose, we introduce the predefined function out, which outputs its first argument (e.g., to a given file or to the standard output) and returns its second argument. This function is implemented as a function *call* (i.e., not as a function application) so that there is no conflict when performing the instrumentation.

Definition 4 (Program Instrumentation). *Let pgm be a flat program. We instrument pgm by replacing each function definition:*

$$\mathsf{f}/k = \mathsf{fun}\; (X_1, \dots, X_k) \to \mathsf{let}\; X = e \;\mathsf{in}\; X$$

with a new function definition of the form

$$\mathsf{f}/k = \mathsf{fun}\; (X_1, \dots, X_k) \to [\![\mathsf{let}\; X = e \;\mathsf{in}\; \mathsf{out}(\text{``}exit(vs, bs, X)\text{''}, X)]\!]_{\mathsf{F}}^{vs, ps}$$

where $vs = [\overline{X_k}]$, $ps = [\mathsf{bv}(e)]$, F is a flag to determine if an exitcase event should be produced when a pattern is reached (see below), and the auxiliary function $[\![\;]\!]$ is shown in Fig. 7.

Let us briefly explain the rules of the instrumentation. First, we add an exit event at the end of each function. An additional bind event is also required when the expression e is neither a function application nor an case expression in order to explicitly bind X to the return expression of e (for function applications and case expressions this is already done in the exit and exitcase events, respectively).

$$\cfrac{\cfrac{\pi_2,\theta_4 \vdash Y \Downarrow_\epsilon [a] \qquad \pi_2,\theta_5 \vdash W_1 \Downarrow_\epsilon [a]}{\cfrac{\pi_2,\theta_4 \vdash \mathsf{let}\ W_1 = \mathsf{case}\dots \Downarrow_{\tau_1} [a] \qquad \pi_2,\theta_6 \vdash [H|W_2] \Downarrow_\epsilon [a,a]}{\cfrac{\pi_2,\theta_3 \vdash \mathsf{let}\ W_2 = \mathsf{apply}\dots \Downarrow_{\tau_2} [a,a] \qquad \pi_2,\theta_7 \vdash W_1 \Downarrow_\epsilon [a,a]}{\pi_2,\theta_2 \vdash \mathsf{let}\ W_1 = \mathsf{case}\dots \Downarrow_{\tau_3} [a,a] \qquad \pi_1,\theta_8 \vdash W \Downarrow_\epsilon [a,a]}}}}{\pi_1,\theta_1 \vdash \mathsf{let}\ W = \mathsf{apply\ app/2}\ (X,X)\ \mathsf{in}\ W \Downarrow_{\tau_4} [a,a]}$$

with

$$\pi_1 = \langle [X],[W] \rangle \quad \text{and} \quad \pi_2 = \langle [X,Y],[W_1,W_2H,T] \rangle$$

$\theta_1 = \{X \mapsto [a]\}$ $\qquad\qquad\qquad\qquad$ $\theta_2 = \{X \mapsto [a], Y \mapsto [a]\}$

$\theta_3 = \{X \mapsto [a], Y \mapsto [a], H \mapsto a, T \mapsto [\ ]\}$ $\quad$ $\theta_4 = \{X \mapsto [\ ], Y \mapsto [a]\}$

$\theta_5 = \{X \mapsto [\ ], Y \mapsto [a], W_1 \mapsto [a]\}$ $\qquad$ $\theta_6 = \{X \mapsto [a], Y \mapsto [a], H \mapsto a, T \mapsto [\ ], W_2 \mapsto [a]\}$

$\theta_7 = \{X \mapsto [a], Y \mapsto [a], W_1 \mapsto [a,a]\}$ $\qquad$ $\theta_8 = \{X \mapsto [a], W \mapsto [a,a]\}$

$\tau_1 = \mathsf{case}([X,Y],[W_1,W_2],1,X,[\ ],[(1,X,[\ ]),(2,X,[H|T])])$
$\qquad +\mathsf{exitcase}([X,Y],[W_1,W_2],W_1,Y)$

$\tau_2 = \mathsf{call}([X,Y],[W_1,W_2],W_2,[T,Y]) + \tau_1 + \mathsf{exit}([X,Y],[W_1,W_2],W_1)$

$\tau_3 = \mathsf{case}([X,Y],[W_1,W_2],2,X,[H|T],[(1,X,[\ ]),(2,X,[H|T])]) + \tau_2$
$\qquad +\mathsf{exitcase}([X,Y],[W_1,W_2],W_1,[H|W_2])$

$\tau_4 = \mathsf{call}([X],[W],W,[X,X]) + \tau_3 + \mathsf{exit}([X,Y],[W_1,W_2],W_1)$

Fig. 6. Example computation with the instrumented semantics

$$[\![e]\!]_\mathsf{F}^{vs,ps} = e \ \text{ if } e \in \mathsf{Pat}$$
$$[\![e]\!]_{\mathsf{T}(p)}^{vs,ps} = \mathsf{out}(\text{``exitcase}(vs,ps,p,e)\text{''}, e) \ \text{ if } e \in \mathsf{Pat}$$
$$[\![\mathsf{let}\ p = \mathsf{apply}\ p_0\ (\overline{p_n})\ \mathsf{in}\ e]\!]_b^{vs,ps} = \mathsf{let}\ p = \mathsf{out}(\text{``call}(vs,ps,p,[p_1,\dots,p_n])\text{''},$$
$$\mathsf{apply}\ p/0\ (p_1,\dots,p_n)\ )$$
$$\mathsf{in}\ [\![e]\!]_b^{vs,ps}$$

$\quad [\![\mathsf{let}\ p = \mathsf{case}\ p_0\ \mathsf{of} \quad = \mathsf{let}\ p = \mathsf{case}\ p_0\ \mathsf{of}$
$\qquad\qquad\qquad p_1 \to e_1; \qquad\qquad\qquad p_1 \to \mathsf{out}(\text{``case}(vs,ps,1,p_0,p_1,alts)\text{''},$
$\qquad\qquad\qquad\qquad\qquad\qquad\qquad\qquad [\![e_1]\!]_{\mathsf{T}(p)}^{vs,ps}\ )$

$\qquad\qquad\qquad \dots \qquad\qquad\qquad\qquad\qquad \dots$

$\qquad\qquad\qquad p_n \to e_n \qquad\qquad\qquad\ p_n \to \mathsf{out}(\text{``case}(vs,ps,n,p_0,p_n,alts)\text{''},$
$\qquad\qquad\qquad\qquad\qquad\qquad\qquad\qquad\quad [\![e_n]\!]_{\mathsf{T}(p)}^{vs,ps}\ )$

$\qquad\qquad\qquad \mathsf{end} \qquad\qquad\qquad\qquad \mathsf{end}$
$\qquad\quad \mathsf{in}\ e]\!]_b^{vs,ps} \qquad\qquad \mathsf{in}\ [\![e]\!]_b^{vs,ps}$

$$[\![\mathsf{let}\ p = e_1\ \mathsf{in}\ e_2]\!]_b^{vs,ps} = \mathsf{let}\ p = [\![e_1]\!]_\mathsf{F}^{vs,ps}\ \mathsf{in}\ \mathsf{out}(\text{``bind}(vs,ps,p,\mathsf{ret}(e_1))\text{''},$$
$$[\![e_2]\!]_b^{vs,ps}\ )$$

$$[\![\mathsf{do}\ e_1\ e_2]\!]_b^{vs,ps} = \mathsf{do}\ [\![e_1]\!]_\mathsf{F}^{vs,ps}\ [\![e_2]\!]_b^{vs,ps}$$
$$[\![e]\!]_b^{vs,ps} = e \ \text{ otherwise}$$

$$\text{where}\ alts = [(p_0,1,p_1),\dots,(p_0,n,p_n)]$$

Fig. 7. Program instrumentation

Then, we also add call and case events in each occurrence of a function application and a case expression, respectively. Finding the value returned by a case expression is a bit more subtle. For this purpose, we introduce a flag that is propagated through the different cases so that only when the expression is the last expression in a case branch (a pattern) we produce an exitcase event. For let expressions, we produce a bind event and continue evaluating both the expression in the right-hand side of the binding and the result. Finally, the *default* case—the last equation in Fig. 7—is only used to ignore the call to the predefined function out/2.

Example 3. Consider again the flat program of Example 2. The instrumented program is shown in Fig. 8.

$$\text{main}/2 = \text{fun } (X) \rightarrow \text{let } W = \text{out}(\text{``call}([X],[W],W,[X,X])\text{''},$$
$$\text{apply app}/2 \ (X,X))$$
$$\text{in out}(\text{``exit}([X],[W],W)\text{''},W)$$

$$\text{app}/2 = \text{fun } (X,Y) \rightarrow$$
$$\text{let } W_1 = \text{case } X \text{ of}$$
$$[\,] \rightarrow \text{out}(\text{``case}([X,Y],[W_1,W_2,H,T],1,X,[\,],alts)\text{''},$$
$$\text{out}(\text{``exitcase}([X,Y],[W_1,W_2,H,T],W_1,Y)\text{''},Y))$$
$$[H|T] \rightarrow \text{out}(\text{``case}([X,Y],[W_1,W_2,H,T],2,X,[H|T],alts)\text{''},$$
$$\text{let } W_2 = \text{out}(\text{``call}([X,Y],[W_1,W_2,H,T],W_2,[T,Y])\text{''},$$
$$\text{apply app}/2 \ (T,Y)))$$
$$\text{in out}(\text{``exitcase}([X,Y],[W_1,W_2,H,T],W_1,[H|W_2])\text{''},$$
$$[H|W_2])$$
$$\text{in out}(\text{``exit}([X,Y],[W_1,W_2,H,T],W_1)\text{''},W_1)$$

$$\text{where } alts = [(1,X,[\,]),(2,X,[H|T])].$$

Fig. 8. Instrumented program

It can easily be shown that the instrumented program produces the same sequence of events of Example 2, e.g., by executing the program in the standard environment of Erlang (together with an appropriate definition of out/2).

The correctness of the program instrumentation is stated in the next result:

Theorem 1. *Let pgm be a flat program and pgm^I its instrumented version according to Definition 4. Given an initial expression, apply main/n $(p_1, \ldots, p_n)$, its execution using pgm and the instrumented semantics (according to Definition 3) produces the same sequence of events as its execution using pgm^I and the standard semantics.*

Proof. We prove that for all program expressions, e, we have that $\langle vs, ps \rangle, \theta \vdash e \Downarrow_\tau p$ implies $\theta \vdash [\![e]\!]_F^{vs,ps} \Downarrow p$ with the standard semantics[2] and, moreover,

---

[2] Here, we consider that the *standard* semantics is that of Fig. 4 without the events labeling the transitions.

it outputs the same sequence of events τ. The claim of the theorem is an easy consequence of this property. We prove the claim by induction on the depth k of the proof tree with the instrumented semantics.

Since the base case $k = 0$ is trivial (the rule to evaluate a pattern is the same in both cases), we now consider the inductive case $k > 0$. We distinguish the following cases depending on the applied rule from the semantics of Fig. 4:

- The first rule of the semantics is not applicable since the depth of the proof is $k > 0$.
- If the applied rule is the second one (to evaluate a function call), then the considered transition has the form

$$\langle vs, ps \rangle, \theta \vdash \text{let } p = \text{apply } p_0 \ (\overline{p_m}) \text{ in } e \Downarrow_\tau p''$$

with $\tau = \text{call}(vs, ps, p, [\overline{p_m}]) + \tau_1 + \text{exit}([\overline{Y_m}], [\text{bv}(e_2)], p_2'') + \tau_2$. The instrumented expression is thus

$$[\![\text{let } p = \text{apply } p_0 \ (\overline{p_m}) \text{ in } e]\!]_b^{vs,ps}$$

Following the rules of Fig. 7, this is transformed to

$$\text{let } p = \text{out}(\text{"call}(vs, ps, p, [\overline{p_m}])\text{"}, \text{apply } p_0 \ (\overline{p_m})) \text{ in } [\![e]\!]_b^{vs,ps}$$

such that the execution of this instrumented code will first output the event $\text{call}(vs, ps, p, [\overline{p_m}])$ similarly to the instrumented semantics. By the induction hypothesis, the evaluation of $p_0, \ldots, p_m$ and e with the instrumented semantics produces the same values and outputs the same events than with their instrumented versions with the standard semantics. Let us now consider that p_0 evaluates to function f/m, whose definition is as follows: $\text{f}/m = \text{fun } \overline{Y_m} \rightarrow \text{let } X = e \text{ in } X$. In the instrumented program, the same function has the form

$$\text{f}/m = \text{fun } (X_1, \ldots, X_m) \rightarrow [\![\text{let } X = e \text{ in out}(\text{"exit}(vs, bs, X)\text{"}, X)]\!]_F^{vs',ps'}$$

$vs' = [\overline{Y_m}]$ and $ps' = [\text{bv}(e)]$. By the induction hypothesis, we know that the sequence of events for let $X = e$ in X in the instrumented semantics, is the same as that of $[\![\text{ let } X = e \text{ in } X]\!]_F^{vs',ps'}$, therefore the claim follows.
- If the applied rule is the second one (to evaluate a function call), then the considered transition has the form

$$\langle vs, ps \rangle, \theta \vdash \text{let } p = \text{case } p_0 \text{ of } \textit{clauses} \text{ end in } e \Downarrow_\tau p_0'$$

with $\textit{clauses} = \overline{p_l \rightarrow e_l}$ and

$$\tau = \text{case}(vs, ps, i, p_0, p_i, alts) + \tau_1 + \text{exitcase}(vs, ps, p, p_i') + \tau_2$$

The instrumented expression is thus

$$[\![\text{let } p = \text{case } p_0 \text{ of } \textit{clauses} \text{ end in } e]\!]_b^{vs,ps}$$

which is transformed to

$$\text{let } p = \text{case } p_0 \text{ of } \textit{clauses}' \text{ end in } [\![e]\!]_b^{vs,ps}$$

with $clauses' = \overline{\text{out}(\text{``}case(vs, ps, l, p_0, p_l, alts)\text{''}, [\![e_l]\!]_{\mathsf{T}(p)}^{vs,ps})}$. By the induction hypothesis, we have that $\langle vs, ps \rangle, \theta \cup \sigma \vdash e_i \Downarrow_{\tau_1} p'_i$ implies $[\![e_i]\!]_{\mathsf{F}}^{vs,ps} \Downarrow p'_i$ outputs the sequence of events τ_1. Therefore, $[\![e_i]\!]_{\mathsf{T}(p)}^{vs,ps} \Downarrow p'_i$ outputs and additional event exitcase, and the claim follows by induction.
- Proving the claim for the two remaining rules is straightforward by the induction hypothesis.

A prototype implementation of the program instrumentation can be found at http://kaz.dsic.upv.es/instrument.html. Here, one can introduce a (restricted) Erlang program that is first transformed to the flat syntax and, then, instrumented (several input examples are provided). Moreover, it is also possible to run the instrumented program and obtain the corresponding sequence of events.

5   Concolic Execution

The relevance of the computed sequences of events is that one can easily reconstruct a symbolic execution that mimics the steps of the concrete execution that produced the sequence of events, as well as to produce alternative bindings for the initial variables so that a different execution path will be followed.

Let us first formalize the reconstruction of the symbolic execution from a sequence of events using the Prolog program shown in Fig. 9. As mentioned before, we should ensure that the elements of τ are renamed apart. In our implementation, the sequence of events is written to a file, that is then consulted as a sequence of *facts* and, thus, their variables are always renamed apart. For simplicity, we do not show these low level details in Fig. 9 but just assume the events in τ have been renamed apart.

$sym(\tau, Res, Vars) \leftarrow eval(\tau, [(Res, Vars, BVars)])$.

$eval([\,],[\,])$.

$eval([call(Vars, BVars, NRes, NVars)|Tau], [(Res, Vars, BVars)|Env]) \leftarrow$
$\quad eval(Tau, [(NRes, NVars, NBVars), (Res, Vars)|Env])$.

$eval([case(Vars, BVars, N, Arg, Pat, Alts)|Tau], [(Res, Vars, BVars)|Env]) \leftarrow$
$\quad Arg = Pat, eval(Tau, [(Res, Vars, BVars)|Env])$.

$eval([exitcase(Vars, BVars, Arg, Pat)|Tau], [(Res, Vars, BVars)|Env]) \leftarrow$
$\quad Arg = Pat, eval(Tau, [(Res, Vars, BVars)|Env])$.

$eval([bind(Vars, BVars, Pat1, Pat2)|R], [(Res, Vars, BVars)|Env]) \leftarrow$
$\quad Pat1 = Pat2, eval(R, [(Res, Vars, BVars)|Env])$.

$eval([exit(Vars, BVars, Pat)|Tau], [(Res, Vars, BVars)|Env]) \leftarrow$
$\quad Res = Pat, eval(Tau, Env)$.

Fig. 9. Prolog procedure for symbolic execution

Let us briefly explain the rules of the procedure. The first clause just calls *eval* and initializes an stack of function environments with $(Res, Vars, BVars)$, where

Res is the result of the evaluation, *Vars* are the variables of the main function, and *BVars* are the bounded variables of the main function. When calling *sym*, all these three variables are unbound.

The first rule of *eval*/2 just finishes the computation when there are no events to be processed.

The next rule deals with *call* events and just pushes a new environment (*NRes,NVars,NBVars*) into the stack of environments. Observe that the names of variables *Vars* and *BVars* occurs twice in the head of the clause—as arguments of the event and as in the current environment—which makes them unify and thus set the right values for them in the current symbolic execution. This is done in all the clauses.

The next rule deals with *case* events and it main purpose is to unify *Arg* and Pat, which represent the case argument and the selected pattern, respectively.

The next rule takes an *exitcase* event and proceeds similarly to the previous one by matching Arg and Pat, now denoting the pattern of a let expression and the result of the evaluation of a case branch.

The next rule deals with a *bind* event in the obvious way by unifying the given patterns Pat_1 and Pat_2.

Finally, the last rule matches *Res* in the current environment (used to store the output of the current function call) with the pattern *Pat* and, moreover, pops the environment (*Res, Vars, BVars*) from the stack of environments.

For example, given the sequence of events of Example 2 and the initial call $sym(\tau, Res, Vars)$, the above program returns:

$$Res = [X, X], \; Vars = [X]$$

which obviously produces less instantiated values than the concrete execution (where we had *Res* = [a,a], *Vars*= [a]).

For concolic testing, though, one is not interested in computing the symbolic execution associated to the concrete execution, but in alternative symbolic executions so that the produced data will give rise to different concrete executions. Luckily, it is easy to extend the previous procedure in order to compute alternative symbolic executions by just replacing the clause for *case* events as follows:

$$eval([case(Vars, BVars, N, Arg, Pat, Alts)|Tau], [(Res, Vars, BVars)|Env])$$
$$\leftarrow member((M, Arg', Pat'), Alts),$$
$$N \neq M, \; Arg' = Pat',$$
$$eval(Tau, [(Res, Vars, BVars)|Env]).$$

By using the call $member((M, Arg', Pat'), Alts)$, this rule nondeterministically chooses all the alternative selections in case expressions, thus producing alternative bindings for the initial call. For instance, for the sequence of events of Example 2, we get three (nondeterministic) answers:

$$Vars = [\,] \; ; \quad Vars = [X] \; ; \quad Vars = [X, Y|R]$$

An implementation of the concolic testing tool has been undertaken. The first stage, flattening and instrumenting the source program has been implemented in Erlang itself, and can be tested at http://kaz.dsic.upv.es/instrument.html. In contrast, the concolic testing algorithm is being implemented in Prolog, since the facilities of this language—unification and nondeterminism—make it very appropriate for dealing with symbolic executions.

6 Discussion

In this paper, we have introduced a transformational approach to concolic execution that is based on flattening and instrumenting the source program—a simple first order, eager functional language—. The execution of the instrumented program gives rise to a stream of events that can then be easily processed in order to compute the variable bindings of the associated symbolic executions, as well as possible alternatives. To the best of our knowledge, our paper proposes the first approach to concolic execution by program instrumentation in the context of functional (or logic) programming. In contrast to using an interpreter-based design, in our approach the instrumented program can be run in any environment, even non-standard ones, which opens the door, for instance, to run the instrumented program in a model checking environment like Concuerror [6] so that its execution would produce the sequences of events for all relevant interleavings.

As a future work, we plan to extend our approach in order to cover a larger subset of Erlang as well as to design a fully automatic procedure for concolic testing (currently, one should manually run the instrumented program and the Prolog procedure for generating alternative bindings). Here, we expect that our transformational approach will be useful to cope with concurrent programs, as mentioned above.

Acknowledgements. We thank the anonymous reviewers and the participants of LOPSTR 2015 for their useful comments to improve this paper.

References

1. Armstrong, J., Virding, R., Williams, M.: Concurrent programming in ERLANG. Prentice Hall, Englewood Cliffs (1993)
2. Carlsson, R.: An Introduction to Core Erlang. In: Proceedings of the PLI 2001 Erlang Workshop (2001). http://www.erlang.se/workshop/carlsson.ps
3. Giantsios, A., Papaspyrou, N.S., Sagonas, K.F.: Concolic testing for functional languages. In: Proceedings of PPDP 2015, pp. 137–148. ACM (2015)
4. Godefroid, P., Klarlund, N., Sen, K.: DART: directed automated random testing. In: Proceedings of PLDI 2005, pp. 213–223. ACM (2005)
5. Godefroid, P., Levin, M.Y., Molnar, D.A.: Sage: whitebox fuzzing for security testing. Commun. ACM **55**(3), 40–44 (2012)

6. Gotovos, A., Christakis, M., Sagonas, K.F.: Test-driven development of concurrent programs using Concuerror. In: Rikitake, K., Stenman, E. (eds.), Proceedings of the 10th ACM SIGPLAN Workshop on Erlang, pp. 51–61. ACM (2011)
7. Kahn, G.: Natural semantics. In: Brandenburg, F.-J., Vidal-Naquet, G., Wirsing, M. (eds.), Proceedings of STACS 1987, pp. 22–39 (1987)
8. King, J.C.: Symbolic execution and program testing. Commun. ACM **19**(7), 385–394 (1976)
9. Mesnard, F., Payet, E., Vidal, G.: Concolic testing in logic programming. Theor. Pract. Logic Program. **15**, 711–725 (2015)
10. Pasareanu, C.S., Rungta, N.: Symbolic PathFinder: symbolic execution of Java bytecode. In: Pecheur, C., Andrews, J., Di Nitto, E. (eds.), ASE, pp. 179–180. ACM (2010)
11. Sen, K., Marinov, D., Agha, G.: CUTE: a concolic unit testing engine for C. In: Proceedings of ESEC/SIGSOFT FSE 2005, pp. 263–272. ACM (2005)
12. Vidal, G.: Towards symbolic execution in Erlang. In: Voronkov, A., Virbitskaite, I. (eds.) PSI 2014. LNCS, vol. 8974, pp. 351–360. Springer, Heidelberg (2015)
13. Vidal, G.: Concolic execution and test case generation in Prolog. In: Proietti, M., Seki, H. (eds.) LOPSTR 2014. LNCS, vol. 8981, pp. 167–181. Springer, Heidelberg (2015)

Memory Policy Analysis for Semantics Specifications in Maude

Adrián Riesco[1], Irina Măriuca Asăvoae[2], and Mihail Asăvoae[3]([⊠])

[1] Universidad Complutense de Madrid, Madrid, Spain
ariesco@fdi.ucm.es
[2] Swansea University, Swansea, UK
I.M.Asavoae@swansea.ac.uk
[3] Inria Paris-Rocquencourt, Paris, France
mihail.asavoae@inria.fr

Abstract. In this paper we propose an approach to the analysis of
formal language semantics. In our analysis we target memory policies,
namely, whether the formal specification under consideration follows a
particular standard when defining how the language constructs work
with the memory. More specifically, we consider Maude specifications of
formal programming language semantics and we investigate these specifications at the meta-level in order to identify the memory elements
(e.g., variables and values) and how the language syntactic constructs
employ the memory and its elements. The current work is motivated by
previous work on generic slicing in Maude, in the pursuit of making our
generic slicing as general as possible. In this way, we integrate the current
technique into an existing implementation of a generic semantics-based
program slicer.

Keywords: Formal semantics · Maude · Slicing · Analysis · Memory
policies

1 Introduction

Static program analysis provides functional and non-functional guarantees with
respect to the program behavior. These guarantees, e.g., invariants, are automatically computed from predefined approximations of the concrete program executions. Examples on standard invariants include pointer behavior in sequential
code, data races in concurrent code, or bounds of execution time/memory usage.

Rewriting logic provides support to define formal and executable language
semantics. A key aspect in a language definition is the memory model—the
set of all semantic entities that are required to describe the storage component of a program execution. Let us consider how memory is organized for two

This research has been partially supported by MICINN Spanish project *Strong-
Soft* (TIN2012-39391-C04-04) and by the Comunidad de Madrid project N-Greens
Software-CM (S2013/ICE-2731).

M. Falaschi (Ed.): LOPSTR 2015, LNCS 9527, pp. 293–310, 2015.
DOI: 10.1007/978-3-319-27436-2_18

languages defined in rewriting logic—an imperative language with functions and input/output support and an assembly language generated from it. For example, the formal definition of the imperative language would require a global memory for (global) program variables, local environment (for locals), call stack for functions, and input/output buffers. A program execution is a sequence of rewrite steps that access one or more of these storages. In a similar fashion, the formal definition of an assembly language relies on a main memory represented as an array of memory cells, each cell stores one value, and a set of (general purpose or specialized) registers for everything else.

Our goal is to design generic program analysis tools based on a meta-level analysis of the programming language semantic definition. This would allow a certain degree of parameterization of the program analysis such that changes in the formal language semantics should not result in the need of adapting the corresponding analyzer, since the analyzer automatically incorporates the modifications. This approach builds on the formal executable language semantics given as a rewriting logic theory [7,19] and on the program to be analyzed. The generic design for program analysis tools based on language semantics comes in two steps. The first step is a meta-analysis of the formal language semantics. The second step is a data dependency analysis of the program. The meta-level analysis is a fixpoint computation of the set of basic language constructs of interest, e.g., side-effect constructs, which is then used to extract safe program slices based on a required criterion. This methodology is instantiated in [3,26] on the classical WHILE language augmented with a side-effect assignment and read/write statements and, respectively, on the WhileF language—an extension of WHILE to allow functions and scope declaration for variables.

An example program in WhileF is in Fig. 1 (left). Note however that both intra- and inter-procedural program slicing methods are based on a less generic assumption: the general memory update operation—the assignment statements—has a fixed destination: its left-hand side. This is not necessarily generic as, for example, the family of the assembly languages uses explicit memory operations (load/store) and arithmetic/logic operations (which update registers), with flexible destination placement in the language syntax. For example:

- in MIPS assembly language, in Fig. 1 (middle), the load instruction lw has a direction right (source) to left (destination), while the store instruction sw has a reverse direction. Moreover, mult multiplies the values in the two registers and writes the result in a special multiplication register.
- in x86 assembly language generated by gas (Fig. 1, right top), which is the GNU assembler and the default back-end of the standard gcc compiler, an instruction like movl 16(%esp), %eax copies into the register %eax the value found at the address referred by the register %esp shifted left by 16, as in Fig. 1 (right top). The update is from left (source) to right (destination).
- in x86 assembly language, in Fig. 1 (right bottom), an instruction like mov eax, DWORD PTR [esp+28] copies into the register eax a word-length from the address found in the register esp shifted to the right with the offset 28. The update is from right (source) to left (destination).

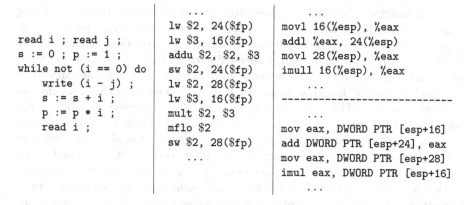

Fig. 1. WhileF program (left) with snapshots of assembly code MIPS IV (middle), x86 - AT&T (right - top) and x86 - Intel (right - bottom)

The direction of the memory update is an example of what we call a memory policy, meaning the way the language constructs make use of the semantic entities that define the memory model in the formal semantics. Moreover, when we infer the direction of the memory update operation we actually address (in a uniform way) a wide range of low-level languages.

In this paper we propose a refinement of a previously introduced technique in [3, 26], where we described a generic intra- and inter-procedural slicing method, respectively. In [26] we focused on inferring the language constructs that produce side-effects from the semantics specification, i.e., language constructs inducing memory updates. In the current work, we infer memory policies, i.e., formal semantics properties about how the language constructs use the memory model defined by the semantics. We particularise the memory policy to *detecting the direction of the data flow in the memory updates*. Namely, given a side-effect construct c in the considered language, we infer which are the sources and which is the destination of the data flow detected in c. For example, in an assignment $x := y + z$ our memory policy detects that y and z are the sources while x is the destination. For inferring this memory policy the meta-analysis tracks down how each element in the construct c is used at the memory level (either read or write) and then we trickle up this information back in the components of c.

Paper Outline. This paper is organized as follows: Sect. 2 covers related work; Sect. 3 introduces rewriting logic and Maude as well as our view on memory polices from the rewriting-logic perspective; Sect. 4 details the algorithm of inferring memory policies; Sect. 5 describes the prototype tool. We conclude in Sect. 6.

2 Related Work

Our goal is to design and implement generic formal semantics-based tools for program analysis in a rewriting logic environment, with focus on memory models. Hence, we relate our approach to static program analysis and rewriting logic.

Static program analysis is a compile-time process for automatically extracting run-time semantic information (i.e., invariants) from programs. Abstract interpretation [8], which systematically derives sound approximations of the concrete semantics, and type systems [22], which define correct programs with respect to typing information, are two of the most used techniques for program analysis.

Program analysis based on abstract interpretation uses abstract domains and abstract semantics. The latter is an abstract re-implementation of (some of) the language operations as well as an abstract memory. From the point of view of the abstract representation of the program memory, the abstract semantics can capture a wide range of properties: functional properties, e.g., pointer and alias analyses [15,23], data race detection [12] on shared memory programs, stack safety [24], automated checks for coding standards [29], or non-functional properties, e.g., computation of safe upper bounds for heap size [1] and stack size [4]. In comparison with these approaches, we propose to infer, via meta-analysis of the formal language semantics, certain information (which we call policies) about the abstract memory system.

Having a formal executable semantics with precise memory models allows verification of both sequential and concurrent code. For example, the encoding of the x86 assembly language semantics in HOL proof assistant [28] allows reasoning about memory consistency in threaded applications while the encoding of the memory model of C language in Coq [18] is suitable for pointer arithmetic reasoning. In general, theorem proving either interactive or automated provides the necessary infrastructure to allow meta-level reasoning for programming language semantics, in a similar fashion with our proposal. These approaches are complemented by the rewriting logic semantics project [21], which focuses on how to define formal semantics of programming languages in rewriting logic and how to construct program analysis tools directly over these semantics. The memory component of a language definition in rewriting logic and its applicability in program analysis is presented in [11,14]. The memory model of [14] is exemplified on a simple imperative language with functions. Also, they define pluggable program analyses by reusing parts of the concrete language semantics. For example, the rewriting logic specification of the Java Memory Model [11] is used for model checking Java programs. Our approach accommodates the concept of pluggable program analysis via meta-level manipulation of the program semantics, as given for program slicing in [3].

The term-slicing aspect of our proposed program slicing technique is rooted into the notions of descendant/ancestor and origin tracking [5,16,17]. Origin tracking, introduced in [17] is a refinement of the descendant/ancestor relationship as it follows the symbols of an expression to their causes in an earlier expression in a rewrite sequence. The origin tracking in first-order term rewriting systems [16] is intrinsic to slicing due to its strategy of reasoning on every reduction from a term to its normal form. The term-slicing uses an extended concept of origin tracking, w.r.t. the aforementioned approaches, because it tracks changes in conditional rewriting rules, as defined in Maude, with a particular emphasis on how to slice through rule conditions. In fact, the proposed notion of term-slicing determines variable dependencies in rules and equations in Maude specifications where the variables are subterms of a certain sort.

In rewriting logic there are several approaches for analysis tools, not necessarily for programs. For example, debugging [2], testing [25], and slicing [2,3,13,26]. The program slicing technique in [13] executes the term representation of a program with the formal semantics and extracts dynamic slices. In comparison, our approach does not execute the formal semantics; the term slicing is based on a meta-level analysis of the semantics. In terms of genericity [13], requires translation steps from a given language semantics into an intermediate language (which is the base for program slicing), whereas our approach works directly on the semantics, as it is defined. The slicing technique in [2] works on generic Maude execution traces. In comparison, we propose a static approach built around a formal semantics and with an emphasis on computing slices for programs and not for execution traces. The work in [25] presents an approach to generate test cases similar to the one presented here in the sense that both use the semantics of programming languages formally specified to extract specific information. However, in [25] the narrowing technique is used on the semantic rules to instantiate the state of the variables in the given program. Matching logic [27] is a program verification technique based on executing a program with a rewriting-based formal semantics, by proving the necessary program invariants. In comparison, our approach is complementary to matching logic as it attempts to compute invariants from the semantics and afterwards, to apply them in program reasoning (e.g., program slicing). Moreover, our approach uses the meta-level capabilities of rewriting logic, which to the best of our knowledge are not available in the matching logic framework.

The technique in the current paper follows our previous work on language-independent program slicing in rewriting logic environment [3]. Actually, the implementation of the current work improves the genericity aspect of the slicing tool developed in [3], since we infer policies about memory updates applied to imperative and assembly languages. The program slicing over the formal semantics S of the language L follows the same two steps as in [3]: (1) an initial meta-analysis of S followed by (2) a program analysis conducted over the programs in L using term slicing.

3 Preliminaries

We present in this section the basic ideas about Maude and memory policies.

3.1 Memory Policies

A formal language semantics consists of the set of all semantic entities that are required to fully specify all possible behaviors of any correct program, i.e., with respect to the semantics definition. Part of the language semantic entities describe the memory system. Examples of such semantic entities are heaps, stacks (e.g., call stack, loop stack), environments, register file, etc. Then, the language constructs interact, directly or indirectly, with the memory system. Our aim is to infer information about this interaction in an automated way.

We achieve our declared goal of designing generic program analysis tools by employing a meta-level analysis of the formal language semantics. Such a meta-analysis extracts semantics level properties, e.g., the sets of language constructs that may induce side-effects or may result into context-updates. From a memory system point of view, these properties are inferred from the semantics specification by following how the language constructs operate on the memory system. We call this kind of properties *memory policies*. For example, in the case of an imperative language semantics as WhileF, i.e., with functions and input-output capabilities, one memory policy could be named as "direction property". This would involve inferring that in the assignment statements the right-hand side is the source and the left-hand side is the destination.

A more formal view on inferring memory policies would require reasoning at the level of sorting relationships of the semantic entities present in the language semantics specification (starting with a given set of memory-related sorts). For assembly languages in Fig. 1, MIPS considers left to right direction for store and right to left for load instructions while the two x86 styles use the same style for both direction, although it is from left to right for one architecture and from right to left in the other. Consequently, if we are to extend our tool for dealing with a larger class of programming languages, we need to incorporate this particular memory policy inference, which automatically deduces from the semantics specification, for the side-effect constructs, what is the direction of the data flow in each such construct. Note that this direction is crucial for the accuracy of the slicing result, as we need to incorporate in the slicing set only the changing points of certain variables, i.e., where those variables are destination.

3.2 Semantics in Maude

Maude modules are executable rewriting logic specifications. Rewriting logic [20] is a logic of change very suitable for the specification of concurrent systems and it is parameterized by an underlying equational logic, for which Maude uses membership equational logic (*MEL*) [6], which, in addition to equations, allows one to state membership axioms characterizing the elements of a sort. Rewriting logic extends *MEL* by adding rewrite rules.

Maude functional modules [7, Chap. 4], introduced with syntax fmod ... endfm, are executable membership equational specifications that allow the definition of sorts (by means of keyword sort(s)); subsort relations between sorts (subsort); operators (op) for building values of these sorts, giving the sorts of their arguments and result, and which may have attributes such as being associative (assoc) or commutative (comm), for example; memberships (mb) asserting that a term has a sort; and equations (eq) identifying terms. Both memberships and equations can be conditional (introduced by the keyword cmb and ceq, respectively). Maude system modules [7, Chap. 6], introduced with syntax mod

... endm, are executable rewrite theories. A system module can contain all the declarations of a functional module and, in addition, declarations for rules (rl) and conditional rules (crl).

Maude has been widely used for specifying the semantics of several languages, such as Java [10] or C [9]. The key idea for specifying semantics is, first, to define the signature by means of declaring sorts and their respective constructors (operators). We illustrate this methodology by presenting a simple assembly language that we will use throughout the rest of the paper. This language uses registers to keep intermediate values and defines standard functions, such as addition and subtraction, over them, and also has a memory where values are stored for later sessions. The specification of this language requires a sort identifying a register (RegId), for the value stored in a register (Register), and for the set of such values (Registers). Note the use of the keyword subsort indicating that Register is a particular case of Registers:

```
sorts RegId Register Registers .        subsort Register < Registers .
```

We define now values for these sorts as follows: RegId are built with the constructor reg, which receives a natural number; a Register is just a pair of a RegId and an integer (underscores are just placeholders); finally, we can have either the empty Registers (mtReg) or the juxtaposition of elements, which is commutative and associative and has mtReg as identity:

```
op reg : Nat -> RegId [ctor] .
op <_,_> : RegId Int -> Register [ctor] .

op mtReg : -> Registers [ctor] .
op __ : Registers Registers -> Registers [ctor assoc comm id: mtReg] .
```

We can also define functions on these sorts. We specify the function _[_] for looking-up a value in the registers (note that it returns 0 if it is not initialized) and update for updating the memory:

```
op _[_] : Registers RegId -> Int .
eq [lu1] : (< R, I > RS) [R] = I .
eq [lu2] : RS [R] = 0 [owise] .

op update : RegId Int Registers -> Registers .
eq [upd_int1] : update(R, I, < R, I' > RS) = < R, I > RS .
eq [upd_int2] : update(R, I, RS) = < R, I > RS [owise] .
```

The sort for the long-term memory, Memory, is defined in a similar way. It is also worth presenting the syntax for instructions and the whole system that will be used when defining the semantics. Instructions have sort Ins and their syntax depends on the specific instructions. For example, the instruction for adding two registers and storing the result in a third one is defined below. We will infer later the direction of this instruction, that is, we identify which one is "the third register." Finally, the complete system has sort System and puts together a list of instructions (`Instructions`), the state of the registers (sort `Registers`), the state of the memory (`Memory`), and the program counter (of a predefined sort `Nat`):

```
op add_,_,_ : RegId RegId RegId -> Ins [ctor] .
op [_|_|_|_] : Instructions Registers Memory Nat -> System [ctor] .
```

Once the signature is established, the semantics are defined by means of rewrite rules. Rewrite rules mimic the behavior specified by the inference rules in the formal semantics by executing the premises in the conditions and the conclusion in the body of the rule. The rule labeled [add] below defines the expected behavior of the add instruction: retrieves the values stored in the second and the third register parametrizing the instruction, adds them, and stores the thus obtained value in the first register:

```
crl [add] : [IIL | RS | M | PC] => [IIL | RS' | M | PC + 1]
   if (add RI, RI', RI'') := getIns(IIL, PC) /\
      I := RS [RI'] /\
      I' := RS [RI''] /\
      RS' := update(RI, I + I', RS) .
```

Note that we use matching conditions (:=) to indicate that the pattern in the lefthand side matches the term in the righthand side, once it has been reduced by means of equations. This condition binds the free variables (that is, the variables that did not appear in the lefthand side of the rule or in previous matching conditions) to the appropriate values.

4   Inferring Memory Policies

We describe next the refinement that extends our previous work on discovering side-effect constructs in a programming language starting with the semantics specification of the considered language [26]. There, we show a generic intraprocedural slicing process where the generic aspect is given by the inference of what we call side-effect language constructs, i.e., the instructions that determine memory changes. To achieve this, we construct a so called *hyper-tree*, whose nodes are sets of rewrite rules and edges are dependencies between these rules. As such, we are able to infer which constructs are going to possibly produce memory updates by following the paths in the hyper-tree from the root to the leaves. We can see our current work as a trickle-up in this hyper-tree. Namely, at the leaves level we extract information regarding the source-destination relation of memory updates and we propagate this relation up in the hyper-tree at the level of the language constructs. Note that the method in [26] produces an over-approximation of the side-effect constructs, which we now refine not in terms of cutting out elements from the resulting set, but by enriching the information contained in this set with data-flow direction.

Hence, we present in this section the ideas underlying our framework, illustrating them on the Maude semantics of an assembly language. The results are equally applicable to other rewriting-based semantics, like the one for WhileF language that we describe in [3]. Next, we elaborate on the semantics of the assembly language and how to infer memory policies like the "direction of a memory update" for programming language constructs.

We assume that the sorts for the memory (`Registers` and `Memory` in the example in Sect. 3) are provided by the user, while the rest of information is inferred by the system. It is important to state that these inferences work under a natural assumption: memory sorts are composed by tuples mapping program variables into values, possibly via addresses.

4.1 Maude Slicing

In order to narrow down the source of the changes, we first apply term-slicing to the equations and rules in the semantics. Slicing for Maude specifications (named in this paper *term-slicing* to differentiate it from the standard program slicing component present in our work) is already used for improving the results from Maude model checker [2]. We use here a simpler approximation of term-slicing that traces back the source of a given set of variables by adding to this set the variables involved in their generation. This approximation is a syntactic procedure for computing dependencies in a single rule/equation by taking into account that variables can be bound in the lefthand side of matching conditions (`:=`) and in the righthand side of rewrite conditions (`=>`). Hence, starting from an initial set of variables of interest $\mathcal{V}$, we traverse the conditions following a bottom-up strategy and, when a variable $v \in \mathcal{V}$ is bound by these conditions we add all the variables in the "opposite" side (hence in the righthand side of matching conditions and the lefthand side of rewrite conditions) to $\mathcal{V}$. For example, let us assume we have a rule as follows

```
crl f(X, Y, Z) => g(h(B, A3), Z)
  if X >= 3 /\
     A1 := aux1(X, Y) /\
     B := other_fun(X, Y, Z) /\
     aux2(Y) => A2 /\
     A3 := aux3(A1, A2) .
```

and we want to trace back `A3`, since it modifies the memory. In the rule above, the condition `A3 := aux3(A1, A2)` indicates that the value in `A3`, the variable in the term-slicing set, depends on the value of both `A1` and `A2` used in function `aux3`, so they are both included in the term-slicing set. The previous condition, `aux2(Y) => A2`, indicates that `A2` depends on `Y`, and hence it is included in the slicing set. Note however that the condition `B := other_fun(X, Y, Z)` does not produce any change in the term-slicing set because `Y` is only used and not changed by this condition, hence `B` is not included in the term-slicing set. The condition `A1 := aux1(X, Y)` adds `X` (and `Y`) into the term-slicing set. Finally, the first condition, `X >= 3`, has no effect because it is not a matching or rewriting condition. From this analysis we find that `A3` depends on `Y` and `X` from the lefthand side of the rule; similarly, we can decide the dependencies of any term.

Data: A specification and the sorts for the memory $\mathcal{M}$.
Result: Set of sorts for values $\mathcal{V}$.
$\mathcal{V} = \emptyset$;
foreach *constructor* $c(s_0, \ldots, s_n, v)$ *of sort* $S, n \geq 0, S \in \mathcal{M}$ **do**
    foreach *function* $f : ar \to v$ **do** // Explicit inference
    | **if** $S \in ar$ **then** $\mathcal{V} = \mathcal{V} \cup \{v\}$;
    end
    foreach *rule* $l \to r$ *if cond* **do** // Implicit inference
    | $vm = varsOfSortMemory(r, \mathcal{M})$;
    | $vs = slicing(l, cond, vm)$;
    | $\mathcal{V} = \mathcal{V} \cup getVarsInConstructor(vs, c)$;
    end
end

Algorithm 1. Algorithm for inferring the sort for the values

4.2   Inferring the Sorts for the Values in the Memory

We now emphasize on the settings characterizing the memory part in the class of language semantics specifications that we consider. As previously mentioned, we assume that the memory component of the specification is connecting the program variables to their current values, either directly as in a simplified memory model, or via a chain of "addresses" as in a more accurate representation of the machine. Note that by "values" we understand those terms building the memory that are used by the semantics to modify the state, while by "addresses" we understand those terms used to access the values. We now show how to obtain the sorts for the values stored in the memory given by the, e.g., Registers sort in the considered language specification.

We present the algorithm for inferring these sorts in Algorithm 1. We traverse the constructors for the sorts specifying the memory and check all the possible outcomes for them. The first inner loop deals with *explicit* access to the memory: functions that receive the memory and return one of the sorts used in the constructor.[1] This case is illustrated by the function look-up (_[_]) in Sect. 3. The look-up function is defined by the equations [lu1] and [lu2] and it extracts a term of sort Int, which is used to build a Register, which is, in turn, a subsort of the sort of a specific part of the memory, i.e., Registers. Since this function is used in the semantics of the language, we infer that Int is the sort of a possible value.

We can also find *implicit* access to the memory: patterns in the lefthand side of rules or in matching/rewrite conditions can be used to retrieve values from the memory, as illustrated in the second inner loop of Algorithm 1. In this case, we trace back the variables modifying the memory and keep only those obtained

---

[1] We have placed the sort v as the last sort in the arity to ease the presentation, but it is not required.

Data: A specification, the sorts for the memory $\mathcal{M}$, and the sorts standing for values $\mathcal{V}$.

Result: Set of functions $\mathcal{F}$ modifying the memory annotated with the variables responsible for the modifications.

$\mathcal{F} = \emptyset$;

foreach *function* $f : s_0, \ldots, s_n \rightarrow s, s \in \mathcal{M}, \exists i.s_i \in \mathcal{V}$ **do** // Explicit inference
   | $\mathcal{F} = \mathcal{F} \cup \{f_{s_i}\}$
end

foreach *rule* $l \rightarrow r \; if \; cond$ **do** // Implicit inference
   | $vm = varsOfSortMemory(r, \mathcal{M})$;
   | $vs = slicing(l, cond, vm)$;
   | $vv = varsOfSortValue(vm, \mathcal{V})$;
   | **if** $vv \neq \emptyset$ **then** $\mathcal{F} = \mathcal{F} \cup \{f_{vv}\}$;
end

Algorithm 2. Algorithm for inferring the functions modifying the memory

from the memory. For example, assume we modify the rule [add] from Sect. 3 to avoid the look-up function, obtaining [addv2].[2]

```
crl [addv2] : [IIL | RS | M | PC] => [IIL | RS'' | M | PC + 1]
  if (add RI, RI', RI'') := getIns(IIL, PC) /\
     < RI', I > < RI'', I' > RS' := RS /\
     RS'' := update(RI, I + I', RS) .
```

In this case, we know that it is possible for the memory to be modified (we have a new variable RS'' for a memory sort), so we consider the term-slicing set to initially contain only this variable. We then trace back its related variables using the technique described in Sect. 4.1, obtaining RI, I, I', RS, IIL, and PC. We can now filter the obtained term-slicing set and retain only the values in the memory (in this case both I and I'), which have the sort Int that we previously inferred. Note that it is possible to use a matching with unrequited information to make the method above to include some sorts that are not proper values. However, this is not a threat for soundness, because our technique computes over-approximations, so adding a sort that is not memory related will just worsen the granularity of the slice computed later.

4.3 Inferring the Functions Modifying the Memory

At this step we look for the functions that introduce new values into the memory. As presented in Algorithm 2, in this case we can also find both *explicit* and *implicit* access to the memory. Note that the algorithm returns the set of functions annotated with the variables responsible for the effects. The explicit case, shown in the first loop, is easy to detect: we just traverse the operators

---

[2] We would need extra rules to take care of non-initialized registers, but this is not relevant for the technique.

looking for those creating/modifying the memory (i.e., the memory appears in the coarity). We then trace the source of this modifications by using the slicing technique in Sect. 4.1 to annotate those arguments responsible for the changes and having one of the sorts annotated in the previous step. For instance, the function update from Sect. 3 modifies the memory by introducing the element of sort Int received as the second parameter.

Note that since update is found to modify the memory, then also the [addv2] rule modifies the memory since it uses update to match the memory variable RS''. This connection between the rules and functions that modify the memory is already presented in our previous work [26]. There we describe the construction of a hyper-tree containing in its nodes rules from the language semantics specification while its arcs are given by relations as the one mentioned above. For instance, [addv2] is a parent of [upd_int1] and [upd_int2] in the hyper-tree because it uses the update function which is described by the two [upd_] rules.

The implicit modifications to the memory, shown in the second loop of Algorithm 2, occur when a rewrite rule modifies the memory directly, i.e., without using any auxiliary function. In this case, we must slice again the rule using the updated memory criteria and keep those variables that have the sort obtained in the previous step. We illustrate this with a third version of the [add] rule from Sect. 3, called [addv3].[3]

```
crl [addv3] : [IIL | RS | M | PC] => [IIL | RS'' | M | PC + 1]
   if (add RI, RI', RI'') := getIns(IIL, PC) /\
      < RI, I > < RI', I' > < RI'', I'' > RS' := RS /\
      RS'' := < RI, I' + I'' > < RI', I' > < RI'', I'' > RS' .
```

In this case, the last matching condition updates the memory onsite by using the values I' and I''. Consequently, I' and I'' are annotated as side-effect sources, i.e., sources of changes in the memory.

4.4   Inferring the Data-Flow Information

By using the results obtained in the previous steps, we have enough information to infer the data-flow relation that is of interest here, i.e., the source-destination relation in the language constructs producing side-effects. As shown in Algorithm 3, we take for each rewrite rule the variables modifying the memory (obtained from either explicit or implicit change) and apply *enriched* slicing to them. This enriched slicing takes into account the assumption stated at the beginning of the section: the memory is composed of cells (tuples) connecting the program variables (or registers in the case of assembly languages) with their values. Hence, when facing a matching condition involving the memory we extend the slicing set to all the elements in the tuple in order to make sure we consider all the "addresses" connecting the program variables with their values. Finally, we need to recognize the instruction being executed. This term is the one that

---

[3] Note that we would need another rule to deal with the case where RI is not initialized, but this does not change the inference.

Data: A specification and the functions modifying the memory $\mathcal{F}$.
Result: Data-flow information $\mathcal{D}$.
$\mathcal{D} = \emptyset$;
foreach *rule* $rl \equiv l \rightarrow r$ *if cond* **do**
$\quad vs = getVarsFromAnnotations(rl, \mathcal{F})$;
$\quad vss = slicing(rl, vs)$;
$\quad ins = getInstruction(rl, vss)$;
$\quad active = getVars(ins) \cap vss$;
$\quad passive = getVars(ins) \setminus vss$;
$\quad \mathcal{D} = \mathcal{D} \cup \{ins : active \mapsto passive\}$
end

Algorithm 3. Algorithm for inferring the data-flow information

fulfills the following properties: (i) must be (or depend on) a subterm that contains the complete state, including the memory and any other sort required by the semantics and (ii) contains all the variables from the slicing set not related with the memory. The variables appearing in this term and in the slicing set are responsible for the modifications we are tracing in the memory. Note that many rules can specify the behavior of the same instruction. In this case, we put together all the possible sources of change.

For example, this method infers for the rule [add] from Sect. 3, that:

1. The term being executed is `add RI, RI', RI''`, since it is not related to the memory and contains `RI'` and `RI''`, which in turn generate `I` and `I'` from the slicing set.
2. The variables `RI'` and `RI''`, which appear in both the term and the slicing set, modify the rest of the variables (`RI`). Hence, this instruction works from right to left.

The same result is easily obtained for `addv2` and `addv3`. Moreover, note that the same approach can be easily followed to analyze the direction of a standard assignment instruction in any imperative language.

5   Prototype

The ideas presented in the previous sections have been used to extend the slicing tool in [3]. It allows us to apply our generic slicing framework to semantics of imperative languages, like the WhileF language in [3], to languages with mixed data-flow policies, like the assembly language presented in this paper, or to "eccentric" semantics, such as, languages with a left to right assignment statement. The source code of the tool, examples, and more explanations are available at http://maude.sip.ucm.es/slicing/.

The tool is started by loading in a Maude session the `slicing.maude` file available at the webpage. This starts an input/output loop where other Maude modules can be introduced and analyzed. We introduce the semantics for the language, e.g., the assembly language partially presented throughout the paper.

One of the tool's features is to infer the data flow information for the basic language constructs. For example, add presented in more detail in Sect. 4 has a right to left direction, i.e., add R1, R2, R3 stores R2 + R3 in R1; the direction of the load instruction is from left to right, i.e., load R1, R2 loads in R2 the data stored in the memory cell indicated by R1; while for the store instruction the tool infers a data flow direction from right to left, i.e., str R1, R2 stores in the register indicated by R1 the value in the cell indicated by R2.

Because we have at hand an executable semantics, we can use the assembly language semantics to execute the program pow, which computes x^y (assuming x and y are stored in the memory cells 0 and 1, respectively) and stores the result in the cell 2:

```
op pow : -> InsList .
eq pow = load R1, R1         *** Load M[0] in R1 (left to right)
             addi R2, R2, 1      *** Add 1 and save it in R2
             load R2, R2         *** Load M[1] in R2
             addi R4, R4, 1      *** Add 1 and save it in R4
     'loop   beq R2, R3, 'out    *** Jump to out when R2 and R3 are equal
             mul R4, R4, R1      *** Store in R4 the result of R4 * R1
                                 *** (function from right to left)
             subi R2, R2, 1      *** Update the counter
             jmp 'loop           *** Jump to loop
     'out    addi R5, R5, 2      *** Add 2 and save it in R5
             str R5, R4          *** Store the value of R4 in M[R5]
                                 *** (function from right to left)
             break .             *** end
```

The execution of the program needs the user's input of initial state, e.g., the function testPow introduces 3 and 5 in the memory cells 0 and 1, respectively:

```
op testPow : -> System .
eq testPow = [ pow | mtReg | [0, 3] [1, 5] | 0 ] .
```

Furthermore, in order to obtain the slicing results, the user introduces the sorts corresponding to the memory with the command:

```
Maude> (set side-effect sorts Memory Registers .)
Memory Registers selected as side effect sorts.
```

Once these sorts are set, we can start the slicing process by indicating the program to slice, e.g., testPow, and the initial slicing set, e.g., the singleton set containing R5 the variable storing the final result of pow:

```
Maude> (slice testPow wrt R5 .)
```

Note that the initial state of the program is not used by the slicer, which performs static analysis in the true sense, i.e., without using any information from the current state of the program. The program's state is there just to exemplify the executing capabilities of the programming language semantics used in our tool. Now, for slicing, the tool analyzes the list of instructions of the program, given in pow, and returns:

– The rules producing side effects, obtained by using the sorts for values and checking which rules modify them:

```
The rules causing side effects are: add addi and load mul muli
                                     str sub subi
```

– The data-flow information for each rule producing side-effects. It is interesting to see the difference between load and store, as discussed above:

```
The inferred data-flow information is:
- For function add RI:RegId,RI':RegId,RI'':RegId :
Variable(s) RI:RegId are modified by RI':RegId RI'':RegId
- For function load RI:RegId,RI':RegId :
Variable(s) RI':RegId are modified by RI:RegId
- For function str RI:RegId,RI':RegId :
Variable(s) RI:RegId are modified by RI':RegId ...
```

– The final slicing set. In this case, the value stored in the position R5 is updated with the contents of R4, which was in turn updated with the contents in R1. Hence, these registers compose the final slicing set:

```
The variables obtained by the slicing process are: R5 R4 R1
```

It is important to remember that the tool works for any programming language whose semantics has been defined in Maude. Hence, we can use the WhileF language from [3] to further test the semantics. Briefly, WhileF is an imperative language with functions and input-output capabilities. Henceforth, the algorithm that infers memory policy information on the WhileF semantics works with the sorts of the underlying memory model: a state sort ST mapping variables to values (the global memory), a sort ESt for the program environment (the local memory), and a sort for the read/write buffer RWBUF. Hence, we can introduce the code from Fig. 1(left) in WhileF as follows:

```
op whileExample : -> Com .
eq whileExample = Read i ; Read j ; s := 0 ; p := 1 ;
                  While Not Equal(i, 0) Do
                  Write (i -. j) ; s := s +. i ; p := p *. i ; Read i .
```

Our tool will traverse the semantics, find the sort of values, and show that the assignment works from left to right. Moreover, it also indicates that the variable related to p is just i, used in the multiplication:

```
Maude> (slice whileExample wrt p .)
The inferred data-flow information is:
- For function X:Var := e:Exp :
Variable(s) X:Var are modified by e:Exp
The variables obtained by the slicing process are: p i
```

6   Concluding Remarks and Future Work

In this paper we used formal language semantics to infer a certain type of memory policy, i.e., data-flow information for language constructs which produce memory updates. This inference has allowed us to improve on the genericity of our slicing technique [3] and to make another step towards a complete design of a automatized semantics-based slicing tool. Moreover, this addition to the slicing tool allowed testing the tool on other class of programming language specifications such as the assembly languages.

We are currently investigating the automatic inference of other slicing prerequisites for interprocedural methods such as the automatic deduction of function call/returns and the inference of their parameter passing patterns. These improvements would further automatize our generic slicing tool as the language designer would roughly need only to define the semantics of the programming language, to give the input program, and the slicing criterion, then our generic slicer will generate all the necessary information for slicing. From a language perspective, we aim to extend the language, for example with pointers and hence, to be able to accommodate more complex memory policies, based on a more refined memory model. Note that the addition of pointers to our framework will allow us to use arrays as well. Finally, our aim is to introduce concurrency in the framework, so we can cover and test out proposed methodology on a larger and significant class of programming languages.

Acknowledgments. We thank the anonymous reviewers for their valuable comments and suggestions, which greatly improved the quality of the paper.

References

1. Albert, E., Genaim, S., Gómez-Zamalloa, M.: Live heap space analysis for languages with garbage collection. In: Proceedings of the International Symposium on Memory Management, ISMM 2009, pp. 129–138. ACM (2009)
2. Alpuente, M., Ballis, D., Frechina, F., Romero, D.: Using conditional trace slicing for improving Maude programs. Sci. Comput. Program. **80**, 385–415 (2014)
3. Asăvoae, I.M., Asăvoae, M., Riesco, A.: Towards a formal semantics-based technique for interprocedural slicing. In: Albert, E., Sekerinski, E. (eds.) IFM 2014. LNCS, vol. 8739, pp. 291–306. Springer, Heidelberg (2014)
4. Baufreton, P., Heckmann, R.: Reliable and precise WCET and stack size determination for a real-life embedded application. In: ISoLA 2007, Workshop On Leveraging Applications of Formal Methods, Verification and Validation, Revue des Nouvelles Technologies de l'Information, pp. 41–48. (2007)
5. Bethke, I., Klop, J.W., de Vrijer, R.C.: Descendants and origins in term rewriting. Inf. Comput. **159**(1–2), 59–124 (2000)
6. Bouhoula, A., Jouannaud, J.-P., Meseguer, J.: Specification and proof in membership equational logic. Theor. Comput. Sci. **236**(1–2), 35–132 (2000)
7. Clavel, M., Durán, F., Eker, S., Lincoln, P., Martí-Oliet, N., Meseguer, J., Talcott, C. (eds.): All About Maude. LNCS, vol. 4350. Springer, Heidelberg (2007)

8. Cousot, P., Cousot, R.: Abstract interpretation: a unified lattice model for static analysis of programs by construction or approximation of fixpoints. In: Conference Record of the Symposium on Principles of Programming Languages, POPL 1977, pp. 238–252. ACM (1977)
9. Ellison, C., Rosu, G.: An executable formal semantics of C with applications. In: Proceedings of the Symposium on Principles of Programming Languages, POPL 2012, pp. 533–544. ACM (2012)
10. Farzan, A., Chen, F., Meseguer, J., Roşu, G.: Formal analysis of Java programs in JavaFAN. In: Alur, R., Peled, D.A. (eds.) CAV 2004. LNCS, vol. 3114, pp. 501–505. Springer, Heidelberg (2004)
11. Farzan, A., Meseguer, J., Roşu, G.: Formal JVM code analysis in JavaFAN. In: Rattray, C., Maharaj, S., Shankland, C. (eds.) AMAST 2004. LNCS, vol. 3116, pp. 132–147. Springer, Heidelberg (2004)
12. Ferrara, P.: A generic static analyzer for multithreaded java programs. Softw., Pract. Exper. 43(6), 663–684 (2013)
13. Field, J., Tip, F.: Dynamic dependence in term rewriting systems and its application to program slicing. Inf. Softw. Technol. 40(11–12), 609–636 (1998)
14. Hills, M., Rosu, G.: A rewriting logic semantics approach to modular program analysis. In: Proceedings of the International Conference on Rewriting Techniques and Applications, RTA 2010, LIPIcs, vol. 6, pp. 151–160. (2010)
15. Hind, M., Pioli, A.: Evaluating the effectiveness of pointer alias analyses. Sci. Comput. Program. 39(1), 31–55 (2001)
16. Huet, G.P., Lévy, J.: Computations in orthogonal rewriting systems, I. In: Computational Logic - Essays in Honor of Alan Robinson, pp. 395–414. (1991)
17. Klop, J.W.: Term rewriting systems from Church-Rosser to Knuth-Bendix and beyond. In: Paterson, M.S. (ed.) Automata, Languages and Programming. LNCS, vol. 443, pp. 350–369. Springer, Heidelberg (1990)
18. Leroy, X., Blazy, S.: Formal verification of a C-like memory model and its uses for verifying program transformations. J. Autom. Reason. 41(1), 1–31 (2008)
19. Martí-Oliet, N., Meseguer, J.: Rewriting logic: roadmap and bibliography. Theor. Comput. Sci. 285(2), 121–154 (2002)
20. Meseguer, J.: Conditional rewriting logic as a unified model of concurrency. Theor. Comput. Sci. 96(1), 73–155 (1992)
21. Meseguer, J., Rosu, G.: The rewriting logic semantics project. Theor. Comput. Sci. 373(3), 213–237 (2007)
22. Pierce, B.C.: Types and Programming Languages. MIT Press, London (2002)
23. Ramalingam, G.: The undecidability of aliasing. ACM Trans. Program. Lang. Syst. 16(5), 1467–1471 (1994)
24. Regehr, J., Reid, A., Webb, K.: Eliminating stack overflow by abstract interpretation. In: Alur, R., Lee, I. (eds.) EMSOFT 2003. LNCS, vol. 2855, pp. 306–322. Springer, Heidelberg (2003)
25. Riesco, A.: Using semantics specified in maude to generate test cases. In: Roychoudhury, A., D'Souza, M. (eds.) ICTAC 2012. LNCS, vol. 7521, pp. 90–104. Springer, Heidelberg (2012)
26. Riesco, A., Asăvoae, I.M., Asăvoae, M.: A generic program slicing technique based on language definitions. In: Martí-Oliet, N., Palomino, M. (eds.) WADT 2012. LNCS, vol. 7841, pp. 248–264. Springer, Heidelberg (2013)
27. Rosu, G., Stefanescu, A.: Matching logic: a new program verification approach. In: Proceedings of the International Conference on Software Engineering, ICSE 2011, pp. 868–871. ACM (2011)

28. Sarkar, S., Sewell, P., Nardelli, F.Z., Owens, S., Ridge, T., Braibant, T., Myreen, M.O., Alglave, J.: The semantics of x86-cc multiprocessor machine code. In: Proceedings of the Symposium on Principles of Programming Languages, POPL 2009, pp. 379–391. ACM (2009)
29. Venkitaraman, R., Gupta, G.: Static program analysis of embedded executable assembly code. In: Proceedings of the International Conference on Compilers, Architecture, and Synthesis for Embedded Systems, CASES 2004, pp. 157–166. ACM (2004)

Smart Environments

SHIP - A Logic-Based Language and Tool to Program Smart Environments

Serge Autexier and Dieter Hutter[✉]

German Research Center for Artificial Intelligence (DFKI), Bremen, Germany
{serge.autexier,dieter.hutter}@dfki.de

Abstract. The increasing availability of smart objects demands for flexible mechanisms to orchestrate different types of these objects to smart environments. As smart objects are typically not aware of each other, an orchestrating platform has to manage common resources, to harmonize the individual behavior of the acting objects, and to combine their activities to an intelligent team work. This paper presents a corresponding framework to implement such an orchestrating platform. It provides a concurrent programming language representing states in Description Logics and state transitions as logical updates enabling deductive support to infer non-explicitly represented knowledge. It uses temporal logic to suspend execution of a process for a particular evolution of the global state that is specified by a LTL formula. Since a process can fork into subprocesses this provides a mechanism for runtime verification by splitting a process into a subprocess executing some critical program and another parallel subprocess monitoring the first one by waiting for the desired evolution of states specified in its LTL formula.

Keywords: Description logic · Programming paradigm · Smart environments

1 Introduction

Smart homes typically comprise various individual components acting almost autonomously. Control systems equipped with sensors and actuators are the classical paradigm to realize such components. Sensors capture more and more data of their environment. Originally used to monitor a device's internal states, their purpose have been diversified (e.g. measuring the health status of a human) reaching the vision of context-aware computing [11]. Actuators operate on these data and actively interfere with the environment. Improving the reasoning capabilities and allowing for connectivity between different systems results in the notion of smart objects behaving autonomously and in an intelligent way in their environment. Typically, these systems are based on different notions of abstractions depending on their application area which results in incompatible middleware infrastructures impeding a universal interoperability on a higher abstraction level [11]. An intelligent media equipment, for instance, runs independently of an intelligent management for indoor climate or for ambient light.

© Springer International Publishing Switzerland 2015
M. Falaschi (Ed.): LOPSTR 2015, LNCS 9527, pp. 313–328, 2015.
DOI: 10.1007/978-3-319-27436-2_19

While an intelligent wheelchair is able to navigate autonomously in a flat using its location sensors, two of them may block each other in a narrow passage because both are not even aware of each other.

To combine smart objects to an intelligent overall system, such objects would have to anticipate and prospectively react to the development in their environment as well as to negotiate and collaborate with other smart objects on common goals. However, such an approach would require that smart objects are aware of each other, their needs and their abilities, which rules out the use of already existing solutions or commercial of the shelf (COTS). Smart homes typically comprise a variety of different tasks that range from rather primitive ones, like detecting motions or remotely control lights, to more sophisticated ones, like to recognize gestures or autonomously drive robotic vehicles. Realizing intelligent behavior in such environments is a major task as it has to go all the way up from protocols to communicate on a low signal level to sophisticated services recognizing, mediating, and planing high level activities.

We propose to use a logic-based programming environment that allows one to easily orchestrate the various processes and smart objects. This environment maintains a consistent view on the overall system, monitors changes in the environment and coordinates the actions of the individual activities. We developed a corresponding programming language SHIP implemented in the SHIP-tool to support the development of programs orchestrating and monitoring heterogeneous processes. The SHIP-tool provides an implementation, simulation, and execution environment for ambient intelligent processes. Thus, it implements an inter-application adaptation to provide thorough task support by coordinating the execution of a distributed application across a set of heterogeneous processes and smart objects. While [2] presented real world ambient intelligence processes developed in the SHIP-tool and [1] described the methodology to model real world states in ontologies in order to allow for an efficient treatment of state changes, this paper defines the programming language and its semantics.

2   Running Example

In this section we introduce a simple running example to illustrate our logic formalism for modeling states (Sect. 3.1) of the real world and also the primitives of the programming language (Sect. 5).

We consider a smart building environment where, among controllable lights, doors, elevators *etc.*, autonomously driving electric wheelchairs operate. The wheelchairs are requested by inhabitants that need a lift to another room or the delivering of some items. Wheelchairs operate along an internal route graph. Vertices denote locations in the environment and edges connect adjacent locations. A wheelchair, requested to drive to a target position, will compute its own path based on this route graph. To orchestrate several wheelchairs operating in an environment, a wheelchair has to communicate its computed route before starting the journey and each reached location during the journey to an orchestrating platform. In-between two locations the wheelchair can deviate from the straight line. Wheelchairs can enter or leave the scope of the platform: e.g. they are switched off for maintenance like recharging the batteries or a person can leave the environment sitting in a wheelchair.

This setting allows for a broad range of assistance processes, such as executing transportation requests, scheduling conflict-free rides of the wheelchairs, assisting individual wheelchairs during their rides by opening doors and turning on lights as needed, supervising the wheelchairs to detect misbehaviors and reacting on it, and many more. As our running example we consider an assistance process represented by (WhChAssist), which interfaces with the real wheelchairs and the scheduler organizing the conflict-free rides, and a monitoring process (MonitorWhChRoute) detecting and reacting on deviations from assigned routes. Moreover, we will illustrate a process (WhChSupervision) reacting on wheelchairs coming and going as well as managing assistance and supervision.

3   Specification and Representation of States

In our programming environment, states are represented in Description Logics (DL). As usual in DL, a state comprises a TBox T and a RBox R (terminological box) describing the properties of the real world as an abstract specification and an ABox A (assertional box) stating the properties of concrete individuals of the specified concepts. All together, they form an ontology $O = \langle T, R, A \rangle$.

The states are modeled as ontologies based on the description logic SROIQ [8], which supports role composition (R), transitive roles (S), inverse roles (I), qualified number restrictions (Q) and nominals concepts build from individuals (O). While full-fledged SROIQ can be used for queries, only its SRIQ fragment is used to define the ontology. For SROIQ, concepts are formed from the predefined atomic concepts top T, bottom F and arbitrary concept names different from T and F, as well as complex concept expressions formed accordingly by union $\sqcup$ and intersection $\sqcap$ of concepts, and role-based existential $\exists r$. C, universal $\forall r$. C, and number restrictions $\{>, =, <\}r$. C, where r is either a role name or the inverse of a role name r^{-1}.

The properties of the concepts are declared by concept inclusion axioms C $\sqsubseteq$ D or by concept definition C = D. Disjointness of concepts is declared by Disjoint. A role r : $D \times R$ is declared by indicating its domain D and range R. Similar to concepts, subroles can be declared by limited complex role inclusion axioms $r_0 \cdot \ldots \cdot r_n \sqsubseteq r$ where $r_i \in R$ and '·' denotes role composition. SHIP supports the standard role properties Sym, Asym, Trans, Ref, Irref, Fun, and FunInv. Furthermore, role names can be defined as the composition of roles $r = r_0 \cdot \ldots \cdot r_n$ or as the reflexive, transitive closure of another role r = r_0*. This is not expressible in description logic and it is translated to $r_0 \sqsubseteq r$, Trans(r), Ref(r) when translated to DL. However, it is important as a meta-property to ensure that nothing else than the transitive closure of r_0 is in r. r is functional iff $\forall (a_1, b_1), (a_2, b_2) \in r.a_1 = a_2 \implies b_1 = b_2$. A role name r directly depends on a role name r' if r' syntactically occurs in the definition r = R of r. A role name r depends on a role name r', if r directly depends on r' or there exists a role name r'' which depends on r' and r directly depends on r'. We denote the dependency relation between role names by $>_r$. Furthermore, we require the induced dependency relation on concept and role names to be irreflexive (i.e. we consider acyclic TBoxes and RBoxes modulo the transitivity of roles).

In addition we provide abbreviations to ease the definition of ontologies in a style inspired by abstract datatypes. For instance, Routes being sequences of Positions can be defined in the following way

```
Route ::= EmptyRoute | NonEmptyRoute(route_next:Position,route_rest:Route)
```

being internally expanded to

```
Route = EmptyRoute ⊔ NonEmptyRoute        route_next:NonEmptyRoute × Position
Disjoint(EmptyRoute, NonEmptyRoute)        Fun(route_next)
NonEmptyRoute                              route_rest:NonEmptyRoute × Route
    ⊑ (∃route_next . Position)             Fun(route_rest)
       ⊓ (∃route_rest . Route)
```

3.1 Representation of States

We interpret the construction of complex concepts similar to specifications of abstract datatypes. An existential quantification as in IDObject⊑ ∃at.AbstPosition specifies a mandatory attribute "at" of "type" AbstPosition in instances of the abstract datatype IDObject. Intersection of concepts combines the attributes of the sub-datatypes while union of concepts resembles the notion of variants (cf. [1] for details). Similar to an initialization of all records in new instances of a datatype we want to enforce a constructive definition of each individual of a complex concept in an ABox. For instance, having an individual d for the concept IDObject above, d must have a position and we demand that we can always deduce the actual position of d from the ABox. Furthermore, since we allow for disjunction of concepts D ⊑ E ⊔ F we also want to know for each individual in D whether it belongs to E or F or both. That means that for any individual d of a (complex) concept the ABox always provides the full information about the composition and settings of the individual. In other words, the ABox provides the individuals necessary to name the values of the various attributes and there is no need to invent new values by introducing Skolem functions.

The same rigor of constructiveness is applied to the specification of roles. SHIP allows for the definition of composed roles, e.g. by defining $r = r_1 \cdot r_2$. Knowing that two individuals a, b are in a role r there must be some individual c such that $(a, c) : r_1$ and $(c, b) : r_2$ holds. We demand that also this witness is specified explicitly, i.e. the ABox contains some individual c and the necessary relations between c and the individuals a and b.

An ontology is *constructive* if it fulfills the described constructiveness properties and refer to [1] for details.

3.2   ABox-Queries

Having an ontological state representation we can query a state for facts derivable from the ontology. Concept queries are built from the standard atomic concepts as before but also allow for nominal concepts formed from sets of individuals as additional atomic concepts; hence we are in the SROIQ fragment. The primitive queries are ABox-queries possibly containing variables for individuals and in addition !i to query whether the individual i exists in the ontology.

4   Updates

4.1   ABox Update

An update is a pair (α, δ), where α is a consistent set of primitive ABox-assertions to be added and δ primitive ABox-assertions to be removed. To guarantee the constructiveness of updates we enforce that they are specified in a non-redundant form by ABox-assertions exclusively over primitive concepts and roles (cf. [1] for details). For a given ontology $O = \langle T, R, A \rangle$ and consistent primitive update (α, δ) the new ABox is determined from the old ABox by

1. If $(\text{i} : \text{C}) \in \alpha$ and $(\text{i} : \text{D}) \in A$ and C and D are disjoint concepts, then $(\text{i} : \text{D})$ is removed;
2. If $((\text{a}, \text{b}) : \text{r}) \in \alpha$, $((\text{a}, \text{c}) : \text{r}) \in A$ and r is functional, then $(\text{a}, \text{c}) : \text{r}$ is removed.

Finally, all assertions from δ are removed. Formally, the result $upd(\alpha, \delta)$ of an update is defined by

$$A' := \alpha \cup (A \setminus (\delta \cup \{(\text{i} : \text{D}) \in A \mid (\text{i} : \text{C}) \in \alpha, \text{C and D are disjoint}\}$$
$$\cup \{((\text{a}, \text{c}) : \text{r}) \in A \mid ((\text{a}, \text{b}) : \text{r}) \in \alpha, \text{r is functional}\})$$

The resulting ontology $\langle T, R, A' \rangle$ may well be inconsistent, for instance if number restrictions are violated. If so, the update is refused and we stick to the previous ontology O. If an action triggered the update, the action fails in the process semantics. If the environment triggered the update, respective repair processes must have been specified to synchronize the SHIP-tool and the environment. If the ontology is consistent, it is not necessarily constructive. To this end we use a procedure to check if the resulting ontology $\langle T, R, A' \rangle$ is constructive. If not, the SHIP-tool can provide information about missing itemization of the ABox. This can be used to statically analyse the effects of actions of the defined processes whether they only contain primitive ABox-assertions and if they are complete enough to preserve constructiveness of the ontology.

Actions (**pre**, **eff**) represent updates initiated in processes. An action is applicable on an ontology O if all its preconditions are satisfied in O. In this case, O is updated to a new ontology O' by applying the effects of the action, including conditional effects which conditions hold in O. If O' is inconsistent, then the action fails and we keep O as the current ontology. Otherwise the action succeeds and O' is the new current ontology. If the preconditions are not satisfied, the action stutters, i.e., waits until it gets applicable. skip is the action which is always applicable and does not modify the ontology.

Using free variables in the effects, new individuals can be added to O, where a fresh name is created at run-time. By annotating a variable with the keyword delete, individuals can be removed from the ontology.

In our running example, there are actions for wheelchairs that are new in the ontology, for instance because the real wheelchair entered the building or was turned on again. For those we need to distinguish if they are under supervision or not and do so by having disjoint concepts Supervised and Unsupervised in the ontology. When a wheelchair gets active again, a new individual w is introduced into the ontology belonging to the concepts WheelChair and Unsupervised. This is caused by the real world (sensors) sending the following update when detecting a new wheelchair: $(\{w:WheelChair, w:Unsupervised\}, \emptyset)$ As soon as a supervision process is started for this wheelchair, we have to bookmark this by calling an action from inside the program. In the SHIP-language actions are parameterized over the name of the individuals. To toggle the supervision status using conditional effects, we define the action for bookmarking as follows:

```
action toggleSupervisionStatus (w) {
  pre = w:WheelChair
  if (w:Unsupervised) w:Supervised
  if (w:Supervised) w:Unsupervised }
```

4.2   Monitor Progression

Monitors constitute predictions on the development of the actual state in the future. They are specified as LTL formulas over ABox-queries (Sect. 3.2), existential/universal quantification over individuals and the classical temporal modalities F (eventually), G (generally), and U (until). Note that X (next) is not included because it is inappropriate for parallel, but interleaved processes.

In our running example we need to monitor the real world to detect when we need to compute a new overall schedule of the rides of all currently operating wheelchairs. This is the case when a currently non-driving wheelchair wants to drive or a currently driving wheelchair gets new driving instructions after having completed its current ride. In SHIP we can express this by the following LTL-formula

$$
\varphi = \left(
\begin{array}{l}
(\exists w:\text{WhChWithEmptyRoute . } F(w:\text{WhChNonEmptyRoute})) \\
\text{or } (\exists w:\text{WhChWithNonEmptyRoute .} \\
\quad F(w:\text{WhChEmptyRoute and } F(w:\text{WhChNonEmptyRoute})))
\end{array}
\right)
$$

Thus, a monitor formula ϕ for a state σ makes demands to the state σ and predicts conditions on its follow-up states. Since the monitor formula represents the conditions on the development of states relative to the actual state, it has to be modified once the system advances to the next state. Each successful ABox-update results in an update of all active monitors. Since a monitor formula predicts the development of the global state in time, the formula changes in each state transition. The *progression* $\Pi(\phi, \sigma)$ (cf. [6]) of a formula propagates its demands to the next state: it checks the demands for the actual state σ and returns a formula specifying the demands for the successor state σ' of σ and the corresponding follow-up states of σ'.

$\Pi(\phi, \sigma) = \text{True}$ [False, resp.] if ϕ is free of temporal modalities $\wedge\, \sigma \models \phi$ [$\sigma \not\models \phi$, resp.]

$\Pi(\phi_1 \text{ and } \phi_2, \sigma) = \Pi(\phi_1, \sigma) \text{ and } \Pi(\phi_2, \sigma)$

$\Pi(\text{G}\phi, \sigma) = \Pi(\phi, \sigma) \text{ and } \text{G}\phi$

$\Pi(\phi_1 \text{ or } \phi_2, \sigma) = \Pi(\phi_1, \sigma) \text{ or } \Pi(\phi_2, \sigma)$

$\Pi(\text{F}\phi, \sigma) = \Pi(\phi, \sigma) \text{ or } \text{F}\phi$

$\Pi(\forall/\exists x : C.\ \phi, \sigma) = \Pi(\phi[x \leftarrow c_1], \sigma) \text{ and/or} \ldots \text{and/or } \Pi(\phi[x \leftarrow c_n], \sigma)$
    if$\{c_1 \ldots c_n\} = \{c \mid \sigma \models c : C\}$

$\Pi(\phi_1 \text{ U } \phi_2, \sigma) = \Pi(\phi_2, \sigma) \text{ or } (\Pi(\phi_1, \sigma) \text{ and } \phi_1 \text{ U } \phi_2)$

For sake of readability, we omitted Boolean simplification rules to simplify the result of Π, but assume, for instance, that Π returns True instead of $\text{True or } \text{F}(\phi)$.

Suppose there are two wheelchairs r1 and r2 in an initial state σ_0 of our running example. The first has an empty route while the second has a non-empty route. Initializing φ in σ_0 we obtain

$$\varphi_0 = \Pi(\varphi, \sigma_0) = \quad \begin{matrix} \texttt{F(r1:WhChNonEmptyRoute) or} \\ \texttt{F(r2:WhChEmptyRoute and F(r2:WhChNonEmptyRoute)))} \end{matrix}$$

Suppose, the situation is updated to a state σ_1 where r2 has now an empty route, then the progression of φ_0 is

$$\varphi_1 = \Pi(\varphi_0, \sigma_1) = \quad \texttt{F(r1:WhChNonEmptyRoute) or F(r2:WhChNonEmptyRoute)))}$$

For any further update where r1 and r2 keep an empty route, the progression leaves the formula invariant. For instance, as soon as r1 gets a non-empty route in some σ_n the progression results in

$$\varphi_n = \Pi(\varphi_1, \sigma_n) = \quad \texttt{True}$$

indicating that the observed trace of world models $\sigma_0, \ldots, \sigma_n$ satisfied the LTL-formula φ.

5 Programming

Programs are constructed on top of actions (representing ABox-updates) and monitor activations (observing the environment) with the help of the process combinators described in Table 1. The last two rows describe syntactic sugar defined in terms of the upper process combinator primitives.

action, proc and monitor allow for the definition of macros for updates, program fragments and monitor formulas. The formal as well as the actual parameters range over individuals such that the execution of a call (of an action, a proc or a monitor) in a program c will simply substitute this call in c by the

Table 1. Process combinators

Name	Syntax	Semantics
complex condition	switch case c_1 => p_1 ... case c_n => p_n	branches according to the specified cases which are checked in order. _ can be used as default case, then, the condition never stutters
iteration	p*	applies p until it fails, always succeeds
sequence	p ; q	applies p then q
monitor start	init m	starts the monitor, continues when the monitor succeeds or fails when the monitor fails
guarded execution	try p catch q	executes p; if p fails, q is executed, but the modifications of p are kept
parallel non-strict	p ⟨|⟩ q	executes p and q in parallel (interleaved), terminates when both p and q terminate, fails only if both fail
bounded parallel non-strict	forall c => p	executes p for all instances matching c in parallel, terminates when all instances terminate, fails only if all fail
parallel strict	p ⋈ q	executes p and q in parallel (interleaved), terminates when both p and q terminate, fails when one of them fails
bounded parallel strict	foralls c => p	executes p for all instances matching c in parallel, terminates when all terminate, fails when one of them fails
	some c => p	switch case c => p
	let x = i.r in p	switch case (x,i):r => p

body of the macro in which the formal parameters have been substituted by the corresponding actual parameters of the call.

We illustrate the process languages with sample processes from our running examples before providing a precise small-step semantics in the next section.

The main process responsible to initiate a supervision for each new wheelchair is encoded in the procedure WhChSupervision:

```
proc WhChSupervision () {
  some w:(WheelChair ⊓ Unsupervised) =>SuperviseWhCh(w) ⟨|⟩ WhChSupervision}
```

It is a parameter-free recursive procedure waiting until we can derive from the actual ontology there is an unsupervised wheelchair. It then starts the supervision process for it (SuperviseWhCh) and in parallel (non-strict) recursively calls itself. The non-strict parallel process combinator ensures that a failure of the supervision process does not cause an abortion of the main process and vice-versa.

The supervision process `SuperviseWhCh` is defined as follows:

```
proc SuperviseWhCh(w) {
  toggleSupervisionStatus(w);
  WhChAssist(w) )( MonitorWhChRoute(w) )( init G(!w) }
```

First, it invokes the action `toggleSupervisionStatus` (cf. Sect. 4.1) marking the wheelchair as supervised. Then it invokes three processes in parallel with the parallel strict process combinator ensuring that all process fail if one of them does. The first process `WhChAssist` mediates between the wheelchair and the route scheduler only allowing the wheelchair to drive on those segments of the route that the scheduler has already cleared for it. The second process `MonitorWhChRoute` monitors if the wheelchair behaves as assumed and follows the travelling directives. The third process is a guard process checking that the wheelchair is still active, i.e. has no failure (e.g., break down, no power) or does not leave the building. In this case the individual representing the wheelchair is removed from the ontology and the global invariant postulating the existence of the individual (`G(!w)`) fails. In this case the monitor fails with an exception. Because of the semantics of strict parallel operators this will cause the interruption of both other processes and finally the stop of the whole supervision process of this wheelchair. However, the main `WhChSupervision` is not interrupted due to the non-strict parallel operator used there.

The procedure `MonitorWhChRoute` oversees the wheelchairs behavior and acts as a kind of run-time verifier which reacts on failures:

```
proc MonitorWhChRoute (w) {
  some w:WhChNonEmptyRoute and (w,n):nextposition and (w,f):routefinalpos
    => try { init MonitorWhChTransition(w) }
      catch {
        switch
        case not(!w) => skip
        case !w =>
            requestWhChTo(w,n);
            try { (MonitorWhChRoute (w) )( init G(not (w,n):at))}
            catch { skip };
            requestWheelChTo(w,f) };
      MonitorWhChRoute (w) }
```

Initially it waits for a wheelchair receiving a non-empty route. Then, it initializes an LTL-formula checking that the wheelchair reports only from designated positions of its specified route.

```
monitor MonitorWhChTransition (w) {
  ∃currentpos:(∃inv(at) . { w }) .
  ∃nextpos:(∃inv(nextposition).{w}) . (w,currentpos):at U (w,nextpos):at}
```

Here we select the current position and the next position of the wheelchair by querying the ontology and initalize the monitoring checking that the wheelchair still reported to be at the starting position until it reaches the next position.

The process `MonitorWhChRoute` observes this behavior. If it could be observed successfully, it recurses invoking the observation for the next segment provided that the route is still not empty. If the postulated behavior has been violated, i.e. the successive formula progression of the formula results in `False`, a failure is raised, which is catched by the surrounding `try-catch` and the catch process-block is invoked to react on it.

If the wheelchair is still there, we issue an explicit request to send the wheelchair back on track. This is monitored by a recursive call of the monitoring process accompanied (strictly) by the invariant that the wheelchair is not back on track. The effect of the invariant is to interrupt the monitoring process as soon as the wheelchair is back on track. After that a new request is issued sending the wheelchair to its original destination and the monitoring is resumed.

5.1   Small-Step Semantics

We now provide a small-step semantics for this programming language on top of ABoxes. We interpret programs as terms representing their abstract syntax trees. For the sake of readability, we write these terms using the concrete syntax.

The *execution* of a program c with respect to a state σ (represented by an ABox) is a sequence $(c, \sigma) = (c_1, \sigma_1) \to \ldots \to (c_n, \sigma_n)$ of pairs (c_i, σ_i) with σ_i being the actual state in step i and c_i the continuation of c still to be done in state i. In general, each c_i represents the continuation of a bundle of interleaved processes arising from forking existing processes with the help of $\bowtie$ and $\langle\!\vert\!\rangle$ combinators. We deal with this concurrency using an interleaving semantics. Hence, given a continuation c_i there are in general various positions in c_i where we can continue the computation. These *evaluation positions* are defined with the help of term access functions. As usual, $\langle\rangle$ denotes the identity, i.e. $t|_{\langle\rangle} = t$ while $\pi \cdot i$ selects the i-th argument of the term accessed by π, i.e. $t|_{\pi \cdot i} = t_i$ if $t|_\pi = f(t_1 \ldots t_n)$. Analogously, $t[\pi \leftarrow t']$ denotes the term emerging from t by replacing the subterm $t|_\pi$ by t'.

Given a continuation c, the set of *evaluable positions* of c is defined as the smallest set $eval(c)$ satisfying the following conditions:

$$\langle\rangle \in eval(c)$$
$$\pi \cdot 1 \in eval(c) \quad \text{if } \pi \in eval(c) \wedge \exists p, q.(c|\pi = p; q \vee c|\pi = \textbf{try } p \textbf{ catch } q),$$
$$\pi \cdot 1, \pi \cdot 2 \in eval(c) \quad \text{if } \pi \in eval(c) \wedge \exists p, q.(c|\pi = p \bowtie q \vee c|\pi = p \langle\!\vert\!\rangle q)$$

We assume that each continuation c is in some normal form obtained by applying the following set of simplification rules exhaustively before each computation step ($\top$ denotes the empty program while $\bot$ refers to a failure of execution):

$$\top; p \Rightarrow p, \quad \bot; p \Rightarrow \bot, \quad \textbf{try } \top \textbf{ catch } p \Rightarrow \top, \quad \textbf{try } \bot \textbf{ catch } p \Rightarrow p,$$
$$\top* \Rightarrow \top, \quad \bot* \Rightarrow \top, \quad \top \bowtie \top \Rightarrow \top, \quad \bot \bowtie p \Rightarrow \bot, \quad p \bowtie \bot \Rightarrow \bot,$$
$$\top \langle\!\vert\!\rangle \top \Rightarrow \top, \quad \top \langle\!\vert\!\rangle \bot \Rightarrow \top, \quad \bot \langle\!\vert\!\rangle \top \Rightarrow \top, \quad \bot \langle\!\vert\!\rangle \bot \Rightarrow \bot$$

These rewriting rules reflect the propagation of $\top$ and $\bot$ inside continuations. E.g. the failure of the execution of the first element p in a sequence p, q results also in a failure of the entire sequence. An iteration $p*$ returns always $\top$. $\bowtie$ and $\langle\!\vert\!\rangle$ denote a fork of a process in subprocesses. While $\bowtie$ binds both subprocesses as the failure of one subprocess causes the failure (termination) of the other, $\langle\!\!|\rangle$ executes both subprocesses independently.

Obviously, the set of these simplification rules is noetherian and locally confluent and thus guarantees the existence of a normal form $c \downarrow$ for all programs c. Additionally, we apply $p* \Rightarrow$ try $p; p * $ catch $\top$ if $p*$ occurs at an evaluable position. Notice that applying this rule in an evaluable position, the resulting position of $p*$ in try $p; p *$ catch $\top$ is never an evaluable one which prevents us from infinite rewriting. Additionally, we open up macro definitions for calls to action, proc and monitor in evaluable positions, as already mentioned above.

In the following, we present the evaluation rules of the small-step semantics. We start with the rules (A1 – A3) for applying an action (pre, eff), which allow us to update the actual state by eff provided that pre holds. Notice that the precondition pre of an action establishes a proof obligation in Description Logic. We use a DL-reasoner like Pellet [13] to find a deducible instance $\rho(\mathtt{pre})$ or to refute the precondition. The execution of an action (i.e. the corresponding process) stutters if its precondition can be neither proved nor refuted.

$$(\sigma, (\mathtt{pre}, \mathtt{eff})) \rightarrow (\sigma', \top) \text{ if } \exists \rho.\ \sigma \models \rho(\mathtt{pre}) \wedge upd(\sigma, \rho(\mathtt{eff})) = \sigma' \quad \text{(A1)}$$
$$(\sigma, (\mathtt{pre}, \mathtt{eff})) \rightarrow (\sigma, \bot) \text{ if } \exists \rho.\ \sigma \models \rho(\mathtt{pre}) \wedge upd(\sigma, \rho(\mathtt{eff})) = \bot \quad \text{(A2)}$$
$$(\sigma, (\mathtt{pre}, \mathtt{eff})) \rightarrow (\sigma, \bot) \text{ if } \forall \rho.\ \sigma \cup \rho(\mathtt{pre}) \models \bot \quad \text{(A3)}$$

The next rules (S1 – S4) define the non-trivial cases for conditions (switch) and process forking ($\bowtie$, $\langle\!\!|\rangle$). Similar to action rules, the evaluation of switch stutters if no case of its cases is applicable and no default operation is specified.

$$(\sigma, \mathtt{switch}\ \mathtt{c_1}\texttt{=>}\ \mathbf{p}_1\ \dots\ \mathtt{c_n}\texttt{=>}\ \mathbf{p}_n\quad \texttt{-=>}\ \mathbf{p}) \rightarrow (\sigma, \rho(\mathbf{p}_j))$$
$$\text{if } \exists j \le n. \exists \rho.\ \sigma \models \rho(\mathbf{c}_j) \wedge \forall i < j.\ \forall \rho.\ \sigma \not\models \rho(\mathbf{c}_i) \quad \text{(S1)}$$
$$(\sigma, \mathtt{switch}\ \mathtt{c_1}\texttt{=>}\ \mathbf{p}_1\ \dots\ \mathtt{c_n}\texttt{=>}\ \mathbf{p}_n\quad \texttt{-=>}\ \mathbf{p}) \rightarrow (\sigma, \mathbf{p})$$
$$\text{if } \forall i \le n. \sigma \not\models \mathbf{c}_i \quad \text{(S2)}$$
$$(\sigma, \mathtt{foralls}\ \mathtt{c} \texttt{=>} \mathbf{p}) \rightarrow (\sigma, \rho_1(\mathbf{p})\ \bowtie\ \dots\ \bowtie\ \rho_n(\mathbf{p}))$$
$$\text{if } \{\rho_1, \dots \rho_n\} = \{\rho\,|\,\sigma \models \rho(\mathbf{c})\} \quad \text{(S3)}$$
$$(\sigma, \mathtt{forall}\ \mathtt{c} \texttt{=>} \mathbf{p}) \rightarrow (\sigma, \rho_1(\mathbf{p})\ \langle\!\!|\rangle\ \dots\ \langle\!\!|\rangle\ \rho_n(\mathbf{p}))$$
$$\text{if } \{\rho_1, \dots \rho_n\} = \{\rho\,|\,\sigma \models \rho(\mathbf{c})\} \quad \text{(S4)}$$

Rules (M1 – M3) are concerned with starting and finishing an LTL-monitor.

$$(\sigma, <\mathtt{init}\ \psi>) \rightarrow (\sigma, \top) \quad \text{if } \Pi(\psi, \sigma) = \mathtt{True} \quad \text{(M1)}$$
$$(\sigma, <\mathtt{init}\ \psi>) \rightarrow (\sigma, \bot) \quad \text{if } \Pi(\psi, \sigma) = \mathtt{False} \quad \text{(M2)}$$
$$(\sigma, <\mathtt{init}\ \psi>) \rightarrow (\sigma, <\mathtt{active}\ \psi>) \text{ else.} \quad \text{(M3)}$$

In general a monitor formula denotes conditions to the actual state and its successor states. If the actual state σ fails to satisfy the corresponding conditions for σ already, the activation of the monitor fails. If the monitor only formulates conditions on σ and these conditions are satisfied, then the monitor stops successfully returning $\top$. Otherwise the monitor is activated to observe the future development of states. Next, we define 'tick' that updates all active monitors in the continuation. Monitors are only active if they occur at evaluation positions.

$\text{tick}(\sigma, <\text{active } \psi>) = \top \, [\bot] \text{ if } \Pi(\psi, \sigma) = \text{True } [\text{False}] \text{ and}$

$\text{tick}(\sigma, <\text{active } \psi>) = <\text{active } \Pi(\psi, \sigma)> \text{ else,}$

$\text{tick}(\sigma, p; q) = \text{tick}(\sigma, p); q \quad \text{and} \quad \text{tick}(\sigma, \text{try } p \text{ catch } q) = \text{try } \text{tick}(\sigma, p) \text{ catch } q,$

$\text{tick}(\sigma, p \, [\![\, q) = \text{tick}(\sigma, p) \, [\![\, \text{tick}(\sigma, q) \quad \text{and} \quad \text{tick}(\sigma, p \, \langle\!|\!\rangle \, q) = \text{tick}(\sigma, p) \, \langle\!|\!\rangle \, \text{tick}(\sigma, q)$

Summing up, we define an execution step in SHIP as follows: Let π a position in a program c in normal form. A rule $R \in \{A1 - A3, M1 - M3, S1 - S4\}$ is *applicable* to a pair (σ, c) at position π iff π is an evaluable position and there is some (σ', c') such that $(\sigma, c|\pi) \rightarrow (\sigma', c'|\pi)$ is an instance of R and $c' = c[\pi \leftarrow c'|\pi]$.

The *result* of the rule application is the pair $(\sigma', \text{tick}(\sigma', c') \downarrow)$ if $R = (A1)$ and $(\sigma, c' \downarrow)$ otherwise.

We illustrate the small-step semantics in our simple scenario. We assume an initial situation, in which no wheelchair is active. Starting the supervision process WhChSupervision, the small-step semantics expands the process body and then stops with the following expression as there is no unsupervised wheelchair around:

```
some w:(WheelChair ⊓ Unsupervised) => SuperviseWhCh(w) ⟨|⟩ WhChSupervision
```

I.e., the process stutters and any update from the real world that does not introduce an unsupervised wheelchair leaves it as it is. Assume a wheelchair r1 gets active, e.g. because it has finished charging its batteries: it triggers the following update to the ontology from the real world ({r1:WheelChair, r1:Unsupervised, (r1, Charger):at},∅). Now, the query w:(WheelChair ⊓ Unsupervised) has an instance r1 and the expression reduces first to

```
SuperviseWhCh(r1) ⟨|⟩ WhChSupervision
```

and further by expansion of the macro bodies we obtain

```
{ toggleSupervisionStatus(r1);
  WhChAssist(r1) [X] MonitorWhChRoute(r1) [X] init G(!r1) }
⟨|⟩ {some w:(WheelChair⊓ Unsupervised)=>SuperviseWhCh(w)⟨|⟩WhChSupervision}
```

Next the action toggleSupervisionStatus is applied which applies the update ({r1:Supervised}, {r1:Unsupervised})) to the ontology. The expression is evaluated further by expanding the macro definitions of WhChAssist and MonitorWhChRoute and activating the monitor init G(!r1). For sake of readability we focus on the evaluation of the second one and ignore for now the expansion and further evaluation of WhChAssist.

```
{ WhChAssist(r1) [X]
  { some r1:WhChNonEmptyRoute and (r1,n):nextposition
         and (r1,f):routefinalpos
    => try { init MonitorWhChTransition(r1) }
       catch {
```

```
                  switch
                  case not(!r1) => skip
                  case !r1 => requestWhChTo(r1,n);
                            (MonitorWhChRoute (r1) ⋈ init G(not (r1,n):at));
                            requestWhChTo(r1,f) };
              MonitorWhChRoute (w) }
        ⋈ active G(!r1) }
⟨⎪⟩ {some w:(WheelChair⊓ Unsupervised)=>SuperviseWhCh(w)⟨⎪⟩ WhChSupervision}
```

At this stage the system stutters until r1 gets a non-empty route. Suppose, a second wheelchair gets active, then the lower part activates another instance analogously to the upper part. Assume, r1 gets a non-empty route to the Kitchen with first target position p1, the expression evaluates further to

```
{ WhChAssist(r1) ⋈
  { try { active (r1,Charger):at U ((r1,p1):at) }
    catch {
    switch
    case not(!w) => skip
    case !w => requestWhChTo(w,n);
              (MonitorWhChRoute (w) ⋈ init G(not (w,n):at));
              requestWhChTo(w,f) };
    MonitorWhChRoute (w) }
  ⋈ active G(!r1) }
⟨⎪⟩ {some w:(WheelChair⊓ Unsupervised)=>SuperviseWhCh(w)⟨⎪⟩ WhChSupervision}
```

Next assume that the wheelchair gets offline due to some mechanical problem, which triggers the update $(\emptyset, \{delete(r1)\})$. It entails the deletion of r1:WheelChair, r1:Supervised, and (r1,Charger):at. The formula progression of active (r1,Charger):at U ((r1,p1):at) reduces to $\bot$, which further reduces the try-expression to the catch part:

```
switch
case not(!r1) => skip
case !r1 => requestWhChTo(r1,p1);
          (MonitorWhChRoute (r1) ⋈ init G(not (r1,p1):at));
          requestWhChTo(r1,Kitchen)
```

Here the first case of the switch-expression evaluates to True causing the execution of skip and reducing that whole expression to MonitorWhChRoute (r1), which gets expanded but stops then on some r1:WhChNonEmptyRoute and (r1,n):nextposition and (r1,f):routefinalpos => .... Furthermore the formula progression for the active monitor active G(!r1) is executed which reduces it to False. This in turn makes the monitoring fail as well as all strictly parallel process, which yields the new process expression

```
some w:(WheelChair ⊓ Unsupervised) => SuperviseWhCh(w) ⟨⎪⟩ WhChSupervision
```

6   Improvements

The language, described so far, provides only updates for the specification of actions. However, in practice some types of updates cannot be stated in a declarative manner, but need to be computed. For instance, consider our running example. If a wheelchair is ordered to a specific location, the route must be computed depending on the route graph resided in the ontology. However, routes and their computation can be more efficiently represented as abstract datatypes (e.g. formalizing the "closest" neighbor of a location is impossible in DL). In Sect. 3 we already emphasized the similarities between constructive ontologies and abstract datatypes. We utilize this similarity to relate parts of the ontology to serializations of datatype instances in a programming language. An example are the declarations of freely generated datatypes using *constructors*. For basic datatypes like sets, lists or maps this can be done uniformly.

These datatypes in programming languages like ML, Haskell or Scala can be automatically serialized into an ontological representation and vice versa. For serialization, the (pointer) structure of a datatype instance can be represented by introducing individuals in the ontology.

For instance, a route in a route graph is a list of positions. The computation of the route takes the current route of a wheelchair as argument, its current position, the target position and the route graph and returns a new route for the wheelchair. To mimic the rewriting of the wheelchair's route to the new route, the subtree of the old route must be replaced by the subtree for the new route. Technically, this can be handled by reusing the same individual at the root of the tree (i.e., keeping the pointer), deleting all individuals in the subtree of the old route and inserting fresh individuals to represent the new subtree. As SHIP is implemented in Scala, we implemented the described connection between the ontological world and Scala datatypes. We extended the action declarations to accommodate the (Scala-)computation of updates as illustrated by

```
action computePlan = {
 pre = r:PlanRequest, (r,w):planrequest_wheelchair,
       (r,src):planrequest_source, (r,trg):planrequest_target
       (w,oldroute):wheelchair_route, g:RouteGraph
 exec = RouteComputation.computeRoute(oldroute,src,trg,r,w,g)}
```

where the Scala-function `computeRoute` is used to update the value of `oldroute` according to the new requirements.

Evaluation. We evaluated the performance of the approach in practice on the described example. The ontology has 89 concepts, 50 roles, 198 TBox/RBox axioms. The ABox has in average 207 individuals and 665 axioms. The ontology updates take in average of 0.37 s (min 0.02 s, max 0.69 s), which is acceptable. Still reaction times of the processes appear sometimes slow, because often there are independent actions in parallel processes that are execute in sequence although they could be executed simultaneously. This is an aspect where a static analysis could help to combine multiple small actions into one large update step.

7   Related Work and Conclusion

We presented a logic-based language to program smart environments featuring the ability to reason about actual states (using DL-reasoner) and also to smartly interleave active (performing updates) and passive (monitoring the environment) phases in a process. Execution monitoring has a long tradition in robotics (cf. [7,12] to detect discrepancies between expected and observed developments. In [5] LTL was used to restrict search in forward-chaining planning. As new states are generated they are incrementally checked against the goal formulated as an LTL formula, which is updated in each step with the help of formula progression. Since then, various approaches combining planning with LTL-monitoring have been proposed (e.g. [3,6,9]). Also logic programming languages have a long tradition, GOLOG [10] being one of the first. In [14] the explicit representation of the environment in a logic allows for the assessment of complex situations and an adaptive behavior. [4] introduces an action formalism on description logics. However to our knowledge, our approach is unique as it combines both features in a programming language controlling real environments: The SHIP-Tool [2], which is a prototypical implementation of an interpreter for the presented programming language, has been successfully tested by implementing intelligent assistance processes for the Bremen Ambient Assisted Living Lab [2].

References

1. Autexier, S., Hutter, D.: Constructive DL update and reasoning for modeling and executing the orchestration of heterogenous processes. In: 26th International Workshop on Description Logics (DL 2013). CEUR, vol. 1014 (2013)
2. Autexier, S., Hutter, D., Stahl, C.: An implementation, execution and simulation platform for processes in heterogeneous smart environments. In: Augusto, J.C., Wichert, R., Collier, R., Keyson, D., Salah, A.A., Tan, A.-H. (eds.) AmI 2013. LNCS, vol. 8309, pp. 3–18. Springer, Heidelberg (2013)
3. Baader, F.: Ontology-based monitoring of dynamic systems. In: 14th International Conference on Principles of Knowledge Representation and Reasoning (KR 2014). AAAI Press (2014)
4. Baader, F., Lutz, C., Milicic, M., Sattler, U., Wolter, F.: Integrating description logics and action formalisms: first results. In: 20th National Conference on Artificial Intelligence, AAAI 2005. AAAI Press (2005)
5. Bacchus, F., Kabanza, F.: Planning for temporally extended goals. Ann. Math. Artif. Intell. **22**, 5–27 (1998)
6. Bauer, A., Falcone, Y.: Decentralised LTL monitoring. Arxiv preprint arXiv:1111.5133 (2012)
7. Doherty, P., Kvarnström, J., Heintz, F.: A temporal logic-based planning and execution monitoring framework for unmanned aircraft systems. Auton. Agents Multi-Agent Syst. **19**(3), 332–377 (2009)
8. Horrocks, I., Kutz, O., Sattler, U.: The even more irresistible SROIQ. In: Knowledge Representation, KR 2006. AAAI Press (2006)
9. Lamine, K.B., Kabanza, F.: History checking of temporal fuzzy logic formulas for monitoring behavioral-based mobile robots. In: IEEE International Conference on Tools with Artificial Intelligence (2000)
10. Levesque, H.J., Reiter, R., Lesperance, Y., Lin, F., Scherl, R.B.: GOLOG: a logic programming language for dynamic domains. The Journal of Logic Programming **31**(1–3), 59–83 (1997)

11. Nakajima, T.: Case study of middleware infrastructure for ambient intelligence environments. In: Nakashima, H., Aghajan, H., Augusto, J.C. (eds.) Handbook of Ambient Intelligence and Smart Environments, pp. 229–256. Springer, Heidelberg (2010)
12. Pettersson, O.: Execution monitoring in robotics: a survey. Robot. Auton. Syst. **53**, 73–88 (2005)
13. Sirin, E., Parsia, B., Grau, B.C., Kalyanpur, A., Katz, Y.: Pellet: a practical owl-dl reasoner. J. Web Sem. **5**(2), 51–53 (2007)
14. Springer, T., Turhan, A.-Y.: Employing description logics in ambient intelligence for modeling and reasoning about complex situations. J. Ambient Intell. Smart Environ. **1**(3), 235–259 (2009)

Program Transformation

Correctness of Context-Moving Transformations for Term Rewriting Systems

Koichi Sato, Kentaro Kikuchi[(⊠)], Takahito Aoto, and Yoshihito Toyama

RIEC, Tohoku University, 2-1-1 Katahira, Aoba-ku, Sendai, Miyagi 980-8577, Japan
{koichi,kentaro,aoto,toyama}@nue.riec.tohoku.ac.jp

Abstract. Proofs by induction are often incompatible with functions in tail-recursive form as the accumulator changes in the course of unfolding the definitions. Context-moving and context-splitting (Giesl, 2000) for functional programs transform tail-recursive programs into non tail-recursive ones which are more suitable for proofs by induction and thus for verification. In this paper, we formulate context-moving and context-splitting transformations in the framework of term rewriting systems, and prove their correctness with respect to both eager evaluation semantics and initial algebra semantics under some conditions on the programs to be transformed. The conditions for the correctness with respect to initial algebra semantics can be checked by automated methods for inductive theorem proving developed in the field of term rewriting systems.

Keywords: Tail-recursion · Program transformation · Term rewriting system · Inductive theorem proving

1 Introduction

Proofs by induction are fundamental in software verification and thus dealt with by many automated theorem provers. An *inductive theorem* of a *term rewriting system* (*TRS* for short) is an equation valid in the initial algebra of the TRS. Inductive theorems correspond to the equations that can be shown by induction on the data structures, and various automated methods have been investigated for proving inductive validity of TRSs [1–3,7,9,10].

Recursive definition is a fundamental tool in various areas. A recursive definition of a function in which the body of the definition is a recursive call of the function itself (with different arguments, typically) is called *tail-recursive*. When evaluating a function call, if the function definition is given in a tail-recursive form, the environment of the function call does not need to be kept to deal with further computations that manipulate the results of its recursive calls. Thus, programs in which function definitions are given in tail-recursive forms are compiled into codes removing extra overheads in function calls. Thus, tail-recursive programs attain both efficiency and readability. However, proofs by induction are often incompatible with tail-recursive definitions, as can be seen in the following example.

© Springer International Publishing Switzerland 2015
M. Falaschi (Ed.): LOPSTR 2015, LNCS 9527, pp. 331–345, 2015.
DOI: 10.1007/978-3-319-27436-2_20

Example 1 (tail-recursion and proofs by induction). Let us consider the following rewrite rules in tail-recursive form computing the addition of natural numbers:

$$R = \{ Add(0, y) \rightarrow y, \ Add(S(x), y) \rightarrow Add(x, S(y)) \}$$

Let us consider proving $Add(x, 0) \doteq x$ by induction on x. In the induction step where $x = S(x')$, one needs to show the equation $Add(x', S(0)) \doteq S(x')$ obtained by unfolding the equation. However, one cannot apply the induction hypothesis $Add(x', 0) \doteq x'$ to this equation, since the second argument is different.

As the second argument y in the rewrite rules in Example 1, a tail-recursive definition usually contains a variable called an *accumulator* which keeps intermediate results of the computation and is passed to the return value at the final recursive call. By unfolding the definition, the value of the accumulator changes step by step in the course of the computation; in proofs by induction, this change of the value makes the application of the induction hypothesis impossible. In this way, proofs by induction are often incompatible with tail-recursive definitions. Most methods for proving inductive theorems of TRSs containing tail-recursive rules (tail-recursive TRSs) suffer a similar difficulty.

On the other hand, "simple" recursive definitions do not suffer such a problem. For example, a "simple" version of the TRS for addition would be the following usual definition.

Example 2 (simple recursion and proofs by induction). Let R' be the following TRS.

$$R' = \{ Add(0, y) \rightarrow y, \ Add(S(x), y) \rightarrow S(Add(x, y)) \}$$

Now let us prove the same equation $Add(x, 0) \doteq x$ of Example 1 using R'. The base step is trivial, and in the induction step, one obtains an equation $S(Add(x', 0)) \doteq S(x')$ by unfolding the definition. This time, one can apply the induction hypothesis $Add(x', 0) \doteq x'$, and thus the proof succeeds.

The TRS R' of Example 2 can be obtained from the TRS R of Example 1 by transforming the rhs of the second rule from $Add(x, S(y))$ to $S(Add(x, y))$, i.e., transforming the rhs of the rewrite rule in such a way that the context $S(\square)$ around the accumulator y is moved outside of the recursive call $Add(x, y)$. Generalizing such a transformation, J. Giesl [4] proposed *context-moving* and *context-splitting* transformations for a particular form of functional programs with eager evaluation. These transformations, under some conditions, transform tail-recursive programs into equivalent "simple" recursive programs more suitable for theorem proving employing proofs by induction.

In a previous paper [8], we formulated context-moving and context-splitting transformations *for TRSs*, and showed their correctness in the case where input TRSs are orthogonal. We also proposed an approach for inductive theorem proving which combines these transformations with *rewriting induction* [7]. It was demonstrated by experiments that the approach is effective for proving inductive theorems of tail-recursive TRSs, compared to other systems based on rewriting induction [3,9,10] (the system of [9] is equipped with lemma generation techniques in [2,11,12]).

In the present paper, we focus on the correctness of the context-moving and context-splitting transformations as formulated in [8] but where input TRSs are more general than orthogonal. To clarify the difference to the approach in [4], we also show the correctness of the context-moving transformation where input and output TRSs are evaluated by a deterministic eager strategy and thus can be seen as a faithful representation of the functional programs discussed in [4].

The sufficient condition of the context-moving transformation for TRSs with eager evaluation is identical to that of [4]. The condition is based on whether two terms are evaluated to the same value by the input TRS (cf. Definition 2). However, this notion does not necessarily coincide with equality in the initial algebra of the TRS, and so is not an inductive theorem in the traditional sense. Hence, the condition cannot in general be verified by automated methods for proving inductive theorems as developed in the field of TRSs. This is an obstacle to implementing our approach to proving inductive theorems of tail-recursive TRSs.

On the other hand, the sufficient conditions of the context-moving and context-splitting transformations in the present paper are precisely equality in the initial algebra, and so can be checked by an inductive theorem prover. Moreover, as consequences of the correctness under the conditions, it turns out that the context-moving and context-splitting transformations preserve equality in the initial algebra, and the terms in each equivalence class have the same normal form with respect to rewriting by the TRSs before and after the transformations.

The contributions of the paper are summarized as follows:

- We present proofs of the correctness of the context-moving transformation for TRSs with respect to both eager evaluation semantics and initial algebra semantics. Moreover, we provide an example to illustrate the usefulness of our result in comparison to [4] (i.e., a transformation for a TRS where the initial algebra semantics differs from the eager evaluation semantics).
- We report on an implementation and experiments of the context-moving and context-splitting transformations for TRSs including non-orthogonal cases. This is novel since [4] does not report on any implementation or experiments.
- Proving the correctness with respect to eager evaluation semantics has not been treated in [8], and our proof of it differs from the one of [4]; we simply use induction on the length of the evaluation while the proof of [4] depends on induction on an unusual ordering (denoted $\succ_f$ in [4]).
- In our proof of the correctness with respect to initial algebra semantics, we do not assume the uniqueness of normal forms nor orthogonality in output TRSs, in contrast to the proofs of the correctness in [4,8]. In the proof for the context-splitting transformation, we introduce a new translation ()$^\bullet$ between the terms of input and output TRSs besides the translation ()$^\circ$ which is the same as one used in [4].

The rest of the paper is organized as follows. Section 2 contains preliminaries. We formulate the context-moving transformation for TRSs and study its correctness in Sect. 3. We briefly discuss the context-splitting transformation for TRSs in Sect. 4. We report on an implementation and experiments in Sect. 5. Section 6 concludes with suggestions for further work.

To save space we omit some of the details in proofs, but a long version of the paper is available at http://www.nue.riec.tohoku.ac.jp/user/kentaro/.

2   Preliminaries

In this section, we fix notations and notions used in the paper.

The set of *terms* over *function symbols* $\mathcal{F}$ and *variables* $\mathcal{V}$ is denoted by $\mathcal{T}(\mathcal{F}, \mathcal{V})$. The set of variables (function symbols) occurring in a term t is denoted by $\mathcal{V}(t)$ (resp. $\mathcal{F}(t)$). We abbreviate a sequence of terms $t_1, t_2, \ldots, t_n$ as $\bar{t}$; we define $\mathcal{V}(\bar{t}) = \bigcup_{i=1}^{n} \mathcal{V}(t_i)$ and $\mathcal{F}(\bar{t}) = \bigcup_{i=1}^{n} \mathcal{F}(t_i)$. A term t is *ground* if $\mathcal{V}(t) = \emptyset$; the set of ground terms is denoted by $\mathcal{T}(\mathcal{F})$. The *root symbol* of a term t is denoted by $root(t)$. A *context* is a term containing precisely one occurrence of each special constant $\square_1, \ldots, \square_n$ (*holes*) for some n. A context C is denoted by $C[\,]$ if $n = 1$. If C is a context with n holes then the term obtained by replacing each $\square_i$ $(1 \leq i \leq n)$ in C with t_i is denoted by $C[t_1, t_2, \ldots, t_n]$. A *substitution* is a function $\theta : \mathcal{V} \to \mathcal{T}(\mathcal{F}, \mathcal{V})$ (we omit the usual condition of substitutions that they have a finite domain to ease the notation). A substitution θ is *ground* if $\theta : \mathcal{V} \to \mathcal{T}(\mathcal{F})$; throughout the paper, θ_g, θ_g', etc. denote ground substitutions.

A *rewrite rule* $l \to r$ satisfies $l \notin \mathcal{V}$ and $\mathcal{V}(l) \supseteq \mathcal{V}(r)$. We assume that variables in rewrite rules are renamed when necessary. A *term rewriting system* (*TRS*, for short) is a finite set of rewrite rules. We call $l \to r$ an *R-rule* if $l \to r \in R$. The set of *defined function symbols* of a TRS R is given by $\mathcal{D} = \{root(l) \mid l \to r \in R\}$ and the set of *constructor symbols* is $\mathcal{C} = \mathcal{F} \setminus \mathcal{D}$. Terms in $\mathcal{T}(\mathcal{C}, \mathcal{V})$ are *constructor terms*; terms in $\mathcal{T}(\mathcal{C})$ are *ground constructor terms*. A TRS R is a *constructor TRS* if for any rewrite rule $f(l_1, \ldots, l_n) \to r \in R$, each l_i $(1 \leq i \leq n)$ is a constructor term. A *ground constructor substitution* is a substitution $\theta : \mathcal{V} \to \mathcal{T}(\mathcal{C})$; throughout the paper, $\theta_{gc}, \theta_{gc}'$, etc. denote ground constructor substitutions.

In this paper, we work with unsorted TRSs for simplicity, but we elaborate lemmas and definitions so that they can be easily adapted to those in the setting of (monomorphic) many-sorted TRSs.

3   Context-Moving Transformation for TRSs

In this section, we formulate the context-moving transformation for TRSs and prove the correctness of the transformation for some classes of TRSs. In the context-moving transformation for functional programs in [4], the context occurring around the accumulator variable is moved outside of the recursive calls in each rule. The context-moving transformation *for TRSs* follows the same idea.

Definition 1 (Context-Moving Transformation for TRSs). A *context-moving transformation from TRS R to TRS R'* is given as:

$$R = R_A \cup R_B \cup R_C \text{ where } R_A = \{f(\bar{l}_i, z) \to f(\bar{r}_i, C_i[z]) \mid 1 \leq i \leq m\}$$
$$R_B = \{f(\bar{l}_j, z) \to C_j[z] \mid m + 1 \leq j \leq n\}$$
$$R_C = \{l_k \to r_k \mid n + 1 \leq k \leq p\}$$
$$R' = R_A' \cup R_B \cup R_C \text{ where } R_A' = \{f(\bar{l}_i, z) \to C_i[f(\bar{r}_i, z)] \mid 1 \leq i \leq m\}$$

Here, R_A, R'_A consist of m recursive f-rules, R_B consists of $(n-m)$ non-recursive f-rules, and R_C consists of $(p-n)$ other (non f-)rules. The function symbol f is the *target* of the transformation, the contexts $C_1[\ ], \ldots, C_n[\ ]$ are the *moving contexts*, and the variable z is the *accumulator*. Furthermore, *it is required that the target f and the accumulator z do not appear anywhere else except the places explicitly indicated*, i.e., (i) $f \notin (\bigcup_{i=1}^m \mathcal{F}(\bar{l}_i, \bar{r}_i, C_i)) \cup (\bigcup_{j=m+1}^n \mathcal{F}(\bar{l}_j, C_j)) \cup (\bigcup_{k=n+1}^p \mathcal{F}(l_k, r_k))$ and (ii) $z \notin (\bigcup_{i=1}^m \mathcal{V}(\bar{l}_i, \bar{r}_i, C_i)) \cup (\bigcup_{j=m+1}^n \mathcal{V}(\bar{l}_j, C_j))$.

Henceforth, we will focus on a context-moving transformation from R to R', and unless otherwise stated, $f, z, R_A, C_i, \bar{l}_i, \ldots$ are supposed to be those specified in the definition above.

Example 3 (context-moving transformation). Let R be the following TRS for multiplication.

$$R = \left\{ \begin{array}{ll} (a)\ Mult(S(x), y, z) \rightarrow Mult(x, y, Add(y, z)), & (b)\ Mult(0, y, z) \rightarrow z \\ (c)\ Add(S(x), y) \rightarrow S(Add(x, y)), & (d)\ Add(0, y) \rightarrow y \end{array} \right\}$$

We apply the context-moving transformation with $Mult$ as the target and z as the accumulator. The rewrite rules of R are partitioned into $R_A = \{(a)\}$, $R_B = \{(b)\}$ and $R_c = \{(c), (d)\}$, and there are two moving contexts, namely $C_1 = Add(y, \square)$ and $C_2 = \square$. Thus, by definition, we obtain

$$R'_A = \left\{ Mult(S(x), y, z) \rightarrow Add(y, Mult(x, y, z)) \right\}$$

Therefore, the following TRS R' is obtained.

$$R' = \left\{ \begin{array}{ll} Mult(S(x), y, z) \rightarrow Add(y, Mult(x, y, z)), & Mult(0, y, z) \rightarrow z \\ Add(S(x), y) \rightarrow S(Add(x, y)), & Add(0, y) \rightarrow y \end{array} \right\}$$

The rest of this section is devoted to the discussion on the correctness of the context-moving transformation.

3.1 Correctness of the Context-Moving Transformation with Respect to Eager Evaluation Semantics

First we discuss the correctness with respect to eager evaluation semantics as considered in [4]. We assume in this subsection that R is a constructor TRS.

An *eager rewrite relation* $\xrightarrow{e}_R$ is a binary relation on $\mathcal{T}(\mathcal{F})$ given by $s \xrightarrow{e}_R t$ iff $s = C[l\theta_{gc}]$ and $t = C[r\theta_{gc}]$ for some $l \rightarrow r \in R$, a context $C[\ ]$ and a ground constructor substitution θ_{gc}. Further, we assume some specific rewrite strategy (e.g. leftmost(-innermost) with rule priority) so that each rewrite step is deterministic. A rewrite step by the deterministic strategy (the *eager evaluation strategy*) is denoted by $s \xrightarrow{ev}_R t$. The reflexive transitive closure of $\xrightarrow{ev}_R$ is denoted by $\xrightarrow{ev}{}^*_R$. A ground term t is said to be *defined in R* if there exists $v \in \mathcal{T}(\mathcal{C})$ such that $t \xrightarrow{ev}{}^*_R v$; in that case, $|t|^{ev}_R$ denotes the length of the reduction sequence from t to v. We use $s \stackrel{ev}{\equiv}_R t$ to mean that for any $v \in \mathcal{T}(\mathcal{C})$, $s \xrightarrow{ev}{}^*_R v$ if and only if $t \xrightarrow{ev}{}^*_R v$. Note that $\stackrel{ev}{\equiv}_R$ is an equivalence relation and if $s \xrightarrow{ev}{}^*_R t$ then $s \stackrel{ev}{\equiv}_R t$.

The following are basic properties of an eager evaluation strategy, which are freely used in the rest of this subsection.

Lemma 1.   1. *If* $s \stackrel{\text{ev}}{\equiv}_R t$ *then* $C[s] \stackrel{\text{ev}}{\equiv}_R C[t]$.
2. *If* $C[s]$ *is defined in* R *and* $s \stackrel{\text{ev}}{\rightarrow}_R t$ *then* $|C[s]|_R^{\text{ev}} = |C[t]|_R^{\text{ev}} + 1$.
3. *If* $C[t]$ *is defined in* R *then so is* t, *and moreover* $|C[t]|_R^{\text{ev}} \geq |t|_R^{\text{ev}}$.

Proof. By induction on $C[\ ]$.   □

Lemma 2. *For any* $l \rightarrow r \in R$ *and ground substitution* θ_g *such that* $\theta_g(x)$ *is defined in* R *for any* $x \in \mathcal{V}(l)$, $l\theta_g \stackrel{\text{ev}}{\equiv}_R r\theta_g$.

Proof. If $l\theta_g$ or $r\theta_g$ is defined in R, then $l\theta_g \stackrel{\text{ev}}{\equiv}_R l\theta_{gc} \stackrel{\text{ev}}{\equiv}_R r\theta_{gc} \stackrel{\text{ev}}{\equiv}_R r\theta_g$ for some θ_{gc}.   □

We require a property concerning the moving contexts $C_1[\ ], \ldots, C_n[\ ]$ to guarantee the correctness of the context-moving transformation. This property is given in a similar way to [4] and is formulated as below.

Definition 2 (Commutativity Law of Moving Contexts). Let $C_1[\ ], \ldots, C_n[\ ]$ be the moving contexts of an instance of the context-moving transformation. The *commutativity law of moving contexts* refers to the following condition:

$$\forall i(1 \leq i \leq m).\forall j(1 \leq j \leq n).\forall \theta_{gc}.C_i[C_j[z]]\theta_{gc} \stackrel{\text{ev}}{\equiv}_R C_j[C_i[z]]\theta_{gc} \qquad (\text{CCOM}^{\text{ev}})$$

Here, we assume that each variable in moving contexts $C_i[\ ], C_j[\ ]$ is renamed so that their variables do not overlap. By Lemma 1, it is seen that the condition $(\text{CCOM}^{\text{ev}})$ is equivalent to the one with θ_g instead of θ_{gc}.

Example 4 (commutativity law of moving contexts). The moving contexts of the transformation in Example 3 are $C_1 = Add(y, \square)$ and $C_2 = \square$ (with $m = 1$ and $n = 2$). As C_2 is a trivial context, the commutativity law of moving contexts is
$\forall \theta_{gc}.Add(x, Add(y, z))\theta_{gc} \stackrel{\text{ev}}{\equiv}_R Add(y, Add(x, z))\theta_{gc}$.

Definition 3 $(R \stackrel{\text{ev} f}{\Rightarrow}_{\text{cm}} R')$. We write $R \stackrel{\text{ev} f}{\Rightarrow}_{\text{cm}} R'$ if R' is obtained from a constructor TRS R by the context-moving transformation such that f is the target and the condition $(\text{CCOM}^{\text{ev}})$ holds.

The commutativity law of moving contexts is essential for guaranteeing the simulation of rewrite sequences from ground terms to ground constructor terms on R by R' and vice versa. The key property to the simulation is the following context-moving lemma.

Lemma 3 (Context-Moving Lemma). *Suppose* $R \stackrel{\text{ev} f}{\Rightarrow}_{\text{cm}} R'$. *Let* $1 \leq i \leq m$, *and let* θ_{gc} *be a ground constructor substitution and* $\bar{t}, u$ *be ground terms.*

1. *If* $C_i\theta_{gc}[f(\bar{t}, u)] \stackrel{\text{ev} *}{\rightarrow}_R v \in \mathcal{T}(\mathcal{C})$ *then* $f(\bar{t}, C_i\theta_{gc}[u]) \stackrel{\text{ev} *}{\rightarrow}_R v$.
2. *If* $f(\bar{t}, C_i\theta_{gc}[u]) \stackrel{\text{ev} *}{\rightarrow}_{R'} v \in \mathcal{T}(\mathcal{C})$ *then* $C_i\theta_{gc}[f(\bar{t}, u)] \stackrel{\text{ev} *}{\rightarrow}_{R'} v$.

Proof. 1. If $C_i\theta_{gc}[f(\bar{t}, u)] \xrightarrow{\text{ev}}_R^* v \in \mathcal{T}(\mathcal{C})$ then $f(\bar{t}, u)$ is defined in R. The claim is proved by induction on $|f(\bar{t}, u)|_R^{\text{ev}}$.

2. By induction on $|f(\bar{t}, C_i\theta_{gc}[u])|_{R'}^{\text{ev}}$.   □

We are now ready to prove the correctness of the context-moving transformation with respect to eager evaluation semantics.

Theorem 1 (Correctness of Context-Moving Transformation). *Let R be a constructor TRS. Suppose $R \xRightarrow{\text{ev} f}_{\text{cm}} R'$. For any ground term s and ground constructor term v, $s \xrightarrow{\text{ev}}_R^* v$ if and only if $s \xrightarrow{\text{ev}}_{R'}^* v$.*

Proof. By induction on the length of the evaluation, using Lemma 3.   □

Remark 1. The proof of the "only if"-part of Theorem 1 given in [4] is based on the converse of Lemma 3(1). For those proofs, induction on an unusual ordering $\succ_f$ is used. In contrast, our proof is based on Lemma 3(2), and it suffices to use induction on the length of the evaluation.

3.2 Correctness of the Context-Moving Transformation with Respect to Initial Algebra Semantics

The correctness theorem in the previous subsection depends on the condition (CCOM^{ev}), which involves a notion of evaluation and does not necessarily correspond to equality in the initial algebra. In this subsection, we show the correctness of the context-moving transformation based on a condition that precisely corresponds to equality in the initial algebra.

First we introduce some standard definitions in term rewriting. A *rewrite relation* $\to_R$ is a binary relation on $\mathcal{T}(\mathcal{F}, \mathcal{V})$ given by $s \to_R t$ iff $s = C[l\theta]$ and $t = C[r\theta]$ for some $l \to r \in R$, a context $C[\ ]$ and a substitution θ. The reflexive transitive closure of $\to_R$ is denoted by $\xrightarrow{*}_R$. If a unique normal form of t exists, then the normal form of t is denoted by $t{\downarrow}_R$. For each substitution θ, the substitution $\theta{\downarrow}_R$ is defined by $\theta{\downarrow}_R(x) = (\theta(x)){\downarrow}_R$, provided that $(\theta(x)){\downarrow}_R$ is defined for any $x \in \mathcal{V}$. We use $\theta_{g\backslash f}$ to denote a ground substitution such that f does not appear in its range. A TRS R is *sufficiently complete* if $\forall s \in \mathcal{T}(\mathcal{F}). \exists v \in \mathcal{T}(\mathcal{C}). s \xrightarrow{*}_R v$ holds [6]; R is *ground confluent* if $\xleftarrow{*}_R \circ \xrightarrow{*}_R \subseteq \xrightarrow{*}_R \circ \xleftarrow{*}_R$ on $\mathcal{T}(\mathcal{F})$. We assume in this subsection that R is a sufficiently complete and ground confluent TRS.[1]

Now we introduce a property on the moving contexts $C_1[\ ], \ldots, C_n[\ ]$ to guarantee the correctness with respect to semantics considered in this subsection.

Definition 4 (Commutativity Law of Moving Contexts). *Let $C_1[\ ], \ldots, C_n[\ ]$ be the moving contexts of an instance of the context-moving transformation. The commutativity law of moving contexts refers to the following condition:*

$$\forall i(1 \leq i \leq m).\forall j(1 \leq j \leq n).\forall \theta_g. C_i[C_j[z]]\theta_g{\downarrow}_R = C_j[C_i[z]]\theta_g{\downarrow}_R \qquad \text{(CCOM)}$$

---

[1] In the case of many-sorted TRSs, we assume sufficient completeness only for the sort of return values of the target f of the context-moving transformation, meaning that any ground term of that sort can be rewritten to a constructor term. Cf. Example 6.

Here, we assume that each variable in moving contexts $C_i[\ ], C_j[\ ]$ is renamed so that their variables do not overlap.

In contrast to the condition (CCOM^{ev}) in Definition 2, the above condition (CCOM) precisely corresponds to equations that are valid in the initial algebra of the input TRS R, i.e. inductive theorems of R, and so may be checked by an inductive theorem prover. (For an actual implementation, see Sect. 5.)

Definition 5 $(R \Rightarrow^f_{\mathrm{cm}} R')$. We write $R \Rightarrow^f_{\mathrm{cm}} R'$ if R' is obtained from a sufficiently complete and ground confluent TRS R by the context-moving transformation such that f is the target and the condition (CCOM) holds.

The condition (CCOM) is essential for guaranteeing the simulation of rewrite sequences from ground terms to ground constructor terms on R by R'. We first show the simulation of rewrite sequences of the form $f(\bar{x}, z)\theta_{g\backslash f} \xrightarrow{*}_R v$ (Lemma 4) and then generalize it to an arbitrary case (Lemma 5).

Lemma 4. *Suppose* $R \Rightarrow^f_{\mathrm{cm}} R'$. *For any ground substitution* $\theta_{g\backslash f}$ *and ground constructor term* v, *if* $f(\bar{x}, z)\theta_{g\backslash f} \xrightarrow{*}_R v$ *then* $f(\bar{x}, z)\theta_{g\backslash f} \xrightarrow{*}_{R'} v$.

Proof. Suppose $f(\bar{x}, z)\theta_{g\backslash f} \xrightarrow{*}_R v$. By the form of the rewrite rules in R, we know that any rewrite sequence α of R from $f(\bar{x}, z)\theta_{g\backslash f}$ to v has the following form:

$$
\begin{aligned}
\alpha : f(\bar{x}, z)\theta_{g\backslash f} \ &= \ f(\bar{l}_{i_1}\theta_1, u_1) \to_{R_A} f(\bar{r}_{i_1}\theta_1, C_{i_1}\theta_1[u_1]) \\
&\xrightarrow{*}_{R_C} f(\bar{l}_{i_2}\theta_2, u_2) \to_{R_A} f(\bar{r}_{i_2}\theta_2, C_{i_2}\theta_1[u_2]) \\
&\qquad\qquad \vdots \\
&\xrightarrow{*}_{R_C} f(\bar{l}_{i_n}\theta_n, u_n) \to_{R_B} C_{i_n}\theta_n[u_n] \xrightarrow{*}_{R_C} v
\end{aligned}
$$

Here $\theta_1, \ldots, \theta_n$ are ground substitutions such that f does not appear in their ranges. Note that rewrite rules of R_A, R_B applicable to any term $f(\bar{t}, u)$ (at root position) are completely specified by $\bar{t}$ regardless of u. Hence, in the rewrite sequence α, the applications of R_A, R_B-rules are not affected even if one postpones the applications of R_C-rules to u_i's. Thus, one can obtain the next rewrite sequence β from α, by distinguishing the applications of R_C-rules to the last argument of f and those to the rest, and postponing the former:

$$
\begin{aligned}
\beta : f(\bar{x}, z)\theta_{g\backslash f} \ &= \ f(\bar{l}_{i_1}\theta_1, u_1) \to_{R_A} f(\bar{r}_{i_1}\theta_1, C_{i_1}\theta_1[u_1]) \\
&\xrightarrow{*}_{R_C} f(\bar{l}_{i_2}\theta_2, C_{i_1}\theta_1[u_1]) \to_{R_A} f(\bar{r}_{i_2}\theta_2, C_{i_2}\theta_2[C_{i_1}\theta_1[u_1]]) \\
&\qquad\qquad \vdots \\
&\xrightarrow{*}_{R_C} f(\bar{l}_{i_n}\theta_n, C_{i_{n-1}}\theta_{n-1}[\cdots C_{i_1}\theta_1[u_1]\cdots]) \\
&\qquad\qquad \to_{R_B} C_{i_n}\theta_n[C_{i_{n-1}}\theta_{n-1}[\cdots C_{i_1}\theta_1[u_1]\cdots]] \xrightarrow{*}_{R_C} v
\end{aligned}
$$

Next we construct a rewrite sequence γ of R' from β (of R). It is easy to observe in the definition of context-moving transformation that for any i and θ, $f(\bar{l}_i, z)\theta \to_{R_A} f(\bar{r}_i, C_i[z])\theta$ implies $f(\bar{l}_i, z)\theta \to_{R'_A} C_i[f(\bar{r}_i, z)]\theta$. Thus, by moving out the contexts $C_{i_j}\theta_j[\ ]$ in each R_A-step, we obtain the corresponding R'_A-step.

Then the next rewrite sequence γ is obtained from β.

$$\gamma : f(\bar{x}, z)\theta_{g\backslash f} = f(\bar{l}_{i_1}\theta_1, u_1) \to_{R'_A} C_{i_1}\theta_1[f(\bar{r}_{i_1}\theta_1, u_1)]$$
$$\xrightarrow{*}_{R_C} C_{i_1}\theta_1[f(\bar{l}_{i_2}\theta_2, u_1)] \to_{R'_A} C_{i_1}\theta_1[C_{i_2}\theta_2[f(\bar{r}_{i_2}\theta_2, u_1)]]$$
$$\vdots$$
$$\xrightarrow{*}_{R_C} C_{i_1}\theta_1[\cdots C_{i_{n-1}}\theta_{n-1}[f(\bar{l}_{i_n}\theta_n, u_1)]\cdots]$$
$$\to_{R_B} C_{i_1}\theta_1[\cdots C_{i_{n-1}}\theta_{n-1}[C_{i_n}\theta_n[u_1]]\cdots]$$

Since R is ground confluent, so is R_C. Thus, by the condition (CCOM) and $f \notin \mathcal{F}(C_{i_1}\theta_1[\cdots C_{i_{n-1}}\theta_{n-1}[C_{i_n}\theta_n[u_1]]\cdots])$, it follows

$$v = C_{i_n}\theta_n[C_{i_{n-1}}\theta_{n-1}[\cdots C_{i_1}\theta_1[u_1]\cdots]]\!\downarrow_{R_C}$$
$$= C_{i_1}\theta_1[\cdots C_{i_{n-1}}\theta_{n-1}[C_{i_n}\theta_n[u_1]]\cdots]\!\downarrow_{R_C}$$

Hence, we obtain $f(\bar{x}, z)\theta_{g\backslash f} \xrightarrow{*}_R C_{i_1}\theta_1[\cdots C_{i_{n-1}}\theta_{n-1}[C_{i_n}\theta_n[u_1]]\cdots] \xrightarrow{*}_{R_C} v$. □

Lemma 5. *Suppose $R \Rightarrow^f_{cm} R'$. For any ground term s and ground constructor term v, if $s \xrightarrow{*}_R v$ then $s \xrightarrow{*}_{R'} v$.*

Proof. By induction on the number of occurrences of f in s, using Lemma 4. □

In contrast to the proof in the previous subsection, a key ingredient of the proof of the correctness here is preservation of two properties of R: sufficient completeness and ground confluence. The former is a direct consequence of Lemma 5.

Lemma 6. *Suppose $R \Rightarrow^f_{cm} R'$. Then R' is sufficiently complete.*

Proof. It follows by Lemma 5 from the sufficient completeness of R.   □

To show preservation of ground confluence, we need the simulation of rewrite sequences from ground terms to ground constructor terms on R' by R, that is, the converse of Lemma 5. To this end, we first prove the following lemma, where we use again the forms of rewrite rules in R and R' and the condition (CCOM).

Lemma 7. *Suppose $R \Rightarrow^f_{cm} R'$. For any ground terms s, s' and ground constructor term v, if $s \xrightarrow{*}_R v$ and $s \to_{R'} s'$ then $s' \xrightarrow{*}_R v$.*

Proof. If $s \to_{R_B \cup R_C} s'$, then $s \to_R s'$ by $R_B \cup R_C \subseteq R$, and hence the claim follows immediately by the ground confluence of R. It remains to prove the case $s \to_{R'_A} s'$. Then one has $s = C[f(\bar{l}_{i_1}\theta_1, u)]$ and $s' = C[C_{i_1}\theta_1[f(\bar{r}_{i_1}\theta_1, u)]]$, and thus, $s = C[f(\bar{l}_{i_1}\theta_1, u)] \to_{R_A} C[f(\bar{r}_{i_1}\theta_1, C_{i_1}\theta_1[u])]$. Furthermore, since $s \xrightarrow{*}_R v$, it follows from the ground confluence of R that $C[f(\bar{r}_{i_1}\theta_1, C_{i_1}\theta_1[u])] \xrightarrow{*}_R v$. Thus, $s = C[f(\bar{l}_{i_1}\theta_1, u)] \to_{R_A} C[f(\bar{r}_{i_1}\theta_1, C_{i_1}\theta_1[u])] \xrightarrow{*}_R v$. Now, as in the proof of Lemma 4, this rewrite sequence looks like:

$$\alpha : s = C[f(\bar{l}_{i_1}\theta_1, u)] \to_{R_A} C[f(\bar{r}_{i_1}\theta_1, C_{i_1}\theta_1[u])]$$
$$\xrightarrow{*}_{R_C} C[f(\bar{l}_{i_2}\theta_2, C_{i_1}\theta_1[u])] \to_{R_A} C[f(\bar{r}_{i_2}\theta_2, C_{i_2}\theta_2[C_{i_1}\theta_1[u]])]$$
$$\vdots$$
$$\xrightarrow{*}_{R_C} C[f(\bar{l}_{i_n}\theta_n, C_{i_{n-1}}\theta_{n-1}[\cdots C_{i_1}\theta_1[u]\cdots])]$$
$$\to_{R_B} C[C_{i_n}\theta_n[C_{i_{n-1}}\theta_{n-1}[\cdots C_{i_1}\theta_1[u]\cdots]]] \xrightarrow{*}_R v$$

Consider the next rewrite sequence β obtained from α by replacing the first R_A-step with an R'_A-step:

$$
\begin{aligned}
\beta : s \; &= \; C[f(\bar{l}_{i_1}\theta_1, u)] \to_{R'_A} C[C_{i_1}\theta_1[f(\bar{r}_{i_1}\theta_1, u)]] \\
&\xrightarrow{*}_{R_C} C[C_{i_1}\theta_1[f(\bar{l}_{i_2}\theta_2, u)]] \to_{R_A} C[C_{i_1}\theta_1[f(\bar{r}_{i_2}\theta_2, C_{i_2}\theta_2[u])]] \\
&\xrightarrow{*}_{R_C} C[C_{i_1}\theta_1[f(\bar{l}_{i_3}\theta_3, C_{i_2}\theta_2[u])]] \to_{R_A} C[C_{i_1}\theta_1[f(\bar{r}_{i_3}\theta_3, C_{i_3}\theta_3[C_{i_2}\theta_2[u]])]] \\
&\quad\quad\quad\vdots \\
&\xrightarrow{*}_{R_C} C[C_{i_1}\theta_1[f(\bar{l}_{i_n}\theta_n, C_{i_{n-1}}\theta_{n-1}[\cdots C_{i_2}\theta_2[u]\cdots])]] \\
&\quad\quad\quad \to_{R_B} C[C_{i_1}\theta_1[C_{i_n}\theta_n[C_{i_{n-1}}\theta_{n-1}[\cdots C_{i_2}\theta_2[u]\cdots]]]]
\end{aligned}
$$

Then, by the condition (CCOM) and ground confluence of R, we have

$$
\begin{aligned}
v &= C[C_{i_n}\theta_n[C_{i_{n-1}}\theta_{n-1}[\cdots C_{i_1}\theta_1[u]\cdots]]]\!\downarrow_R \\
&= C[C_{i_1}\theta_1[C_{i_n}\theta_n[C_{i_{n-1}}\theta_{n-1}[\cdots C_{i_2}\theta_2[u]\cdots]]]]\!\downarrow_R
\end{aligned}
$$

Since $s' = C[C_{i_1}\theta_1[f(\bar{r}_{i_1}\theta_1, u)]] \xrightarrow{*}_R C[C_{i_1}\theta_1[C_{i_n}\theta_n[C_{i_{n-1}}\theta_{n-1}[\cdots C_{i_2}\theta_2[u]\cdots]]]]$ (in β), we conclude $s' \xrightarrow{*}_R v$. □

The rewrite step $s \to_{R'} s'$ in the above lemma can be generalized to $s \xrightarrow{*}_{R'} s'$.

Lemma 8. *Suppose* $R \Rightarrow^f_{cm} R'$. *For any ground terms* s, s' *and ground constructor term* v, *if* $s \xrightarrow{*}_R v$ *and* $s \xrightarrow{*}_{R'} s'$ *then* $s' \xrightarrow{*}_R v$.

Now we can prove the converse of Lemma 5.

Lemma 9. *Suppose* $R \Rightarrow^f_{cm} R'$. *For any ground term* s *and ground constructor term* v, *if* $s \xrightarrow{*}_{R'} v$ *then* $s \xrightarrow{*}_R v$.

Proof. By sufficient completeness of R, there exists a ground constructor term v' such that $s \xrightarrow{*}_R v'$. By Lemma 8, we have $v \xrightarrow{*}_R v'$, and thus $v = v'$ as v is a constructor term. Hence, $s \xrightarrow{*}_R v$. □

Now we arrive at the preservation of ground confluence.

Lemma 10. *Suppose* $R \Rightarrow^f_{cm} R'$. *Then* R' *is ground confluent.*

Proof. Let t be a ground term and suppose that $t \xrightarrow{*}_{R'} t_1$ and $t \xrightarrow{*}_{R'} t_2$. Since R' is sufficiently complete by Lemma 6, there exist ground constructor terms v_1, v_2 such that $t_1 \xrightarrow{*}_{R'} v_1$ and $t_2 \xrightarrow{*}_{R'} v_2$. By Lemma 9, we have $t \xrightarrow{*}_R v_1$ and $t \xrightarrow{*}_R v_2$. Then by ground confluence of R, we obtain $v_1 = v_2$. Hence R' is ground confluent. □

We are now ready to show the main theorem of this subsection, which implies that the context-moving transformation preserves equality in the initial algebra and the terms in each equivalence class have the same normal form by R and R'.

Theorem 2 (Correctness of Context-Moving Transformation). *Let* R *be a sufficiently complete and ground confluent TRS. Suppose* $R \Rightarrow^f_{cm} R'$. *Then for any ground term* s, $s\!\downarrow_R = s\!\downarrow_{R'}$.

Proof. By sufficient completeness and ground confluence of R and R', $s\downarrow_R$ and $s\downarrow_{R'}$ are unique constructor ground terms. By Lemma 5, we have $s \xrightarrow{*}_{R'} s\downarrow_R$. Thus, $s\downarrow_R = s\downarrow_{R'}$. □

Example 5 (context-moving transformation for non-orthogonal system). Let R be the following non-orthogonal TRS for a list calculation.

$$R = \begin{cases} (a) & Minlist(Cons(x, xs), z) \to Minlist(xs, Min(x, z)) \\ (b) & Minlist(Nil, z) \to z \\ (c) & Min(S(x), S(y)) \to S(Min(x, y)) \\ (d) & Min(0, y) \to 0, \quad (e) \quad Min(x, 0) \to 0 \end{cases}$$

where we assume that it is many-sorted with sorts *Nat* and *NatList* in an appropriate way. We apply the context-moving transformation with *Minlist* as the target and z as the accumulator. We have $R_C = \{(c), (d), (e)\}$ and there are two moving contexts, namely $C_1 = Min(x, \square)$ and $C_2 = \square$. Then we have

$$\forall \theta_g. Min(x, Min(y, z))\theta_g\downarrow_R = Min(y, Min(x, z))\theta_g\downarrow_R$$

and thus, $R \Rightarrow^{Minlist}_{cm} R'$, where

$$R' = \{Minlist(Cons(x, xs), z) \to Min(x, Minlist(xs, z))\} \cup \{(b)–(e)\}$$

Example 6. In this example, we use a many-sorted TRS R with sorts *Nat* and *NatStream*, where ":" of sort $Nat \times NatStream \to NatStream$ is the only constructor symbol for terms of sort *NatStream*.

$$R = \begin{cases} (a) & Sum(S(x), \alpha, z) \to Sum(x, Tl(\alpha), Add(Hd(\alpha), z)) \\ (b) & Sum(0, \alpha, z) \to z \\ (c) & Hd(x : \alpha) \to x, \quad (d) \quad Tl(x : \alpha) \to \alpha \\ (e) & Inc \to 0 : Succ(Inc), \quad (f) \quad Succ(x : \alpha) \to S(x) : Succ(\alpha) \\ (g) & Add(S(x), y) \to S(Add(x, y)), \quad (h) \quad Add(0, y) \to y \end{cases}$$

Here we have sufficient completeness for sort *Nat*, which is the sort of return values of the target *Sum*. Then, all the arguments for the correctness of the context-moving transformation follows for terms of sort *Nat*.[2] We have $R_C = \{(c)–(h)\}$ and there are two moving contexts, namely $C_1 = Add(Hd(\alpha), \square)$ and $C_2 = \square$. Then we have

$$\forall \theta_g. Add(Hd(\alpha), Add(Hd(\beta), z))\theta_g\downarrow_R = Add(Hd(\beta), Add(Hd(\alpha), z))\theta_g\downarrow_R$$

Thus, we obtain $R \Rightarrow^{Sum}_{cm} R'$, where

$$R' = \{Sum(S(x), \alpha, z) \to Add(Hd(\alpha), Sum(x, Tl(\alpha), z))\} \cup \{(b)–(h)\}$$

Note here that, for terms of sort *Nat*, normal forms may not be reached by the eager evaluation strategy because of the rule for *Inc*.

---

[2] For terms of sort *NatStream*, we do not seek the correctness of the context-moving transformation in the style of Theorem 2.

4   Context-Splitting Transformation for TRSs

In this section, we formulate the context-splitting transformation for TRSs and prove the correctness of the transformation. In the context-splitting transformation for functional programs in [4], the context occurring around the accumulator variable is required to be split into a "common" part and an "own" part, where the "common" part needs to be common to all f-rules. Then, in each rule, the context is moved outside of the recursive calls, moving its own part and removing the accumulator. Furthermore, the target f is replaced with a new function symbol f', obtained by removing the accumulator argument of f. The context-splitting transformation *for TRSs* follows the same idea.

Definition 6 (Context-Splitting Transformation for TRSs). The *context-splitting transformation from TRS R to TRS R'* is given as:

$$R = R_A \cup R_B \cup R_C \quad \text{where} \quad R_A = \{f(\bar{l}_i, z) \to f(\bar{q}_i, C_i[z]) \mid 1 \le i \le m\}$$
$$R_B = \{f(\bar{l}_j, z) \to C_j[z] \mid m+1 \le j \le n\}$$
$$R_C = \{l_k \to r_k \mid n+1 \le k \le p\}$$

For each i ($1 \le i \le n$), it is required that either $C_i[\ ] = C[r_i, \square]$ or $C_i[\ ] = \square$.

$$R' = R'_A \cup R'_B \cup R_C \quad \text{where} \quad R'_A = \{f'(\bar{l}_i) \to C'_i[f'(\bar{q}_i)] \mid 1 \le i \le m\}$$
$$R'_B = \{f'(\bar{l}_j) \to r'_j \mid m+1 \le j \le n\}$$

Here, for each i ($1 \le i \le m$) and j ($m+1 \le j \le n$), the context $C'_i[\ ]$ and the term r'_j are given like this:

$$C'_i[\ ] = \begin{cases} C[\square, r_i] & \text{if } C_i[\ ] = C[r_i, \square] \\ \square & \text{if } C_i[\ ] = \square \end{cases} \qquad r'_j = \begin{cases} r_j & \text{if } C_j[\ ] = C[r_j, \square] \\ e & \text{if } C_j[\ ] = \square \end{cases}$$

The function symbol f is the *target* of the transformation, the variable z is the *accumulator*, the context C is the *common context*, and the term e is the *unit*. Here, the common context C should be a ground context such that $f \notin \mathcal{F}(C)$ and the unit e should be a ground constructor term. Furthermore, *it is required that the target f and the accumulator z do not appear anywhere else except the places explicitly indicated.*

Example 7 (context-splitting transformation). Let R be the following TRS for list concatenation. Here we assume that it is many-sorted in an appropriate way.

$$R = \begin{cases} (a) \ Cat(LCons(x, xs), z) \to Cat(xs, App(z, x)), & (b) \ Cat(LNil, z) \to z \\ (c) \ App(Cons(x, xs), y) \to Cons(x, App(xs, y)), & (d) \ App(Nil, y) \to y \end{cases}$$

We apply the context-splitting transformation with Cat as the target and z as the accumulator. The TRS R is partitioned like this: $R_A = \{(a)\}$, $R_B = \{(b)\}$ and $R_C = \{(c), (d)\}$. We remark that the common context is $C = App(\square_2, \square_1)$ and we have $C_1[\ ] = App(\square, x)$ and $C_2[\ ] = \square$. The unit is $e = Nil$. We construct R'_A, R'_B from R_A, R_B as follows:

$$R'_A = \{ Cat'(LCons(x, xs)) \to App(x, Cat'(xs)) \} \qquad R'_B = \{ Cat'(LNil) \to Nil \}$$

Thus, we obtain $R' = R'_A \cup R'_B \cup \{(c), (d)\}$.

4.1 Correctness of the Context-Splitting Transformation with Respect to Initial Algebra Semantics

In the context-moving transformation, the commutativity laws of moving contexts played an important role. In the context-splitting transformation, we require two conditions instead: "associativity law of common context" and "unit law of common context"; they are defined again following [4] but in the forms that correspond to equality in the initial algebra as (CCOM) in Definition 4.

Definition 7 (Associativity Law of Common Context). Let C be the common context of an instance of the context-splitting transformation. The *associativity law of common context* for the transformation refers to the following condition:

$$\forall \theta_g.C[C[x,y],z]\theta_g\!\downarrow_R = C[x,C[y,z]]\theta_g\!\downarrow_R \qquad \text{(CASSOC)}$$

Definition 8 (Unit Law of Common Context). Let C be the common context and e be the unit of an instance of the context-splitting transformation. The *unit law of common context* for the transformation refers to the following condition:

$$\forall \theta_g.C[x,e]\theta_g\!\downarrow_R = C[e,x]\theta_g\!\downarrow_R = \theta_g(x) \qquad \text{(CUNIT)}$$

Definition 9 ($R \Rightarrow_{cs}^{f} R'$). We write $R \Rightarrow_{cs}^{f} R'$ if R' is obtained from a sufficiently complete and ground confluent TRS R by the context-splitting transformation such that f is the target and the conditions (CASSOC) and (CUNIT) hold.

Definition 10 (Translations $()^{\circ}, ()^{\bullet}$). Let C be the common context and e be the unit of an instance of the context-splitting transformation. We recursively define the term $t^{\circ} \in \mathcal{T}(\mathcal{F}', \mathcal{V})$ for each term $t \in \mathcal{T}(\mathcal{F}, \mathcal{V})$ and the term $t^{\bullet} \in \mathcal{T}(\mathcal{F}, \mathcal{V})$ for each term $t \in \mathcal{T}(\mathcal{F}', \mathcal{V})$ as follows:

$$t^{\circ} = \begin{cases} C[f'(\bar{t}^{\circ}), u^{\circ}] & \text{if } t = f(\bar{t}, u) \\ g(\bar{t}^{\circ}) & \text{if } t = g(\bar{t}), g \neq f \\ t & \text{if } t \in \mathcal{V} \end{cases} \qquad t^{\bullet} = \begin{cases} f(\bar{t}^{\bullet}, e) & \text{if } t = f'(\bar{t}) \\ g(\bar{t}^{\bullet}) & \text{if } t = g(\bar{t}), g \neq f' \\ t & \text{if } t \in \mathcal{V} \end{cases}$$

Here, for each sequence $\bar{t} = t_1, \ldots, t_n$, we let $\bar{t}^{\star} = t_1^{\star}, \ldots, t_n^{\star}$ for $\star \in \{\circ, \bullet\}$.

We first show two kinds of simulation of rewrite sequences from ground terms to ground constructor terms on R by R'.

Lemma 11. *Suppose $R \Rightarrow_{cs}^{f} R'$. For any ground term s and ground constructor term v, (i) if $s \xrightarrow{*}_R v$ then $s^{\circ} \xrightarrow{*}_{R'} v$ and (ii) if $s^{\bullet} \xrightarrow{*}_R v$ then $s \xrightarrow{*}_{R'} v$.*

Using Lemma 11, we can show the correctness of the context-splitting transformation. As in the case of the context-moving transformation, a key ingredient of the proof is preservation of sufficient completeness and ground confluence.

Lemma 12. *Suppose $R \Rightarrow_{cs}^{f} R'$. Then R' is sufficiently complete.*

Lemma 13. *Suppose $R \Rightarrow_{cs}^{f} R'$. Then R' is ground confluent.*

Theorem 3 (Correctness of Context-Splitting Transformation). *Let R be a sufficiently complete and ground confluent TRS. Suppose $R \Rightarrow_{cs}^{f} R'$. Then for any ground term s, $s\!\downarrow_R = s^{\circ}\!\downarrow_{R'}$.*

5 Automating the Context-Moving and Context-Splitting Transformations

In this section, we report on an implementation and experiments of the context-moving and context-splitting transformations for TRSs presented in this paper. A key feature of our implementation is to employ inductive theorem proving to verify the commutative law of moving contexts, etc. to guarantee the correctness of the transformations.

An equation $s \doteq t$ is an *inductive theorem* of a TRS R ($R \models_{\text{ind}} s \doteq t$) if $s\theta_g \overset{*}{\leftrightarrow}_R t\theta_g$ for any ground substitution θ_g. It is known that these equations coincide with the equations that are valid in the initial algebra of R. The next lemma follows immediately from the definition.

Lemma 14. *Let R be a sufficiently complete and ground confluent TRS. Then $R \models_{\text{ind}} s \doteq t$ iff for any ground substitution θ_g, $s\theta_g{\downarrow}_R = t\theta_g{\downarrow}_R$.*

Thus, the commutative law of moving contexts and the associative and unit laws of common context, in the forms of (CCOM) in Definition 4, (CASSOC) in Definition 7 and (CUNIT) in Definition 8, are guaranteed if one succeeds in proving the following conditions (C), (A) and (U), respectively.

(C) $\forall i(1 \leq i \leq m)\forall j(1 \leq j \leq n).R \models_{\text{ind}} C_i[C_j[z]] \doteq C_j[C_i[z]]$

(A) $R \models_{\text{ind}} C[C[x,y],z] \doteq C[x,C[y,z]]$

(U) $R \models_{\text{ind}} C[x,e] \doteq x,\ R \models_{\text{ind}} C[e,x] \doteq x$

Here, $C_1,\ldots,C_m$ are the moving contexts, C is the common context, and e is the unit of the transformation.

We have implemented a TRS transformation procedure with the context-moving and context-splitting transformations using Standard ML of New Jersey. We employed rewriting induction [7] for proving conditions (C), (A) and (U). Since one generally needs to deal with non-orientable equations for proving the condition (C), we have used rewriting induction for non-orientable equations [1].

We have tested context-moving transformations, context-splitting transformations, and their combinations. Among 21 examples, the context-moving transformations succeeded at 15 examples and the context-splitting transformations succeeded at 10 examples. There are 6 examples which succeeded in both of the transformations. Failure of 3 examples in context-moving transformations and 4 in context-splitting transformations are due to failure of rewriting induction.

All details of the experiments are available on the webpage http://www.nue.riec.tohoku.ac.jp/tools/experiments/lopstr15/.

6 Conclusion

We have presented proofs of the correctness of context-moving and context-splitting transformations for TRSs. First we gave a proof of the correctness of the context-moving transformation with respect to eager evaluation semantics as considered in [4]. Then we gave proofs of the correctness of the context-moving and context-splitting transformations with respect to initial algebra semantics, where the conditions of the transformations precisely correspond to equality in the initial algebra and so can be checked by an inductive theorem prover.

The context-moving transformation for TRSs with eager evaluation as well as the transformations in [4] allows input programs where a term may not be evaluated to a ground constructor term either because it is not terminating under the evaluation strategy or because evaluation gets stuck at a non-constructor term. To deal with such programs in general (i.e., to prove their properties and to check the conditions for the correctness of the transformations), one needs methods for induction proofs with partial functions as studied in [5]. Also, the correctness of the transformations for programs with other evaluation strategies, e.g. lazy evaluation, is to be investigated. These problems and their implementation are left as future work.

Acknowledgements. We are grateful to the anonymous referees for valuable comments. This research was supported by JSPS KAKENHI Grant Numbers 25330004, 25280025 and 15K00003.

References

1. Aoto, T.: Designing a rewriting induction prover with an increased capability of non-orientable equations. In: Proceedings of 1st SCSS, volume 08–08 of RISC Technical report, pp. 1–15 (2008)
2. Aoto, T.: Sound lemma generation for proving inductive validity of equations. In: Proceedings of 28th FSTTCS, LIPIcs, vol. 2, pp. 13–24. Schloss Dagstuhl (2008)
3. Bouhoula, A., Kounalis, E., Rusinowitch, M.: Automated mathematical induction. J. Logic Comput. **5**(5), 631–668 (1995)
4. Giesl, J.: Context-moving transformations for function verification. In: Bossi, A. (ed.) LOPSTR 1999. LNCS, vol. 1817, pp. 293–312. Springer, Heidelberg (2000)
5. Giesl, J.: Induction proofs with partial functions. J. Autom. Reasoning **26**(1), 1–49 (2001)
6. Kapur, D., Narendran, P., Zhang, H.: On sufficient-completeness and related properties of term rewriting systems. Acta Informatica **24**(4), 395–415 (1987)
7. Reddy, U.S.: Term rewriting induction. In: Stickel, M.E. (ed.) CADE 1990. LNCS, vol. 449, pp. 162–177. Springer, Heidelberg (1990)
8. Sato, K., Kikuchi, K., Aoto, T., Toyama, Y.: Automated inductive theorem proving using transformations of term rewriting systems. JSSST Comput. Softw. **32**(1), 179–193 (2015). In Japanese
9. Shimazu, S., Aoto, T., Toyama, Y.: Automated lemma generation for rewriting induction with disproof. JSSST Comput. Softw. **26**(2), 41–55 (2009). In Japanese
10. Stratulat, S.: A general framework to build contextual cover set induction provers. J. Symbolic Comput. **32**, 403–445 (2001)
11. Urso, P., Kounalis, E.: Sound generalizations in mathematical induction. Theor. Comput. Sci. **323**, 443–471 (2004)
12. Walsh, T.: A divergence critic for inductive proof. J. Artif. Intell. Res. **4**, 209–235 (1996)

Constraint Solving and Programming

Why CP Portfolio Solvers Are (under)Utilized? Issues and Challenges

Roberto Amadini[1]($\boxtimes$), Maurizio Gabbrielli[1], and Jacopo Mauro[2]

[1] Department of Computer Science and Engineering/Laboratory Focus INRIA,
University of Bologna, Bologna, Italy
{amadini,gabbri}@cs.unibo.it
[2] Department of Informatics, University of Oslo, Oslo, Norway
jacopom@ifi.uio.no

Abstract. It is well recognized that a single, arbitrarily efficient solver can be significantly outperformed by a *portfolio solver* exploiting a combination of possibly slower on-average different solvers. Despite the success of portfolio solvers within the context of solving competitions, they are rarely used in practice. In this paper we give an overview of the main limitations that hinder the practical adoption and development of portfolio solvers within the Constraint Programming (CP) paradigm, discussing also possible ways to overcome them and potential extensions outside the CP field.

1 Introduction

Solving combinatorial search problems is hard, and there exist nowadays plenty of techniques and constraint solvers for performing this task. It has become clear that different solvers are better when solving different problem instances, even within the same problem class. It has also been shown that a single, arbitrarily efficient solver can be significantly outperformed by using a *portfolio* of possibly on-average slower solvers.

Algorithm portfolios [25] can be seen as instances of the more general *Algorithm Selection* problem [57] where, as reported in [42], the algorithm selection is performed case-by-case for each problem to solve. Within the context of constraint solving, a portfolio approach enables to combine a number $m > 1$ of different constituent solvers $s_1, \ldots, s_m$ in order to create a globally better constraint solver, dubbed a *portfolio solver*. When a new, unseen problem p comes, the portfolio solver tries to predict the best constituent solver(s) $s_{i_1}, \ldots, s_{i_k}$ (with $1 \leq i_j \leq m$ for $j = 1, \ldots, k$) for solving p and then runs them on p. Properly selecting and scheduling the solvers is a crucial step for the performance of a portfolio solver, and it is usually performed by exploiting *Machine Learning* techniques based on features extracted from the problem p to solve.

We can safely say that portfolio approaches have proven to be particularly effective within the context of solving challenges. For instance, the SAT portfolio solvers 3S [38] and CSHC [46] won gold medals in the SAT Competition

Supported by the EU project FP7-644298 *HyVar: Scalable Hybrid Variability for Distributed, Evolving Software Systems*.

© Springer International Publishing Switzerland 2015
M. Falaschi (Ed.): LOPSTR 2015, LNCS 9527, pp. 349–364, 2015.
DOI: 10.1007/978-3-319-27436-2_21

2011 and 2013 respectively. SATZilla [72] won the SAT Challenge 2012. CPHydra [54] was the winner of the International Constraint Solver Competition 2008. The ASP portfolio solver claspfolio [17] was gold medalist in different tracks of the ASP Competition 2009 and 2011. ArvandHerd [70] and IBaCoP [14] won some tracks in the Planning Competition 2014, where 29 out of 67 solvers were portfolio-based. Surprisingly enough, despite the remarkable results achieved in such challenges, portfolio solvers have been in general poorly adopted in the real word. So, a question naturally arises: why portfolio approaches are scarcely used outside the walls of solving competitions?

In this paper we tackle this problem by focusing in particular on portfolio approaches within the *Constraint Programming* (CP) paradigm, where the goal is to model and solve *Constraint Satisfaction Problems* (CSPs) as well as the more general *Constraint Optimisation Problems* (COPs). From this perspective the state of the art of CP portfolio solvers is still a raw fruit if compared, e.g., to the SAT field where a number of effective portfolio approaches have been developed and tested. As an example, the first and the only portfolio solver that won a MiniZinc Challenge [67] —the only still active competition for evaluating CP solvers— has been sunny-cp [6] in 2015. There are certainly a number of difficulties because of which many users prefer to take refuge in a more classical "single-solver" approach, rather than relying on portfolio solvers. However, we believe that in a not negligible number of cases a proper combination of different solvers might significantly improve the solving process. Our goal is therefore trying to reduce the obstacles that hinder the practical adoption and development of portfolio approaches. Among the various issues, we identified four main challenges for the future of CP portfolio solvers:

- *prediction model* (Sect. 2): what are the scientific and engineering issues that arise when building or using the prediction model responsible for the solver selection;
- *optimisation problem* (Sect. 3): how to apply the portfolio theory to COPs, being the state of the art in this field still in an embryonic stage;
- *parallelisation* (Sect. 4): how to exploit different processing units, possibly running in parallel more than one constituent solver;
- *utilisation* (Sect. 5): how to facilitate the practical use of portfolio solvers for solving generic CP problems.

In the rest of the paper we will explain in more detail these issues, by discussing possible ways to overcome them and providing also proposals for future directions, such as for example the extension of portfolio solving outside the CP field.

2   Prediction Model

With the term *"prediction model"* we refer to the set of data, knowledge and algorithms required to predict and run the best solver(s) for solving a new CP problem. In this section we focus in particular on three key components of the

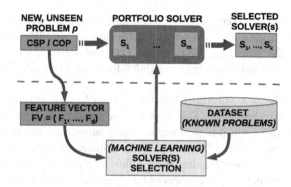

Fig. 1. Basic components of the prediction model.

prediction model: the dataset of problems used to make (and test) the predictions, the features used to characterize each problem, and the algorithms used to perform the solver selection (Fig. 1).

The basic framework of a prediction model is summarized in Fig. 2. When a new, unseen problem p needs to be solved, the *feature vector* $FV = (F_1, \ldots, F_d)$ of p is firstly computed. Broadly speaking, FV is a collection of $d \geq 0$ numerical attributes that characterize the problem p (e.g., statistics over the problem structure). Then, a subset $s_{i_1}, \ldots, s_{i_k}$ of $1 \leq i_j \leq m$ solvers of the portfolio $\{s_1, \ldots, s_m\}$ is selected and executed according to FV and to a dataset of already known problems on which the portfolio solver is trained. Note that, although the solver selection is usually performed by means of Machine Learning algorithms, the use of Machine Learning is not strictly necessary. For example, we could define a purely static prediction model that for every instance p always runs a schedule of solvers which is pre-computed *a priori*, regardless of p. In this case no prediction is needed.

The prediction should be transparent for the end user, i.e., the user should run the portfolio solver on p just like a regular, individual solver without worrying about the underlying structure of the model.

2.1 Dataset

The performance of a portfolio solver is strongly dependent on the choice of the set of problem instances used to perform the solver selection. The difficulties of choosing a suitable dataset are well recognized [57]. If we restrict ourselves to the CP field, the first issue is the lack of a standard language. Differently from SAT, ASP, and Planning, no standard format exists for specifying a CP problem. This problem affects individual CP solvers and, *a fortiori*, represents a major obstacle for defining and comparing portfolio solvers. Lately the CP community seems to converge on *MiniZinc* language [52] but, despite more than 8000 MiniZinc instances are publicly available, the standardisation process looks far from over. Other formats like XCSP [59] and Essence [20] are still in use, and *ad hoc* solver-specific languages are widely adopted. Even the natural language is used for the problem specification (e.g., see the well-known CSPLib library [23]).

Clearly we propose to converge to a common language, whether it be MiniZinc or any other. This should foster the definition of compilers/interpreters for switching from the standard language to the preferred target language (e.g., in [2] we introduced the xcsp2mzn compiler for converting an XCSP model to MiniZinc).

Assuming to have a standard language, even having standard datasets of problems is a desirable goal. This can be useful not only for making better predictions, but also for a fair performance comparison of different solvers. The *Algorithm Selection Library* [13] is currently addressing this task by collecting and standardizing datasets coming from different algorithm selection scenarios. A good starting point for the creation of CP standard benchmarks might consist in using the instances of the MiniZinc Challenge, but also other choices might be equally justifiable (e.g., using the more extensive benchmarks of the International Constraint Solver Competition 2008/2009).

Having a too large dataset may also be counterproductive: including without criteria all the available instances is a poor choice, since it may create noise and hinder the predictions accuracy. A reasonable approach to construct a dataset consists in grouping the problems by their nature, difficulty, and origin and add few representatives per group into the dataset. Since such a classification may require a considerable human effort, a promising direction for future works concerns the automation of the dataset construction. For instance, in [31] a dataset of SAT problems is automatically generated by means of a clustering algorithm.

2.2   Features

The concept of feature is crucial for algorithm selection. Features are instance-specific attributes that characterize a given problem. Early as 1976, Rice stated that *"The determination of the best (or even good) features is one of the most important, yet nebulous, aspects of the algorithm selection problem"* [57]. Features can be categorized in static and dynamic [42]. In the context of CP solving, static features are computed off-line by parsing the input problem (e.g., statistics over the variables and the constraints of the problem). Dynamic features are instead collected by retrieving information at runtime (e.g., the number of propagation performed or nodes explored in the search tree).

A weakness of dynamic features regards the limited portability. Running a solver for short runs makes the features dependent on the architecture on which the solver is executed. This may distort the predictions, which may change when performed by different machines on the same problem.

From a problem specification it is possible to collect hundreds of features (see for instance [2,9,19,36]) but, as for the dataset construction, care should be taken for avoiding to retrieve redundant and noisy knowledge. As also shown in [19,44] usually a very small subset of features is really needed.

There is an extensive literature concerning the problem of *feature selection* [27], i.e., the problem of selecting the most significant features for a prediction model. Indeed, a common issue for many portfolio approaches consists in using features that do not properly characterize a given problem. Feature selection can be a very costly task that might result in negligible gains—or even deterioration—of performance when not properly performed. The selection procedure can be safely performed off-line, but turns out to be almost infeasible when the behaviour of the portfolio solver can not be simulated. For instance, in a COP setting the side effects of bounds communication are not predictable in advance [7] and thus can not be safely simulated. In these cases it is desirable to use a selection algorithm the more robust as possible w.r.t. redundant features.

The cost of feature extraction plays an important role. The time needed to compute the features should be minimized, since every second of feature computation is a second taken from the portfolio solver execution. For instance, extracting features based on constraint graphs can be very time and space consuming. Beyond the "classic" scientific challenges relating to the feature identification and selection—which are typical of the Machine Learning field—as a future challenge we also propose to not overlook more practical problems like the computational tractability and the usability of features. In this respect we developed the feature extractor `mzn2feat-1.0` [51], that improves a previous version of `mzn2feat` [2] with the aim of being more portable, light-weight, flexible, and independent from the particular machine on which it is run. To the best of our knowledge, this is the only publicly available tool able to extract features from a generic MiniZinc model: of course, we welcome every other analogous tool for retrieving and selecting new, significant features.

2.3 Solver Selection

Selecting the best solver to run on a given problem is a critical issue, clearly related to the available dataset and the features considered for each problem of the dataset. In this context, classification techniques appear to be more robust than regression ones for the runtime prediction (e.g., see [48,55,71]). A common drawback of portfolio approaches is that typically a prediction model is built by first running each solver of the portfolio on every instance of the dataset. This task typically requires weeks of computations, and it is not very flexible and portable. For example, having a new (version of a) solver means re-running such solver on all the problems. Moreover, the runtime information computed during the training phase on a given machine may be no longer significant when the portfolio solver is run on another machine. An interesting direction to follow is shown in [65], where the prediction model is built by using only short runs of the constituent solvers on the training instances.

Another issue concerns the explicit construction of a prediction model. As pointed out also in [61], portfolio solvers usually require a complex off-line training phase for selecting the solver(s) to run. For example, SATzilla [71] uses a weighted Random Forest machine learning approach while CHSC [46] clusters the instances of the training set. Despite the proven effectiveness of these approaches, we think that a major challenge is to lighten as much as possible the training phase. In [3,54,61] the authors show that also "training-less" approaches can be competitive w.r.t. those that build an explicit prediction model. Another interesting direction to consider when building a prediction model is to use alternatives to the supervised learning approach. Techniques such as self-training and co-training already used for example for QBF [56] and ASP [47] may simplify the construction of the prediction model.

Of course, even the time needed for the on-line selection of the solvers to run must be considered. For example, a prediction model with a negligible building cost might be totally useless if selecting the solver(s) to run for a new instance takes an unreasonable amount of time. Also having a huge dataset may involve a time dilation in the solvers selection process. For instance, the time required by using an approach based on *k-Nearest Neighbour* [18] classification depends on the size of the dataset since it requires the scan of the whole dataset.

Note that being able to reduce the effort of building the prediction model allows one to quickly adapt the prediction to new unseen problems or to exploit new available solvers. This is of particular importance since real life applications usually focus on solving a specific class of problems that may be not fully known in advance. Moreover, as the results of the Learning track of the Planning Competition 2014 [37] show, learning from new incoming instances may dramatically increase the performance.

3 Optimisation Problems

Optimisation problems are of great interest in many real life applications where we are interested in finding an optimal (or good enough) solution. If CSP portfolio solvers can draw inspiration from SAT approaches—possibly through an encoding into SAT—in a straightforward way, when dealing with COPs the matter becomes more complicated. Unlike CSPs, here the dichotomy solved/not solved is no longer suitable since a COP solver can provide *sub-optimal solutions* without finding the optimal one (or proving its optimality).

The first issue here is the lack of a universally accepted metric for measuring the performance of a COP solver, and therefore for building COP prediction models. Since for hard combinatorial problems it is often very difficult to complete the search in reasonable time, it is clear that the solution quality must be taken into account. What is less clear is *how* to do it. There are plenty of metrics, used in well known solving competitions, for evaluating the performance of COP solvers. In [49,62] the solvers are ranked by using a lexicographic order over the solution quality, the number of the solved instances, and the solving time. The purse score described in [12] was used in the SAT Competition 2005, while a metric exploiting results aggregation and pair-wise comparisons between solvers is proposed in [22]. The MiniZinc Challenge uses instead a *Borda* count voting system: problems are treated like voters who rank the solvers. This approach is surely reasonable, but in our opinion has a disadvantage: it could overestimate small time differences in case of easy instances, as well as underrate big time differences in case of medium and hard instances. In [5,7] we proposed and evaluated alternative metrics that take into account both the solution quality (i.e., the score metric) and the anytime performance of the solvers (i.e., the area metric) without relying on cross-comparisons between the solvers.

The possibility of producing sub-optimal solutions is a key factor worth investigating. We argue that collaborative strategies can be successfully adopted by COP portfolio solvers. Indeed, some solvers quickly find good sub-optimal solutions but fail to improve later, while others are slower but in the end find better solutions. In this setting a solver can exploit the objective function bounds found by another solver to reduce its search space, as done for instance in [7]. As an example, consider the behaviours of the solvers s_1 and s_2 in Fig. 2 within a timeout of $T = 1000$ s. The best value $v^* = 10$ is found by s_2 after 900 s, but it takes 800 s to find its first solution ($v = 45$). Meanwhile, s_1 finds a better value ($v = 40$) after just 10 s and even better values in just 100 s. So, the question is: what happens if we "inject" the upper bound 40 from s_1 to s_2? Considering that starting from $v = 45$ the solver s_2 is able to find v^* in 100 s (from 800 to 900), hopefully starting from any better (or equal) value $v' \leq 45$ the time needed by s_2 to find v^* is no more than 100 s. From a graphical point of view, this means in some way to "shift" the curve of s_2 towards the left from $t = 800$ to 10, by exploiting the fact that after 10 s s_1 can suggest to s_2 the upper bound $v = 40$. The cooperation between s_1 and s_2 would thereby reduce by $\Delta t = 790$ s the time needed to find v^*, and moreover would allow to exploit the remaining Δt s for finding better solutions or even proving the optimality of v^*. However, note that this *virtual* behaviour may not occur: it may be that s_2 calculates important information in the first 800 s required to find the solution $v^* = 10$, and therefore the injection of $v = 40$ could be useless (if not harmful).

The decision of switching between the solvers can be made statically, as done in [7], but also dynamically at run time. It may be counterproductive to stop a solver if it is actively producing new solutions while it is likely that it will not produce solutions if no solutions are produced so far. Of course, the behaviour of a solver depends on its nature and on the problem to be solved. We believe that

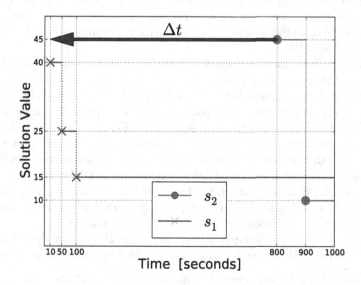

Fig. 2. Example of bound communication from s_1 to s_2.

Table 1. Solvers outcomes on a bin-packing instance.

	CPX	FD	LazyFD	MIP	par-folio	obj-folio
Best Value Found	No Value	No Value	10	10	10	10
Optimisation Time [s]	Timeout	Timeout	Timeout	182.59	341.49	**16.71**

interesting patterns can emerge by studying the problems and the solvers in a more "qualitative" way rather than performing only "quantitative" observations. From a practical point of view, *monitoring* the sub-optimal solutions found by the running solvers might be the starting point for the definition of more dynamic portfolios, where decisions are made during the search instead of relying on costly prediction models. We are currently examining this approach, that we consider promising especially in parallel settings.

Making tons of experiments on a large amount of data is certainly significant from a statistical point of view, but somehow hinders the understanding of what we are experimenting. Indeed, as mentioned in [42], often we observe and evaluate the performance of different algorithms without being able to give a full *explanation* for such performance. Hopefully, looking at the COP solvers behaviour more in depth could give us some explanation and hints on how to better combine the different solvers.

4   Parallelisation

Having a finite portfolio, its parallelisation would seem a trivial issue: you only need to run in parallel all the solvers. Unfortunately, often the number of the constituent solvers exceeds the number of available cores. Furthermore, even assuming to have fewer solvers than cores, it is likely that—due to synchronisation and memory contention issues—running in parallel all the solvers on the same multicore machine is actually different from running the same solvers on different machines [60].

In the SAT field parallel portfolios have been extensively studied. Usually, different configurations of the same solver are run simultaneously by enabling the sharing of learned clause between solvers [10,28,58]. Conversely, the parallelisation of CP solvers does not appear currently so fruitful. For example, in the MiniZinc Challenges 2014/15 the possibility of multiprocessing did not lead in general to remarkable performance gains despite the availability of 8 logical cores: the overall best single solver was the single-threaded solver Chuffed [15]. Except for some preliminary investigations done for the CPHydra and Numberjack solvers [29,73] we are not aware of parallel CP portfolio solvers. This issue gives rise to interesting research perspectives. Specifically, parallelisation seems to be highly promising when applied to optimisation problems.

Let us consider as an example an instance of the *Bin Packing* problem taken from the `minizinc-1.6` benchmarks[1] when solved by using a portfolio of the

---

[1] The model file is `2DPacking.mzn` while the data file is `Class7_40_3.dzn`.

solvers coming with the G12 MiniZinc distribution (viz., CPX, FD, LazyFD, and MIP) on a quadcore architecture within a timeout of 500 s. Table 1 shows the results achieved by the single solvers against two portfolio solvers: par-folio that just runs all the solvers in parallel and obj-folio that runs all the solver in parallel, restarting the solvers with the new bound of the objective function every time a better solution is found. The best solver of the portfolio is MIP, which is able to complete the search in 182.59 s. LazyFD finds in few seconds the optimal value 10, but never proves its optimality. FD and CPX instead do not find any solution. par-folio is remarkably worse than MIP (it takes almost twice the time to prove the optimality) witnessing that in practice running all the solvers in parallel does not mimic the sequential execution of the single solvers. obj-folio instead significantly *outperforms* the best solver: it quickly proves the optimality because in this case MIP is restarted by exploiting the value 10 found by LazyFD after few seconds.

Another interesting point concerns the solvers scheduling when the portfolio size m exceeds the number n of available cores. In the above example $m = n = 4$, but what if $n < m$? Is it better to select $k < n$ solvers for reducing the processor load, or to choose $k = n$ solvers to be run on all the available cores, or even scheduling $k > n$ solvers by properly splitting the solving time window? Furthermore, is it better to use a static approach, where the solvers to run are decided in advance, or a dynamic one, where solvers are selected on-line according to the instance to be solved?

A major challenge is also predicting if, and when, restarting a solver is beneficial. Particular care must be taken in restarting a solver with new objective bounds, since interrupting the solver search means to lose the knowledge gained during the computation. This may be harmful for solvers that accumulate information during the search, e.g., lazy clause generation solvers [53].

Techniques like nogood learning and lazy clause generation have proven to be very effective, and in a parallel setting can gain additional benefits. Unfortunately, only few CP solvers use nogoods and there is no standard API to retrieve this knowledge. A standard protocol for extracting and sharing nogoods is hence desirable, since often the portfolio solver views its constituent solvers as "black-boxes" on which it has a very limited control. However, even without the control over the constituent solvers, it is possible to work directly on the problem to be solved. For instance, one can adopt work splitting techniques for dividing the original problem into a number of sub-problems, each of which assigned to a different solver.

5   Utilisation

As pointed out also in [61], a key reason for the lack of common adoption of portfolio solvers is their poor usability. The effort required to set up a portfolio solver is typically much higher than the cost of installing its constituent solvers. Building the prediction model is a hard work, partially justifiable by the fact that this process is performed off-line.

We believe that a major challenge for CP portfolio solvers relies in simplifying their installation and use, and that the best way to overcome this problem is to encourage the dissemination of (possibly open-source) portfolio solvers to be downloaded, installed, and used just like a regular individual solver.

For easing the everyday use, the installation of a portfolio solver should not be a nightmare. Wrapping all the software needed in a minimal virtual machine might be an idea: in this way the user just need to start the virtual machine and run the portfolio solver, instead of dealing with the installation of all the necessary components. Furthermore, following the current trend of cloud computing, it would be interesting to develop an *"online"* portfolio solver as a service to be installed and run in a public or private cloud. Some preliminary works for testing portfolios on the cloud have already been done [24,43] and, as underlined in [40], this solution may have some advantages. The solving process is transparent for the end-user, which only needs the API for communicating with the server. Moreover, the internals of the portfolio solver can be studied and maintained directly in the cloud, by taking advantage of the emerging capabilities of cloud computing. An "immutable service" approach [21,50] would enable to use the cloud resources to concurrently solve the incoming problems and update the prediction models.

The diffusion of CP portfolio solvers could have positive implications also for the individual solvers. Aside from the "image return" for a solver belonging to a successful portfolio, there are also technical aspects. For example, we realized that a lot of solvers we tested have some bugs (e.g., only considering the MiniZinc Challenge 2014 there have been 24 wrong answers given by 5 different solvers). Portfolio solvers can be used for checking the reliability of a solver, by comparing its answer on a given problem against the answers given by each other solver of the portfolio. We believe that a portfolio solver should take into account the unreliability of its constituent solvers. Getting rid of a buggy solver may be too penalizing since it is often the case that the most promising solvers to include in a portfolio are the experimental ones, usually maintained by few people and not extensively tested. Where it is not possible to fix the bug, the verification *a posteriori* of the solution is preferable: the constituent solvers of the portfolio can be used for double-checking the solution. Unfortunately, the verification of a solution is sometimes computationally infeasible, especially when it comes to prove the unsatisfiability or the optimality. An alternative idea might be to encode the reliability in the prediction model, e.g., by associating to each solver a trust level.

5.1   SUNNY and sunny-cp

In order to facilitate and encourage the practical usage of CP portfolio solvers we developed sunny-cp [4], a parallel portfolio solver built on top of the SUNNY algorithm [3] able to solve generic CP problems encoded in MiniZinc language.

SUNNY is a lazy portfolio approach which exploits instances similarity to guess the best solver(s) to use. For a given problem instance p, SUNNY uses a k-Nearest Neighbours (k-NN) algorithm to select from a training set of known instances the subset $N(p, k)$ of the k problems closer to p. According to the $N(p, k)$ instances, SUNNY relies on three heuristics: h_{sel}, for *selecting* the most promising solvers to run; h_{all}, for *allocating* to each solver a certain runtime (the more a solver is promising, the more time is allocated); and h_{sch}, for *scheduling* the sequential execution of the solvers according to their presumed speed. These heuristics depend on the application domain. For example, for CSPs h_{sel} selects the smallest sub-portfolio $S \subseteq \Pi$ that solves the most instances in $N(p, k)$, by using the solving time for breaking ties. h_{all} allocates to each $s_i \in S$ a time t_i proportional to the instances that S can solve in $N(p, k)$, while h_{sch} sorts the solvers by increasing solving time in $N(p, k)$. For COPs the approach is analogous, but different performance metrics are used [5].

The first version of sunny-cp was sequential [6] and relied on eight solvers, viz. Chuffed, CPX, G12/CBC, G12/FD, G12/LazyFD, G12/Gurobi, Gecode, and MinisatID.[2] Then, we significantly improved it by adding more solvers (viz. Choco, iZplus, HaifaCSP, and OR-Tools) to its portfolio and especially by allowing their simultaneous execution and cooperation on multiple cores. This allowed sunny-cp to win the gold medal in the open category of MiniZinc Challenge 2015.

6   Related Work

The interest in algorithm selection and configuration is quite general and growing. It is outside the scope of the current paper to give a global overview of the plethora of portfolio approaches tried in the literature. For more comprehensive surveys, we refer the interested reader to [36,42,64].

As said earlier, portfolio solvers have proven their effectiveness in many international solving competitions. The SAT portfolio solvers 3S [38] and CSHC [46] won gold medals in SAT Competition 2011 and 2013 respectively. SATZilla [72] won the SAT Challenge 2012, CPHydra [54] the Constraint Solver Competition 2008, the ASP portfolio solver claspfolio [17] was gold medallist in different tracks of the ASP Competition 2009 and 2011, ArvandHerd [70] and IBaCoP [14] won some tracks in the Planning Competition 2014.

Apart from CPHydra and SUNNY, there are only few other approaches that can deal with CSPs. In [8,9] Machine Learning techniques are used to enhance the performances of a single CSP solver by dynamically adapting its search heuristics. These works lists an extensive set of features to train and improve the heuristics model through Support Vector Machines. Proteus [33] is a recent CSP portfolio approach that does not rely purely on CSP solvers, but may decide to encode a CSP problem instance into SAT, by selecting an appropriate encoding and a corresponding SAT solver.

Regarding optimisation problems, we can say that COP portfolios are mostly developed just for some specific optimisation problems like Knapsack, Most Probable Explanation, Set Partitioning, Travel Salesman Problem [26,36,69].

---

[2] sunny-cp attended the MiniZinc Challenge 2014 with respectable results (4th out of 18). It has also been awarded with an *honourable mention* by the challenge organizers.

The only COP solver we are aware of is presented in [7] using an adaptation of the SUNNY algorithm. An empirical evaluation of different portfolio approaches applied to COPs was performed in [5].

Surprisingly enough, only a few portfolio solvers are parallel and even fewer are the dynamic ones selecting on-line the solvers to run. We are aware of only two dynamic and parallel portfolio solvers that attended a solving competition, namely p3S [45] (in the SAT Challenge 2012) and IBaCoP2 [14] (in the Planning Competition 2014). Apart from a preliminary investigation about CPHydra parallelisation [73], the only parallel and dynamic CP portfolio solver able to deal with also COPs is sunny-cp [4]. The parallelisation of portfolio solvers is a hot topic which is drawing some attention in the community. For instance, parallel extensions of well-known sequential portfolio approaches are studied in [32], while in [30] ASP techniques are used for computing a static schedule of solvers which can even be executed in parallel.

Finally, a number of tools are being developed in order to improve portfolio solvers usability. snappy [61] is a simple and training-less algorithm portfolio which relies on a nearest neighbours prediction mechanism. LLAMA (Leveraging Learning to Automatically Manage Algorithm) [41] is instead a framework that facilitates the exploration of different portfolio techniques on any problem domain, by supporting the most common solver selectors and possibly combining them.

7   Conclusions and Extensions

Portfolio approaches have been extensively studied, and successfully used in solving competitions. In this paper we discussed the main challenges that, in our view, need to be tackled for spreading the use of portfolio approaches in Constraint Programming. We identified in particular four main aspects: the prediction model used for the solver selection, the treatment of optimisation problems, the parallelisation of execution, and the actual usability of CP portfolio solvers.

We already performed some preliminary investigations, and we are currently working on the implementation of some ideas we proposed. In particular, we are working to further improve the sunny-cp solver.

Clearly, portfolio solvers are not a *panacea* and there are contexts in which their use is unnecessary. For instance, when a given solver of the portfolio strongly dominates all the others it might be preferable switching to other related techniques such as *Algorithm Configuration* [34,35,39] for properly tuning the parameters of the dominant solver. Scenarios like this are not uncommon in real life applications, where the focus is on solving a specific problem (or class of problems) rather than different problems disparate in their nature.

Having more and better datasets and solvers is of course welcome for our purposes. We would like to encourage the CP community to submit new problems and solvers to international solving competitions like the MiniZinc Challenge. To advance the state of the art and bridge the current gaps, it would be nice to have a number of CP portfolio entrants (maybe running in a dedicated track). This somehow would go against the—surprising in our opinion—direction taken in the SAT competition 2014, where only portfolio approaches consisting of at most two core algorithms were allowed.

We conclude the paper by discussing some possible extensions of the portfolio approach beyond the CP paradigm. As mentioned, Algorithm Portfolios can be viewed as particular instances of the Algorithm Selection framework where we are interested in predicting case-by-case the best algorithm (not necessarily a constraint solver) for any new, unseen problem to be solved. The generality of this framework allows its instantiation to different paradigms such as Boolean Satisfiability (SAT), Answer-Set Programming (ASP), Quantified Boolean Formula (QBF), or even for solving different instances of the same problem, e.g., the Container Pre-marshalling Problem [13].

A natural, yet unexplored, target for portfolio solvers is certainly the *Constraint Logic Programming* (CLP) field. On the one hand, from a CLP specification is possible to derive a CP problem to be solved by CP portfolio solvers like for instance what done in [16]. On the other hand, a CLP solver can provide interfaces for dealing with different CP problem specifications (e.g., the Zinc library of SICStus Prolog [63] that allows to solve both FlatZinc and MiniZinc models) and therefore be embedded into a portfolio solver.

The portfolio framework also enables to consider program transformation techniques that may speed up the solving process. A possible approach consists in splitting the input problem into different, maybe overlapping sub-problems and to assign the sub-problems to the different constituent solvers. This might be advantageous especially when solvers are running simultaneously. Furthermore, the input model can even be enriched by adding redundant constraints (e.g., bounds, nogoods or other clauses learned by a solver during the search) for narrowing the search space. We remark that program transformation techniques are not uncommon in constraint solving and in particular there exists a lot of work proposing different techniques for encoding a CP problem into a SAT problem [1,11,33,66,68]. In this setting, portfolio approaches can be used for predicting whether and how to compile a CP model into SAT.

References

1. Abío, I., Stuckey, P.J.: Encoding linear constraints into SAT. In: O'Sullivan, B. (ed.) CP 2014. LNCS, vol. 8656, pp. 75–91. Springer, Heidelberg (2014)
2. Amadini, R., Gabbrielli, M., Mauro J.: An enhanced features extractor for a portfolio of constraint solvers. In: SAC, pp. 1357–1359. ACM (2014)
3. Amadini, R., Gabbrielli, M., Mauro, J.: SUNNY: a lazy portfolio approach for constraint solving. TPLP 14(4–5), 509–524 (2014)
4. Amadini, R., Gabbrielli, M., Mauro J.: A multicore tool for constraint solving. In: IJCAI, pp. 232–238. AAAI Press (2015)
5. Amadini, R., Gabbrielli, M., Mauro, J.: Portfolio approaches for constraint optimization problems. In: AMAI, pp. 1–18 (2015)
6. Amadini, R., Gabbrielli, M., Mauro, J.: SUNNY-CP: a sequential CP portfolio solver. In: SAC, pp. 1861–1867. ACM (2015)
7. Amadini, R., Stuckey, P.J.: Sequential time splitting and bounds communication for a portfolio of optimization solvers. In: O'Sullivan, B. (ed.) CP 2014. LNCS, vol. 8656, pp. 108–124. Springer, Heidelberg (2014)

8. Arbelaez, A., Hamadi, Y., Sebag, M.: Online heuristic selection in constraint programming. In: SoCS (2009)
9. Arbelaez, A., Hamadi, Y., Sebag, M.: Continuous search in constraint programming. In: ICTAI, pp. 219–243. IEEE Computer Society (2010)
10. Audemard, G., Hoessen, B., Jabbour, S., Lagniez, J-M., Piette, C.: PeneLoPe, a parallel clause-freezer solver. In: SAT Challenge, pp. 43–44 (2012)
11. Barahona, P., Hölldobler, S., Nguyen, V.-H.: Representative encodings to translate finite CSPs into SAT. In: Simonis, H. (ed.) CPAIOR 2014. LNCS, vol. 8451, pp. 251–267. Springer, Heidelberg (2014)
12. Le Berre, D., Simon, L.: Preface to the special volume on the SAT 2005 competitions and evaluations. In: JSAT, 2(1–4), (2006)
13. Bischl, B., Kerschke, P., Kotthoff, L., Lindauer, M.T., Malitsky, Y., Fréchette, A., Hoos, H.H., Hutter, F., Leyton-Brown, K., Tierney, K., Vanschoren, J.: Aslib: A benchmark library for algorithm selection. CoRR, abs/1506.02465 (2015)
14. Cenamor, I., de la Rosa, T., Fernández, F.: IBACOP and IBACOP2 Planner (2014). http://www.plg.inf.uc3m.es/~icenamor/files/IBaCoPPlanner.pdf
15. Chu, G., de la Banda, M.G., Stuckey, P.J.: Automatically exploiting subproblem equivalence in constraint programming. In: Lodi, A., Milano, M., Toth, P. (eds.) CPAIOR 2010. LNCS, vol. 6140, pp. 71–86. Springer, Heidelberg (2010)
16. Cipriano, R., Dovier, A., Mauro, J.: Compiling and executing declarative modeling languages to gecode. In: Garcia de la Banda, M., Pontelli, E. (eds.) ICLP 2008. LNCS, vol. 5366, pp. 744–748. Springer, Heidelberg (2008)
17. claspfolio: http://www.cs.uni-potsdam.de/claspfolio/
18. Duda, R.O., Hart, P.E., Stork, D.G.: Pattern Classification, 2nd edn. Wiley-Interscience, New York (2000)
19. Fawcett, C., Vallati, M., Hutter, F., Hoffmann, J., Hoos, H.H., Leyton-Brown, K.: Improved features for runtime prediction of domain-independent planners. In: ICAPS. AAAI (2014)
20. Frisch, A.M., Harvey, W., Jefferson, C., Martínez-Hernández, B., Miguel, I.: Essence: a constraint language for specifying combinatorial problems. Constraints 13(3), 268–306 (2008)
21. FullContact. How We Used Immutable Servers to Simplify Our Cloud Infrastructure. (2014) http://www.fullcontact.com/blog/immutable-servers-benefits/
22. Van Gelder, A.: Careful ranking of multiple solvers with timeouts and ties. In: Sakallah, K.A., Simon, L. (eds.) SAT 2011. LNCS, vol. 6695, pp. 317–328. Springer, Heidelberg (2011)
23. Gent, I.P., Walsh, T.: CSPlib: a benchmark library for constraints. In: Jaffar, J. (ed.) CP 1999. LNCS, vol. 1713, pp. 480–481. Springer, Heidelberg (1999)
24. Geschwender, D., Hutter, F., Kotthoff, L., Malitsky, Y., Hoos, H.H., Leyton-Brown, K.: Algorithm configuration in the cloud: a feasibility study. In: Pardalos, P.M., Resende, M.G.C., Vogiatzis, C., Walteros, J.L. (eds.) Learning and Intelligent Optimization. Lecture Notes in Computer Science, pp. 41–46. Springer, Heidelberg (2014)
25. Gomes, C.P., Selman, B.: Algorithm portfolios. Artif. Intell. 126(1–2), 43–62 (2001)
26. Guo, H., Hsu, W.H.: A machine learning approach to algorithm selection for NP-hard optimization problems: a case study on the MPE problem. Ann. OR 156(1), 61–82 (2007)
27. Guyon, I., Elisseeff, A.: An introduction to variable and feature selection. J. Mach Learn. Res. 3, 1157–1182 (2003)
28. Hamadi, Y., Jabbour, S., Sais, L.: ManySAT: a parallel SAT solver. JSAT 6(4), 245–262 (2009)

29. Hebrard, E., O'Mahony, E., O'Sullivan, B.: Constraint programming and combinatorial optimisation in numberjack. In: Lodi, A., Milano, M., Toth, P. (eds.) CPAIOR 2010. LNCS, vol. 6140, pp. 181–185. Springer, Heidelberg (2010)

30. Hoos, H.H., Kaminski, R., Lindauer, M.T., Schaub, T.: aspeed: solver scheduling via answer set programming. In: TPLP (2015)

31. Hoos, H.H., Kaufmann, B., Schaub, T., Schneider, M.: Robust benchmark set selection for boolean constraint solvers. In: Nicosia, G., Pardalos, P. (eds.) LION 7. LNCS, vol. 7997, pp. 138–152. Springer, Heidelberg (2013)

32. Lindauer, M., Hoos, H., Hutter, F.: From sequential algorithm selection to parallel portfolio selection. In: Jourdan, L., Dhaenens, C., Marmion, M.-E. (eds.) LION 9 2015. LNCS, vol. 8994, pp. 1–16. Springer, Heidelberg (2015)

33. Hurley, B., Kotthoff, L., Malitsky, Y., O'Sullivan, B.: Proteus: a hierarchical portfolio of solvers and transformations. In: Simonis, H. (ed.) CPAIOR 2014. LNCS, vol. 8451, pp. 301–317. Springer, Heidelberg (2014)

34. Hutter, F., Hoos, H.H., Leyton-Brown, K.: Sequential model-based optimization for general algorithm configuration. In: Coello, C.A.C. (ed.) LION 2011. LNCS, vol. 6683, pp. 507–523. Springer, Heidelberg (2011)

35. Hutter, F., Hoos, H.H., Leyton-Brown, K., Stützle, T.: ParamILS: an automatic algorithm configuration framework. J. Artif. Intell. Res. (JAIR) **36**, 267–306 (2009)

36. Hutter, F., Xu, L., Hoos, H.H., Leyton-Brown, K.: Algorithm runtime prediction: the state of the art. CoRR, abs/1211.0906 (2012)

37. International planning competition: http://ipc.icaps-conference.org/

38. Kadioglu, S., Malitsky, Y., Sabharwal, A., Samulowitz, H., Sellmann, M.: Algorithm selection and scheduling. In: Lee, J. (ed.) CP 2011. LNCS, vol. 6876, pp. 454–469. Springer, Heidelberg (2011)

39. Kadioglu, S., Malitsky, Y., Sellmann, M., Tierney, K.: ISAC - instance-specific algorithm configuration. In: Coelho, H., Studer, R., Wooldridge, M. (eds.) ECAI, Frontiers in Artificial Intelligence and Applications, vol. 215 pp. 751–756. IOS Press (2010)

40. Kiziltan, Z., Mauro, J.: Service-oriented volunteer computing for massively parallel constraint solving using portfolios. In: Lodi, A., Milano, M., Toth, P. (eds.) CPAIOR 2010. LNCS, vol. 6140, pp. 246–251. Springer, Heidelberg (2010)

41. Kotthoff, L.: LLAMA: leveraging learning to automatically manage algorithms. CoRR, abs/1306.1031 (2013)

42. Kotthoff, L.: Algorithm selection for combinatorial search problems: a survey. AI Mag. **35**(3), 48–60 (2014)

43. Kotthoff, L.: Reliability of computational experiments on virtualised hardware. J. Exp. Theor. Artif. Intell. **26**(1), 33–49 (2014)

44. Kroer, C., Malitsky, Y.: Feature filtering for instance-specific algorithm configuration. In: ICTAI, pp. 849–855. IEEE (2011)

45. Malitsky, Y., Sabharwal, A., Samulowitz, H., Sellmann, M.: Parallel SAT solver selection and scheduling. In: Milano, M. (ed.) CP 2012. LNCS, vol. 7514, pp. 512–526. Springer, Heidelberg (2012)

46. Malitsky, Y., Sabharwal, A., Samulowitz, H., Sellmann, M.: Algorithm portfolios based on cost-sensitive hierarchical clustering. In: IJCAI. AAAI (2013)

47. Maratea, M., Pulina, L., Ricca, F.: Multi-engine ASP solving with policy adaptation. J. Logic Comput. 1–22 (2013)

48. Maratea, M., Pulina, L., Ricca, F.: A multi-engine approach to answer-set programming. TPLP **14**(6), 841–868 (2014)

49. Max-SAT 2013: http://maxsat.ia.udl.cat/introduction/

50. Morris, K.: Immutable server web page (2013). http://martinfowler.com/bliki/ImmutableServer.html
51. mzn2feat-1.0 web page: http://www.cs.unibo.it/amadini/mzn2feat-1.0.tar.bz2
52. Nethercote, N., Stuckey, P.J., Becket, R., Brand, S., Duck, G.J., Tack, G.: MiniZinc: towards a standard CP modelling language. In: Bessière, C. (ed.) CP 2007. LNCS, vol. 4741, pp. 529–543. Springer, Heidelberg (2007)
53. Ohrimenko, O., Stuckey, P.J., Codish, M.: Propagation via lazy clause gener. Constraints 14(3), 357–391 (2009)
54. O'Mahony, E., Hebrard, E., Holland, A., Nugent C., O'Sullivan, B.: Using case-based reasoning in an algorithm portfolio for constraint solving. In: AICS 2008 (2009)
55. Pulina, L., Tacchella, A.: A multi-engine solver for quantified boolean formulas. In: Bessière, C. (ed.) CP 2007. LNCS, vol. 4741, pp. 574–589. Springer, Heidelberg (2007)
56. Pulina, L., Tacchella, A.: A self-adaptive multi-engine solver for quantified boolean formulas. Constraints 14(1), 80–116 (2009)
57. Rice, J.R.: The algorithm selection problem. Adv. Comput. 15, 65–118 (1976)
58. Roussel, O.: ppfolio. http://www.cril.univ-artois.fr/~roussel/ppfolio/
59. Roussel O., Lecoutre, C.: XML representation of constraint networks: format XCSP 2.1. CoRR, abs/0902.2362 (2009)
60. Sabharwal, A., Samulowitz, H.: Insights into parallelism with intensive knowledge sharing. In: O'Sullivan, B. (ed.) CP 2014. LNCS, vol. 8656, pp. 655–671. Springer, Heidelberg (2014)
61. Samulowitz, H., Reddy, C., Sabharwal, A., Sellmann, M.: Snappy: a simple algorithm portfolio. In: Järvisalo, M., Van Gelder, A. (eds.) SAT 2013. LNCS, vol. 7962, pp. 422–428. Springer, Heidelberg (2013)
62. SAT Challenge 2012: http://baldur.iti.kit.edu/SAT-Challenge-2012/
63. Zinc Interface library(zinc). https://sicstus.sics.se/sicstus/docs/4.1.0/html/sicstus/lib_002dzinc.html#lib_002dzinc
64. Smith-Miles, K.: Cross-disciplinary perspectives on meta-learning for algorithm selection. ACM Comput. Surv. 41(1) (2009)
65. Stojadinovic M., Maric, F.: Instance-based selection of CSP solvers using short training. In: Pragmatics of SAT (2014)
66. Stojadinovic, M., Maric, F.: meSAT: multiple encodings of CSP to SAT. Constraints 19(4), 380–403 (2014)
67. Stuckey, P.J., Becket, R., Fischer, J.: Philosophy of the miniZinc challenge. Constraints 15(3), 307–316 (2010)
68. Tamura, N., Taga, A., Kitagawa, S., Banbara, M.: Compiling finite linear CSP into SAT. Constraints 14(2), 254–272 (2009)
69. Telelis, O., Stamatopoulos, P.: Combinatorial optimization through statistical instance-based learning. In: ICTAI, pp. 203–209 (2001)
70. Valenzano, R.A., Nakhost, H., Müller, M., Schaeffer, J., Sturtevant, N.R.: Arvand-Herd: parallel planning with a portfolio. In: ECAI, pp. 786–791 (2012)
71. Xu, L., Hutter, F., Shen, J., Hoos, H., Leyton-Brown, K.: SATzilla2012: improved algorithm selection based on cost-sensitive classification models. In: SAT Challenge (2012)
72. Lin, X., Hutter, F., Hoos, H.H., Leyton-Brown, K.: SATzilla: portfolio-based algorithm selection for SAT. JAIR 32, 565–606 (2008)
73. Yun, X., Epstein, S.L.: Learning algorithm portfolios for parallel execution. In: Hamadi, Y., Schoenauer, M. (eds.) LION 2012. LNCS, vol. 7219, pp. 323–338. Springer, Heidelberg (2012)

CHR in Action

Arwa Ismail$^{(\boxtimes)}$, Nada Sharaf, and Slim Abdennadher

German University in Cairo, New Cairo, Egypt
{arwa.sayed-ismail,nada.hamed,slim.abdennadher}@guc.edu.eg

Abstract. Constraint Handling Rules (CHR) has expanded its application range over the past few years to include different algorithms rather than only constraint solvers. Animation of algorithms has been used over the past few decades to aid the understanding of programming languages and how they are processed. In this work, we present a generic form of animating CHR programs using source-to-source transformation. The transformation converts CHR programs into their equivalent CHR programs enhanced with animation features, in an automated manner.

Keywords: Constraint Handling Rules · Animation · Source-to-source transformation · XPCE

1 Introduction

Over the past years, Constraint Handling Rules (CHR), a high-level constraint-based language [10], has been used with a wide range of fields. CHR was initially introduced for writing constraint solvers. It has developed into a general purpose language due to its rule-based declarative nature.

Generally, the human brain is capable of imagining scenarios shown through images in motion, better than being presented only with written text or static pictures. The animation tool presented in [15] depended on creating and displaying a well-animated and easy-to-follow representation of an algorithm in order to aid one's understanding of its functionality. The reason is that a static drawing or a typed piece of code is not sufficient to describe its behaviour.

Due to the advantages of visualizing algorithms [13], various attempts have been made to visualize CHR programs. In [3,4], new approaches for visualizing the execution of CHR programs were presented. The visualization, however, focused on showing which rules are being executed. [17,18] presented the first approach towards a generic CHR program animation tool. The tool used source-to-source transformation to interface the CHR programs with another tool offering visual objects. The idea presented was to allow users to specify the interesting constraints/rules of a program. Such constraints were thus associated with visual objects. On removing/adding the constraints, the visual objects are changed thus animating the program. The platform presented in [17,18] used an external visualization tool for producing the animations. Specifically, Jawaa [14] was used for proof of concept. Jawaa is a visualization tool that offers different types of visual objects in addition to some actions to move objects and

© Springer International Publishing Switzerland 2015
M. Falaschi (Ed.): LOPSTR 2015, LNCS 9527, pp. 365–383, 2015.
DOI: 10.1007/978-3-319-27436-2_22

change their graphical characteristics. The problem, however, was that without the available actions, provided by Jawaa, it would have not been possible to have such animated visualizations. The users in that case would only see visual objects being added and removed. This would thus provide no animation of the changes happening to the objects such as a translation from one position to another. In addition, the actions available to the user depend on the tool being used. Thus if another visualization tool with different actions was used, the user might miss on some needed animations.

The new tool is thus a standalone Prolog-based platform. It is able to provide users with any needed action without using or depending on an external visualization tool. This isolates the platform from any changes in such external tools. The platform only uses CHR rules to add new animation actions into a normal CHR program. The platform solely depends on Prolog. This is due to the fact that CHR is, eventually, converted to Prolog for execution. Animation was also done using a Prolog graphical toolkit (XPCE). Conceptually, each constraint is represented by a graphical object with certain characteristics which is drawn onto the visualized environment. The animation takes place as a result of detecting a change in one or more of the object's characteristics, removing the object from the visualization, and adding it back again as a new object with a new set of characteristics. This implies that each time an object's characteristic (such as its position) is changed, the change is simulated in a step-by-step manner. The new CHR programs are thus able to automatically detect changes in the characteristics of an "interesting constraint". As a result, a graphical tool providing only basic objects (such as XPCE) can be instructed by the new program to produce an animation of the change such as moving an object from one position to another. Figure 3a shows how the tool was used to animate a sorting algorithm. The tool is available from http://met.guc.edu.eg/chr/chrinaction.

The paper is organized as follows: Sect. 2 presents previous related work and the contribution of this work accordingly. Sects. 3 and 4 introduce the needed background information. Section 5 introduces the needed steps for the source-to-source transformation of the input files. In Sect. 6, the CHR animation is discussed. Section 7 shows an example of how the transformed file works. We finalize with conclusions and directions of future work.

2   Related Work

Over the years, there has been a lot of work into visualizing and animating algorithms. For logic programming, however, the focus was on the program regardless of the implemented algorithm. In some of the systems such as the system presented in [16], logic programs were represented using a variation of cyclic AND/OR graphs representing the structure of the program. Dynamic graphs showed steps of the solution. A set of binding dependency graphs were also used to show how the values of the variables were generated. In [8], Augmented AND/OR trees (AORTA) were used as a means for a tracing and debugging facility. Another set of tools [19,20] focused on visualizing the search space and its changes over the course of execution.

Unlike the previous systems, the focus in our system is to animate the execution of different types of algorithms. There were various attempts to provide animations of algorithms in general. [5] provides a 30-minute video showing the animation of different sorting algorithms. The platform presented in [12] provides animations of different types of algorithms such as sorting, tree and graph algorithms instead of only sorting algorithms. However, both systems lacked generality. They only had built in animation facilities for well-known algorithms.

The system presented in this work aims at providing a general animation for the execution of CHR programs. The system was first introduced in [17]. Similar to BALSA [6] and Zeus [7], the notion of interesting events was used. The main difference is the different type of interesting events offered in the new system. It is also simpler to use. Unlike Balsa and Zeus no algorithm animators are required to write the views and specify how animation should work. However, one of the main drawbacks of the system presented in [17] is that the animation was outsourced to an external tool. This was done to have a generic animation platform depending on the basic graphical objects provided by animation tools such as Jawaa. As a consequence, the animation depends on the external tool. Thus any change of the tool could affect the needed animations. That is why the work presented in this paper aims at having the animation detected and handled automatically by the CHR program. In addition, a Prolog graphical toolkit (XPCE) was used as a proof of concept. The tool originally only provided a set of basic graphical objects (circle, rectangle, . . . etc.

3 Constraint Handling Rules

As previously mentioned, CHR [9,10] is a rule-based language. A CHR program contains a sequence of "simpagation" rules. The rules act upon constraints in the store until they are solved. There are two types of constraints in a program: user-defined/CHR constraints and built-in constraints handled by the host language. The general format of a simpagation rule is:

$$optional\_rule\_name \ @ \ H_K \setminus H_R \ \Leftrightarrow \ G \mid B.$$

The head constraints (H_K, H_R) contain a conjunction of CHR constraints. A conjunction is represented by a comma. The guard (G) contains built-in constraints only. The body (B) could contain CHR and built-in constraints. The rule is executed if the constraint store contains constraints matching the head constraints and if the guard (G) is satisfied. Once the rule is executed, the head constraints H_K are kept in the store while H_R are removed from the store. In the case where H_K is empty, the rule is a simplification rule of the form:

$$optional\_rule\_name \ @ \ H_R \ \Leftrightarrow \ G \mid B.$$

On the other hand, if H_R is empty, the rule is a propagation rule of the form:

$$optional\_rule\_name \ @ \ H_K \ \Rightarrow \ G \mid B.$$

Furthermore, the body of the rule (B) is executed. This results in having the body constraints (if any) thrown into the constraint store in their order of appearance.

In order to provide a better understanding for how CHR works, the following CHR program is introduced:

Listing 3.1. Sorting constraints in a descending manner.

```
sortingRule @ element(I1,V1),element(I2,V2) <=> I1<I2,
    V1<V2 | element(I2,V1), element(I1,V2).
```

The program sorts `element/2` constraints in descending order through the simplification rule `sortingRule`. The arguments of a constraint `element(I,V)` are respectively as follows: I represents the index of the element within the order and V corresponds to its value. For every two `element/2` constraints where the guard is satisfied, their indices are swapped. The guard states that the first constraint is at an index before the second constraint's index, but the value of the first constraint is smaller than the second one.

When querying the program with the following: `element(1,30)`, `element(2,100)` and `element(3,50)` respectively, the execution will proceed as follows:

1. `element(1,30)`, `element(2,100)` get matched to the rule `sortingRule`. They are removed from the constraint store and the following constraints are thrown into the constraint store respectively:
   (a) `element(2,30)` which gets matched to `sortingRule` again along with `element(3,50)`. Similarly, they are removed from the constraint store, and the following constraints are thrown respectively: `element(3,30)` and `element(2,50)` which do not match to any rule.
   (b) `element(1,100)` is then thrown into the constraint store.
2. The execution stops since there are no more constraints that can be matched to the `sortingRule`. Eventually, we are left with the following constraints: `element(1,100)`, `element(2,50)` and `element(3,30)`.

4   XPCE

XPCE[1] is a platform portable tool-kit used for developing GUIs where the platforms covered are UNIX/X11 and Windows (Windows-NT/2000/XP/Vista).

Assuming one's familiarity with Prolog syntax, the simplest way to introduce XPCE is through the four predicates it adds to Prolog introduced in [1,21]. The following are the three most important ones that have been used throughout the implementation of this work:

---

[1] An introduction to XPCE can be found in [2].

- new(?Ref, +Class(...Arg...)): Creates an object Ref as an instance of a class Class using the given argument(s) Arg, which is a set of arguments containing the characteristics of the instance created. The reference Ref can either be a Prolog variable or an XPCE object reference displayed as @ref where the name after the "@" can be any name.
- send(+Ref, +Method(...Arg...)): Similar to new/2, send/2 invokes the method Method with arguments Arg on the already existing object Ref.
- free(+Ref): destructs the referenced object Ref. As a result, it disappears from the visual environment.

Note that the above XPCE predicates have extended implementations which allow taking more arguments as input [1] (as can be seen in Listing 3.8).

5   Source-to-Source Transformation

In order to have a CHR program augmented with animation features, a source-to-source transformation is used as presented in Fig. 1. The transformation process feeds the input CHR program (P) to a parser. Next, the parser outputs a query (Q) to be introduced to the implemented CHR transformer (T) which in turn formulates the output transformed CHR program (P^T). The output of the parser is similar to the relational normal form presented in [11]. This form represents the different constituents of the original CHR program using special constraints. Initially, the user marks which constraints are to be animated. Each of these constraints are referred to as C^A. For each C^A, the user chooses a graphical object to represent it. This implementation supports the following graphical objects: circle, box, ellipse, line, text and image. For each object, a set of appropriate characteristics can be manipulated such as x-coordinate, y-coordinate, size, color, etc. The user is then prompted to specify the values of such characteristics through a GUI as presented in [17].

For any C^A constraint, it is augmented with an extra argument that represents a unique ID. Each ID maps to a specific object/constraint. In order to provide a correct animation and avoid manual edits to the original program, the user is prompted to mark any desired relations between a C^A head constraint and a corresponding C^A body constraint in a rule. A relation indicates that the

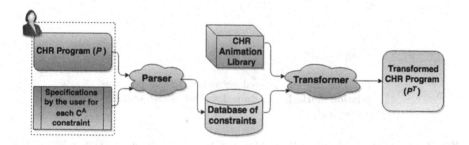

Fig. 1. Generic transformation process.

body constraint is an update of the head constraint. Such piece of information is important to detect the change (if any) of the object corresponding to such a C^A constraint. The detection of change will be further explained through Sect. 6. The two constraints involved in a relation have to be the same C^A constraint; same name and arity. For the sorting example in Listing 3.1, we use the implemented tool to mark the relationships desired for this program as shown in Fig. 2. For this program, we would like to see the translation of a certain element. Thus, we would like to link the head constraint `element(I1,V1)` with the body constraint `element(I2,V1)` to indicate that V1's position moved from I1 to I2 (provided that the user annotated an element/2 constraint's index to correspond to the graphical object's x-position). This is done through assigning the same ID to both constraints. Similarly, the head constraint `element(I2,V2)` is linked with the body constraint `element(I1,V2)`. In general, after all rules have been presented for the user to mark any desired relations, any C^A constraint not involved in a relation is assigned a unique ID.

Afterwards, the original CHR program is parsed to produce a set/database of CHR constraints carrying information about the original CHR program.

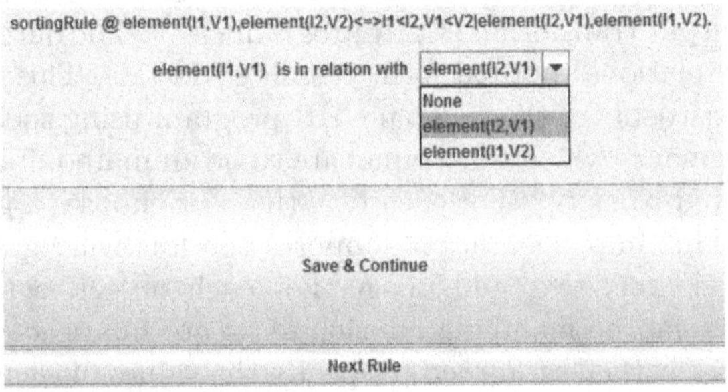

(a) Marking the relation for the first element. The drop-down list shows all possible body constraints for a relation.

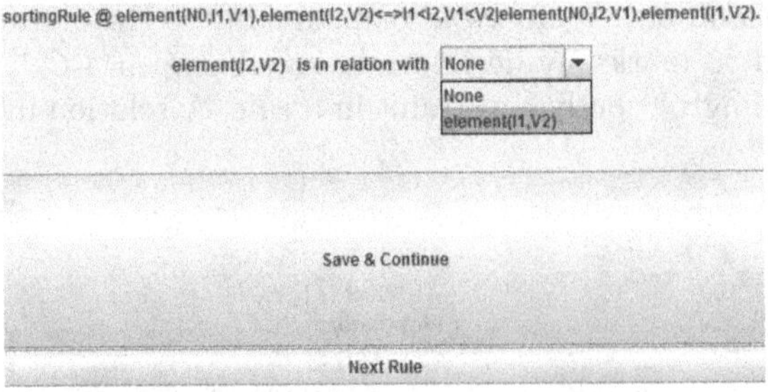

(b) The rule is updated according to the previous relation marked. The drop-down list is also updated accordingly.

Fig. 2. Screen-shots of the tool showing the user's input for marking relationships between head-body constraints for the sorting program.

The database also contains constraints carrying information about the characteristics of each C^A constraint (as per the user's input). Subsequently, the query (Q) includes all of the constraints within the database, constraints that would trigger rules in the transformer (T) to formulate the (transformed) rules of P^T as well as constraints that initiate and describe the animation environment.

In general, the transformed CHR program is structured as follows:

1. CHR rules responsible for linking C^A constraints to their corresponding graphical objects
2. Transformed rules of the original CHR program P to accommodate the extended graphical features of the transformed program P^T. The transformation of the rules follows the formalized form presented in Sect. 6.
3. A set of animation rules that constitutes the implemented CHR animation library. This library provides different animation features regarding the type/shape of objects to be visualized as well as different animation actions to be performed for an object. The rules of the implemented library are generic and are augmented to the end of any P^T to perform the animation actions.

6   Animation Using CHR

In order to activate the animation feature for any CHR program, the query is augmented with start/0. The predicate start/0 is used to create a window where the animation would be presented. It is also responsible for throwing three important constraints that aid the animation process.

– priority/2 is responsible for simulating a graphical queue which controls and maintains correct order of animation execution. In other words, it ensures that a currently running animation is not interrupted by any introduction of a new C^A constraint. For a constraint priority(PrA,PrD), the last executed action throughout the animation is of priority PrD. Whereas the next available position in the queue for an action to be executed is PrA. Initially, it is introduced as priority(1,0).
– order/1, initialized as order(0), is used as a counter for the characteristics of objects generated. This helps keeping track of the order of the characteristics of a certain object as per their appearance. For an object, given two instances of the same characteristic, we are able to determine which is the old or new one based on the order (or time) it appeared in.

As previously mentioned, every C^A constraint appearing in the constraint store or in a rule is augmented with an extra argument that acts as its identifier. This is done to ensure the distinction between different constraint objects to be animated. For example, given that an element(I,V) constraint is a C^A, it should be updated and used as follows: element(N,I,V) where N is a unique identifier corresponding to the corresponding graphical object of the constraint element. Hence, for the upcoming sections of this paper, a C^A constraint refers to the updated format with the augmented argument.

The following sections present an overview of the implemented approach.

6.1   Linking C^A Constraints to Graphical Objects

A constraint obj(N,Type,Ch) in the constraint store represents the initial declaration of a C^A constraint, with identifier N, according to the graphical part of the program. The argument Type is the graphical object type (circle, box, ..., etc.) specified through the user's input for the corresponding constraint C^A. Ch is a list of the object's characteristics (width, height, ..., etc.) also provided by the user.

Given an obj(N,Type,Ch) constraint, further extraction of the graphical information of N is necessary to create its corresponding graphical object. This is performed through the execution of the rule genericAnimRule shown in Listing 3.2. The following constraints are produced by the rule.

- maxChange/2 determines the maximum number of characteristics that can be modified for a certain object. The number of modifications depend on the object type and is specified through the fact max/2. max(Type,X) is true iff the number of changes possible for the object of type Type is X. For example, max(box,5) means that an object of type box, has 5 possible characteristics to be updated. The transformed CHR program (P^T) is augmented with a predefined set of max/2 facts covering all object types.
- a constraint obj2/2 which carries the set of characteristics of an object. For obj2(N,Ch), Ch is a list of characteristics of the object N.
- object/3 which introduces the corresponding graphical object of the C^A constraint. object(N1,Type1,St), is the constraint responsible for adding the object N1 of object graphical type Type1 to the queue for animation. St represents the status of the object; kept, removed or unknown. In general, a kept status indicates that the C^A constraint has appeared as an H_K constraint. Conversely, a removed status indicates that the C^A constraint has appeared as an H_R constraint. For the case of a newly introduced object, its status is unknown.

Listing 3.2. Generic animation rule augmented at the beginning of P^T

```
genericAnimRule @ obj(N1,Type1,[H|T])==>max(Type1,MN1) |
    maxChange(N1,MN1),obj2(N1,[H|T]),object(N1,Type1,unknown).
```

Whenever a C^A constraint appears in the constraint store, we would like to create its equivalent graphical object. The characteristics of the graphical object depends on the user's input. Hence, we introduce two rules to P^T for each C^A constraint:

1. **Rule C^A2:**

$$C^A2 @ C^A(N,..) \Rightarrow X_0, ..., X_i, obj(N,T,Ch). \tag{1}$$

where $i \geq 0$.

The propagation rule C^A2 is matched for each newly introduced constraint C^A that does not have an already existing object in the graphical environment. The obj/3 constraint corresponds to the initial declaration of such a constraint in the graphical environment of P^T. $X_0, ..., X_i$ is a set of constraints which the user can specify about the desired animated objects. For the sorting example, if we would like the I argument in an element(N,I,V) constraint to determine the x-position of its corresponding object in the graphical environment, then we could add it as a constraint on the object. Accordingly, the x-position characteristic in the list Ch would carry the value I rather than a fixed value. In other words, we create a dynamic relation between the animated object and its related constraint through mapping variables of the C^A constraint to control values of the characteristics of the object.

2. **Rule C^A1:**

$$C^A1@C^A(N,..)\backslash checkAnimation(N) \Leftrightarrow X_0, ..., X_i, obj2(N, Ch). \qquad (2)$$

where $i \geq 0$.

The existence of a checkAnimation(N) constraint in the constraint store denotes that a relation between two C^A constraints has been detected. In other words, it indicates the detection of two constraints of the same C^A, carrying the same ID (N). The checkAnimation/1 constraint triggers the rule C^A1 for a C^A constraint. The rule throws a constraint obj/2 with the characteristics of the new object. obj/2 in turn triggers the set of rules for extracting the characteristics of the object as presented in Subsect. 6.3. The extracted characteristics are used to further draw the object as well as detect any change that could occur for the object through the execution of the program.

6.2   Transforming the Rules of the Original Program P

In order to accommodate for the link between the constraints of the program P and the graphical abilities of the transformation, the rules of P are transformed accordingly. An initial transformation is required to serve the augmentation of the extra argument for every C^A constraint corresponding to its identifier as previously mentioned. We will refer to a constraint C^A in the head of the rule as H^A. A constraint C^A that is found in the body of a rule carrying the same identifier as a H^A will be referred to as B^A and is called a "successor".

Further transformation of the rules is only applied to the rules where a constraint H^A appears. For every H^A constraint at position i-1 with identifier N , the constraint obj/3 is concatenated to the head of the rule at position i and referred to as O_i. This is done to force the graphical object to act the same as its corresponding constraint. A constraint object(N,Type,St) is added to the body of the rule with the same identifier as its corresponding H^A with the same N, where Type refers to the graphical object type of N. The third argument of object/3 is set to be kept or removed. This depends on whether the H^A is a kept constraint H^A_K or a removed constraint H^A_R. In this case the status of the object is said to be *known*. object/3 is later used to add the object N to a graphical queue to initiate a dynamic animation sequence for N (if detected) as will be further explained through Subsect. 6.3.

Next, we follow the `object/3` constraint with another one. The new constraint added depends on whether there is a successor constraint B^A to the H^A constraint or not.

- If the H^A has a successor, the constraint `checkAnimation/1` is added. It has one argument carrying the same identifier of the constraint H^A. This constraint is responsible for triggering a set of rules to perform the dynamic animation. The pair of `object/3` and `checkAnimation/1` constraints are referred to as OS. For every B^A in position $k+1$, OS is added at position k. An example of such rule transformation can be seen through Listing 3.7.
- If a constraint H^A with identifier N does not have a successor, the constraint `noSuccessor(N)` is added. This constraint triggers a set of rules that correspond to deciding the final state of an animated object. For each H^A without a successor, the pair of constraints `object/3` and `noSuccessor/1` are added to the beginning of the body and are referred to as ON.

Below is the formalized form of thr transformed simpagation rule shown in Eq. 3. The case where a successor is found is presented in Eq. 4 and accordingly OS is augmented. Whereas Eq. 5 shows the case with no successor detected where ON can be observed. Without loss of generality, we will assume that the head and body constraints that are associated with the animation come before the rest. However, in the reality, the head and body constraints can have any order and the transformer will keep the constraints in the corresponding positions.

- Simpagation:

$$H_{K_1}^A, ..., H_{K_n}^A, H_{K_{n+1}}, ..., H_{K_m} \backslash H_{R_1}{}^A, ..., H_{R_l}^A, H_{R_{l+1}}, ..., H_{R_o}$$
$$\Leftrightarrow G \mid B_1^A, ..., B_i^A, B_{i+1}, ..., B_j. \tag{3}$$

- Simpagation (with successor):

$$H_{K_1}^A, O_{K_1}, ..., H_{K_n}^A, O_{K_n}, H_{K_{n+1}}, ..., H_{K_m} \backslash H_{R_1}{}^A, O_{R_1} ..., H_{R_l}^A, O_{R_l}, H_{R_{l+1}}, ..., H_{R_o}$$
$$\Leftrightarrow G \mid OS_1, B_1^A, ..., OS_i, B_i^A, B_{i+1}, ..., B_j. \tag{4}$$

- Simpagation (without a successor):

$$H_{K_1}^A, O_{K_1}, ..., H_{K_n}^A, O_{K_n}, H_{K_{n+1}}, ..., H_{K_m} \backslash H_{R_1}{}^A, O_{R_1} ..., H_{R_l}^A, O_{R_l}, H_{R_{l+1}}, ..., H_{R_o}$$
$$\Leftrightarrow G \mid ON_1, ..., ON_{heads}, B_1^A, ..., B_i^A, B_{i+1}, ..., B_j. \tag{5}$$

where $i, j, k, l, m, n, o \geq 1$, $i = n + l$ and $heads = n + l$.
For the no successor case as shown in Eq. 5, note that the B^A constraints appearing are not successors to any of the head constraints H^A.

6.3   Implemented CHR Animation Library

The CHR animation library implemented is generic and does not depend on the user input. It provides different functions to enable the graphical animation of the program, given the link provided through the rules C^A1 and C^A2 for every C^A, genericAnimRule, as well as the transformed rules of the program P.

- **Extraction of an Object's Characteristics:**
  This is done through the constraint obj2(N,Ch) which triggers the set of rules responsible for extracting the characteristics of an object N from the list Ch. As a result, a pair of constraints for each characteristic char of the list Ch is introduced to the constraint store: charC(N,X,O) and char(N,X), where X is the value of the characteristic, and O is the order in which the characteristic of the object N appeared in. The constraint order/1 previously introduced is responsible for feeding in the value of the argument O. For example for the characteristic xpos(x-position of an object), the two new constraints introduced to the constraint store: xPosC(N, X,O) and xPos(N,X).
- **Giving Priority to an Object:**
  In order to provide a correctly ordered graphical representation without interruptions of newly introduced constraints to the constraint store, a virtual queue form is implemented through the animation library. Whenever a C^A constraint requires dynamic animation, it is added to the queue which operates in a first come first serve manner. Adding an object to the queue and giving it the current highest priority, ensures that the animation of the object would be performed before allowing any entry of a constraint, resulting from an application of a rule from the transformed program rules, to the store. Without such a feature, it is possible to have a constraint that is irrelevant to the current animation introduced to the store whilst executing the animation sequence of an object. This would cause the newly introduced constraint to be active and thus could possibly match and cause the execution of further rules. As a result, the animation could be interrupted by another animation sequence, which eventually would not map to the execution sequence of the original program. Thus, the transformation would lose the credibility of mapping the program execution to the graphical environment.

  The queue is managed through the constraint priority/2 previously introduced. The values of the arguments of this constraint are updated appropriately whenever a C^A constraint is queued for animation (update the first argument) or animation of a certain C^A has been performed successfully (update the second argument).

  The target is to produce a corresponding obj(ID,N,Type,St,Pr) constraint for a C^A constraint N of object type Type and status St. The obj/5 constraint is responsible for triggering a set of rules that creates and controls a graphical object whose identifier is ID and position in the queue is Pr. This constraint is thrown into the constraint store as a result of triggering the appropriate rule from Listing 3.3.

Listing 3.3. Adding an object to the graphical queue

```
givingPrKept @ object (N, Type , kept) ,  priority (Pr ,D) <=> var
    (ID)  | PrN is  Pr +1,  priority (PrN,D) ,checkUnknown (N) ,
    obj (ID ,N, Type , kept , Pr) ,  kept (ID) .
givingPrRemoved @ object (N, Type , removed) ,  priority (Pr ,D)
    <=> var (ID)  | PrN is  Pr +1,  priority (PrN,D) ,
    checkUnknown (N) ,  obj (ID ,N, Type , removed , Pr) .
givingPrUnknown @ object (N, Type , unknown) <=> var (ID)  | obj
    (ID ,N, Type , unknown) .
```

An `object/3` constraint in the store, whose status is known, denotes that a B^A constraint has been detected to be a successor to a constraint H^A (refer to Subsect. 6.2). Thus, is it to be added to the queue to ensure finalizing the animation of the object N without having any interruptions that could be caused by external rule executions. The functionality of each of the two rules `givingPrKept` and `givingPrRemoved` is to update the status of the object N from `unknown` to be known (`kept` or `removed`). This is done through `checkUnknown(N)` constraint which triggers a set of rules that removes any previous definitions of N as an `unknown` object. Afterwards, a new constraint `obj/5` is thrown to the store to represent the object with the updated status St.

On the other hand, an object of `unknown` status denotes that it is still new and no animation is detected yet. Thus, there is no need to include it in the queue as it is only required to draw the object onto the graphical environment without further animation. Hence, executing the rule `givingPrUnknown` results in throwing an `obj(ID,N,Type,St)` constraint into the store which carries the same first four arguments as `obj/5`.

– **Drawing an Object:**
  Given a `obj/5`, or `obj/4` for an object with `unknown` status, a set of rules are provided and are responsible for the direct communication between CHR and XPCE in order to draw an object. The rules require the existence of the set of constraints corresponding to the characteristics of the object (a set of `char/2` constraints). For example, for the object shape `circle`, the below rules in Listing 3.4 are provided. For an object whose status is known, the corresponding rule is not executed unless it is the turn of the object in the queue which is ensured through the guard of the rule given the `priority/2` constraint. The correct turn of an object is ensured by checking that the Pr value of the object is the next one following the last executed turn PrD. Furthermore, after executing an action for the object, the queue's PrD argument is updated accordingly.

Listing 3.4. CHR animation library rules responsible for communicating drawing commands to XPCE - circle shape.

```
drawRCircle @ xPos(N,X) , yPos(N,Y) , color(N,Color) , radius(N,Radius)
    \ obj(ID,N,circle ,removed,Pr) , priority(PrA,PrD) <=> NPr is PrD
    +1 , Pr = NPr, var(ID) | PrN is PrA +1, draw(ID,circle ,[X,Y,
    Color ,Radius]) , time1 , freeObj(ID) ,obj(ID,N,circle ,removed,PrA)
    , priority(PrN,Pr).
drawKCircle @ xPos(N,X) , yPos(N,Y) , color(N,Color) , radius(N,Radius) ,
    obj(ID,N,circle ,kept,Pr) \ priority(PrA,PrD) <=> NPr is PrD +1
    , Pr = NPr, var(ID) | PrN is PrA +1, draw(ID,circle ,[X,Y,Color ,
    Radius]) , time1 ,obj(ID,N,circle ,kept,PrA) , priority(PrN,Pr).
drawUCircle @ obj(ID,N,circle ,unknown) , xPos(N,X) , yPos(N,Y) , color(
    N,Color) , radius(N,Radius)==> var(ID) |draw(ID,circle ,[X,Y,Color
    ,Radius]) , time1 .
```

Regardless of the object's status, execution of any of the above three rules triggers the predicate `draw/3`. In general, a predicate `draw(Id,Type,ChList)` is responsible for sending XPCE commands to the graphical window to create a new graphical object of shape `Type` with characteristics values represented in the list `ChList`. The `draw/3` predicate responsible for drawing a circle object is shown in Listing 3.5. A `time1/0` constraint, as observed in the rules of Listing 3.4, is responsible for causing a small delay after drawing an object onto the graphical environment to allow us to view the animation.

In general, for an object shape `Shape`, the CHR animation library provides the three rules `drawRShape,drawKShape` and `drawUShape` similar to the ones in Listing 3.4 where the appropriate characteristics constraints are used to draw the object. Furthermore, a `draw/3` predicate is provided for each shape.

Listing 3.5. CHR animation library: Predicate for drawing a circle object using XPCE commands

```
draw(Id , circle ,[X,Y,Color ,Radius]):-
    send(@p, display , new(Id , circle(Radius)) , point(X,Y)) ,
    updateFillC(Id ,circle ,Color) ,
    send(@p, flush ).
```

– **Detecting Change in an Object:**
Whenever a successor is detected for a C^A constraint whose identifier is N, its characteristics are extracted through the execution of rule $C^A 1$ (refer to Subsect. 6.1) and the series of rule executions caused by the generated `obj2/2` constraint. Given that there would already exist a set of constraints corresponding to the original object's characteristics, this results in the existence of two sets of characteristics constraints for the constraint N; old and new ones (successor characteristics). In order to animate such change, the object undergoes multiple successive checks to detect the change that occurred. For an object's characteristic `char`, the following steps take place:

1. Compare old(X1) and new (X2) values of char which is done through the third argument of the constraint charC/3. Equation 6 shows the abstracted rule detect for detecting a change for a characteristic char.

$$detect @ charC(N, X2, O2) \setminus charC(N, X1, O1), maxChange(N, M) \Leftrightarrow$$
$$O1 < O2 | M2 \, is \, M - 1, animate(N, char, X1, X2), maxChange(N, M2).$$
$$(6)$$

The animate constraint is responsible for triggering animation rules to animate the change of N from X1 to X2 with respect to the characteristic char. Furthermore, the second argument for the maxChange/2 constraint is decremented by one to denote that a characteristic has been examined for change.

2. Detecting the kind of change whether increasing, decreasing or null.
3. Remove any old char-related constraints.
4. Animate the change between the old value and the new one. This is done through successive drawing and removal of the graphical object with intermediate changes to simulate the dynamic animation effect. An abstract rule animateChar is presented below and is responsible for updating intermediate values for a characteristic char of an object N. The constraint getNext computes the intermediate value XN from X1 towards X2.

$$animateChar @ animate(N, char, X1, X2), obj(Id, N, Type, St, Pr),$$
$$char(N, X1), priority(PrA, PrD) \Leftrightarrow NPr \, is \, PrD + 1, Pr = NPr$$
$$|PrN \, is \, PrA + 1, getNextValue(X1, X2, XN), char(N, XN),$$
$$obj(Id2, N, Type, St, PrA), priority(PrN, Pr),$$
$$animate(N, char, XN, X2).$$
$$(7)$$

The rule execution stops as soon as it is detected that the current char value reached the new target value. This is detected through the stopping (abstracted) rule finalizeChar whenever the values of the characteristic char in the animate constraint reach the same value.

$$finalizeChar @ obj(Id, N, Type, St, Pr), animate(N, char, X, X),$$
$$priority(PrA, PrD) \Leftrightarrow NPr \, is \, PrD + 1, Pr = NPr | PrN \, is \, PrA + 1,$$
$$obj(Id2, N, Type, St, PrA), priority(PrN, Pr).$$
$$(8)$$

The above four steps are repeated for every char of the constraint N until all changes have been applied (if any). This is detected through the second argument of maxChange/2 reaching zero, denoting that there are no more characteristics for N to be inspected for change. Finally, the object is removed from the queue and depending on the status St of it, it is removed from or kept in the graphical environment. The abstracted rule changeDone shown in Eq. 9

is responsible for such a process. Also, the store is cleared from any unnecessary constraints associated with the object N. For the successor constraint, maxChange/2 is then reset/refilled to hold the number of characteristics available for N's object. This step is necessary to allow future animation detection for N.

$$changeDone @ obj(ID, N, Type, St, Pr), maxChange(N, 0),$$
$$priority(PrA, PrD) \Leftrightarrow NPr \; is \; PrD + 1, Pr = NPr \, | \, priority(PrA, Pr).$$

$$(9)$$

7   Animation Example: Sorting

This section presents the execution of the previously presented sorting example in Sect. 3 given the same query provided, as well a applying the transformation presented in Sect. 6.

Initially, we mark element/2 constraint to be a C^A. Thus the two customized rules in Listing 3.6 are added to the output program $sorting^T$ where it is observed that element/2 is transformed into element/3 to accommodate the ID argument.

Listing 3.6. Customized animation rule augmented to $sorting^T$

```
element1 @ element(N,I,V)\checkAnimation(N) <=> IN is I*50,obj2(N,[
    height(V),xPos(IN),yPos(300),width(20), color(blue)]).
element2 @ element(N,I,V)==>IN is I*50,obj(N ,box,[height(V),xPos(IN),
    yPos(300),width(20), color(blue)]).
```

Here, we chose that each element(N,I,V) constraint is to be visualized as a box, its x-coordinate is determined according to the constraint's second argument I and the height is determined through its third argument V. The y-coordinate has a constant value of 300, the width is equal to 20 and the color of the box is blue.

As a second step to the transformation, the original rule of the program presented in Listing 3.1 is transformed according to the formalized transformation presented in Sect. 6. The transformation of the rule shown in Listing 3.7 also depends on the user's input regarding any head-body relations.

Listing 3.7. Sorting example rule transformed

```
sortingRule @ element(N1,I1,V1),obj(N1,TypeN1,_),element(N2,
    I2,V2),obj(N2,TypeN2,_)<=>I1<I2,V1<V2| object(N1,TypeN1,
    removed),checkAnimation(N1),element(N1,I2,V1),object(N2,
    TypeN2,removed),checkAnimation(N2),element(N2,I1,V2).
```

Finally, the query is also transformed as follows: element(a,1,30), element(b,2,100) and element(c,3,50), where unique values are augmented onto every C^A constraint to act as identifiers.

1. **Introduction of a Constraint** C^A

   Initially, element(a,1,30) is thrown into the constraint store which triggers the rule element2. As a result, the constraint obj(a,box,[height(30),xPos (50),yPos(300),width(20), color(blue)] is thrown into the constraint store.

2. **Extraction of Object Characteristics**

   Next, genericAnimRule is triggered by the obj/3 constraint and throws the corresponding obj2/2 constraint which in turn extracts the characteristics of element a and produce the following constraints: height(a,30), heightC(a,30,0),xPos(a,50),xPosC(a,50,1),yPos(a,300), yPosC(a, 300,2), width(a,20), widthC(a,20,3), color(a,blue), colorC (a,blue,4). Also generi cAnimRule throws the constraint object(a,box, unknown) into the store.

3. **Creation of a New (Unknown) Object:**

   The object/3 constraint accordingly triggers the rule givingPrUnknown and generates the constraint obj(ID,a,box,unknown) where ID is a non-bound variable. Given the obj/4 constraint as well as the set of characteristic constraints of element a, the rule drawUBox is executed which in turn triggers the corresponding draw/3 predicate for drawing a box shape. Both rules are shown in Listing 3.8.

   Listing 3.8. Creating an object of type box with an unknown status and drawing it with XPCE commands

   ```
   drawUBox @ obj(ID,N,box,unknown), xPos(N,X), yPos(N,Y),
       color(N,Color), width(N,Width), height(N,Height)  ==>
       var(ID) | draw(ID,box,[X,Y,Color,Width,Height]), time1.

   draw(Id,box,[X,Y,Color,Width,Height]):-
   send(@p, display, new(Id, box(Width,Height)),point(X,Y)),
   updateFillC(Id,box,Color),
   send(@p, flush).
   ```

   The same process is executed for element(b,2,100) until the box object corresponding to element b is drawn as well. Figure 3a shows the graphical window with both elements in the constraint store drawn.

4. **Checking for Applicable CHR Rules:**

   Given that there are two element/3 constraints in the store that can satisfy the guard of the rule sortingRule, the rule can be applied. As a result, the constraint object(a,box,removed) is thrown into the constraint store which replaces the previous obj/4 constraint for element a with an unknown status with the constraint obj(ID,a,box,unknown,Pr) where Pr is the currently available turn in the queue. Such replacement is done through applying the rule givingPrRemoved shown in Listing 3.3.

5. **Inspecting Change in an Object:**

   Next, checkAnimation(a) constraint is thrown into the store through the rule sortingRule, which triggers the rule element1 (refer to Listing 3.6) as soon as the new `element(a,2,30)` is thrown into the store as well. As a result, an `obj2/2` constraint is generated carrying the new characteristics of `element` a. The following set of characteristic constraints are generated accordingly: `height(a,30)`,`heightC(a,30,10)`,`xPos(a,100)`,`xPosC(a,100,11)`,`yPos` `(a,300)`,`yPosC(a,300,12)`,`width(a,20)`,`widthC(a,20,13)`,`color(a,` `blue)`, `colorC(a,blue,14)`. Thus, the change of characteristics for `element` a is detected according to the process previously explained through Sect. 6. As a result, we observe the graphical object corresponding to `element` a moving to its new position as shown in Fig. 3b and c.

Furthermore, the same process takes place for `element` b until it reaches its target position.

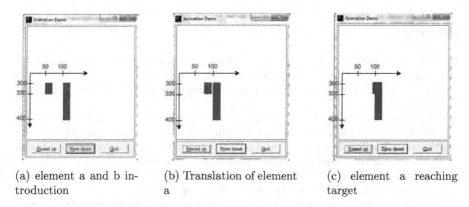

(a) element a and b introduction

(b) Translation of element a

(c) element a reaching target

Fig. 3. Screen-shots of the step-by-step animation of the sorting example (edited with guiding axes).

Next, `element` c is introduced into the store and the graphical environment and further checks for applicable rules are applied. The animation stops whenever the program execution stops. For the full execution of the transformed program using the query examined above, please refer to this Youtube video link: https://youtu.be/fXjwg5XLr0w. In general, the above five steps formulate the path of a C^A constraint when it enters the constraint store.

8   Conclusion and Future Work

In conclusion, this paper presented a generic transformation that can be applied to a CHR program. The transformation augments the program with actions. Various animation actions are supported such as translating, resizing and changing colors of constraint objects. The transformation was applied to different CHR programs including a simple chemical reactions program as shown in http://youtu.be/LcTlLiOsuKk. The advantage of this approach is that it does not depend on a visualization tool with embedded animated actions. It is able to infer and apply the actions automatically from the program. The tool can be used for educational purposes to provide a better visual experience for students.

For future work, a tool will be used to enable the user to define new objects and actions to be supported. Furthermore, grouping of several objects to represent a single constraint will be investigated. The different types of annotations supported in [17] will be also included. Different visualization tools other than XPCE should also be tested. Also, the tool will be enhanced with debugging features that could enable the user to optimize his program. The aim, for now, was generality rather than efficiency. The focus was to be able to encode animations using CHR only without having to use the possibly efficient ready animation tools that could change or die over time. However, efficiency of the tool is part of the future work.

References

1. SWI-Prolog - Manual. http://www.swi-prolog.org/download/xpce/doc/userguide/userguide.pdf. Accessed 8 June 2015
2. SWI-Prolog GUI - Manual. http://www.swi-prolog.org/packages/xpce/. Accessed 8 June 2015
3. Abdennadher, S., Saft, M.: A visualization tool for constraint handling rules. In: Proceedings of 11th Workshop on Logic Programming Environments (2001)
4. Abdennadher, S., Sharaf, N.: Visualization of CHR through source-to-source transformation. In: Technical Communications of the 28th International Conference on Logic Programming (ICLP 2012). Leibniz International Proceedings in Informatics (LIPIcs), vol. 17, pp. 109–118. Dagstuhl, Germany (2012)
5. Baecker, R.: Sorting out sorting: a case study of software visualization for teaching computer science. Softw. Vis. Program. Multimedia Experience 1, 369–381 (1998)
6. Brown, M.H., Sedgewick, R.: A system for algorithm animation. SIGGRAPH Comput. Graph. 18(3), 177–186 (1984)
7. Brown, M.H.: Zeus: a system for algorithm animation and multi-view editing. In: Proceedings of the IEEE Workshop on Visual Languages, pp. 4–9, October 1991
8. Eisenstadt, M., Brayshaw, M.: Graphical debugging with the transparent PROLOG machine (TPM). In: McDermott, J.P. (ed.) Proceedings of the 10th International Joint Conference on Artificial Intelligence, pp. 83–86. Morgan Kaufmann, August 1987
9. Frühwirth, T.: Theory and practice of constraint handling rules, special issueon constraint logic programming. J. Logic Program. 37(1–3), 95–138 (1998)
10. Frühwirth, T.: Constraint Handling Rules. Cambridge University Press, New York (2009)
11. Frühwirth, T., Holzbaur, C.: Source-to-source transformation for a class of expressive rules. In: Buccafurri, F. (ed.) APPIA-GULP-PRODE, pp. 386–397 (2003)
12. Halim, S., Koh, Z.C., Loh, V., Halim, F.: Learning algorithms with unified and interactive web-based visualization. Olympiads Inform. 6, 53–68 (2012)
13. Hundhausen, C., Douglas, S., Stasko, J.: A meta-study of algorithm visualization effectiveness. J. Vis. Lang. Comput. 13(3), 259–290 (2002)
14. Pierson, W.C., Rodger, S.H.: Web-based animation of data structures using JAWAA. In: Proceedings of the Twenty-Ninth SIGCSE Technical Symposium on Computer Science Education, SIGCSE 1998, pp. 267–271. ACM, New York (1998)

15. Rößling, G., Schüer, M., Freisleben, B.: The animal algorithm animation tool. In: ACM SIGCSE Bulletin, vol. 32, pp. 37–40. ACM (2000)
16. Senay, H., Lazzeri, S.G.: Graphical representation of logic programs and their behavior. In: Proceedings of the 1991 IEEE Workshop on Visual Languages, Japan, 8–11 October 1991, pp. 25–31. IEEE Computer Society (1991)
17. Sharaf, N., Abdennadher, S., Frühwirth, T.: CHRAnimation: an animation tool for constraint handling rules. In: Proietti, M., Seki, H. (eds.) LOPSTR 2014. LNCS, vol. 8981, pp. 92–110. Springer, Heidelberg (2015)
18. Sharaf, N., Abdennadher, S., Frühwirth, T.W.: Visualization of constraint handling rules. CoRR, abs/1405.3793 (2014)
19. Simonis, H., Aggoun, A.: Search-tree visualisation. In: Deransart, P., Hermenegildo, M.V., Małuszynski, J. (eds.) DiSCiPl. LNCS, vol. 1870, pp. 191–208. Springer, Heidelberg (2000)
20. Smolka, G.: The definition of kernel oz. In: Podelski, A. (ed.) Constraint Programming: Basics and Trends. LNCS, vol. 910, pp. 251–292. Springer, Heidelberg (1994)
21. Wielemaker, J., Anjewierden, A.: An architecture for making object-oriented systems available from prolog. arXiv preprint cs/0207053 (2002)

Author Index

Printed in the United States
By Bookmasters